HANDBOOK

of

RECEPTORS

and

CHANNELS

G PROTEIN COUPLED RECEPTORS

Edited by
Stephen J. Peroutka

CRC Press
Boca Raton   Ann Arbor   London   Tokyo

Library of Congress Cataloging-in-Publication Data

Catalog information is available from the Library of Congress.

ISBN 0-8493-8321-8

Preface

Two major advances have occurred within the past few years in the field of receptor analysis as a result of molecular biological studies. First, receptor diversity within a species is far greater than might have been predicted. Second, significant pharmacological variability may occur between species, despite high percentages of homology, indicating that pharmacological and toxicological studies in non-human species may be misleading. At present, more than 400 receptors have been cloned and sequenced. This book series represents an attempt to compile this rapidly expanding data in a practical and useful format.

Series Format

The coding, sequencing, and expression of a variety of membrane receptors and channels indicates the existence of at least four "superfamilies" of molecular structures that mediate signal transduction. This book series is meant to be a compendium of data derived from recent studies on each receptor "superfamily". Each volume will contain complete amino acid sequence information on all cloned receptors as well as relevant pharmacological information.

Volumes (Editors)

1. G Protein-Coupled Receptors
   Stephen J. Peroutka, M.D., Ph.D.

2. Ligand- and Voltage-Gated Ion Channels
   R. Alan North, M.D., Ph.D.

Chapter Headings

Chapter headings will be consistent throughout the series. For example, a review of "Glutamate Receptors" will always be located in Chapter 13 if they have been identified in the relevant "superfamily". If "Glutamate Receptors" have not been identified within a receptor "superfamily" (e.g., within the Tyrosine Kinase family), Chapter 13 will be absent in that particular volume.

Chapters

1. Acetylcholine

2. Adenosine

3. Adrenergic

4. Atrial Natriuetic Factor

5. Calcium

6. Cannabinoids

7. Cytokines

8. Dopamine

9. Endothelins

10. Epidermal Growth Factor-Urogastrone

11. GABA

12. Glucose

13. Glutamate

14. Histamine

15. Hormone

16. 5-Hydroxytryptamine

17. Insulin

18. Neurotrophic Factors

19. Neuropeptides

20. Odorants

21. Opioids

22. Platelet-Derived Growth Factor

23. Potassium

24. Rhodopsins

25. Sodium

26. Thrombin

99. Miscellaneous

Species Codes

Chapters will be divided into subheadings based upon individual receptor clones. An additional subheading will be used to identify the species from which the clone was obtained. Once again, the number system used for individual species will be consistent throughout the series, as shown below.

Species Codes

1. Human

2. Gorilla

3. Monkey

4. Bovine

5. Sheep

6. Canine

7. Porcine

8. Rabbit

9. Guinea pig

10. Rat

11. Mouse

12. Hamster

13. Opposum

14. Turkey

15. Chicken

16. Frog

17. Snake

18. Octopus

19. Lamprey

20. Xenopus

21. Drosophila

22. Fish

23. Slime mold

24. Yeast

25. Virus

Volume I
G Protein-Coupled Receptors

Chapters

Acknowledgments

I thank Jean M. Peroutka for editorial assistance. This work was supported in part by the Kleiner Family Foundation and NIH Grant NS 25360-06. I also thank Karoly Nikolics, Ph.D. and Colin Watanabe for their invaluable assistance and patience in assisting me with the receptor sequence analyses.

Dedication

To Jean, Laura, and Michael for their love, encouragement, and willingness to share me with a word processor.

The Editor

Stephen J. Peroutka, M.D., Ph.D. is the Director of Neuroscience at the Palo Alto Institute for Molecular Medicine and the President and Founder of Spectra Biomedical, Inc.

Dr. Peroutka received his A.B. degree from Cornell University in 1975. He obtained his M.D. and Ph.D. degrees in 1979 and 1980, respectively, from the Johns Hopkins University School of Medicine. After completing an internship in Internal Medicine at Stanford University in 1981, he was a Resident and Fellow in the Department of Neurology at the Johns Hopkins Hospital. Dr. Peroutka was an Assistant Professor of Neurology and Pharmacology at Stanford University from 1984 to 1990 and was Chief, Neurology Service, at the Palo Alto Veteran's Administration Hospital from 1988 to 1990. In 1990, he joined Genentech, Inc. where he was the Director of Neuroscience from 1991 to 1993.

Dr. Peroutka is a member of the Society for Neuroscience, Society for Neurochemistry, American College of Neuropsychopharmacology, American Association for the Study of Headache and the American Academy of Neurology.

He has been the recipient of the Sandoz Award for excellence in the field of Pharmacology (1979), Alfred P. Sloan Fellowship (1985), John A. and George L. Hartford Fellowship (1985), McKnight Foundation Scholar Award (1986), Japan Society for the Promotion of Science Fellowship (1987), Syntex Award for Receptor Pharmacology (1993), Harold G. Wolff Award from the American Society for the Study of Headache (1993), and was the First Irvine H. Page Lecturer (1990).

Dr. Peroutka has been the recipient of research grants from the National Institutes of Health, the Veteran's Administration, the National Headache Foundation, the Kleiner Family Foundation, and private industry. He has presented over 25 lectures at international meetings and approximately 100 guest lectures at universities and institutes. He has published more than 150 papers and edited 3 other books. His current research interests are the characterization of 5-hydroxytryptamine receptor subtypes and their relationship to human disease states such as migraine.

Contributors

Jean-Michel Arrang, Ph.D.
Unité de Neurobiologie & Pharmacologie
Centre Paul Broca de L'Inserm
Paris, France

Avi Ashkenazi, Ph.D
Genentech, Inc.
Department of Molecular Biology
South San Francisco, California

Paul Blount
Department of Anatomy & Neurobiology
Washington University School of Medicine
St. Louis, Missouri

Tom I Bonner, Ph.D.
Department of Cell Biology
National Institute of Mental Health
Bethesda, Maryland

J. Diaz
Unité de Neurobiologie & Pharmacologie
Centre Paul Broca de L'Inserm
Paris, France

Carola Eva
Instituto di Farmacologia e Terapia Sperimentale
University of Torino
Torino, Italy

Christopher J. Evans, Ph.D.
Department of Psychiatry & Biobehavioral
Sciences
UCLA School of Medicine
Los Angeles, California

Susan R. George, Ph.D.
Addiction Research Foundation
Primary Mechanisms
Department of Pharmacology
University of Toronto
Toronto, Ontario, Canada

Xiao-Ming Guan, Ph.D.
Department of Molecular Pharmacology &
Biochemistry
Merck Research Laboratories
Rahway, New Jersey

Richard Horuk, Ph.D.
Department of Protein Chemistry
Genentech, Inc.
South San Francisco, California

Charles W. Hutchins
Department of Structural Biology
Pharmaceutical Products Division
Abbott Laboratories
Abbott Park, Illinois

Marlene A. Jacobson
Department of Pharmacology
Merck Research Laboratories
West Point, Pennsylvania

James E. Krause, Ph.D.
Department of Anatomy & Neurobiology
School of Medicine
Washington University
St. Louis, Missouri

R. Leurs
Unité de Neurobiologie & Pharmacologie
Centre Paul Broca de L'Inserm
Paris, France

Frederic Libert, Ph.D.
IRIBHN
Université Libre de Bruxelles
Bruxelles, Belgium

Joel Linden, Ph.D.
Department of Internal Medicine and Molecular
Physiology and Biological Physics
University of Virginia
Charlottesville, Virginia

Lisa A. Matsuda, Ph.D.
Department of Psychiatry and Behavioral
Sciences
Medical University of South Carolina
Charleston, South Carolina

Brian O'Dowd, Ph.D.
Department of Pharmacology
University of Toronto
Toronto, Ontario, Canada

Marc Parmentier, M.D., Ph.D.
IRIBHN
Université Libre de Bruxelles
Bruxelles, Belgium

David J. Pepperl
Department of Pharmacology and Toxicology
University of Arizona
Tucson, Arizona

Ernest Peralta, Ph.D.
Deparment of Biochemistry and Molecular
Biology
Harvard University
Cambridge, Massachusetts
Genentech, Inc.
South San Francisco, California

Stephen J. Peroutka, M.D., Ph.D.
Palo Alto Institute for Molecular Medicine
Spectra Biomedical, Inc.
Palo Alto, California

John W. Regan, Ph.D.
Department of Pharmacology and Toxicology
College of Pharmacy
University of Arizona
Tucson, Arizona

Marial Ruat
Unité de Neurobiologie & Pharmacologie
Centre Paul Broca de L'Inserm
Paris, France

Bruce S. Sachais
Department of Anatomy & Neurobiology
Washington University
School of Medicine
St. Louis, Missouri

Thoms Sakmar
Howard Hughes Medical Institute
Rockefeller University
New York, New York

Darryle D. Schoepp, Ph.D.
CNS Research Division
Eli Lilly and Company
Indianapolis, Indiana

Stephane Schurmans, M.D.
IRIBHN
Université Libre de Bruxelles
Bruxelles, Belgium

Jean-Charles Schwartz
Unité de Neurobiologie & Pharmacologie
Centre Paul Broca de L'Inserm
Paris, France

Philip Seeman, Ph.D.
Deparment of Pharmacology
University of Toronto
Toronto, Ontario, Canada

Rolf Sprengel
ZMBH
Universitat Heidelberg
Heidelberg, Germany

Joel Tardivel-Lacombe
Unité de Neurobiologie & Pharmacologie
Centre Paul Broca de L'Inserm
Paris, France

Elisabeth Traiffort
Unité de Neurobiologie & Pharmacologie
Centre Paul Broca de L'Inserm
Paris, France

Pierre Vanderhaeghen, Ph.D.
IRIBHN
Université Libre de Bruxelles
Bruxelles, Belgium

Gilbert Vassart, M.D., Ph.D.
Department of Medical Genetics
Université Libre de Bruxelles
Bruxelles, Belgium

Michael Williams, Ph.D.
Department of Neuroscience Discovery
Abbott Laboratories
Abbott Park, Illinois

Table Of Contents

Chapter 6
Cannabinoid Receptors 79
Tom I. Bonner and Lisa A. Matsuda

Chapter 7
Cytokine Receptors 87
Richard Horuk

Chapter 19
Olfactory Receptors...237
Marc Parmentier, Stéphane Schurmans, Frédéric Libert, Pierre Vanderhaeghen, and Gilbert Vassart

Chapter 20
Opioid And Opiate Receptors 251
Chris Evans

Chapter 22
Opsins 257
Thomas P. Sakmar

Chapter 24
Tachykinin Receptors 277
James E. Krause, Bruce S. Sachais, and Paul Blount

Chapter 99
Xiao-Ming Guan

1

Muscarinic Acetylcholine Receptors

Avi Ashkenazi and Ernest G. Peralta

1.1.0   Introduction

The neurotransmitter actions of acetylcholine are mediated by two distinct types of receptor: muscarinic and nicotinic (see Table 1). In vertebrates, muscarinic receptors are located on heart and smooth muscle cells, on secretory gland cells, and, together with nicotinic receptors, on autonomic ganglion cells and on neurons of the central nervous system. In contrast, nicotinic receptors are located mainly at the motor endplate of skeletal muscle. These two types of acetylcholine receptor were identified in 1914 by Sir Henry Dale, on the basis of acetylcholine-like effects of plant alkaloids: those receptors which are activated by muscarine and blocked by atropine were defined as muscarinic, whereas those which are activated by nicotine and blocked by curare were defined as nicotinic.

Despite their activation by a common neurotransmitter, muscarinic and nicotinic receptors constitute quite different molecular entities. Nicotinic acetylcholine receptors (nAChRs) are in fact ligand-activated ion channels which mediate a rapid, depolarizing response to acetylcholine in target tissues such as striated muscle. In contrast, muscarinic acetylcholine receptors (mAChRs), like many other neurotransmitter receptors, hormone receptors, odorant receptors, and the visual rhodopsins, belong to a large class of integral membrane glycoproteins which contain seven putative α helical transmembrane domains (reviewed by Kobilka, 1992). Each of these "seven transmembrane receptors" transduces an agonist or sensory stimulus across the plasma membrane via the activation of heterotrimeric guanine nucleotide binding (G) proteins (reviewed by Kaziro et al., 1991). Activation of specific G proteins by mAChRs results in various biochemical and electrophysiological effects which lead to biological responses at the level of cells, tissues, and the whole organism. The biochemical effectors regulated by mAChRs include the enzymes adenylyl cyclase, phospholipase C, D, and A_2 (PLC, PLD, PLA_2), and guanylyl cyclase, which control the biosynthesis of second messengers such as cyclic adenosine monophosphate (cAMP), inositol 1,4,5-trisphosphate, diacylglycerol, arachidonic acid, and cyclic guanosine monophosphate (cGMP). The electrophysiological effectors regulated by mAChRs include K^+, Ca^{2+}, and Cl^- ion channels.

The biological responses evoked by mAChRs of the parasympathetic nervous system include inhibition of the rate and force of heart contraction, stimulation of

Table 1.

Receptor	Second messenger	Species cloned	Chromosomal location	a.a. Sequence	Accession number	Primary reference
m1	+PI	Human		460	P11229	Peralta et al., 1987a
		Porcine		460	P04761	Kubo et al., 1986a
		Rat		460	P08482	Bonner et al., 1987
		Mouse		460	P12657	Shapiro et al., 1988
m2	–AC	Human		466	P08172	Peralta et al., 1987b
		Porcine		466	P06199	Peralta et al., 1987a
		Rat		466	P10980	Gocayne et al., 1987
		Chicken		466	A40972	Tietje et al., 1991
m3	+PI	Human		590	P20309	Peralta et al., 1987b
		Porcine		590	P11483	Akiba et al., 1988
		Rat		589	P08483	Bonner et al., 1987
m4	–AC	Human	11p11.2-p12	479	P08173	Peralta et al., 1987b
		Rat		478	P08485	Bonner et al., 1987
		Chicken		490	P17200	Tietje et al., 1990
		Xenopus		484	X65865	Olate (unpublished)
m5	+PI	Human		532	P08912	Bonner et al., 1988
		Rat		531	P08911	Bonner et al., 1988
m		Drosophila		722	P16395	Shapiro et al., 1989

airway and smooth muscle contraction, and stimulation of exocrine and endocrine secretion. In the central nervous system, mAChRs are thought to be involved in functions such as learning, memory, and motor control (reviewed by Taylor, 1990). In addition, mAChRs may regulate the growth of astroglial cells in an age-dependent manner during perinatal brain development (Ashkenazi et al., 1989c), and can act as conditional oncogenes when expressed in cells capable of proliferation (Gutkind et al.,

Table 2. Expression and Pharmacological Properties of Muscarinic Acetylcholine Receptor Subtypes*

Receptor	mRNA expression	Ligands with relative selectivity
m1	Brain, parotid gland, lacrimal gland	Pirenzepine, m1 > m3, m4, m5 > m2, (+)-telenzepine
m2	Heart, brain, urinary bladder, large intestine, small intestine, parotid gland, lung (trachea)	AF-DX 116, m2 > m4 > m3 > m1 > m5 methocarmine, m2 > m1 > m4, m5 > m3, himbacine
m3	Brain, pancreas, lacrimal gland, parotid gland, submandibular gland, urinary bladder, small intestine, large intestine, lung (trachea)	Hexahydrosiladifenidol, m3 > m1, m4, m5 > m2, p-fluorohexahydro-siladifenidol
m4	Brain, lung embryonic chick heart	N/A
m5	Brain	N/A

* See text for references; N/A, not available.

1991). Finally, mAChRs may regulate the release of the amyloid precursor protein which contributes to the formation of amyloid deposits in the brain of individuals with Alzheimer's disease (Nitsch et al., 1992).

Each type of acetylcholine receptor consists of a family of related polypeptide gene products, termed receptor subtypes. Initial studies on the binding of classical antagonists such as atropine and N-methylscopolamine to mAChRs from different tissues suggested a single, homogeneous population of binding sites (Hammer et al., 1980). However, differences were reported in the ability of some drugs to antagonize muscarinic responses in different tissue preparations, suggesting that mAChRs are not identical in different tissues (Barlow et al., 1976). Indeed, the development of the antagonist pirenzepine enabled the pharmacological distinction of different mAChR subtypes in different tissues, since pirenzepine bound with significantly higher affinity to neuronal mAChRs, as compared with mAChRs from heart and smooth muscle (Hammer et al., 1980). Other drugs were developed as well, which exhibit different mAChR subtype selectivities. For example, AF-DX 116 (Giachetti et al., 1986; Giraldo et al., 1987) and himbacine (Gilani et al., 1986) are relatively selective for cardiac mAChRs, whereas 4-DAMP (4-diphenylacetoxy-N-methylpiperidine methiodide) (Barlow et al., 1976) and hexahydrosiladifenidol (Mutschler et al., 1987) are relatively more selective for neuronal mAChRs (reviewed by Hulme et al., 1990).

Since the initial pharmacological classification of mAChR subtypes was limited by the degree of selectivity of drugs available, efforts were made to identify mAChR subtypes at the genetic level by molecular cloning. These studies have confirmed the existence of multiple mAChR subtypes, and demonstrated that they are encoded by distinct genes. In addition, whereas pharmacological classification suggested the existence of three mAChR subtypes (M_1 to M_3), genetic studies revealed at least five (now termed m1 to m5, as adopted by the Fourth Symposium on Subtypes of Muscarinic Receptors, 1989).

At the level of amino acid sequence, the mAChRs are most homologous (~30%) to the α_2-adrenergic receptors and least homologous (~20%) to G protein-coupled receptors that bind peptidic ligands, and to visual rhodopsins. Alignment of the 5 mAChR subtypes reveals ~43% amino acid sequence identity overall, and ~65% identity in the predicted transmembrane domains and short connecting loops (Figure 1). This supports the notion that the binding site for the common ligand, acetylcholine, is located mainly within the lipid bilayer. In contrast, the predicted large cytoplasmic loop joining the 5th and 6th transmembrane domains is highly divergent amongst vertebrate mAChRs and ranges in size between 156 amino acids for the m1 receptor and 241 residues for the m3 subtype. Based on overall amino acid sequence homology, the mAChR subtype family can be divided into two classes, one containing the m1, m3, and m5 subtypes, and the other containing the m2 and m4 subtypes. Notably, the m1, m3, and m5 receptors share high sequence homology, particularly in regions of the large cytoplasmic loop that are adjacent to the fifth and sixth transmembrane segments, as do the m2 and m4 mAChRs.

Homologs of various mAChR subtypes have been cloned from several species, including man, rat, pig, mouse, chicken, and *Drosophila* (Figure 1). There is high sequence conservation for individual receptor subtypes across different mammalian species. For example, the human and porcine m1 receptors are 98.9% identical throughout their 460 amino acids. Residues in the third cytoplasmic domain are the least highly conserved between variants of the same receptor subtype from different species. Phylogenetic tree analysis of mAChR sequences from several species supports further the division into the two branches of "odd" and "even" numbered mAChR subtypes (Figure 2).

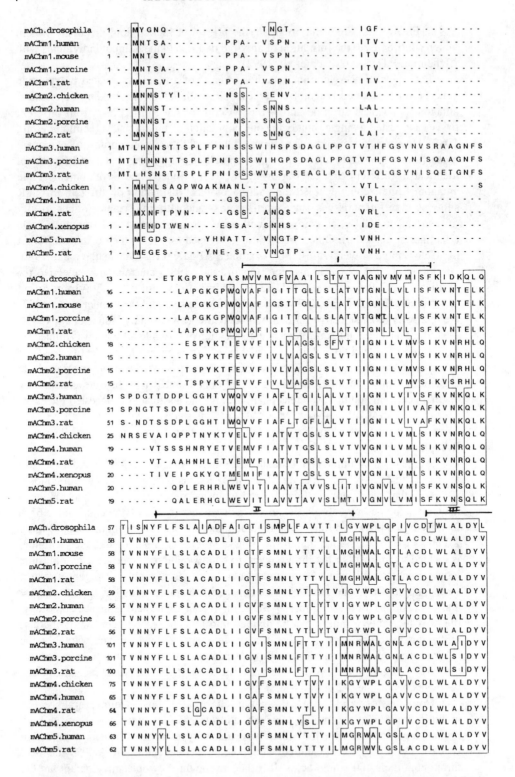

FIGURE 1. Alignment of the amino acid sequences of muscarinic acetylcholine receptors. The sequences were obtained from commercial protein databases such as GenBank and EMBL. The references for each sequence are cited in the text.

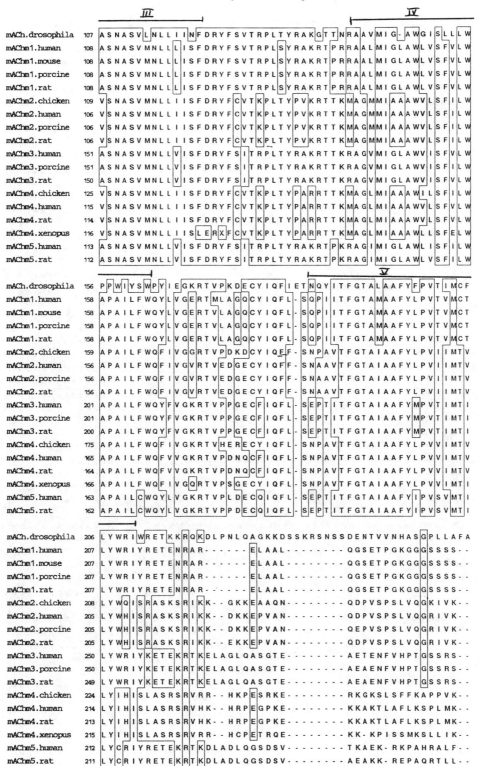

FIGURE 1b.

```
mACh.drosophila  256  QVGGNDHDTWRR P RSESSADAESVYMTNMVIDSGYHGMHSRKSSIKSTNT
mAChm1.human     241  - - - - - SERSQPG- - -AEGSP- - - - - - - - - - - - - -ETPPGRCCRCCRAPR
mAChm1.mouse     241  - - - - - SERSQPG- - -AEGSP- - - - - - - - - - - - - -ESPPGRCCRCCRAPR
mAChm1.porcine   241  - - - - - SERSQPG- - -AEGSP- - - - - - - - - - - - - -ETPPGRCCRCCRAPR
mAChm1.rat       241  - - - - - SERSQPG- - -AEGSP- - - - - - - - - - - - - -ESPPGRCCRCCRAPR
mAChm2.chicken   246  - - - - - PNNNNI P - - - -TSSD- - - - - - - - - - - GLEHN K VQNGKTTGE
mAChm2.human     243  - - - - - PNNNNM P - - - -SSDD- - - - - - - - - - - GLEHN K IQNGKAPRD
mAChm2.porcine   243  - - - - - PNNNNM P - - - -GSDE- - - - - - - - - - - ALEHN K IQNGKAPRD
mAChm2.rat       243  - - - - - PNNNNM P - - - -GGDG- - - - - - - - - - - GLEHN K IQNGKAPRD
mAChm3.human     290  - - - - - CSSYELQQQSMKRSNRRKY- - -GRCHFWFTTKSW K PSSEQMDQD
mAChm3.porcine   290  - - - - - CSSYELQQQSLKRSARRKY- - -GRCHFWFTTKSW K PSAEQMDQD
mAChm3.rat       289  - - - - - CSSYELQQQGVKRSSRRKY- - -GRCHFWFTTKSW K PSAEQMDQD
mAChm4.chicken   262  - - - - - QNNNNS P KRAVEVKE- - - - - - - - - - -EVRNGKVDDQPSAQT
mAChm4.human     252  - - - - - QSVKKP P - PGEAARE- - - - - - - - - - - -ELRNG K LEEAPPPAL
mAChm4.rat       251  - - - - - PSIKKP P - PGGASRE- - - - - - - - - - - -ELRNG K LEEAPPPAL
mAChm4.xenopus   252  - - - - - QT-KNI P KQDAGDKV- - - - - - - - - - -VEKKNGVSNGKIEKS
mAChm5.human     251  - - - - - RSCLRC P RPTLAQRERNQA- - -S- - - -WSSSRRSTSTTGKPSQA
mAChm5.rat       250  - - - - - RSFFSC P RPSLAQRERNQA- - -S- - - -WSSSRRSTSTTGKTTQA


mACh.drosophila  306  IKKSYTCFGSIKEWCIAW W HSGREDSDDFAY E QEEPSDLGYATPVTIET P
mAChm1.human     268  LLQAYS- - - - - - - - - - - W KEE- - - - - - - - E EEDEGS M ES L TSS E GEEP
mAChm1.mouse     268  LLQAYS- - - - - - - - - - - W KEE- - - - - - - - E EEDEGS M ES L TSS E GEEP
mAChm1.porcine   268  LLQAYS- - - - - - - - - - - W KEE- - - - - - - - E EEDEGS M ES L TSS E GEEP
mAChm1.rat       268  LLQAYS- - - - - - - - - - - W KEE- - - - - - - - E EEDEGS M ES L TSS E GEEP
mAChm2.chicken   272  SV MENC- - - - - - - - - - -VQGE- - - - - - - - E KDSSND S T SV S VVPSNTK
mAChm2.human     269  PVTENC- - - - - - - - - - -VQGE- - - - - - - - E KESSND S T SV S AVASNMR
mAChm2.porcine   269  AVTENC- - - - - - - - - - -VQGE- - - - - - - - E KESSND S T SV S AVASNMR
mAChm2.rat       269  GVTETC- - - - - - - - - - -VQGE- - - - - - - - E KESSND S T S S AAVASNMR
mAChm3.human     331  HSSSDS- - - - - - - - - - - W NNN- - - - - - - -DAAASLENSASS S DE E DIGS
mAChm3.porcine   331  HSSSDS- - - - - - - - - - - W NNN- - - - - - - -DAAASLENSASS S DE E DIGS
mAChm3.rat       330  HSSSDS- - - - - - - - - - - W NNN- - - - - - - -DAAASLENSASS S DE E DIGS
mAChm4.chicken   292  EATGQQ- - - - - - - - - - -EEK- - - - - - - - E TSNESS T V S MTQTTKDK P
mAChm4.human     281  PPPPRP- - - - - - - - - - -VADK- - - - - - - -DTSNESS S G S ATQNTKER P
mAChm4.rat       280  PPPPRP- - - - - - - - - - -VPDK- - - - - - - -DTSNESS S G S ATQNTKER P
mAChm4.xenopus   281  MTNLQT- - - - - - - - - - -AEEK- - - - - - - - E TSNESS A S L SHNPPEKQ
mAChm5.human     288  TGPSAN- - - - - - - - - - - W AKA- - - - - - - - E QLTTCS S YPS S EDE D -KP
mAChm5.rat       287  TDLSAD- - - - - - - - - - - W EKA- - - - - - - - E QVTTCS S YPS S EDE A -KP


mACh.drosophila  356  LQSSVSRCTSMNVMRDNYSMGGSVSGVRPPSILLSDVSPTPLPRPPPLAS I
mAChm1.human     297  - - - - - - - -GS E VVIKM- - - - - - -PMVDPE- - - - - - - - - - - - - - - -A
mAChm1.mouse     297  - - - - - - - -GS E VVIKM- - - - - - -PMVDPE- - - - - - - - - - - - - - - -A
mAChm1.porcine   297  - - - - - - - -GS E VVIKM- - - - - - -PMVDPE- - - - - - - - - - - - - - - -A
mAChm1.rat       297  - - - - - - - -GS E VVIKM- - - - - - -PMVDSE- - - - - - - - - - - - - - - -A
mAChm2.chicken   301  - - - - - - - -EDE AAKDA- - - - - -SQISASQD- - - -HLK- -VENSKLTC I
mAChm2.human     298  - - - - - - - -DD E ITQDE- - - - - -NTVSTSLG- - - -HSK- -DENSKQTC I
mAChm2.porcine   298  - - - - - - - -DD E ITQDE- - - - - -NTVSTSLG- - - -HSK- -DENSKQTC I
mAChm2.rat       298  - - - - - - - -DD E ITQDE- - - - - -NTVSTSLD- - - -HSR- -DDNSKQTC I
mAChm3.human     360  - - - - - - - - - -ETRAIYSIVLKLPG-HSTILNSTKLPSSDNLQVPEEELGMV
mAChm3.porcine   360  - - - - - - - - - -ETRAIYSIVLKLPG-HSTILNSTKLPSSDNLQVPEEELGTV
mAChm3.rat       359  - - - - - - - - - -ETRAIYSIVLKLPG-HSSILNSTKLPSSDNLQVSNEDLGTV
mAChm4.chicken   320  - - - - - - - - -TTE ILPAG- - - - -QGQSPA- - - - - - - -HPR-VNPTSKWSK I
mAChm4.human     310  - - - - - - - - -ATE -LSTT- - - - - -EATTPAMPAPPLQPRALNPASRWSK I
mAChm4.rat       309  - - - - - - - - -PTE -LSTA- - - - - -EATTPALPAPTLQPRTLNPASKWSK I
mAChm4.xenopus   310  - - - - - - - - -PLSEASSG- - - - -VVLAPTQSMPPLPAK-ANTASKWSK I
mAChm5.human     316  - - - - - - - - -ATDPVLQVVYKSQGKESPGEEFSAEETEETFVKAETEKSDY
mAChm5.rat       315  - - - - - - - - -TTDPVFQMVYKSEAKESPGKESNTQETKETVVNTRTENSDY
```

FIGURE 1c.

```
mACh.drosophila  406  S Q L Q E M S A V T A S T T A N V N T S G N G N G A I N N N N N A S H N G N G A V N G N G T G N G S
mAChm1.human     312  Q A P T K Q P P R S S P N T V K - - - - - - - - - - - - - - - - - - - - - - - - - - - - - - - ;
mAChm1.mouse     312  Q A P T K Q P P K S S P N T V K - - - - - - - - - - - - - - - - - - - - - - - - - - - - - - -
mAChm1.porcine   312  Q A P A K Q P P R S S P N T V K - - - - - - - - - - - - - - - - - - - - - - - - - - - - - - -
mAChm1.rat       312  Q A P T K Q P P K S S P N T V K - - - - - - - - - - - - - - - - - - - - - - - - - - - - - - -
mAChm2.chicken   329  R I V T K S Q K G D C C A P T N - - - - - - - - - - - - - - - - - - - - - - - - - - - - - - -
mAChm2.human     326  R I G T K T P K S D S C T P T N - - - - - - - - - - - - - - - - - - - - - - - - - - - - - - -
mAChm2.porcine   326  K I V T K T Q K S D S C T P A N - - - - - - - - - - - - - - - - - - - - - - - - - - - - - - -
mAChm2.rat       326  K I V T K A Q K G D V Y T P T S - - - - - - - - - - - - - - - - - - - - - - - - - - - - - - -
mAChm3.human     400  D L E R K A D K L Q A Q K S V D - - - - - - - - - - - - - - - - - - - - - - - - - - - - - - -
mAChm3.porcine   400  D L E R K A S K L Q A Q K S M D - - - - - - - - - - - - - - - - - - - - - - - - - - - - - - -
mAChm3.rat       399  D V E R N A H K L Q A Q K S M G - - - - - - - - - - - - - - - - - - - - - - - - - - - - - - -
mAChm4.chicken   347  K I V T K Q T G T E S V T A I E - - - - - - - - - - - - - - - - - - - - - - - - - - - - - - -
mAChm4.human     343  Q I V T K Q T G N E C V T A I E - - - - - - - - - - - - - - - - - - - - - - - - - - - - - - -
mAChm4.rat       342  Q I V T K Q T G N E C V T A I E - - - - - - - - - - - - - - - - - - - - - - - - - - - - - - -
mAChm4.xenopus   343  K I V T K Q T G N E C V T A I E - - - - - - - - - - - - - - - - - - - - - - - - - - - - - - -
mAChm5.human     357  D T P N Y L L S P A A A H R P K - - - - - - - - - - - - - - - - - - - - - - - - - - - - - - -
mAChm5.rat       356  D T P K Y F L S P A A A H R L K - - - - - - - - - - - - - - - - - - - - - - - - - - - - - - -

mACh.drosophila  456  G I G L G T T G N A T H R D S R T L P V I N R I N S R S V S Q D S V Y T I L I R L P S D G A S S N A
mAChm1.human     328  - - - - - - - - - - - - - - - - - - - - - - - - - - - - - - - - - - - - - - - R P T K K - - - - - -
mAChm1.mouse     328  - - - - - - - - - - - - - - - - - - - - - - - - - - - - - - - - - - - - - - - R P T K K - - - - - -
mAChm1.porcine   328  - - - - - - - - - - - - - - - - - - - - - - - - - - - - - - - - - - - - - - - R P T R K - - - - - -
mAChm1.rat       328  - - - - - - - - - - - - - - - - - - - - - - - - - - - - - - - - - - - - - - - R P T K K - - - - - -
mAChm2.chicken   345  - - - - - - - - - - - - - - - - - - - - - - - - - - - - - - - - - - - - - - - T T V E I V G T - - -
mAChm2.human     342  - - - - - - - - - - - - - - - - - - - - - - - - - - - - - - - - - - - - - - - T T V E V V G S S G Q
mAChm2.porcine   342  - - - - - - - - - - - - - - - - - - - - - - - - - - - - - - - - - - - - - - - T T V E L V G S S G Q
mAChm2.rat       342  - - - - - - - - - - - - - - - - - - - - - - - - - - - - - - - - - - - - - - - T T V E L V G S S G Q
mAChm3.human     416  - - - - - - - - - - - - - - - - - - - - - - - - - - - - - - - - - - D G G S F P K S F S K L P I Q L E S A V D
mAChm3.porcine   416  - - - - - - - - - - - - - - - - - - - - - - - - - - - - - - - - - - D G G S F Q K S F S K L P I Q L E S A V D
mAChm3.rat       415  - - - - - - - - - - - - - - - - - - - - - - - - - - - - - - - - - - D G D N C Q K D F T K L P I Q L E S A V D
mAChm4.chicken   363  - - - - - - - - - - - - - - - - - - - - - - - - - - - - - - - - - - - - - I V P A K A G A S D H
mAChm4.human     359  - - - - - - - - - - - - - - - - - - - - - - - - - - - - - - - - - - - - - I V P A T P A - - - -
mAChm4.rat       358  - - - - - - - - - - - - - - - - - - - - - - - - - - - - - - - - - - - - - I V P A T P A - - - -
mAChm4.xenopus   359  - - - - - - - - - - - - - - - - - - - - - - - - - - - - - - - - - - - - - I V P E C A I P L P E
mAChm5.human     373  - - - - - - - - - - - - - - - - - - - - - - - - - - - - - - - - - S Q K C V A Y K F - R L V V K A D G N Q E
mAChm5.rat       372  - - - - - - - - - - - - - - - - - - - - - - - - - - - - - - - - - S Q K C V A Y K F - R L V V K A D G T Q E

mACh.drosophila  506  A N G G G G G P G A G A A A S A S L S M Q G D C A P S I K M I H E D G P T T T A A A A P L A S A A A
mAChm1.human     333  - - - G R D - - - - - - - - - - - - - - - - - - - - - - - - - - - - - - - - - - - - - - - - - - -
mAChm1.mouse     333  - - - G R D - - - - - - - - - - - - - - - - - - - - - - - - - - - - - - - - - - - - - - - - - - -
mAChm1.porcine   333  - - - G R E - - - - - - - - - - - - - - - - - - - - - - - - - - - - - - - - - - - - - - - - - - -
mAChm1.rat       333  - - - G R D - - - - - - - - - - - - - - - - - - - - - - - - - - - - - - - - - - - - - - - - - - -
mAChm2.chicken   353  N - - G D E - - - - - - - - - - - - - - - - - - - - - - - - - - - - - - - - - - - - - - - - - - -
mAChm2.human     353  N - - G D E - - - - - - - - - - - - - - - - - - - - - - - - - - - - - - - - - - - - - - - - - - -
mAChm2.porcine   353  N - - G D E - - - - - - - - - - - - - - - - - - - - - - - - - - - - - - - - - - - - - - - - - - -
mAChm2.rat       353  S - - G D E - - - - - - - - - - - - - - - - - - - - - - - - - - - - - - - - - - - - - - - - - - -
mAChm3.human     437  T A K T S D V N S S V G K S T A T L - - - - - - - - - - - - - - - - - - - - - - - - - - - - - -
mAChm3.porcine   437  T A K A S D V N S S V G K T T A T L - - - - - - - - - - - - - - - - - - - - - - - - - - - - - -
mAChm3.rat       436  T G K T S D T N S S A D K T T A T L - - - - - - - - - - - - - - - - - - - - - - - - - - - - - -
mAChm4.chicken   374  N S L S N S - - - - - - - - - - - - - - - - - - - - - - - - - - - - - - - - - - - - - - - - - -
mAChm4.human     366  - - - G M R - - - - - - - - - - - - - - - - - - - - - - - - - - - - - - - - - - - - - - - - - - -
mAChm4.rat       365  - - - G M R - - - - - - - - - - - - - - - - - - - - - - - - - - - - - - - - - - - - - - - - - - -
mAChm4.xenopus   370  Q - - A N N - - - - - - - - - - - - - - - - - - - - - - - - - - - - - - - - - - - - - - - - - - -
mAChm5.human     393  T N N G C H - - - K V K I M P C P F - - - - - - - - - - - - - - - - - - - - - - - - - - - - - - -
mAChm5.rat       392  T N N G C R - - - K V K I M P C S F - - - - - - - - - - - - - - - - - - - - - - - - - - - - - - -
```

FIGURE 1d.

```
mACh.drosophila  556  T R R P L P S R D S E F S L P L G R R M S H A Q H D A R L L N A K V I P K Q L G K A G G G A A G G G
mAChm1.human     336  - - - - - - R A G K G Q K P R G - K - - E - - - - - - - - - - - - - - - - - - - - - - - - - - - - -
mAChm1.mouse     336  - - - - - - R G G K G Q K P R G - K - - E - - - - - - - - - - - - - - - - - - - - - - - - - - - - -
mAChm1.porcine   336  - - - - - - R A G K G Q K P R G - K - - E - - - - - - - - - - - - - - - - - - - - - - - - - - - - -
mAChm1.rat       336  - - - - - - R G G K G Q K P R G - K - - E - - - - - - - - - - - - - - - - - - - - - - - - - - - - -
mAChm2.chicken   357  - - - - - - K Q N S V A R K I V - K M T K - - - - - - - - - - - - - - - - - - - - - - - - - - - - -
mAChm2.human     357  - - - - - - K Q N I V A R K I V - K M T K - - - - - - - - - - - - - - - - - - - - - - - - - - - - -
mAChm2.porcine   357  - - - - - - K Q N I V A R K I V - K M T K - - - - - - - - - - - - - - - - - - - - - - - - - - - - -
mAChm2.rat       357  - - - - - - K Q N V V A R K I V - K M P K - - - - - - - - - - - - - - - - - - - - - - - - - - - - -
mAChm3.human     455  - - - P L S F K E A T L A K R F A L K T R S - - - - - - - - - - - - - - - - - - - - - - - - - - - -
mAChm3.porcine   455  - - - P L S F K E A T L A K R F A L K T R S - - - - - - - - - - - - - - - - - - - - - - - - - - - -
mAChm3.rat       454  - - - P L S F K E A T L A K R F A L K T R S - - - - - - - - - - - - - - - - - - - - - - - - - - - -
mAChm4.chicken   380  - - - - - - R P A N V A R K F A - S I A R - - - - - - - - - - - - - - - - - - - - - - - - - - - - -
mAChm4.human     369  - - - - - - P A A N V A R K F A - S I A R - - - - - - - - - - - - - - - - - - - - - - - - - - - - -
mAChm4.rat       368  - - - - - - P A A N V A R K F A - S I A R - - - - - - - - - - - - - - - - - - - - - - - - - - - - -
mAChm4.xenopus   374  - - - - - - R P V N V A R K F A - S I A R - - - - - - - - - - - - - - - - - - - - - - - - - - - - -
mAChm5.human     408  - - - P V A - K E P S - T K G L N P N P S H - - - - - - - - - - - - - - - - - - - - - - - - - - - -
mAChm5.rat       407  - - - P V S - K D P S - T K G P D P N L S H - - - - - - - - - - - - - - - - - - - - - - - - - - - -
```

```
                                                                                              VI
mACh.drosophila  606  C G A H A L M N P N A A K K K K K S Q E K R Q E S K A A K T L S A I L L S F I I T W T P Y N I L V L
mAChm1.human     348  - - - - - - - - - Q L A K R K T F S L V K - E K K A A R T L S A I L L A F I L T W T P Y N I M V L
mAChm1.mouse     348  - - - - - - - - - Q L A K R K T F S L V K - E K K A A R T L S A I L L A F I L T W T P Y N I M V L
mAChm1.porcine   348  - - - - - - - - - Q L A K R K T F S L V K - E K K A A R T L S A I L L A F I V T W T P Y N I M V L
mAChm1.rat       348  - - - - - - - - - Q L A K R K T F S L V K - E K K A A R T L S A I L L A F I L T W T P Y N I M V L
mAChm2.chicken   371  - - - - - - - - - Q P A K K K P P - P S R - E K K V T R T I L A I L L A F I I T W T P Y N V M V L
mAChm2.human     371  - - - - - - - - - Q P A K K K P P - P S R - E K K V T R T I L A I L L A F I I T W A P Y N V M V L
mAChm2.porcine   371  - - - - - - - - - Q P A K K K P P - P S R - E K K V T R T I L A I L L A F I I T W A P Y N V M V L
mAChm2.rat       371  - - - - - - - - - Q P A K K K P P - P S R - E K K V T R T I L A I L L A F I I T W A P Y N V M V L
mAChm3.human     474  - - - - - - - - - Q I T K R K R M S L V K - E K K A A Q T L S A I L L A F I I T W T P Y N I M V L
mAChm3.porcine   474  - - - - - - - - - Q I T K R K R M S L I K - E K K A A Q T L S A I L L A F I I T W T P Y N I M V L
mAChm3.rat       473  - - - - - - - - - Q I T K R K R M S L I K - E K K A A Q T L S A I L L A F I I T W T P Y N I M V L
mAChm4.chicken   394  - - - - - - - - - S Q V R K K R Q M A A R - E K K V T R T I F A I L L A F I L T W T P Y N V M V L
mAChm4.human     383  - - - - - - - - - N Q V R K K R Q M A A R - E R K V T R T I F A I L L A F I L T W T P Y N V M V L
mAChm4.rat       382  - - - - - - - - - N Q V R K K R Q M A A R - E R K V T R T I F A I L L A F I L T W T P Y N V M V L
mAChm4.xenopus   388  - - - - - - - - - N Q V R K K R Q M A A R - E K K V T R T I F A I L L A F I I T W T P Y N V M V L
mAChm5.human     425  - - - - - - - - - Q M T K R K R V V L V K - E R K A A Q T L S A I L L A F I I T W T P Y N I M V L
mAChm5.rat       424  - - - - - - - - - Q M T K R K R M V L V K - E R K A A Q T L S A I L L A F I I T W T P Y N I M V L
```

```
                                                                        VII
mACh.drosophila  656  I K P L T T C S D C I P T E L W D F F Y A L C Y I N S T I N P M S Y A L C N A T F R R T Y V R I L T
mAChm1.human     387  V S T F - - C K D C V P E T L W E L G Y W L C Y V N S T I N P M C Y A L C N K A F R D T F R L L L L
mAChm1.mouse     387  V S T F - - C K D C V P E T L W E L G Y W L C Y V N S T V N P M C Y A S C N K A F R D H F R L L L L
mAChm1.porcine   387  V S T F - - C K D C V P E T L W E L G Y W L C Y V N S T I N P M C Y A L C N K A F R D T F R L L L L
mAChm1.rat       387  V S T F - - C K D C V P E T L W E L G Y W L C Y V N S T V N P M C Y A L C N K A F R D T F R L L L L
mAChm2.chicken   409  I N S F - - C A S C I P G T V W T I G Y W L C Y I N S T I N P A C Y A L C N A T F K K T F K H L L M
mAChm2.human     409  I N T F - - C A P C I P N T V W T I G Y W L C Y I N S T I N P A C Y A L C N A T F K K T F K H L L M
mAChm2.porcine   409  I N T F - - C A P C I P N T V W T I G Y W L C Y I N S T I N P A C Y A L C N A T F K K T F K H L L M
mAChm2.rat       409  I N T F - - C A P C I P N T V W T I G Y W L C Y I N S T I N P A C Y A L C N A T F K K T F K H L L M
mAChm3.human     513  V N T F - - C D S C I P K T F W N L G Y W L C Y I N S T V N P V C Y A L C N K T F R T T F K M L L L
mAChm3.porcine   513  V N T F - - C D S C I P K T Y W N L G Y W L C Y I N S T V N P V C Y A L C N K T F R T T F K M L L L
mAChm3.rat       512  V N T F - - C D S C I P K T Y W N L G Y W L C Y I N S T V N P V C Y A L C N K T F R T T F K T L L L
mAChm4.chicken   433  I N T F - - C E T C V P E T V W S I G Y W L C Y V N S T I N P A C Y A L C N A T F K K T F K H L L M
mAChm4.human     422  V N T F - - C Q S C I P D T V W S I G Y W L C Y V N S T I N P A C Y A L C N A T F K K T F R H L L L
mAChm4.rat       421  V N T F - - C Q S C I P E R V W S I G Y W L C Y V N S T I N P A C Y A L C N A T F K K T F R H L L L
mAChm4.xenopus   427  I N T F - - C Q T C I P E T I W Y I G Y W L C Y V N S T I N P A C Y A L C N A T F K K T F K H L L M
mAChm5.human     464  V S T F - - C D K C V P V T L W H L G Y W L C Y V N S T V N P I C Y A L C N R T F R K T F K M L L L
mAChm5.rat       463  V S T F - - C D K C V P V T L W H L G Y W L C Y V N S T I N P I C Y A L C N R T F R K T F K L L L L
```

FIGURE 1e.

```
mACh.drosophila   706   C K W H T R - - - - - - - - - N R E G M V R G V Y N - - -
mAChm1.human      435   C R W D K R R W R K I - - - - P K R P G S V H R T P S R Q C
mAChm1.mouse      435   C R W D K R R W R K I - - - - P K R P G S V H R T P S R Q C
mAChm1.porcine    435   C R W D K R R W R K I - - - - P K R P G S V H R T P S R Q C
mAChm1.rat        435   C R W D K R R W R K I - - - - P K R P G S V H R T P S R Q C
mAChm2.chicken    457   C H Y - - - - - - - - - - - - K N I G A T R - - - - - - -
mAChm2.human      457   C H Y - - - - - - - - - - - - K N I G A T R - - - - - - -
mAChm2.porcine    457   C H Y - - - - - - - - - - - - K N I G A T R - - - - - - -
mAChm2.rat        457   C H Y - - - - - - - - - - - - K N I G A T R - - - - - - -
mAChm3.human      561   C Q C D K K K R R K Q Q Y Q Q R Q S V I F H K R A P E Q A L
mAChm3.porcine    561   C Q C D K R K R R K Q Q Y Q Q R Q S V I F H K R V P E Q A L
mAChm3.rat        560   C Q C D K R K R R K Q Q Y Q Q R Q S V I F H K R V P E Q A L
mAChm4.chicken    481   C Q Y - - - - - - - - - - - - R N I G T A R - - - - - - -
mAChm4.human      470   C Q Y - - - - - - - - - - - - R N I G T A R - - - - - - -
mAChm4.rat        469   C Q Y - - - - - - - - - - - - R N I G T A R - - - - - - -
mAChm4.xenopus    475   C Q Y - - - - - - - - - - - - K S I G T A R - - - - - - -
mAChm5.human      512   C R W K K K K V E E K L Y - - W Q G - - - N S K L P - - - -
mAChm5.rat        511   C R W K K K K V E E K L Y - - W Q G - - - N S K L P - - - -
```

FIGURE 1f.

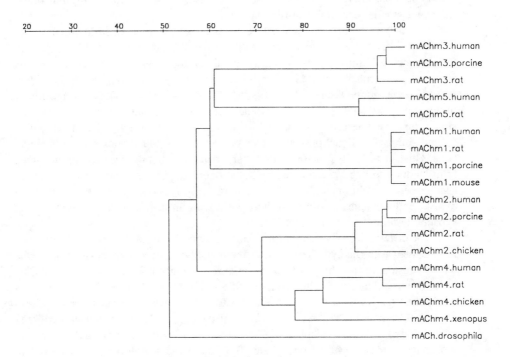

FIGURE 2. Phylogenetic tree of acetylcholine receptors. The phylogenetic tree was constructed according to the method of Feng and Doolittle (1990). The length of each "branch" of the tree correlates with the evolutionary distance between receptor sub-populations.

Prior to molecular cloning of mAChR cDNAs, analysis of radioligand binding to tissue sections was the primary method used to determine the location of mAChR subtypes in different tissues. This approach has been particularly useful for mapping mAChR sites in various brain regions. However, due to the limited subtype selectivity of existing muscarinic ligands, such studies were not conclusive. The isolation of mAChR subtype genes has allowed the generation of subtype-specific probes for mapping mAChR mRNA expression. Studies with such probes have shown that peripheral tissues express a subset of mAChR subtypes, while regions of the brain have a more complex, multiple subtype pattern of expression (Kubo et al., 1986a,b; Peralta et al., 1987b, Bonner et al., 1987; Maeda et al., 1988). For example, only the m2 subtype is expressed in mammalian heart and only m3 is expressed in pancreas. In contrast, exocrine tissues such as lacrimal and parotid glands express both the m1 and m3 subtypes, while in smooth muscle tissue, both m2 and m3 are expressed. All mAChR subtypes, except m5, are detected readily by Northern blot analysis of whole brain mRNA. However, the relative abundance of mAChR subtype mRNA in subregions of the brain varies significantly. The m1 and m4 subtypes are most abundant in the cerebral cortex, although m3 expression is also detectable. In contrast, m2 expression is prevalent in the cerebellum and brainstem. *In situ* hybridization studies (Brann et al., 1988; Weiner et al., 1990) have demonstrated that m1 and m4 are expressed in the cerebral cortex, hippocampus, and striatum. Expression of m3 is found in similar regions, as well as in brainstem regions. In contrast, m2 expression is detected most clearly in brainstem regions, such as the pons, but also in regions of the frontal cortex; m5 expression is seen in the substantia nigra.

Complexity in mAChR subtype expression is observed even at the level of individual cells. For example, a study with a panel of brain-derived cell lines showed that certain cell lines express one, while others coexpress two or more mAChR subtypes (Ashkenazi et al., 1988). Similar complexity was observed in individual neurons of rat brain (Weiner et al., 1990; Bernard et al., 1992). The coexpression of multiple mAChR subtypes may contribute to the elicitation of complex signals within single cells. It is interesting that mAChRs of similar structure, e.g., m2 and m4, which exhibit similar functional properties (see below), can be found coexpressed in single cells. This suggests the possibility that there is overlap, but not identity, in the signal transduction mechanism employed by homologous mAChR subtypes coexpressed in a single cell. In addition, it is possible that homologous subtypes interact with different cytoskeletal elements, resulting in their positioning in distinct cellular regions, e.g., synaptic terminals vs. cell bodies.

The differential expression of mAChR subtypes in various tissues and cell types is consistent with the hypothesis that different subtypes have distinct functions. This was suggested, in fact, by biochemical studies that preceded the cloning of mAChR genes (reviewed by Birdsall and Hulme, 1983 and by Nathanson, 1987). These studies established that mAChRs in brain tissue and certain neuronal cell lines activate a phosphoinositide (PI)-specific phospholipase C (PLC), which hydrolyzes phosphatidylinositol 4,5 bisphosphate to produce two important second messenger molecules: inositol 1,4,5-trisphosphate (IP3) and 1,2-diacylglycerol (DAG). IP3 in turn mediates the release of Ca^{2+} from intracellular stores, while DAG activates the protein kinase C (PKC) family of serine/threonine kinases (reviewed by Berridge and Irvine, 1989). In contrast, it was found that activation of mAChRs in other tissues or cells, notably heart or GH3 pituitary cells, leads to inhibition of adenylyl cylcase activity. In addition, it was observed that activation of mAChRs in endothelial cells or in N1E-115 neuroblastoma cells leads to stimulation of guanylyl cyclase activity (Nathanson, 1987). These studies demonstrated that mAChRs from different cells or tissues regulate several second messenger pathways; however, the contribution of individual subtypes to each response was not discernable, since subtype identification was based on ligands with partial selectivity.

More recently, expression of cloned mAChR subtypes in cells that lack endogenous mAChRs has enabled investigators to assign specific functions to individual mAChR subtypes. The feasibility of this approach was established in studies with recombinant porcine m2 mAChR transfected into Chinese hamster ovary (CHO) cells (Ashkenazi et al., 1987). These studies demonstrated that a transfected exogenous receptor can regulate the signal transduction machinery of a host cell, through interaction with endogenous G proteins. The results of several studies, cited in the sections below, which have characterized the effector-coupling properties of individual mAChR subtypes, suggest the following principles. First, subtypes with similar primary structures employ similar signal transduction pathways. For example, the m1, m3, and m5 subtypes couple primarily to PLC-mediated PI hydrolysis; whereas, the m2 and m4 subtypes couple primarily to adenylyl cyclase inhibition. Second, a given receptor subtype can regulate multiple effector systems, either in different cellular contexts or even within a single cell. The first example of this principle was the observation that the cloned m2 mAChR, when expressed in CHO cells, can mediate both the inhibition of adenylyl cyclase and the activation of PLC (Ashkenazi et al., 1987). Similarly, the m1, m3, or m5 subtypes, when expressed in certain cells, can mediate the activation of PLC, PLA_2, PLD, and adenylyl cyclase (Peralta et al., 1988; Conklin et al., 1988; Lechleiter et al., 1991b). Recent studies suggest that at least in some cases (i.e., m2 mAChR), simultaneous coupling of a single receptor to two effector systems is achieved by activation of a single G protein whose released subunits, in turn, regulate different effector enzymes (i.e., $G_{\alpha i}$ inhibits adenylyl cyclase while $G_{\beta \gamma}$ activates $PLC_{\beta 2}$ (Katz et al., 1992).

Studies on the electrophysiological effects of individual mAChR subtypes have yielded results which are consistent with the notion that mAChR subtypes of related structure carry out similar functions. For example, cloned m1 and m3, but not m2 or m4, mAChRs expressed in NG108-15 neuroblastoma x glioma cells couple to the inhibition of a time- and voltage-dependent K^+ current, termed the M-current, by an unknown mechanism (Fukuda et al., 1988; Pfaffinger et al., 1988). In contrast, m2 mAChRs in pacemaker and atrial cells activate an inwardly rectifying K^+ channel, leading to hyperpolarization of the cell membrane and to inhibition in the rate of heart contraction (Brown and Birnbaumer, 1990). Regulation of this channel involves a unique mechanism which does not require classical second-messenger molecules such as cAMP or Ca^{2+}. Experiments with atrial patches perfused with purified G protein subunits indicate that the G_α subunit may mediate a more direct activation of the channel, while activation by $G_{\beta \gamma}$ subunits may occur through generation of lipophilic channel activators by stimulation of PLA_2 (reviewed by Brown and Birnbaumer, 1990). Similar to m2 mAChRs, m4 receptors are capable of activating an inward rectifying K^+ conductance in the GH3 and AtT20 cell lines (Codina et al., 1987).

The principal mAChR sequences controlling the specificity of G protein activation have been identified by studying hybrid receptors derived from the functionally distinct m2 and m3 subtypes (Lechleiter et al., 1990). Agonist stimulation of m1 or m3 mAChRs expressed in *Xenopus* oocytes results in a strong and rapid activation of Ca^{2+} dependent chloride currents, while the m2 and m4 subtypes generate a weak, delayed, and oscillatory activation of these currents. Activation of a pertussis toxin (PTX)-insensitive G protein pathway, leading to a rapid and transient release of intracellular Ca^{2+} characteristic of the m3 receptor, could be specified by the exchange of as few as nine amino acids from the m3 to the m2 receptor. In a reciprocal manner, transfer of no more than 21 residues from the m2 to the m3 receptor was sufficient to specify activation of a PTX-sensitive G protein, coupled to the slow and oscillatory Ca^{2+} release pathway typical of the m2 mAChR subtype. In each case, the critical residues conferring G protein selectivity occur within the region of the third cytoplasmic loop adjacent to the fifth transmembrane segment. Other mutagenesis studies suggest that the regions of 9 through 11 amino acids adjacent to both the fifth or the sixth transmembrane segments of the m1 mAChR are important for coupling to PI hydrolysis (Arden et al., 1992). Experiments with chimeric m1 and m2 receptors expressed in COS-7 or HEK 293 cells identified the same region of the third cytoplasmic loop as a critical determinant for G protein specificity, and showed that other regions, perhaps the second cytoplasmic domain, also are involved but are less important (Wong et al., 1990; Lai et al., 1992). In addition, synthetic peptides based on the C-terminal region of the third cytoplasmic loop of the human m4 mAChR (residues 382 to 400) had the capacity to stimulate G_i and G_o proteins at nanomolar concentrations (Okamoto and Nishimoto, 1992). Interestingly, m3 mAChR-mediated activation of Ca^{2+} influx in A9 fibroblasts, which is independent of PI hydrolysis and mobilization of intracellular Ca^{2+}, was found to involve a region of the m3 receptor outside of the third cytoplasmic loop (Felder et al., 1992).

In summary, the family of mAChR subtypes exemplifies many of the themes common to G protein-coupled receptors in general. First, there exist multiple subtypes of the receptor for a given neurotransmitter. These subtypes have homologous but distinct amino acid sequences and can be divided into subgroups of subtypes with greater sequence similarity. Second, each receptor subtype has a unique pattern of expression in tissues and organs and in subregions therein. Some cell types express a single receptor subtype, while others express multiple subtypes. Third, each receptor subtype couples to a specific set of G proteins and effector enzymes and/or ion channels. While the effector-coupling properties of each subtype are distinct, there is significant overlap, especially between subtypes that are more similar in sequence. Therefore, the effect of a neurotransmitter such as acetylcholine in a particular cell or tissue will depend on the particular make-up of receptor subtypes and cognate G proteins and effectors present, and the particular physiology of that cell or tissue. For example, in mammalian heart muscle cells, acetylcholine will activate an m2 receptor subtype, which, via a specific G protein, will inhibit the activity of an adenylyl cyclase effector enzyme. In addition, the m2 receptor will activate an inward rectifying K^+ channel via another G protein. The result would be inhibition in the rate and force of contraction of the heart muscle. Thus, the multiplicity of receptor subtypes, G proteins and effectors employed by a given neurotransmitter or hormone enables a great diversity in the repertoire of biological responses that it can evoke.

1.1.1   m1 Muscarinic Acetylcholine Receptors

The M_1 mAChR subtype was defined pharmacologically based on relatively high affinity for pirenzepine (K_D = 25 to 50 nM), and was reported to be present mainly in neuronal tissue (Hammer et al., 1980). Efforts to isolate cDNA and genomic clones encoding mAChRs began with the purification of mAChRs from porcine cerebrum, using affinity chromatography (Haga and Haga, 1985). Probes based on partial amino acid sequence of the purified receptor were used to screen a porcine cerebrum cDNA library, from which clones encoding the mAChR were isolated (Kubo et al., 1986a). The cDNA sequence predicted a polypeptide of 460 amino acid residues and a calculated M_r of ~51,000. This could be reconciled with the M_r of ~70,000 seen with the purified mAChR on SDS-polyacrylamide gels by the presence of two potential sites for N-linked glycosylation near the predicted amino terminus.

Hydropathicity analysis of the sequence showed the presence of seven hydrophobic segments, similar to bacterial and mammalian rhodopsins and to the β-adrenergic receptor. Sequence homology with those receptors was also apparent, especially within the hydrophobic segments. Therefore, it was proposed by Kubo et al. that the mAChR assumes a membrane topology similar to that of bacteriorhodopsin, which had been shown by electron-diffraction studies to contain seven α-helical segments that span the membrane (Ovchinnikov, 1982).

Northern blot analysis of mRNA derived from different rat brain regions and from rat atrium using the m1 receptor cDNA as a probe showed expression in the brain in cerebral cortex and corpus striatum, but not in medulla pons, and not in the atrium. This mRNA localization was consistent with the tissue-distribution of the M1 subtype observed in radioligand binding studies, thus supporting the notion that the cloned receptor was identical to the pharmacologically defined M_1 subtype.

1.1.1.1 Human m1 mACh Receptors

The human m1 mAChR gene was isolated (along with genes encoding other mAChR subtypes) from a human genomic library using a hybridization probe based on the porcine m2 mAChR (Peralta et al., 1987b). This approach took advantage of the observation that the related porcine m2 gene did not contain introns within the coding region (Peralta et al., 1987a). Indeed, the human m1 gene proved also to be encoded by a single exon. The DNA sequence predicted a 460 amino acid polypeptide, with 2 potential N-linked glycosylation sites in the N-terminal region.

Due to the scarcity of human tissue samples, virtually no studies have been done to map m1 receptor expression in human tissues. Northern blot analysis of a panel of cell lines derived from human brain tumors, including H4 neuroglioma, TE671 medulloblastoma, A172, U87, U138, and U373 glioblastoma, SK-N-MC and SK-N-SH neuroblastoma, and 1321N1 neuroblastoma, failed to detect expression of m1 mRNA (although expression of other mAChR subtype mRNAs was detected). These results, taken together with the results of studies with animal brain tissue, suggest a highly specific pattern of expression of the m1 subtype in the human central nervous system.

The pharmacological properties of the cloned human m1 mAChR were investigated by transfection of the cloned DNA into human embryonic kidney (HEK) 293 cells (Peralta et al., 1987b). The human m1 receptor exhibited relatively high affinity for the M1-selective antagonist pirenzepine (25-fold higher than human m2 mAChR), and relatively low affinity for the M2-selective antagonist AF-DX 116 (>100-fold lower than human m2 mAChR).

The functional properties of the cloned human m1 mAChR have been investigated in transfected HEK 293 cells (Peralta et al., 1988). These studies have demonstrated efficient coupling of the human m1 mAChR to activation of PLC-mediated PI hydrolysis. No inhibition of adenylyl cyclase was observed; in fact, cAMP levels were elevated upon m1 activation. A similar elevation in cAMP levels mediated by m1 mAChRs was observed in transfected murine A9 L cells, and it was proposed that this effect is secondary to PI hydrolysis (Felder et al., 1989). However, recent studies in human SK-N-SH neuroblastoma cells (reviewed by Baumgold, 1992) suggest that m1-mediated adenylyl cyclase activation is independent of Ca^{2+}-mobilization or PKC stimulation, which result from PI hydrolysis. This led to the hypothesis that adenylyl cyclase activation by PLC-coupled mAChRs occurs by a G protein activation that leads directly to activation of adenylyl cyclase, by the released G protein $\beta\gamma$ subunit complex (Baumgold, 1992). In addition to the responses listed above, human m1 mAChRs expressed in HEK 293 cells exhibit efficient activation of PLD (Sandman et al., 1991).

Studies with human m1 mAChR expressed in CHO cells (Ashkenazi et al., 1989a) confirmed the efficient coupling of this subtype to PI hydrolysis. Pertussis toxin (PTX), an agent which uncouples certain G proteins from their cognate receptors by adenosine diphosphate (ADP) ribosylation of the G_α subunit, has little effect on this response, suggesting the involvement of a PTX-insensitive G protein in m1 coupling to PLC. Other studies in CHO cells showed human m1 mAChR-mediated mobilization of Ca^{2+} from internal stores (Lechleiter et al., 1989). Like the PI hydrolysis response, the Ca^{2+} response, which is secondary to PI hydrolysis, was mostly unaffected by PTX. Similarly, m1-mediated activation of PLD in HEK 293 cells was mostly resistant to PTX (Sandman et al., 1991). Reconstitution experiments with purified proteins indicate that the human m1 mAChR can activate PI hydrolysis by PLC-β_1 via the PTX-insensitive G protein $G_{q/11}$ (Berstein et al., 1992).

Studies with human m1 mAChR in CHO cells showed efficient coupling to PI hydrolysis; in addition, induction of protooncogene *c-fos* mRNA and activation of DNA synthesis were observed, which may be secondary responses to PI hydrolysis (Ashkenazi et al., 1989b,c). Muscarinic activation of DNA synthesis was observed also in primary cultures of perinatal rat astroglial cells (Ashkenazi et al., 1989c; Zohar and Salomon, 1992), suggesting a potential involvement of m1 or m1-like mAChRs in the regulation of astroglial cell growth in the developing brain. Human m1 mAChRs expressed in NIH 3T3 cells can act as conditional oncogenes, further suggesting that PI-coupled mAChRs may be involved in regulating the growth of certain proliferating cells (Gutkind et al., 1991). Human m1 mAChRs expressed in CHO cells also were found to inhibit endocytosis and endosome trafficking (Haraguchi and Rodbell, 1991). The m1-regulated endosomes appeared to contain increased levels of G proteins, suggesting that mAChRs may regulate cellular signal transduction elements at the level of trafficking, in addition to the classical regulation of second messenger systems. Finally, human m1 mAChR expressed in HEK 293 cells was found to stimulate the release of the amyloid precursor protein, suggesting that m1 mAChRs may be involved in regulating events leading to the formation of amyloid deposits in the brain of patients with Alzheimer's disease (Nitsch et al., 1992).

1.1.1.7 Porcine m1 mACh Receptors

As described in Section 1.1.1, the porcine m1 subtype was the first mAChR to be cloned (Kubo et al., 1986a). Northern blot analysis of mRNA from porcine tissues revealed m1 expression within the brain in the cerebrum and cortex but not in corpus striatum or medulla pons (Kubo et al., 1986a; Maeda et al., 1988). In addition, m1 mRNA expression was detected in porcine lacrimal and parotid glands, but not in small and large intestines, trachea, urinary bladder, or atrium.

The cloned porcine m1 mAChR was investigated by expressing mRNA derived from its cDNA in *Xenopus* oocytes (Kubo et al., 1986a). Binding experiments indicated high affinity for pirenzepine (K_D = 25 to 50 nM), suggesting an M_1 subtype pharmacology. Functional analysis demonstrated an ability of the porcine m1 mAChR to regulate electrophysiological activity, primarily of chloride channels, in the injected oocytes. Additional studies with the porcine m1 mAChR expressed in NG108-15 neuroblastoma x glioma cells have demonstrated its ability to mediate the activation of PI hydrolysis and of intracellular Ca^{2+} mobilization, the activation of a Ca^{2+}-dependent K^+ current, and the inhibition of a voltage-sensitive K^+ current (M-current) (Fukuda et al., 1988; Neher et al., 1988).

1.1.1.10   Rat m1 mACh Receptors

Rat m1 mAChR cDNA was isolated, among other mAChR cDNAs, from a rat cerebral cortex library (Bonner et al., 1987), using a hybridization probe based on the previously cloned cDNA of the porcine cerebral m1 mAChR (Kubo et al., 1986a). The cDNA sequence predicted a 458-amino acid protein with 2 amino-terminal potential glycosylation sites and 98% identity with the porcine m1 receptor.

Northern hybridization studies of rat tissues indicated m1 mRNA expression in certain regions of the brain, but not heart or pancreas (Bonner et al., 1987; Peralta et al., 1987b). *In situ* hybridization studies showed that in the rat brain, m1 receptors are most abundant in cerebral cortex, striatum, and hippocampus (Brann et al., 1988; Weiner et al., 1990). At the cellular level, m1 mRNA was observed in ~80% of cholinergic neurons of rat striatum, coexpressed with m2 and m4 receptors (Bernard et al., 1992). Northern blot analysis of mRNA from NG108-15 mouse neuroblastoma x rat glioma cells and from PC-12 rat adrenal pheochromocytoma cells failed to detect expression of the m1 mAChR (Ashkenazi et al., 1988).

Pharmacological characterization of the rat m1 receptor expressed in transfected COS-7 cells (Bonner et al., 1987) indicated relatively high affinity for pirenzepine. Similar results were obtained upon expression of the rat m1 mAChR in CHO-K1 cells (Buckley et al., 1989). Biochemical studies with the rat m1 receptor expressed in murine A9 L cells (Conklin et al., 1988) have yielded similar findings to those obtained with human m1 receptors expressed in HEK 293 cells. Rat m1 mAChRs coupled efficiently to activation of PLC, PLA_2, as well as to adenylyl cyclase. In murine A9 L cells expressing rat m1 mAChRs, inhibition of DNA synthesis was observed (Conklin et al., 1988). This is in contrast to the activation of DNA synthesis observed for human m1 mAChRs expressed in CHO cells, and for endogenous mAChRs in certain brain-derived cell lines, in primary rat astrocytes (Ashkenazi et al., 1989c; Zohar and Salomon, 1992), and in primary parotid gland cells (Kikuchi et al., 1987). The inhibition of DNA synthesis also appears to be in contrast to the conditional oncogenic activity of human m1 mAChRs in NIH 3T3 cells (Gutkind et al., 1991). Growth factor-like effects such as stimulation of neurite outgrowth also have been reported for rat m1 mAChRs expressed by transfection in rat pheochromocytoma PC12 cells (Pinkas-Kramarski et al., 1992). The apparent discrepancy between the stimulatory and inhibitory growth factor-like effects of m1 mAChRs may be attributable to the different cellular contexts in which these effects were observed. Differential effects of cloned m1 mAChRs expressed in different cell types also have been observed with respect to adenylyl cyclase: in HEK 293 cells or in murine A9 L cells, m1 mAChRs mediate elevation of cAMP levels (Peralta et al., 1988; Conklin et al., 1988), whereas in RAT-1 cells, inhibition of adenylyl cyclase is observed (Stein et al., 1988).

1.1.1.11   Mouse m1 mACh Receptors

The mouse m1 mAChR gene was isolated from a mouse genomic library, using a full length cDNA of the porcine m2 receptor as a hybridization probe (Shapiro et al., 1988). When expressed in mouse Y1 cells, the cloned mouse m1 mAChR exhibited high affinity for pirenzepine ($K_D = 10$ nM) and efficient activation of PI hydrolysis, which was not inhibited by PTX.

1.1.2   m2 Muscarinic Acetylcholine Receptors

The pharmacological M_2 subtype was defined based on low affinity for pirenzepine ($K_D = 1$ to $5 \mu M$), and was reported to be detectable mainly in heart and smooth muscle tissue (reviewed by Birdsall and Hulme, 1983). The cardiac m2 mAChR cDNA was isolated from a porcine atrial cDNA library by screening with hybridization probes based on partial amino acid sequence of purified atrial receptor protein (Peralta et al., 1987a). A similar cDNA was isolated from the same porcine cerebrum cDNA library used initially to isolate the m1 mAChR cDNA (Kubo et al., 1986a). The predicted amino acid sequence of the cardiac receptor shared 38% overall homology with the porcine m1 mAChR, and hydropathicity analysis suggested a similar seven transmembrane helix topology. Greater sequence homology with the m1 mAChR was observed in the predicted transmembrane domains (51 to 91%) and the small connecting loops (44 to 67%).

Northern blot analysis of mRNA derived from different porcine brain regions and from rat atrium using the m2 cDNA as a probe showed the most expression in atrium and significant expression in the medulla pons (Kubo et al., 1986a). This mRNA localization was consistent with the previously documented cardiac localization of M_2 radioligand binding sites, and suggested that m2 mAChRs may be expressed also in certain brain regions. Pharmacological characterization of the cloned m2 receptor expressed in CHO cells showed low affinity for pirenzepine, confirming its identification as an M_2 receptor (Peralta et al., 1987a). Southern blot analysis of porcine and human genomic DNA using the m2 cDNA as a hybridization probe revealed that the m2 receptor was encoded by a single gene. One to three additional weakly hybridizing bands were observed, suggesting the presence of genes encoding additional mAChR subtypes or related receptors.

1.1.2.1   Human m2 mACh Receptors

The human m2 mAChR gene was isolated (along with genes encoding other mAChR subtypes) from human genomic libraries (Bonner et al., 1987; Peralta et al., 1987b). This mAChR, like the m1 mAChr, proved to be encoded by a single exon. The DNA sequence predicted a polypeptide containing 466 amino acids sharing 97.4% identity with the porcine cardiac m2 mAChR.

Direct studies are lacking on the location of m2 receptors in human tissues. Northern blot analysis of a panel of human cell lines derived from brain tumors, indicated m2 mRNA expression in the SK-N-SH neuroblastoma cell line (together with higher levels of m3 mAChR mRNA), but not in the H4 neuroglioma, TE671 medulloblastoma, A172, U87, U138, and U373 glioblastoma, SK-N-MC neuroblastoma, and 1321N1 neuroblastoma cell lines (Ashkenazi et al., 1988).

The pharmacological properties of the cloned human m2 mAChR were investigated by transfection of the DNA into HEK 293 cells (Peralta et al., 1987b). The human m2 receptor exhibited relatively high affinity for the M_2-selective antagonist AF-DX 116 and low affinity for the M_1-selective antagonist pirenzepine. Similar results were obtained upon expression of human m2 mAChR in CHO-K1 cells; in addition, the human m2 mAChR exhibited relatively high affinity for methoctramine (Buckley et al., 1989).

The functional properties of the cloned human m2 mAChR also were investigated in HEK 293 cells (Peralta et al., 1988; Lechleiter et al., 1989). These studies demonstrated efficient m2 mAChR coupling to inhibition of adenylyl cyclase activity. In addition, weak but significant activation of PI hydrolysis and of intracellular Ca^{2+} release was seen. The magnitude of PI hydrolysis and Ca^{2+} release was much smaller than that induced by m1 mAChRs in HEK 293 cells. In addition, m2-mediated PI hydrolysis was inhibited completely by PTX, whereas the m1 response was mostly insensitive to PTX (Ashkenazi et al., 1989b; Lechleiter et al., 1989). These results suggested the involvement of distinct G proteins and/or PLC isozymes in the coupling of m2 or m1 receptors to PLC. This notion is supported by the recent observation of m2-mediated PTX-sensitive activation of $PLC_{\beta 2}$ via the $\beta\gamma$ subunit complex of a G_i protein (Katz et al., 1992) and of m1-mediated activation of $PLC_{\beta 1}$ via the PTX-insensitive α subunit of the G_q protein (Berstein et al., 1992). Activation of PLD by human m2 mAChR also was observed in HEK 293 cells; this response was markedly weaker than that evoked by human m1 or m3 mAChRs (Sandman et al., 1991). The coupling of human m2 mAChRs to intracellular Ca^{2+} release has been investigated also in *Xenopus* oocytes injected with mAChR mRNAs. In this system, Ca^{2+}-dependent chloride channels were activated differentially by human m2 vs. m3 mAChRs (Lechleiter et al., 1990). Confocal microscopy demonstrated that the pathways of Ca^{2+} release mediated by the m2 and m3 mAChRs in the oocyte system are distinguishable in the spatial domain (Lechleiter et al., 1991a). At submaximal acetylcholine activation, both receptors stimulate pulses of focal Ca^{2+} release. However, maximal stimulation of m2 receptors increases the number of focal release sites, while m3 receptors evoke a Ca^{2+} wave propagating rapidly beneath the plasma membrane. The focal Ca^{2+} release pattern regulated by m2 receptors is completely inhibited by PTX, while the Ca^{2+} waves evoked by m3 are insensitive to the toxin, indicating that these responses are mediated by different G proteins. Heparin, an inhibitor of IP3-gated intracellular Ca^{2+} channels, blocks both Ca^{2+} responses, indicating the involvement of PLC-mediated formation of IP3 in both the m2 and m3 responses (Lechleiter et al., 1991a). Analysis of the mechanisms underlying these distinct patterns of Ca^{2+} release led to the discovery of novel regenerative Ca^{2+} signals, which behave according to chemical wave theory (Lechleiter et al., 1991b).

1.1.2.7 Porcine m2 mACh Receptors

The porcine m2 mAChR cDNA was cloned from a porcine cerebrum cDNA library (Kubo et al., 1986b) and from a porcine atrial cDNA library (Peralta et al., 1987a). The cDNA sequence predicted a polypeptide of 466 amino acid residues and a caluulated M_r of 51,700. Northern blot analysis of porcine tissue mRNA indicated m2 expression in atrium, urinary bladder, trachea, small and large intestines, and parotid gland (Kubo et al., 1986b; Maeda et al., 1988). Within the porcine brain, significant expression of m2 mRNA was detected in cerebrum and medulla pons, and little expression in cortex and corpus striatum.

The pharmacology of the porcine m2 receptor is described in Section 1.1.2 above. Functional characterization of the cloned receptor expressed in NG108-15 cells showed lack of activation of PI hydrolysis and of a Ca^{2+} dependent K^+ current and lack of inhibition of the M-current (Fukuda et al., 1988). Studies in transfected CHO cells showed efficient coupling of the porcine m2 mAChR to inhibition of adenylyl cyclase, as well as weak, but significant activation of PI hydrolysis (Ashkenazi et al., 1987). Recent studies involving cotransfection of m2 with various G protein subunits and PLC isozymes into COS-7 cells indicate that simultaneous activation of two effector systems by m2 is mediated by the α and the $\beta\gamma$ subunits of a G_i protein, which activate adenylyl cyclase or $PLC_{\beta 2}$, respectively (Katz et al., 1992). The physiological significance of the activation of PI hydrolysis by m2 mAChRs is not known. The induction of the second messengers IP3 and Ca^{2+} mediated by m2 differs in magnitude, kinetics, and subcellular location from that evoked by m1, m3, or m5 mAChRs. Thus, m2-

mediated activation of PI hydrolysis may be a means of attaining greater plasticity in the response to acetylcholine, e.g., by brain cholinergic neurons involved in learning and memory.

1.1.2.10   Rat m2 mACh Receptors

The rat m2 mAChR cDNA was isolated from a rat heart cDNA library by screening with a hybridization probe based on the porcine m2 mAChR sequence (Gocayne et al., 1987). The DNA sequence predicted a polypeptide of 466 amino acids with a caluclated M_r of 51,500, sharing 97.6% homology with the porcine m2 mAChR.

Northern blot hybridization studies showed m2 mRNA expression in rat heart and whole brain (Peralta et al., 1987b). *In situ* hybridization studies with rat brain sections indicated m2 mRNA expression mostly in brain stem regions such as the pons, but also in regions of the frontal cortex (Brann et al., 1988). Other studies have demonstrated m2 expression in cholinergic neurons of the rat striatum (Bernard et al., 1992).

1.1.2.15   Chicken m2 mACh Receptors

The chicken m2 mAChR gene was isolated from a chicken genomic library by screening with a hybridization probe based on the chicken m4 mAChR (Tietje and Nathanson, 1991). The intronless genomic sequence predicted a 466 amino acid polypeptide, sharing 92% identity with the porcine m2 mAChR. Northern blot hybridization analysis of mRNA from embryonic chick tissue showed predominant m2 expression in heart and a lower, although significant, expression in whole brain. Interestingly, unlike mammalian heart tissue, which expresses only the m2 mAChR, the chick heart expresses both m2 and m4 mAChR mRNA (Tietje et al., 1990; Tietje and Nathanson, 1991).

The pharmacological properties of the chicken m2 mAChR were investigated by expression in CHO cells and in Y1 adrenal carcinoma cells (Tietje and Nathanson, 1991). These studies revealed that the chicken m2 mAChR resembles mammalian m2 receptors in that it binds the M_2-selective antagonist AF-DX 116 with high affinity. However, in contrast to mammalian m2 mAChRs, the chicken m2 receptor exhibits high affinity for the M_1-selective antagonist pirenzepine. Given the high sequence identity between the chicken and mammalian m2 mAChRs, it is likely that this pharmacological difference arises from limited amino acid changes (Tietje and Nathanson, 1991). Pharmacological differences between species homologs of individual receptor subtypes have been found in other families of G protein-coupled receptors. For example, a single amino acid difference causes a profound pharmacological variation between human and rodent 5-hydroxytryptamine 1B receptors (Oksenberg et al., 1982). Such observations suggest that caution is advisable when extrapolating pharmacological properties of a given receptor subtype across species, even in the context of high amino acid sequence homology.

Characterization of the effector-coupling properties of the chicken m2 mAChR showed efficient inhibition of adenylyl cyclase in CHO and in Y1 cells (Tietje and Nathanson, 1991). In addition, significant activation of PI hydrolysis was seen in CHO cells, as seen with porcine m2, but not in Y1 cells. These results confirm the ability of the m2 mAChR to activate PLC in certain cellular contexts.

1.1.3   m3 Muscarinic Acetylcholine Receptors

The existence of a third mAChR subtype had been suggested by pharmacological and functional studies long before the cloning of mAChR genes (reviewed by Birdsall and

Hulme, 1983). This receptor was referred to as the M_3 or the M_2-glandular subtype. Indeed, the availability of DNA sequences encoding the m1 and m2 mAChR enabled the identification of additional mAChR-encoding genes or cDNAs (Bonner et al., 1987; Peralta et al., 1987b; Braun et al., 1987; Akiba et al., 1988). One of these genes encoded a mAChR subtype which exhibited M_3-like pharmacology; this subtype is now known as the m3 mAChR.

1.1.3.1 Human m3 mACh Receptors

The human m3 mAChR gene (originally designated HM4) was isolated from a human genomic library by low-stringency hybridization, with probes based on homologous regions of the porcine m1 and m2 mAChRs (Peralta et al., 1987b). The m3 DNA sequence predicted a relatively large polypeptide of 590 amino acids, an M_r of ~66,100, and 5 potential N-linked glycosylation sites in the N-terminal region. Amino acid sequence comparisons indicated that the human m3 receptor is significantly more homologous to the human m1 (43%) than human m2 receptor (35%).

As with other mAChR subtypes, virtually no studies have been done to map m3 receptor expression in human tissues. Northern blot analysis of a panel of cell lines derived from human brain tumors indicated m3 mRNA expression in H4 neuroglioma, TE671 medulloblastoma, A172 glioblastoma, SK-N-SH neuroblastoma, and 1321N1 astrocytoma cells, but not in U87, U138, and U373 glioblastoma cells (Ashkenazi et al., 1988). In some of these cell lines, coexpression of m3 with other mAChRs was observed.

Pharmacological characterization upon expression of the cloned human m3 mAChR in HEK 293 cells indicated a pirenzepine affinity which was intermediate between those of human m1 and m2 receptors and comparable to that of human m4 mAChR; in addition, the m3 receptor exhibited AF-DX 116 affinity which was lower than that of m2 and comparable to that of m1 and m4 mAChRs (Peralta et al., 1987b). Similar results were observed with cloned human m3 mAChRs expressed in CHO-K1 cells; in addition, the affinity of m3 for the muscarinic antagonist hexahydrosiladifenidol was significantly higher than that of other mAChR subtypes (Buckley et al., 1989).

The functional properties of the cloned human m3 mAChR were investigated in transfected HEK 293 and CHO cells (Peralta et al., 1988; Ashkenazi et al., 1989a). The properties observed were similar to those of cloned human m1 mAChRs expressed in these cells, i.e., efficient coupling to PI hydrolysis, and no inhibition of adenylyl cyclase (in fact, activation was seen in HEK 293 cells). In addition to these responses, human m3 mAChRs, like human m1 receptors, exhibited efficient activation of PLD when expressed in HEK 293 cells (Sandman et al., 1991). Other responses shared by human m3 and m1 receptors are efficient mobilization of Ca^{2+} from intracellular stores in *Xenopus* oocytes (34) and in CHO cells (Lechlieter et al., 1989), induction of *c-fos* mRNA (Ashkenazi et al., 1989b), of DNA synthesis (Ashkenazi et al., 1989c), of conditional oncogenic activity (Gutkind et al., 1991) and of amyloid precursor protein release (Nitsch et al., 1992), and inhibition of endosome trafficking (Kazutaka and Rodbell, 1991) in CHO cells. Muscarinic activation of PI hydrolysis and of DNA synthesis in human SK-N-SH neuroblastoma cells and in human 1321N1 astrocytoma cells probably is mediated by endogenous m3 mAChRs (Ashkenazi et al., 1989c). The coupling of m3 mAChRs to many of the above responses appears to be mediated primarily by a PTX-insensitive G protein(s).

1.1.2.7 Porcine m3 mACh Receptors

The porcine m3 mAChR gene was isolated from a porcine genomic library (Akiba et al., 1988). The DNA sequence predicted a polypeptide of 590 amino acid residues and a calulated M_r of ~66,000, sharing 96 and 94% identity with the human and rat m3 mAChRs, respectively. Northern blot analysis of mRNA from porcine tissues indicated m3 expression in brain cerebrum, lacrimal and parotid glands, small and large intestines, trachea and urinary bladder, but not in atrium (Maeda et al., 1988).

Expression of the cloned porcine m3 mAChR in *Xenopus* oocytes showed pharmacological properties similar to those described above for human and rat m3 mAChRs (Akiba et al., 1988). Studies in transfected NG108-15 cells showed efficient coupling of the porcine m3 mAChR to activation of PI hydrolysis and of a Ca^{2+}-dependent K^+ current and to inhibition of the M-current, similar to results obtained with the porcine m1 mAChR (Fukuda et al., 1988).

1.1.3.10　Rat m3 mACh Receptors

The rat m3 mAChR gene was isolated from a rat genomic library (Bonner et al., 1987). The rat m3 mAChR cDNA was isolated from a rat forebrain cDNA library (Braun et al., 1987). The predicted polypeptide was 589 amino acids long, with an Mr of ~66,150, and sharing 45 and 39% homology with porcine m1 and m2 mAChRs, respectively.

Northern blot analysis of mRNA from rat tissues showed m3 mAChR expression in whole brain and in cerebral cortex, as well as in the pancreas (Peralta et al., 1987b). *In situ* hybridization studies in rat brain demonstrated m3 mRNA expression in the inner and outer layers of cerebral cortex and in the olfactory bulb and pyramidal cell layer of hippocampus, but low abundance in the dentate gyrus; in addition, m3 mRNA was observed in some thalamic and brain-stem nuclei (Bonner et al., 1987; Braun et al., 1987; Buckley et al., 1988).

Pharmacological characterization of rat m3 mAChRs expressed in COS-7 cells showed a pirenzepine affinity which was lower than that of m1 or m4 and higher than that of m2 mAChRs (Bonner et al., 1987). Analysis of cloned rat m3 receptors expressed in RAT-1 cells showed higher affinity for pirenzepine than for AF-DX 116, and yet higher affinity for the muscarinic antagonist 4-DAMP (Pinkas-Kramarski et al., 1988).

Biochemical studies with the cloned rat m3 receptor in murine A9 L cells yielded similar findings to those obtained with rat m1 receptors (Conklin et al., 1988). Rat m3 mAChRs coupled efficiently to activation of PLC, PLA_2, as well as to activation of adenylyl cyclase. Studies with cloned rat m3 mAChRs expressed in RAT-1 cells indicate efficient activation of PI hydrolysis and inability to inhibit adenylyl cyclase (Pinkas-Kramarski et al., 1988).

1.1.4　m4 Muscarinic Acetylcholine Receptors

The m4 mAChR subtype was difficult to distinguish from the m2 receptor, due to similar pharmacology and function. Therefore, the m4 subtype was not known prior to the molecular cloning of its DNA. Indeed, the m4 receptor was found to be significantly more homologous in amino acid sequence to m2 than to any other mAChR subtype, particularly in the predicted third cytoplasmic loop.

1.1.4.1　Human m4 mACh Receptors

The human m4 mAChR gene was isolated from human genomic libraries by screening with hybridization probes based on previously known mAChR DNA sequences (Bonner et al., 1987; Peralta et al., 1987b). The m4 gene predicted a polypeptide of 478 or 479 amino acids and an M_r of ~53,000, with three potential N-linked glycosylation sites in the N-terminal region. A significantly greater amino acid sequence homology was observed with human m2 than m1 or m3 mAChRs (55 vs. 40 or 37%, respectively). The human m4 gene has been mapped by *in situ* hybridization to the short arm of chromosome 11 (11p11.2-p12) (Bonner et al., in preparation) and shown to be genetically linked to both 11q and 11p markers that are located near the centromere (Grewal et al., 1992)

Northern blot analysis studies with human cell lines indicated m4 mRNA expression in H4 neuroglioma, TE671 medulloblastoma, A172 glioblastoma, SK-N-MC, and SK-N-SH neuroblastoma cells (Ashkenazi et al., 1988). The m4 subtype was found to be expressed alone in SK-N-MC cells, in combination with m3 and m2 in SK-N-SH cells, and with m3 in the other three cell lines.

Pharmacological charcterization of the cloned human m4 receptor in HEK 293 cells showed properties similar to those of human m3 mAChR, i.e., pirenzepine affinity lower than m1 but higher than m2, and AF-DX 116 affinity lower than m2 but higher than m1 mAChRs (Peralta et al., 1987b). A similar pharmacology was observed upon expression of the human m4 mAChR in CHO-K1 cells, with the exception that the AF-DX 116 affinity was similar to that of m1 (Buckley et al., 1989).

Biochemical studies of human m4 mAChR expressed in HEK 293 cells showed effector coupling properties similar to those of the m2 subtype, i.e., efficient inhibition of adenylyl cyclase and weak but significant activation of PI hydrolysis; both responses were inhibited by PTX (Peralta et al., 1988). Similar results were obtained in CHO cells (Ashkenazi et al., 1989b). In addition, Ca^{2+} mobilization was observed in CHO cells, which was similar to the m2-mediated response, and distinct from m1 or m3-mediated responses (Lechleiter et al., 1989). In HEK 293 cells, human m4 mAChR activated PLD as well; this response was significantly weaker than PLD-activation by human m1 or m3 mAChRs in this cell line (Sandman et al., 1991). In CHO-K1 cells, stimulation of human m4 mAChR with low concentrations of agonist-decreased cAMP levels, but at high agonist concentrations, cAMP levels were elevated (Jones et al., 1991). PTX treatment blocked the decrease and, instead, caused a marked increase in cAMP levels. Less than twofold induction of PI hydrolysis was observed in CHO-K1 cells, which was not affected by PTX, suggesting that PLC was not involved in the elevation of cAMP levels.

1.1.4.10   Rat m4 mACh Receptors

The rat m4 mAChR cDNA was isolated from a rat cerebral cortex cDNA library (Bonner et al., 1987). The sequence predicted a 478 amino acid polypeptide with an M_r of ~53,000. Northern hybridization analysis of rat tissues indicated m4 mRNA expression in whole brain and in cerebral cortex and cerebrum, but not in heart, pancreas, submandibular gland, small intestine, trachea, or urinary bladder (Peralta et al., 1987b; Maeda et al., 1988). In addition, m4 mRNA expression was detected in the NG108-15 mouse neuroblastoma x rat glioma cell line and in the rat PC12 adrenal pheochromocytoma cell line (Ashkenazi et al., 1988). Northern hybridization studies with mRNA from different rat brain subregions revealed m4 expression in cerebral cortex, striatum, hippocampus, and cerebellum (Buckley et al., 1988). *In situ* hybridization studies revealed m4 mRNA expression in the olfactory bulb and pyramidal cell layer of the hippocampus and low levels in the dentate gyrus; m4 was also detected in the caudate putamen (Buckley et al., 1988).

Expression of the cloned rat m4 mAChR cDNA in COS 7 cells indicated high affinity for pirenzepine, comparable to that of m1 mAChR (Bonner et al., 1987). The functional properties of rat m4 mAChRs were studied in the NG108-15 cell line, which expresses endogenous m4 receptors (Fukuda et al., 1988). These studies showed little or no effect of m4 receptor activation on PI hydrolysis, Ca^{2+} mobilization, or electrophysiological activity. In contrast, electrophysiological studies with the GH3 and AtT20 cell lines, in which endogenous m4 mAChR is found, showed inhibition of inwardly rectifying K^+ channels (Codina et al., 1987), similar to the inhibition by m2 mAChR in pacemaker and atrial cells (Brown and Birnbaumer, 1990). Recent studies using antisense oligonucleotides to inhibit the expression of endogenous G protein subunits in GH3 cells showed that the $G_{\alpha o1}$ and $G_{\beta 3}$ isoforms couple the endogenous m4 mAChR to the inhibition of voltage-gated calcium channels (Kleuss et al., 1991a,b).

1.1.4.15   Chicken m4 mACh Receptors

The chicken m4 mAChR gene was isolated from a chicken genomic library, using a full length porcine m2 mAChR cDNA as a hybridization probe (Tietje et al., 1990). The intronless genomic DNA sequence predicted a 490 amino acid polypeptide, although an alternative prediction with a more optimal initiation of translation sequence was of a 477 amino acid polypeptide. The shorter receptor shared 83% amino acid sequence identity with rat m4 mAChR and 70% with the porcine m2 mAChR in regions not including the highly divergent large cytoplasmic loop. In this region, 40, 15, 4, and 10% identity was observed with m2, m1, m3 and m5 mAChRs, respectively. Taken together, these results suggested that the cloned chicken gene indeed encoded an m4 mAChR.

Northern blot analysis indicated expression of m4 mAChR mRNA in embryonic chick heart and brain, but not in liver (Tietje et al., 1990). More recent studies showed that in both embryonic chick heart and brain, m4 and m2 mAChRs are coexpressed, at a ratio of about 1:15 (Tietje et al., 1991). Thus, unlike mammalian heart, in which m2 mAChR alone is expressed, in embryonic chick heart both m2 and m4 mAChRs are found. It remains to be determined whether m4 expression occurs in adult chicken heart and, conversely, whether coexpression of m2 and m4 occurs in embryonic mammalian heart.

Expression of the cloned chicken m4 mAChR in CHO and Y1 adrenal carcinoma cells (Tietje et al., 1990) revealed high affinity for pirenzepine and low affinity for AF-DX 116. Effector coupling studies showed efficient inhibition of adenylyl cyclase activity by the m4 receptor in both cell lines. In addition, a weak but significant activation of PI hydrolysis was seen in CHO cells, comparable to that mediated by human m4 mAChR in HEK 293 cells. This response was not seen in Y1 cells.

1.1.4.20   *Xenopus* m4 mACh Receptors

The putative m4 receptor from *Xenopus* has been cloned (Olate, unpublished data).

1.1.5   m5 Muscarinic Acetylcholine Receptors

Southern blot analysis of human genomic DNA with mAChR probes suggested the existence of related sequences in the genome in addition to those of the m1 through m4 mAChRs (Bonner et al., 1987; Peralta et al., 1987b). Indeed, an additional mAChR sequence, designated m5, was isolated subsequently from human and rat genomic and cDNA libraries (Bonner et al., 1988; Liao et al., 1989).

1.1.5.1 Human m5 mACh Receptors

The human m5 mAChR gene was isolated from a human genomic library by screening with a rat m1 receptor DNA probe (Bonner et al., 1988). The genomic DNA sequence predicted a 532 amino acid polypeptide, with 2 potential N-linked glycosylation sites in the N-terminal region, and with significant homology to the other mAChR subtypes. Comparison of the amino acid sequence excluding the amino and carboxyl termini and the large cytoplasmic loop, which are the most diverged between mAChR subtypes, indicated 85, 79, 73, and 68% identity of human m5 with human m3, m1, m4, and m2 mAChRs, respectively.

Pharmacological characterization of the cloned human m5 mAChR expressed in CHO cells showed higher pirenzepine affinity than that displayed by m1, m2, m3, or m4 mAChRs, and an AF-DX 116 affinity which was lower than that of m2, but comparable to that of m1, m3, and m4 (Bonner et al., 1988).

Functional studies of the human m5 mAChR in transfected CHO cells indicated properties similar to those of m1 and m3 mAChRs, i.e., efficient activation of PI hydrolysis and elevation, rather than inhibition, of cAMP levels (Bonner et al., 1988).

1.1.5.10 Rat m5 mACh Receptors

The rat m5 mAChR gene was isolated from a rat genomic library, using a portion of the human m5 gene as a hybridization probe (Bonner et al., 1988). The genomic DNA sequence predicted a 531 amino acid protein, sharing 89% identity with the human m5 mAChR. An identical cDNA sequence was cloned independently from a rat cDNA library (Liao et al., 1989).

Northern blot and *in situ* hybridization studies failed to detect m5 mRNA expression in various rat tissues, including brain, atrium, and submandibular and sublingual glands (Bonner et al., 1988). However, the isolation of m5 cDNA from a rat brain cDNA library (Liao et al., 1989) indicates at least some level of m5 mRNA expression in brain. Indeed, more recently, m5 mRNA expression was demonstrated by *in situ* hybridization in the substantia nigra and par compacta of the rat brain (Weiner et al., 1990). Expression of the cloned rat m5 mAChR cDNA in murine L cells indicated efficient activation of PI hydrolysis (Liao et al., 1989). No effect was observed on cAMP levels, in contrast to the elevation mediated by human m5 mAChR in CHO cells.

1.1.11 Invertebrate Muscarinic Acetylcholine Receptors

1.1.11.21 *Drosophila melanogaster* Muscarinic Acetylcholine Receptors

Molecular cloning efforts heretofore have identified a single mAChR gene in *Drosophila melanogaster* (Shapiro et al., 1989). The *Drosophila* mAChR appears to be most homologous to the vertebrate m3 and m5 mAChRs, although it is significantly longer. Much of the additional sequence occurs in the large cytoplasmic domain of the *Drosophila* receptor. Barring the possibility that additional mAChR subtypes exist in *Drosophila melanogaster*, the identification of only a single mAChR gene in this species suggests that the multiplicity of mAChR subtypes may have arisen as a requirement for the relatively complex functions of vertebrate nervous systems.

ACKNOWLEDGMENTS

We thank Drs. John Winslow and J. Ramachandran for their contributions to many of our studies. We thank Dr. Daniel Capon especially, for his guidance, encouragement, and ideas. Some of the research described in this review was supported by grants from the NIH and Searle Scholars Program to E.G.P.

REFERENCES

Akiba, I., Kubo, T., Maeda, A., Bujo, H., Nakai, J., Mishina, M., and Numa, S. (1988) Primary structure of porcine muscarinic acetylcholine receptor III and antagonist binding studies. *FEBS Lett.* 235, 257–261.

Arden, J. R., Nagata, O., Shockley, M. S., Philip, M., Lameh, J., and Sadee, W. (1992) Mutational analysis of third cytoplasmic loop domains in G-protein coupling of the Hm1 muscarinic receptor. *Biochem. Biophys. Res. Commun.* 188, 1111–1115.

Ashkenazi, A., Winslow, J. W., Peralta, E. G., Peterson, G. L., Schimerlik, M., Capon, D. J., and Ramachandran, J. (1987) An M2 muscarinic receptor subtype coupled to both adenylyl cyclase and phosphoinositide turnover. *Science* 238, 672–675.

Ashkenazi, A., Peralta, E. G., Winslow, J. W., Ramachandran, J., and Capon, D. J. (1988) Functional role of muscarinic acetylcholine receptor subtype diversity. *Cold Spring Harbor Symp. Quant. Biol.* LIII, 263–272.

Ashkenazi, A., Peralta, E. G., Winslow, J. W., Ramachandran, J., and Capon, D. J. (1989a) Functionally distinct G proteins selectively couple different receptors to PI hydrolysis in the same cell. *Cell* 56, 487–493.

Ashkenazi, A., Peralta, E. G., Winslow, J. W., Ramachandran, J., and Capon, D. J. (1989b) Functional diversity of muscarinic receptor subtypes in cellular signal transduction and growth. *Trends Pharmacol. Sci.* (Suppl. Subtypes of Muscarinic Receptors IV). pp.16–22.

Ashkenazi, A., Ramachandran, J., and Capon, D.J. (1989c) Acetylcholine analogue stimulates DNA synthesis in brain-derived cells via specific muscarinic receptor subtypes. *Nature* 340, 146–150.

Barlow, R. B., Berry, K. J., Glenton, P. A. M., Nikolau, N. M., and Soh, S. (1976) A comparison of affinity constants for muscarine-sensitive acetylcholine receptors in guinea-pig atrial pacemaker cells at 29°C and in ileum at 29°C and 37°C. *Br. J. Pharmacol.* 58, 613–620.

Baumgold, J. (1992) Muscarinic receptor-mediated stimulation of adenylyl cyclase. *Trends Pharmcol. Sci.* 13, 339–340.

Bernard, V., Normand, E., and Bloch, B. (1992) *Neuroscience* (in press).

Berridge, M. J. and Irvine, R. F. (1989) Inositol phosphates and cell signalling. *Nature* 341, 197–205.

Berstein, G., Blank, J. L., Smrcka, A. V., Higashijima, T., Sternweis, P. C., Exton, J. H., and Ross, E. M. (1992) Reconstitution of agonist-stimulated phosphatidylinositol 4,5-bisphosphate hydrolysis using purified m1 muscarinic receptor, $G_{q/11}$, and phospholipase C-b$_1$. *J. Biol. Chem.* 267, 8081–8088.

Birdsall, N. J. M. and Hulme, E. C. (1983) Muscarinic receptor subclasses. *Trends Pharmacol. Sci.* 4, 459–463.

Bonner, T. I., Buckley, N. J., Young, A. C., and Brann, M. R. (1987) Identification of a family of muscarinic acetylcholine receptor genes. *Science* 237, 527–532.

Bonner, T. I., Young, A. C., Brann, M. R., and Buckley, N. J. (1988) Cloning and expression of the human and rat m5 muscarinic acetylcholine receptor genes. *Neuron* 1, 403–410.

Brann, M. R., Buckley, N. J., and Bonner, T. I. (1988) The striatum and cerebral cortex express different muscarinic receptor mRNAs. *FEBS Lett.* 230, 90–94.

Braun, T., Schofield, P. R., Shivers, B. D., Pritchett, D. B., and Seeburg, P. H., (1987) A novel muscarinic receptor identified by homology screening. *Biochem. Biophys. Res. Commun.* 149, 125–132.

Brown, A. M. and Birnbaumer, L. (1990) Ionic channels and their regulation by G protein subunits. *Annu. Rev. Physiol.* 52, 197–213.

Buckley, N. J., Bonner, T. I., and Brann, M. R. (1988) Localization of a family of muscarinic receptor mRNAs in rat brain. *J. Neurosci.* 8, 4646–4652.

Buckley, N. J., Bonner, T. I., Buckley, C. M., and Brann, M. R. (1989) Antagonist binding properties of five cloned muscarinic receptors expressed in CHO-K1 cells. *Mol. Pharmacol.* 35, 469–476.

Codina, J., Grenet, D., Yatani, A., Birnbaumer, L., and Brown, A. M. (1987) Hormonal regulation of pituitary GH_3 cell K^+ channels by G_k is mediated by its a-subunit. *FEBS Lett.* 216, 104–106.

Conklin, B. R., Brann, M. R., Buckley, N. J., Ma, A. L., Bonner, T. I., and Axelrod, J. (1988) Stimulation of arachidonic acid release and inhibition of mitogenesis by cloned genes for muscarinic receptor subtypes expressed in A9 L cells. *Proc. Natl. Acad. Sci. U.S.A.* 85, 8698–8702.

Felder, C. C., Kanterman, R. Y., Ma, A. L., and Axelrod, J. (1989) A transfected m1 muscarinic acetylcholine receptor stimulates adenylate cyclase via phosphatidylinositol hydrolysis. *J. Biol. Chem.* 264, 20356–20362.

Felder, C. C., Poulter, M. O., and Wess, J. (1992) Muscarinic receptor-operated Ca^{2+} influx in transfected fibroblast cells is independent of inositol phosphates and release of intracellular Ca^{2+}. *Proc. Natl. Acad. Sci. U.S.A.* 89, 509–513.

Feng, D. F. and Doolittle, R. F. (1990) Progressive alignment and phylogenetic tree construction of protein sequences. *Methods Enzymol.* 183, 375–387.

Fukuda, K., Higashida, H., Kubo, T., Maeda, A., Akiba, I., Bujo, H., Mishina, M., and Numa, S. (1988) Selective coupling with K^+ currents of muscarinic acetylcholine receptor subtypes in NG108–15 cells. *Nature* 335, 355–358.

Giachetti, A., Micheletti, R., and Montagna, E. (1986) Cardioselective profile of AF-DX 116, a muscarinic M_2 receptor antagonist. *Life Sci.* 38, 1663–1672.

Gilani, S. A. H. and Cobbin, L. B. (1986) The cardioselectivity of himbacine: a muscarine receptor antagonist. *Naunyn-Schmiedebergs Arch. Pharmacol.* 332, 16–20.

Giraldo, E., Hammer, R., and Ladinsky, H. (1987) Distribution of muscarinic receptor subtypes in rat brain as determined in binding with AF-DX 116 and pirenzepine. *Life Sci.* 40, 833–840.

Gocayne, J., Robinson, D. A., FitzGerald. M. G., Chung, F. Z., Kerlavage, A. R., Lentes, K. U., Lai, J. Wang, C. D., Fraser, C., and Venter, J. C. (1987) Primary structure of rat cardiac β-adrenergic and muscarinic cholinergic receptors obtained by automated DNA sequence analysis: further evidence for a multigene family. *Proc. Natl. Acad. Sci. U.S.A.* 84, 8296–8300.

Grewal, R. P., Martinez, M., Hoehe, M., Bonner, T. I., Gershon, E. S., and Detera-Wadleigh, S. (1992) Genetic linkage mapping of the m4 human muscarinic receptor (CHRM4). *Genomics* 13, 239–240.

Gutkind, J. S., Novotny, E. A., Brann, M. R., and Robbins, K. C. (1991) Muscarinic Acetylcholine receptor subtypes as agonist-dependent oncogenes. *Proc. Natl. Acad. Sci. U.S.A.* 88, 4703–4707.

Haga, K. and Haga, T. (1985) Purification of the muscarinic acetylcholine receptor from porcine brain. *J. Biol. Chem* 260, 7927–7935.

Hammer, R., Berrie, C. P., Birdsall, N. J. M., Burgen, A.S.V., and Hulme, E. C. (1980) Pirenzepine distinguishes between subclasses of muscarinic receptors. *Nature* 283, 90–92.

Haraguchi, K. and Rodbell, M. (1991) Carbachol-activated muscarinic (M1 and M3) receptors transfected into Chinese hamster ovary cells inhibit trafficking of endosomes. *Proc. Natl. Acad. Sci. U.S.A.* 88, 5964–5968.

Hulme, E. C., Birdsall, N. J. M., and Buckley, N. J. (1990) Muscarinic receptor subtypes. *Annu. Rev. Pharmacol. Toxicol.* 30, 633–673.

Jones, S. V. P., Heilman, C. J., and Brann, Mark, R. (1991) Functional responses of cloned muscarinic receptors expressed in CHO-K1 cells. *Mol. Pharmacol.* 40, 242–247.

Katz, A., Wu, D., and Simon, M. I. (1992) Subunits βγ of heterotrimeric G protein activate β2 isoform of phospholipase C. *Nature* 360, 686–689.

Kaziro, Y., Itoh, H., Kozasa, T., Nakafuku, M., and Satoh, T. (1991) Structure and function of signal-transducing GTP-binding proteins. *Annu. Rev. Biochem.* 60, 349–400.

Kikuchi, K., Nishino, M., and Inoue, H. (1987) Effects of sialagogues on the syntheses of polyamines and DNA in murine parotid gland. *Biochem. Biophys. Res. Commun.* 144, 1161–1166.

Kleuss, C., Heschler, J., Ewel, C., Rosenthal, W., Schultz, G., and Wittig, B. (1991) Assignment of G-protein subtypes to specific receptors inducing inhibition of calcium currents *Nature* 353, 43–48.

Kleuss, C., Scherubl, H., Heschleler, J., Schultz, G., and Wittig, B. (1992) Different β-subunits determine G-protein interaction with transmembrane receptors. *Nature* 358, 424–427.

Kobilka, B. (1992) Adrenergic receptors as models for G protein-coupled receptors. *Annu. Rev. Neurosci.* (1992) 15, 87–114.

Kubo, T., Fukuda, K., Mikami, A., Maeda, A., Takahashi, H., Mishina, M., Haga, K., Ichiyama, A., Kangawa, K., Kojima, M., Matsuo, H., Hirose, T., and Numa, S. (1986a) Cloning, sequencing and expression of complementary DNA encoding the muscarinic acetylcholine receptor. *Nature* 323, 411–416.

Kubo, T., Maeda, A., Sugimoto, K., Akida, I., Mikami, A., Takahashi, H., Haga, K., Ichiyama, A., Kangawa, K., Matsuo, H., Hirose, T., and Numa, S. (1986b) Primary structure of porcine cardiac muscarinic acetylcholine receptor deduced from the cDNA sequence. *FEBS Lett.* 209, 367–373.

Lai, J., Nunan, L., Waite, S. L., Ma, S. W., Bloom, J. W., Roeske, W. R., and Yamamura, H. I. (1992) Chimeric M1/M2 muscarinic receptors: correlation of ligand selectivity and functional coupling with structural modifications. *J. Pharmacol. Exp. Therapeu.* 262, 173–180.

Lechleiter, J., Peralta, E. G., and Clapham, D. (1989) Diverse functions of muscarinic acetylcholine receptor subtypes. *Trends Pharmacol. Sci.* (Suppl. Subtypes of Muscarinic Receptors IV). pp.34–38.

Lechleiter, J., Hellmiss, R., Duerson, K., Ennulat, D., David, N., Clapham, D., and Peralta, E. (1990) Distinct sequence elements control the specificity of G protein activation by muscarinic acetylcholine receptor subtypes. *EMBO J.* 9, 4381–4390.

Lechleiter, J., Girard, S., Clapham, D., and Peralta, E. G. (1991a) Subcellular patterns of calcium release determined by G protein-specific residues of muscarinic receptors. *Nature* 350, 505–508.

Lechleiter, J., Girard, S., Peralta, E., and Clapham, D. (1991b) Spiral calcium wave propagation and annihilation in *Xenopus laevis* oocytes. *Science* 252, 123–126.

Liao, C. F., Themmen, A. P. N., Joho, R., Barberis, C., Birnbaumer, M., and Birnbaumer, L. (1989) Molecular cloning and expression of a fifth muscarinic acetylcholine receptor. *J. Biol. Chem.* 264, 7328–7337.

Maeda, A., Kubo, T., Mishna, M., and Numa, S. (1988) Tissue distribution of mRNAs encoding muscarinic acetylcholine receptor subtypes. *FEBS Lett.* 239, 339–342.

Mutschler, E. and Lambrecht, G. (1984) Selective muscarinic agonists and antagonists in functional tests. *Trends Pharmacol. Sci.* 5 (Suppl. Subtypes of Muscarinic Receptors I). pp.39–44.

Nathanson, N. M. (1987) Molecular properties of the muscarinic acetylcholine receptor. *Annu. Rev. Neurosci.* 10, 195–236.

Neher, E., Marty, A., Fukuda, K., Kubo, T., and Numa S. (1988) Intracellular calcium release mediated by two muscarinic receptor subtypes. *FEBS Lett.* 240, 88–94.

Nitsch, R. M., Slack, B. E., Wurtman, R. J., and Growdon, J. H. (1992) Release of Alzheimer amyloid precursor derivatives stimulated by activation of muscarinic acetylcholine receptors. *Science* 258, 304–307.

Oksenberg, D., Masters, S. A., O'Dowd, B. F., Jin, H., Havlik, S., Peroutka, S. J., and Ashkenazi, A. (1992) A single amino acid difference confers major pharmacological variation between human and rodent 5-hydroxytryptamine receptors. *Nature* 360, 161–163.

Ovchinnikov, Y. A. (1982) Rhodopsin and bacteriorhodopsin: structure-function relationships. *FEBS Lett.* 148, 179–191.

Peralta, E. G., Winslow, J. W., Peterson, G. L., Smith, D. H., Ashkenazi, A., Ramachandran, J., Schimerlik, M., and Capon, D. J. (1987a) Primary structure and biochemical properties of an M_2 muscarinic receptor. *Science* 236, 600–605.

Peralta, E. G., Ashkenazi, A., Winslow, J. W., Smith, D. H., Ramachandran, J., and Capon, D. J. (1987b) Distinct primary structures, ligand binding properties and tissue-specific expression of four human muscarinic acetylcholine receptors. *EMBO J.* 6, 3923–3929.

Peralta, E. G., Ashkenazi, A., Winslow, J. W., Ramachandran, J., and Capon, D. J. (1988) Differential regulation of PI hydrolysis and adenylyl cyclase by muscarinic receptor subtypes. *Nature* 334, 434–437.

Pfaffinger, P. J., Leibowitz, M. D., Subers, E. M., Nathanson, N. M., Almers, W., and Hille, B. (1988) Agonists that supress M current elicit phosphoinositide turnover and Ca^{+2} transients but these events do not explain M-current supression. *Neuron* 1, 477–484.

Pinkas-Kramarski, R., Stein, R., Lindenboim, L., and Sokolovsky, M. (1992) Growth factor-like effects mediated by muscarinic receptor in PC12M1 cells. *J. Neurochem.* 59, 2158–2166.

Pinkas-Kramarski, R., Stein, R., Zimmer, Y., and Sokolovsky, M. (1988) Cloned rat M3 muscarinic receptors mediate phosphoinositide hydrolysis but not adenylate cyclase inhibition. *FEBS Lett.* 239, 174–178.

Sandmann, J., Peralta, E. G., and Wurtman, R. J. (1991) Coupling of transfected muscarinic receptor subtypes to phospholipase D. *J. Biol. Chem.* 266, 6031–6034.

Shapiro, R. A., Scherer, N. M., Habecker, B. A., Subers, E. M., and Nathanson, N. M. (1988) Isolation, sequence, and functional expression of the mouse M1 muscarinic acetylcholine receptor gene. *J. Biol. Chem.* 263, 18397–18403.

Shapiro, R., Wakimoto, B., Subers, E., and Nathanson, N. (1989) *Proc. Natl. Acad. Sci. U.S.A.* 86, 9039–9043.

Stein, R., Pinkas-Kramarski, R., and Sokolovsky, M. (1988) Cloned M1 muscarinic receptors mediate both adenylate cyclase inhibition and phosphoinositide turnover. *EMBO J.* 7, 3031–3035.

Taylor, P. (1990) Cholinergic agonists. In *The Pharmacological Basis of Therapeutics,* Gilman, A., G., Rall, T. W., Nies, A. S., and Taylor, P., Eds., Pergamon, New York, pp. 122–130.

Tietje, K. M., Goldman, P. S., and Nathanson, N. M. (1990) Cloning and functional analysis of a gene encoding a novel muscarinic acetylcholine receptor expressed in chick heart and brain. *J. Biol. Chem.* 265, 2828–2834.

Tietje, K. M. and Nathanson, N. M. (1991) Embryonic chick heart expresses multiple muscarinic acetylcholine receptor subtypes. *J. Biol. Chem.* 266, 17382–17387.

Weiner, D. M., Levey, A., and Brann, M. R. (1990) Expression of muscarinic acetylcholine and dopamine receptor mRNAs in rat basal ganglia. *Proc. Natl. Acad. Sci. U.S.A.* 87, 7050–7054.

Wong, S. K. F., Parker, E. M., and Ross, E. M. (1990) Chimeric Muscarinic Cholinergic: β-adrenergic receptors that activate Gs in response to muscarinic agonists. *J. Biol. Chem.* 265, 6219–6224.

Zohar, M. and Salomon, Y. (1992) Melanocortins stimulate proliferation and induce morphological changes in cultured rat astrocytes by distinct transducing mechanisms. *Brain Res.* 576, 49–58.

2

Adenosine Receptors

Joel Linden, Marlene A. Jacobson, Charles Hutchins, and Michael Williams

1.2.0   Introduction

Recognition sites for the purine nucleoside neuromodulator, adenosine, comprise three main classes of molecular structure (Table 1): G protein-linked receptors , ligand-gated ion channels (Volume 2) and transporters (Volume 3). In addition, given the ubiquitous role of the purine in intracellular energy processes (Arch and Newsholme, 1978; Meghi, 1991), there are a variety of enzymes that recognize adenosine and its phosphorylated derivatives, AMP, ADP, ATP, and Ap4A. Distinct receptors for these purine nucleotides have been described (Burnstock and Kennedy, 1985; Hoyle, 1990; Pintor et al., 1991) and are discussed briefly in Section 1.2.99.

Prior to the cloning and expression of adenosine receptors, their initial classification was based on their pharmacological and functional properties; specifically, their sensitivity to blockade by aryl and alkylxanthines (Sattin and Rall, 1970; Daly, 1982) and their ability to modulate G-protein-linked adenylate cyclase activity (Londos et al., 1980; van Calker et al., 1979). The initial classification by Burnstock (Burnstock, 1978) defined receptors sensitive to adenosine as P_1 receptors while those sensitive to ATP and related nucleotides, were classified as P_2 receptors.

The A_1 subclass of the P_1 receptor was originally defined on the basis of the ability of stable adenosine analogs to inhibit cAMP formation while the A_2 receptor subclass was defined on the basis of the ability of adenosine analogs to increase cAMP formation (Londos et al., 1980; van Calker et al., 1979). Since it has become evident that there are multiple second messenger systems for adenosine (Morgan, 1991), the delineation of A_1 and A_2 receptors on the basis of the modulation of adenylate cyclase activity has been replaced by a pharmacologically based classification (Hamprecht and van Calker, 1985). The A_2 receptor has been further subdivided into A_{2a} and A_{2b} subclasses based on adenosine agonist affinity - A_{2a} - high affinity and A_{2b}- low affinity-, anatomical distribution in the brain and limited pharmacological analysis (Bruns, 1980; Daly et al., 1983; Bruns and Pugsley, 1986). Selective agonists for the A_1 receptor (Bruns, 1980; Bruns and Snyder, 1980; Williams et al., 1986) and the A_{2a} subclass of the A_2 receptor (Hutchison et al., 1989; Ueeda et al., 1991; Ueeda et al., 1991) have thus provided convenient pharmacological tools to define receptor function.

While cloning of adenosine receptors has been a primary focus of purinergic research efforts over the past 2 years, the A_{2a} receptor was serendipitously identified

Table 1. Overview of Adenosine and Purine Nucleotide Purinoceptors

G Protein-coupled receptors		Selective ligands
A_1:		CHA, CPA, CCPA
A_2 "Family":	A2a	CGS 21680
	A2b	None
A_3		?
A_4		CV 1808 (?)
P_{2t}		2-Methylthio-ADP
P_{2u}/P_{2n}		UTP, ATP
(Nucleotide receptor)		
P_{2y}		2-Methylthio-ATP
Ligand-gated channels		
P_{2x}		α,β-Methyene ATP
P_{2z}		ATP^{4-}

as part of a family of G-protein linked receptors cloned from canine thyroid gland as the clone, RDC8 on the basis of *in situ* hybridization (Schiffmann et al., 1990), positive coupling to cAMP and the selective binding of the A_{2a} selective agonist, [³H]-CGS 21680 (Maenhaut et al., 1990). Like other G-protein linked receptors, the cloned A_{2a} receptor had the typical pattern of seven transmembrane hydrophobic α-helices of 20 to 28 amino acid residues but had only about 30% homology with other members of this class. Further examination of a closely related "orphan" clone from canine thyroid led to the identification of RDC7 as the canine A_1 receptor (Libert et al., 1991).

RDC7 is among the smallest members of the G-protein linked superfamily with 326 amino acids (Libert et al., 1989; Figure 2) but shares several common features with other members including small N-termini (less than 12 residues) and landmark amino acids in the transmembrane segments (GN in TM1; LXXXD in TM2; DRY at the border of TM3; WXP in TM6 and NP in TM7; Figure 2). Both the cloned A_1 and A_{2a} receptors lack the aspartate residue in TM3 that is important in muscarinic, adrenergic, serotonergic, and dopaminergic receptor cDNAs for cationic amine binding (Linden et al., 1991).

A_1, A_{1a}, and A_{2b} adenosine receptors have been cloned and expressed from a variety of mammalian species including human brain (Libert et al., 1992; Salavatore et al., 1992) and a novel adenosine receptor has been cloned and designated as the A_3 receptor (Zhou et al., 1992). Within the group of G protein-coupled adenosine receptors (Table 2), the evolutionary relationships between the known subtypes were determined by a phylogenetic tree analysis (Figure 2; Feng and Doolittle, 1990). The aligned sequences of all identified G protein-coupled adenosine receptors can be compared and a concensus phylogenetic tree constructed (Peroutka, 1992). The length of each "branch" (Figure 1) correlates with the evolutionary distance between receptor subpopulations. This includes all the adenosine receptors that have been cloned to date and indicates that the sequence of the adenosine A_3 receptor that is abundant in rat testis (Zhou et al., 1992) is clearly distinct from both A_1 and A_2 receptors.

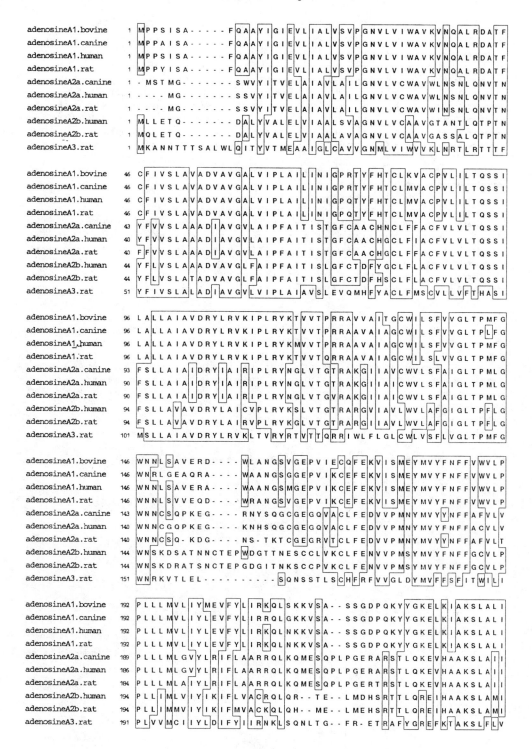

FIGURE 1. Phylogenetic tree of adenosine receptors.

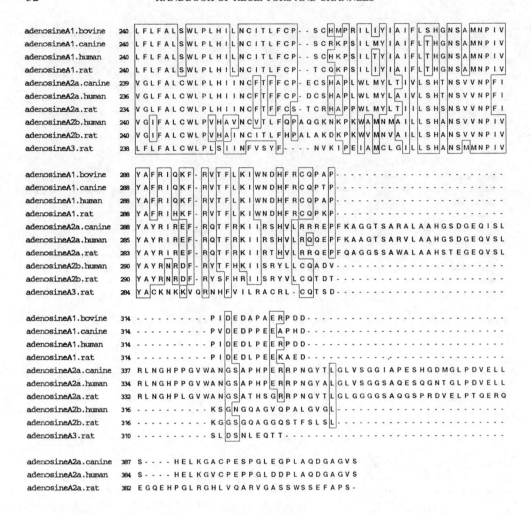

FIGURE 1b.

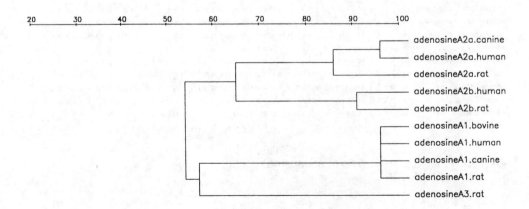

FIGURE 2. Alignment of cloned adenosine receptors.

Table 2.   Molecular Biological Characteristics of G-Protein Coupled Adenosine Receptors

Receptor	Second messenger	Species cloned	Chromosomal location	Amino acids	GenBank accesion	Primary reference
A1	Inhibition of adenylate cyclase	Human	20q 11.2 -13.1	326	545235	26, 27
		Canine		326	P11616	24
		Rat		326	M69054	36, 37
		Bovine		326	X63592	38, 39
A2a	Stimulation of adenylate cyclase	Human	11 q 11 - 13	412	M97370	27, 40, 42, NIH unpublished[a]
			22 q 11	412		45
		Rat		410	JT0607	46, 47
		Canine		411[b]	P11617	22
A2b	Stimulation of adenylate cyclase	Human		332	M97759	27, 53
		Rat		332	M91466	54
A3	Inhibition of adenylate cyclase	Rat		320[c]	S42407	28, 56
		Human		318	L22607	unpublished
A4 (putative)	Activation of potassium channels					58

[a] Tiffany, H.L. and Murphy, P.M., unpublished Genebank Accession M 97370.
[b] The data published on the canine A2a receptor reported 411 residues that ended in DGAGV. The form deposited in the Genebank was 412 residues ending in DGAGVS.
[c] The data published on the rat A3 receptor reported 320 residues that ended in EQTTE. The form deposited in the Genebank was 319 residues ending in EQTT.

Comparison of the cladistic tree for adenosine receptors with those of other G-protein linked receptors for cationic amines (adrenergics, histamine, dopamine, serotonin, and muscarinic cholinergic) indicates that adenosine receptors thus comprise a *family* of G protein coupled receptors that can be grouped by species or by subtypes. The differences between species are much smaller than differences between subtypes with similarity scores for A_1 receptors across species ranging from 94.5 to 96.9%. By contrast, similarity scores for receptor subtypes within a single species are much lower; in the rat A_1–A_{2a} = 67.3%; A_1–A_{2b} = 67.7%; A_1–A_3 = 68.7%; A_{2a}–A_{2b} = 77.8%; A_{2a}–A_3 = 64.4%; and A_{2b}–A_3 = 62.9%. The similarities between G protein coupled receptors in general, and adenosine receptors in particular, are far greater in the transmembrane (TM) regions than in the amino and carboxy termini or in the loops connecting the TM regions. This can be illustrated for the adenosine receptors by plotting the similarity scores of aligned receptor amino acids for all 11 of the adenosine receptors that have been cloned to date. There are seven *peaks* of high similarity that correspond to the seven transmembrane regions. Among the transmembrane segments there is a region of low similarity only in the carboxy terminal half of TM4. The similarities in individual transmembrane amino acids among 11 adenosine receptors are shown in Figure 3. An examination of the bottom portions of the consensus amino acid sequences of TM2, TM3, and TM7 reveals regions of very high similarity corresponding to probable α-helical surfaces. Given this striking conservation among adenosine receptors and the fact that TM3 and TM7 have been implicated in the binding of several cationic ligands to their respective ligands, it may be speculated that these conserved surfaces may form a binding pocket responsible for ligand recognition.

Due to the great similarities in the transmembrane regions of G protein coupled receptors, it is possible to accurately align nucleotide sequences of the TM regions of even relatively distantly related G proteins coupled receptors. An alignment of just the transmembrane regions provides a more precise way of judging relatedness among the family members than does alignment of the entire nucleotide sequences.

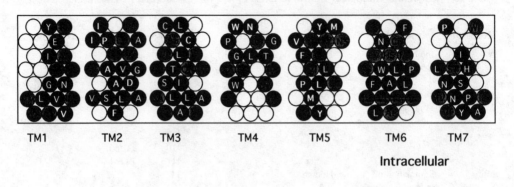

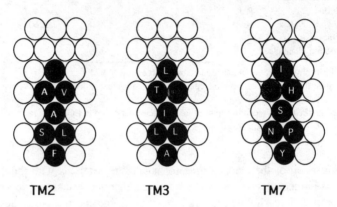

FIGURE 3. Consensus amino acid sequence of adenosine receptors. Top: The putative transmembrane segments of 11 adenosine receptors are indicated. Amino acid identity among all the receptors is indicated by black circles. Positions where only two amino acids are found among the 11 receptors are indicated in grey, both amino acids are indicated with the most common listed first. Bottom: Highly conserved putative alpha-helical surfaces that may correspond to ligand recognition sites are indicated.

The identification by Bruns (Bruns and Fergus, 1990; Bruns et al., 1990) of some novel ligands that had no direct interaction with classical A_1 receptor binding sites but were able to modulate the binding of ligands to the receptors has raised the possibility of there being an allosteric site, interaction of ligands with which can modify the normal binding parameters of adenosine receptors. Daly et al. (Daly et al., 1992) have identified a 33-amino acid peptide termed adenoregulin that can also modulate adenosine receptor binding. To date, there is no molecular evidence for an allosteric site on the A_1 receptor nor for that matter any G protein-linked, 7-transmembrane helix-type receptor. The mechanism for these allosteric effects remains to be determined and may possibly involve transduction as well as recognition elements of the receptor. The 3-benzoylthiophene, PD 81,723, and related compounds (Bruns et al., 1990) are prototypic ligands for the proposed allosteric site.

1.2.1 Adenosine A_1 Receptors

A_1 receptor binding sites originally labeled by Bruns and co-workers with [^{3}H]-cyclohexyladenosine (CHA) have only provided definitive evidence for a single class of A_1 receptor (Bruns et al., 1980; Jacobson et al., 1992) which may exist on low- and high-affinity states. While A_{1a} and A_{1b} receptors have been proposed on the basis of limited pharmacology (Gustaffson et al., 1990), there has been no consistent evidence from binding, functional, or physiological studies to date to suggest that such subtypes exist.

The A_1 receptor has high affinity for N^6-substituted adenosine analogs like CHA, the cyclopentyl analog, CPA (Bruns et al., 1980; Williams et al., 1986), and the 2-chloro analog of CPA, CCPA (Lohse et al., 1988). To date, there have been no reports of any agonists for the adenosine receptor that are not based on the purine nucleoside pharmacophore (Jacobson et al., 1992; Olsson and Pearson, 1990; van Galen et al., 1992).

1.2.1.1   Human A_1 Receptors

The human adenosine A_1 receptor has been cloned from cDNA libraries from human hippocampus (Libert et al., 1992) ventricle, brain, and kidney (Salavatore et al., 1992). Oligonucleotide probes designed from the 5′ and 3′ coding region of RDC7 (Libert et al., 1991) were used to isolate the cDNA clones. The human A_1 receptor cDNA encodes a protein of 326 amino acids (Libert et al., 1992; Salavatore et al., 1992) and is 94% homologous at the amino acid level with the canine, bovine, and rat A_1 receptors (Libert et al., 1991; Reppert et al., 1991; Mahan et al., 1991; Olah et al., 1991; Tucket et al., 1992) differing by 20, 19, and 18 amino acids, respectively (Figure 2).

A comparison of the human A_1 receptors from heart, brain, and kidney showed no tissue specific differences in either the coding or untranslated regions. The human A_1 receptor transfected in the CHO cell line had a Kd value of 2 nM using [^{3}H]CPX (cyclopentylxanthine) as ligand (Libert et al., 1992). That expressed in COS cells exhibited saturable binding of the selective A_1 agonist, [^{3}H]-CHA with a Kd value of 4 nM (Salavatore et al., 1992). The rank order of activity of adenosine agonists was CPA > R-PIA > S-PIA consistent with an A_1 receptor subtype. Northern blot analysis of the human poly $^+$A RNA derived from brain, heart, kidney, and lung identified a 2.9-kb transcript in all four tissues with an apparent higher abundancy in brain. A second transcript of 4.3 kb was also detected in the brain. The human A_1 receptor has been mapped to chromosome 22 q11.2-q 13.1 (Libert et al., 1991).

1.2.1.4   Bovine A_1 Receptors

The bovine A_1 receptor has been cloned by two groups (Olah et al., 1991; Tucker et al., 1991). Antisense oligonucleotide probes based on nucleotide sequences encoding TM3 of RDC7 were used to screen a size selected (2 to 4 kb) bovine brain cDNA library to isolate nucleotide sequences encoding an A_1 receptor of 326 amino acids (Olah et al., 1991; Tucker et al., 1991). Transmembrane segments 1, 2, and 6 of the bovine A_1 receptor showed 100% homology with those in canine and rat A1 receptors. The bovine A_1 receptor differs by 7 (rat) and 8 (dog) amino acids in TM 3, 4, 5, and 7. The majority of differences between the receptors exist in extracellular loops 2 and 3 and the carboxy terminus. The bovine A_1 receptor, unlike RDC7, has two potential glycosylation sites in the second extracellular loop and has a third extracellular loop consisting of only 34 amino acids. The third extracellular loop has potential phosphorylation sites for protein kinases (PK) A and C while a single concensus sequence for PKA is present in the second intracellular loop. Transient expression of the bovine A_1 receptor in COS-7 cells showed high-affinity binding (Kd = 0.37 nM) for the antagonist ligand [^{3}H]-XAC consistent with data from the bovine A_1 receptor present in brain membranes.

The transfected COS7 receptor shares the unique pharmacology previously described (Ferkany et al., 1986) for the bovine A_1 receptor in that the rank order activity was R-PIA ≥ S-PIA > NECA as contrasted to the "classic" A_1 binding profile, R-PIA ≥ NECA > S-PIA.

Tucker et al. (1992) adopted a similar strategy in cloning the bovine A_1 receptor, using 27- and 44-mers encoding TM3 of RDC7 and the ZAP bovine brain cDNA library. A 2.8-kb cDNA clone designated pBOV13 encoding a protein of 326 amino acids showing greater than 90% homology with RDC7 was expressed in COS cells. High-affinity (Kd = 0.24 n*M*), saturable binding of the adenosine antagonist ligand [^{125}I]-BWA844 was observed.

The role of histidine residues in ligand binding has been investigated using site-directed mutagenesis in the COS-7 cell line (Olah et al., 1991). Replacement of His-278 in TM7 with a leucine decreased both agonist ([^{3}H]-APNEA) and antagonist ([^{3}H]-XAC) binding by greater than 90%. A similar mutant at His-251 in TM6 decreased affinity of antagonist (Kd = 0.17 n*M* to 0.65 n*M*) but not agonist radioligand even though the number of receptors labeled by either type of ligand (Bmax) was decreased by 74%. The authors concluded that the His-251 to Leu-251 mutation retained G-protein coupling and was more important than His-278 in ligand binding.

1.2.1.6   Canine A_1 Receptors

As noted, the canine A_1 receptor was identified using PCR from the canine thyroid gland (Libert et al., 1989) as RDC7 (Libert et al., 1991) and contains 326 amino acids (Figure 2). In COS-7 cells transfected with RDC7, [^{3}H]-CHA binds with a Kd value of 10 n*M*. When expressed in CHO cells, RDC7 confers CPA-mediated inhibition of forskolin-stimulated adenylate cyclase activity. Northern blot analysis demonstrated RDC7 transcripts in brain, thyroid gland, kidney, heart, and testis. *In situ* hybridization (Libert et al., 1991) showed preferential localization of RDC7 mRNA in cortical regions of the brain.

1.2.1.10   Rat A_1 Receptors

Using degenerate concensus primers from the transmembrane regions of the canine thyroid G protein family (Reppert et al., 1991; Mahan et al., 1991), the rat brain A_1 receptor has been cloned using a PCR fragment by two groups. The clone designated RFL7 (Reppert et al., 1991) shows 91% homology with the canine A_1 receptor and consists of 326 amino acids (Figure 2). Transient expression in COS-6M and NIH 3T3 cell lines established pharmacology consistent with an A_1 receptor for the clone using binding and functional assessment (inhibition of adenylate cyclase activity). A Kd value of 0.75 n*M* for the receptor was determined using [^{3}H]-CCPA. The rat A_1 receptor cloned and expressed from a rat striatum cDNA library consisted of 326 amino acids and had a Kd value ([^{3}H]-CPX) of 0.53 n*M* in transfected A9-L cells (Mahan et al., 1991) and had greater than 90% homology with RDC7 (Figure 1). Agonist ligand binding to this receptor transfected to CHO cells, unlike that reported for RFL7, indicated the presence of high and low affinity sites.

Northern blot analysis of RFL7 (Reppert et al., 1991) revealed hybridizing transcripts of 3.5 and 5.5 kb in brain, spinal cord, testis, and white adipose tissue. For the rat analog of RDC7 (Mahan et al., 1991), two species of mRNA of approximately 3.1 and 5.6 kb were detected. The former was the predominant form in peripheral rat tissues while highest expression of both was observed in cortex, cerebellum, and hippocampus. Retina predominately expressed the 5.6-kb species.

In situ hybridization using an antisense cRNA probe of the PCR cDNA fragment, PCR7 showed A_1 receptor mRNA to be widely distributed in the rat CNS (Reppert et al., 1991). A discrepancy was noted however with the previously reported A_1 receptor distribution determined by radioligand binding in the cerebellar granule cell layer, striatum, and cranial nerve nuclei.

1.2.2 Adenosine A_{2a} Receptors

1.2.2.1 Human A_{2a} Receptors

Two groups have reported cloning human A_{2a} receptor cDNA (Salavatore et al., 1992; Furlong et al., 1992). A third human A_{2a} receptor gene designated as M97370 has been deposited in the Genebank by Tiffany and Murphy from the NIH but has yet to be published. The cDNA cloned by Salavatore et al. (1992) from heart ventricle and striatal cDNA libraries encodes a protein of 412 amino acids (Figure 2) and has a higher homology (93%) to the canine A_{2a} receptor, RDC8, with 28 amino acid differences than the rat A_{2a} receptor (82% homology) with 71 amino acid differences. The greatest difference in sequence exists between the human and rat receptors in the carboxy terminus. A similar difference has been observed for the rat and canine A_{2a} receptor sequences. Like the human A_1 receptor, a comparision of sequences derived from human heart and brain cDNA clones showed no tissue-specific differences.

Furlong et al. (1992) cloned the A_{2a} receptor from a human neuroblastoma cDNA library. A cDNA clone H3.1 encoding a protein of 412 amino acids had 93% homology with RDC8 with 30 mismatches in the carboxy terminal region. The human A_{2a} receptor expressed in COS cells exhibited saturable binding with the selective A_{2a} receptor agonist, [^{3}H]-CGS 21680 (Jarvis et al., 1989) with a Kd value of 17 nM. Competition studies with agonists showed a rank order of activity NECA > R-PIA > CPA > S-PIA consistent with the labeling on an A_{2a} receptor.

The positive coupling of the human A_{2a} receptor was established through transient expression of mRNA in oocytes. Intracellular cAMP levels were found to be increased in the absence of exogenously added adenosine, an effect that could be blocked by the antagonist, 8-(*p*-sulfophenyl)-theophylline. The observed stimulation of adenylate cyclase activity, like that observed for RDC8 expressed in oocytes, was probably due to leakage of endogenous adenosine from the injected oocytes into the medium.

HEK293 cells transfected with the H3.1 cDNA bound [^{3}H]-CGS 21680 but not [^{3}H]CCPA (Furlong et al., 1992) with a Kd value of 10 nM. A 6-fold CGS 21680-evoked increase in cAMP formation was consistent with the data of Salavatore et al. (1992).

The human A_{2a} receptor is expressed as a 2.9-kb transcript in brain, heart, kidney, and lung (Salavatore et al., 1992). Human A_{2a} receptor mRNA has been localized to the medium-sized neurons in caudate, putamen, and nucleus accumbens (Schiffman et al., 1991), similar to that observed for the A_{2a} receptor in dog and rat brain (Ueeda et al., 1991).

The human A_{2a} receptor has been mapped to both chromosome 11 q11-q13 (Libert et al., 1991) and chromosome 22 q11(Peterfruend et al., 1992), a difference that remains to be resolved. A minor hybridizing peak on chromosome 10 q 25.3-q 26.3 could also be detected with the A_{2a} probe and was suggested to be a putative A_2-like receptor (Libert et al., 1991).

1.2.2.6 Canine A_{2a} Receptors

As already discussed, screening of a canine thyroid library with a human thyroid PCR amplified cDNA fragment with homology to G-protein linked receptors led to the identification of RDC8. The clone expressed in COS cells exhibited characteristics of the high-affinity A2a receptor with a Kd value of 26 nM for CGS 21680 and an agonist profile, NECA > 2-CADO > CPA (Maenhaut et al., 1990). Antipeptide antibodies against the second extracellular loop and carboxy terminal domain of RDC8 have provided evidence for a distinct A_{2a} receptor in dog liver that may be distinct from the receptor found in striatum (Palmer et al., 1992).

1.2.2.10   Rat A_{2a} Receptors

The rat A_{2a} receptor sequence was initially identified as RA2 from a rat brain cDNA library (Chern et al., 1992). Fink et al. (1992) have also cloned the A_{2a} receptor and characterized the pharmacology. The two sequences are similar with the exception of four amino acid differences (Figure 2). The rat A_{2a} receptor has 82% homology with the canine A2a receptor with differences in 71 amino acids.

The rat A_{2a} receptor (Fink et al., 1992) has high affinity for [³H]-CGS 21680 (Kd = 37 nM) with an agonist profile: NECA > R-PIA > CPA > S-PIA. *In situ* hybridization showed that the rat A_{2a} receptor mRNA was expressed in the same striatal neurons as mRNA for the dopamine D_2 receptor with no colocalization of A_{2a} receptor mRNA with mRNA for D_1 receptors (Fink et al., 1992) and finding independently confirmed by Schiffman et al. (1993).

These observations supported previous pharmacological studies showing that adenosine receptor antagonists could mimic the effect of exogenous dopamine agonists in preclinical models of dopamine hypofunction (Jarvis and Williams, 1987; Ferre et al., 1991; Fuxe et al., 1992; Ferre and Fuxe, 1992). The possibility that xanthine adenosine antagonists, like the putative A_{2a} receptor selective 8-dimethoxystyrene analog, KW 17837 (Suzuki et al., 1992), may have potential use in the treatment of Parkinson's disease is under investigation (Fuxe et al., 1992; Ferre and Fuxe, 1992; Suzuki et al., 1992).

1.2.3   Adenosine A_{2b} Receptors

The A_{2b} receptor was first designated as a low-affinity adenosine receptor that stimulated cAMP formation in VA 13 fibroblasts (Bruns, 1980) and brain slices (Daly et al., 1983) and was further delineated on the basis of binding studies (Bruns and Pugsley, 1986). In contrast to A_{2a} receptors that are localized to the striatum, A_{2b} receptors are widely distributed in the brain (Daly et al., 1983; Bruns and Pugsley, 1986). There are no known selective ligands for the A_{2b} receptor. This has limited studies related to its pharmacology and physiological function.

1.2.3.1   Human A_{2b} Receptors

The human A_{2b} receptor has been cloned from frontal cortex (Salavatore et al., 1992) and hippocampal (Pierce et al., 1992) cDNA libraries. That isolated from cortex is 68% homologous with the human A2a receptor and 58% homologous with the human A_1 receptor at the nucleotide level.

The human cortex A_{2b} receptor cDNA encodes a protein of 332 amino acids and is 46 and 58% homologous to the human A_1 and A_{2a} receptor sequences, respectively (Figure 2). It is 86% homologous to the rat A_{2b} receptor (Salavatore et al., 1992) with 48 amino acid differences between species. Of these, all but 9 are located outside of the transmembrane domains. The receptor isolated from hippocampus showed 45% homology with A_1 and A_{2a} receptors and is 328 amino acids in length (Pierce et al., 1992).

The identity of the cortical human A_{2b} receptor was confirmed on the basis of the pharmacological profile determined on the receptor expressed in COS cells (Salavatore et al., 1992). As compared to the human A_{2a} receptor, the A_{2b} receptor had lower affinity for NECA and higher affinity for the xanthine adenosine antagonists, DPX and CPT, consistent with the activities established for the A_{2b} receptor (Bruns, 1980; Daly et al., 1983). Transient expression of the human A_{2b} receptor in oocytes showed an increase in intracellular cAMP only upon addition of adenosine (Salavatore et al., 1992; Pierce et al., 1992) with no effect of CCPA or CGS 21680 (Suzuki et al., 1992). The need for exogenous adenosine for A_{2b} receptor activation as compared to the effects of endogenous adenosine seen with the human A_{2a} receptor is most likely due to differences in receptor subtype affinity for adenosine (Salavatore et al., 1992).

Northern blot analysis of human A_{2b} transcript expression in heart, brain, kidney, and lung revealed two transcripts of 1.7 and 2.1 kb in all the tissues except kidney, where no expression was observed.

1.2.3.10 Rat A_{2b} Receptors

A cDNA encoding a protein of 332 amino acids designated as RFL9 was cloned from a rat brain cDNA library using degenerate olignucleotide primers based on conserved regions of TM3 and TM7 (Stehle et al., 1992; Figures 1 and 2). The amino acid sequence of RFL9 showed 46 and 45% homology with rat A_1 and A_2 receptors, respectively, with the carboxyl terminus and the second extracellular loop being the most divergent. Homologies of 62 and 73% were found in the transmembrane regions with respect to the rat A_1 and A_2 receptors, respectively. RFL9 also lacked the aspartic acid residue in TM3, a common feature with adenosine receptors (Linden et al., 1989).

Transient expression of RFL9 in COS-6M cells resulted in cells capable of binding the nonselective adenosine receptor agonist, [³H]-NECA but neither the A_1 selective agonist, [³H]-CCPA nor the A_{2a} selective agonist, [³H]-CGS 21680. RFL9 transfected cells also showed NECA-stimulated adenylate cyclase activation with a maximal 50-fold increase at a concentration of 10 μM that could be blocked by millimolar concentrations of aminophylline. CGS 21680 (10 μM) had no effect on cAMP production in these cells.

Northern blot analysis of RFL9 revealed two hybridizing transcripts of 1.8 and 2.2 kb that were predominately expressed in large intestine, urinary bladder, and cecum with less expression in lung and brain. Antisense cDNA probes to RFL9 showed that RFL9 mRNA in brain was restricted to to the hypophyseal pars tuberalis. Based on the affinity for NECA and the stimulation of adenylate cyclase, RFL9 appears to be a member of the A_2 family of receptors, potentially the A_{2b}. The lack of effect of CGS 21680 is consistent with its inability to interact with the A_{2b} subclass (Jarvis et al., 1989).

Adenosine-modulated cAMP responses in CHO cells transfected with RFL9 (Rivkees and Reppert, 1992) were pharmacologically similar to those seen in VA13 fibroblasts (Bruns, 1980). RFL9 was detected in the fibroblasts (Rivkees and Reppert, 1992) along with cDNA fragments of both the rat A_1 and A_{2a} receptors. No evidence for multiple forms of RFL9 was noted.

1.2.4   Adenosine A_3 Receptors

A new adenosine receptor demonstrating 58% sequence homology with the RDC7 and RDC8 receptors has been cloned from a rat brain cDNA library (Zhou et al., 1992). The clone, R226, has been reported to encode 320 amino acids and has the same amino acid sequence as the previously reported G-linked protein, *tgpcr*1 (Meyerhof et al., 1991). The sequence deposited in the Genebank was, however, one amino acid residue shorter ending in EQTT rather than EQTTE. R226 has been designated as the A_3 receptor (Zhou et al., 1992). Expression in COS-7 and CHO cell lines showed that the transfected rat A_3 receptor was able to bind the A_1 receptor selective agonist [^{125}I]-APNEA and the nonselective agonist [^{3}H]-NECA but not the A_1 selective antagonists [^{3}H]-CPX or [^{3}H]-XAC or the A_{2a} receptor selective agonist, [^{3}H]-CGS 21680. Binding of [^{125}I]-APNEA (Kd = 17 n*M)* was inhibited by a number of adenosine agonists and nucleotides with a rank order of potency: R-PIA = NECA > S-PIA > adenosine > ATP = ADP. Adenosine agonists can inhibit forskolin-stimulated adenylate cyclase activity by 50 to 60% in R 226 transfected CHO cells with the following order of potency: R-PIA (18 n*M)* > NECA (23 n*M)* > CGS 21680 (144 n*M)*. Northern blot analysis showed that the highest expression of the A_3 receptor, like that of *tgpcr*1, was in testis with low levels in lung, kidney, heart, cortex, striatum, and olfactory bulb. Unlike the majority of adenosine receptors identified to date, agonist activation of the A_3 receptor is not blocked by prototypic xanthine adenosine antagonists.

1.2.99   Other Adenosine Receptors

1.2.99.1 A_4 Receptors

A new adenosine receptor that has been provisionally designated as A_4 by the IUPHAR Subcommittee on Purine Receptor Nomenclature (Abbraccio et al., 1993) has been described in rat striatum using [^{3}H]-CV 1808 (2-phenylaminoadenosine; 2-PAA) as a radioligand (Cornfield et al., 1992). This receptor has not yet been cloned.

Having many of the characteristics of the A_{2a} receptor, the A_4 receptor, like the A_3 receptor, does not bind CGS 21680 and is insensitive to xanthine antagonists. The nonxanthine adenosine antagonist, CGS 15943A (60; Figure 5) and a series of 2-substituted adenosine analogs interact potently with this binding site. [^{3}H]-CV 1808 binding to the A_4 site has a Kd value of 16 to 24 n*M* depending on the brain region studied. The putative A_4 receptor appears to be linked to potassium channels rather than adenylate cyclase (Cornfield et al., 1992).

1.2.99.2 P_2/ATP Receptors

A growing number of P_2 receptors, sensitive to ATP and other purine nucleotides, have been described over the past 5 years (Burnstock and Kennedy, 1985; Hoyle, 1990; Pintor et al., 1991; O'Connor et al., 1991; Table 1). Those initially described, the P_{2x} and P_{2y} (Burnstock and Kennedy, 1985) have been joined by the P_{2t} and P_{2z} receptors and a receptor sensitive to both ATP and UTP that has been designated as the P_{2u} receptor. A site labeled by the diadenosine polyphosphate, Ap4A, a compound that has been described as an "alarmone" (Varshavsky, 1983), has been identified in membranes from adrenal chromaffin cells (Pintor et al., 1991) and other tissues (Hilderman et al., 1991) and may represent the P_{2y} receptor (Pintor et al., 1991).

[^{3}H] α, β methylene ATP (Bo and Burnstock, 1989) and ADP-β-S (Cooper et al., 1989) have been used to label P_{2x} and P_{2y} receptors, respectively, and may prove useful as probes to clone these receptors (Bo et al., 1992). The cloning of P2 receptors will be a useful aid to developing a more systematic nomenclature for this class of purinoceptor (Abbracchio et al., 1993).

Note Added in Proof

Recently several purinergic receptor have been cloned: the sheep A_3 adenosine receptor (1), the human A_3 adenosine receptor (2), the mouse P_{2u} receptor (3), and the chicken P_{2y} receptor (4).

Acknowledgments

The authors would like to thank Dr. Stephen J. Peroutka and William Pearson for constructing the dendograms and Kevin Lynch for aligning the receptor TM segments in Figure 3.

REFERENCES

Abbracchio, M. P., Cattabeni, F., Fredholm, B. B., and Williams, M. (1993) Purine receptor nomenclature: a status report. *Drug Dev. Res.* 28, in press.

Arch J. R. S. and Newsholme E. A. (1978) The control of the metabolism and the hormonal role of adenosine. *Essays Biochem.* 14, 82–123.

Bo, X., Simon, J., Burnstock, G., and Barnard, E. A. (1992) Solubilization and molecular size determination of the P2x-purinoceptor from rat vas deferens. *J. Biol. Chem.,* in press.

Bo, X. and Burnstock, G. (1989) [^{3}H]-a, b-methylene ATP, a radioligand labelling P2-purinoceptors. *J. Autonom. Nervous Sys.* 28, 85–88.

Bruns R. F. (1980) Adenosine receptor activation in human fibroblasts: nucleoside agonists and antagonists. *Can. J. Physiol. Pharmacol.* 58, 673–691.

Bruns R. F., Daly J. W., and Snyder S. H. (1980) Adenosine receptors in brain membranes: binding of N^6-cyclohexyl-[^{3}H]-adenosine and 1,3-diethyl-8-[^{3}H]phenylxanthine. *Proc. Natl. Acad. Sci. U.S.A.* 77, 5547–5551.

Bruns R. F., Lu G. H., and Pugsley T. A. (1986) Characterization of the A_2 adenosine receptor labeled by [^{3}H]NECA in rat striatal membranes. *Mol. Pharmacol.* 29, 331–346.

Bruns, R. F. and Fergus, J. H. (1990) Allosteric enhancement of adenosine A_1 receptor binding and function by 2-amino-3-benzoylthiophenes. *Mol. Pharmacol.* 38, 939–949.

Bruns, R. F., Fergus, J. H., Coughenour, L. L., Courtland G., Pugsley, T. A., Dodd, J. H., and Tinney F. J. (1990) Structure-activity relationships for enhancement of adenosine A_1 receptor binding by 2-amino-3-benzoylthiophenes. *Mol. Pharmacol.* 38, 950–958.

Burnstock G. (1978) A basis for distinguishing two types of of purinergic receptor. In *Cell Membrane Receptors for Drugs and Hormones,* Bolis, L. and Straub, R. W., Eds., Raven, New York, pp. 107–118.

Burnstock, G. and Kennedy C. (1985) Is there a basis for distinguishing two types of P_2-purinoceptor? *Gen. Pharmacol.* 16, 433–440.

Chern, Y., King, K., Lai, H.-L., and Lai, H.-T. (1992) Molecular cloning of a novel adenosine receptor gene from rat brain. *Biochem. Biophys. Res. Commun.* 185, 304–309.

Cooper, C. L., Morris, A. J., and Harden, T. K. (1989) Guanine nucleotide-sensitive interaction of a radiolabeled agonist with a phospholipase C-linked P_{2y}-purinergic receptor. *J. Biol. Chem.* 264, 6202–6206.

Cornfield, L. J., Hu, S., Hurt, S. D., and Sills, M. A. (1992) [^{3}H] 2-Phenylaminoadenosine ([^{3}H] CV 1808) labels a novel adenosine receptor in rat brain, *J. Pharmacol. Exp. Ther.* 263, 552–561.

Daly, J. W. (1982) Adenosine receptors: targets for future drugs. *J. Med. Chem.* 25, 197–207.

Daly J. W., Butts-Lamb, P., and Padgett, W. (1983) Subclasses of adenosine receptors in the central nervous system. Interaction with caffeine and related methylxanthines. *Cell. Mol. Neurobiol.* 36, 69–80.

Daly, J. W., Caceres, J., Moni, R. W., Gusovsky, F., Moos, Jr., M., Seamon, K. B., Milton, K., and Myers, C. W. (1992) Frog secretions and hunting magic in the upper Amazon: identification of a peptide interacting with an adenosine receptor. *Proc. Natl. Acad. Sci. U.S.A.,* in press.

Feng, D. F. and Doolittle, R. F. (1990) Progressive alignment and phylogenetic tree construction of protein sequences. *Meth. Enzymol.* 183, 375–387.

Ferkany, J. W., Valentine, H. L., Stone, G. A., and Williams, M. (1986) Adenosine A_1 receptors in mammalian brain: species differences in their interactions with agonists and antagonists. *Drug Dev. Res.* 9, 85–93.

Ferre, S. and Fuxe, K. (1992) Dopamine denervation leads to an increase in the intramembrane interaction between adenosine A_2 and dopamine D_2 receptors in the neostriatum. *Brain Res.* 594, 124–130.

Ferre, S., Von Euler, G., Johansson, B., Fredholm B. B., and Fuxe, K. (1991) Stimulation of high-affinity adenosine A_2 receptors decreases the affinity of D2 receptors in rat striatal membranes. *Proc. Natl. Acad. Sci. U.S.A.* 88, 7238–7241.

Fink, J. S., Weaver, D. R., Rivkees, S. A., Peterfreund, R. A., Pollack, A. E., Adler, E. M., and Reppert, S. M. (1992) Molecular cloning of the rat A_2 adenosine receptor: selective co-expression with D2 dopamine receptors in rat striatum. *Mol. Brain Res.* 14, 186–195.

Furlong, T. J., Pierce, K. D., Selbie, L. A., and Shine, J. (1992) Molecular characterization of a human brain adenosine A_2 receptor. *Mol. Brain Res.* 15, 62–66.

Fuxe, K., Ferre, S., von Euler, G., Johansson, B., and Fredholm, B. B. (1992) Antagonistic A2a/D2 receptor interactions in the striatum as a basis for adenosine/dopamine interactions in the central nervous system. *Int. J. Purine Pyrimidine Res.* 3, 38.

Gustaffson, L. E., Wiklund, C. U., Wiklund, N. P., and Stelius, L. (1990) Subclassification of neuronal adenosine receptors. In *Purines in Cell Signalling. Targets for New Drugs,* Jacobson, K. A., Daly, J. W., and Manganiello, V., Eds., Springer-Verlag, New York, pp. 200–205.

Hamprecht, B. and van Calker, D. (1985) Nomenclature of adenosine receptors. *Trends Pharmacol. Sci.* 6, 153–154.

Hilderman, R. H., Martin, M., Zimmerman, J. K., and Pivorun, E. B. (1991). Identification of a unique membrane receptor for adenosine 5′, 5‴-P1. P4 tetraphosphate. *J. Biol. Chem.* 266, 6915–6918.

Hoyle, C. H. V. (1990) Pharmacological activity of adenine dinucleotides in periphery: possible receptor classes and transmitter function. *Gen. Pharmacol.* 21, 827–831.

Hutchison, A. J., Webb, R. L., Oei, H. H., Ghai, G. R., Zimmerman, M. B., and Williams, M. (1989) CGS 21680C, an A_2 selective adenosine receptor agonist with preferential hypotensive activity. *J. Pharmacol. Exp. Ther.* 251, 47–55.

Jacobson, K. A., van Galen, P. J. M., and Williams, M. (1992) Adenosine receptors: pharmacology, structure-activity relationships, and therapeutic potential. *J. Med. Chem.*, 35, 407–415.

Jarvis, M. F. and Williams, M. (1987) Adenosine and dopamine function in the CNS. *Trends Pharmacol. Sci.* 8, 330–331.

Jarvis, M. F., Schulz, R., Hutchison, A. J., Do, U. H., Sills, M. A., and Williams, M. (1989). [^{3}H] CGS 21680, a selective A_2-adenosine receptor agonist ligand directly labels A_2-receptors in rat brain. *J. Pharmacol. Exp. Ther.* 251, 888–893.

Libert, F., Parmentier, M., Lefort, A., Dinsart, C., Van Sande, J., Maenhaut, C., Simons, M.-J., Dumont, J. E., and Vassart, G. (1989) Selective amplification and cloning of four new members of the G protein-coupled receptor family. *Science* 244, 569–572.

Libert, F., Schiffmann, S., Lefort, A., Parmentier, M., Gerard, C., Dumont, J. E., Vanderhaegen, J. J., and Vassart, G. (1991) The orphan receptor cDNA RDC7 encodes an A_1 adenosine receptor. *EMBO J.* 10, 1677–1682.

Libert, F., Van Sande, J., Lefort, A., Czernilofsky, A., Dumont, J. E., Vassart, G., Ensinger, H. A., and Mendla, K. D. (1992) Cloning and functional charactertization of a human A_1 adenosine receptor. *Biochem. Biophys. Res. Commun.* 187, 919–926.

Libert, L., Passage, E., Parmentier, M. J., Simons, M., Vassart, G., and Mattei, M. G. (1991) Chromosomal mapping of A_1 and A_2 adenosine receptors, VIP receptor and a new subtype of serotonin receptor. *Genomics* 11, 225–227.

Linden, J., Tucker, A. l., and Lynch, K. R. (1991) Molecular cloning of adenosine A_1 and A_2 receptors. *Trends Pharmacol. Sci.* 12, 326–328.

Lohse, M. J., Klotz, K. N., Schwabe, U., Crisatalli, G., Vittori, S., and Grifantini, M. (1988) 2-Chloro-N^6-cyclopentyladenosine: a highly selective agonist at A_1 adenosine receptors. *Naunyn-Schiemedbergs' Arch. Pharmacol.* 337, 687–689.

Londos, C., Cooper, D. M. F., and Wolff, J. (1980) Subclasses of external adenosine receptors. *Proc. Natl. Acad. Sci. U.S.A.* 77, 2551–2554.

Maenhaut, C., van Sande, S. J., Libert, F., Abramowicz, M., Parmentier, M., Vanderhaegen, J. J., Dumont, J. E., Vassart, G., and Schiffmann, S. (1990) RDC8 codes for an adenosine A_2 receptor with physiological constitutive activity. *Biochem. Biophys. Res. Commun.* 173, 1169–1178.

Mahan, L. C., McVittie, L. D., Smyk-Randall, E. M., Nakata, H., Monsma, F. H., Jr., Gerfen, C. R., and Sibley, D. R. (1991) Cloning and expression of an A_1 adenosine receptor from rat brain. *Mol. Pharmacol.* 40, 1–7.

Meghji, P. (1991) Adenosine production and metabolism. In *Adenosine and the Nervous System*, Stone, T. W., Ed., Academic Press, London, pp. 25–42.

Meyerhof, W., Muller-Brechelin, R., and Richler, D. (1991) Molecular cloning of a novel G-protein coupled receptor expressed during rat spermiogenesis. *Biochem. Biophys. Res. Commun.* 284, 155–160.

Morgan, P. F. (1991) Post-receptor mechanisms. In *Adenosine and the Nervous System*, Stone, T. W., Ed., Academic Press, London, pp. 119–136.

O' Connor, S. E., Dainty, I. A., and Leff, P. (1991) Further subclassification of ATP receptors based on agonist studies. *Trends Pharmacol. Sci.* 12, 137–141.

Olah, M. E., Ren, H., Ostrowski, J., Jacobson, K. A., and Stiles, G. (1991) Cloning, expression, and characterization of a unique bovine A_1 adenosine receptor. *J. Biol. Chem.* 267, 10764–10770.

Olsson, R. A. and Pearson, J. D. (1990) Cardiovascular purinoceptors. *Physiol. Rev.* 70, 761–845.

Palmer, T. M., Jacobson, K. A., and Stiles, G. L. (1992) Immunological identification of A_2 adenosine receptors by two antipeptide antibody preparations. *Mol. Pharmacol.* 42, 391–397.

Peroutka, S. J. (1992) Phylogenetic tree analysis of G protein-coupled 5-HT receptors: implications for receptor nomenclature. *Neuropharmacology* 31, 609–612.

Peterfruend, R. A., MacCollin, M., MacDonald, M., Lekanne Deprez, R. H., Zwarthoff, E. C., Gusella, J., and Fink, J. S. (1992) Adenosine-2 receptor sequences on human chromosome 22. *Abstr. Soc. Neurosci.* 18, 999.

Pierce, K. D., Furlong, T. J., Selbie, L. A., and Shine, J. (1992) Molecular cloning and expression of an adenosine A_{2b} receptor from human brain. *Biochem. Biophys. Res. Commun.* 187, 86–93.

Pintor, J., Torres, M., Castro, E., and Miras-Portugal, M. T. (1991) Characterization of diadenosine tetraphosphate (Ap4A) binding sites in cultured chromaffin cells: evidence for a P_{2y} site. *Br. J. Pharmacol.* 103, 1980–1984.

Reppert, S. M., Weaver, D. R., Stehle, J. H., and Rivkees, S. A. (1991) Molecular cloning and characterization of a rat A_1-adenosine receptor that is widely expressed in brain and spinal cord. *Mol. Endocrinol.* 5, 1037–1048.

Rivkees, S. A. and Reppert, S. M. (1992) RFL9 encodes an A_{2b}-adenosine receptor. *Mol. Endocrinol.* in press.

Salavatore, C. A., Luneau, C. J., Johnson, R. G., and Jacobson, M. A. (1992) Cloning from multiple tissues and characterization of three human adenosine receptor subtypes. *Int. J. Purine Pyrimid. Res.* 3, 82.

Sattin, A. W. and Rall, T. W. (1970) The effect of adenosine and adenine nucleotides on the cyclic adenosine 3¢-5¢-phosphate content of guinea pig cerebral cortex slices. *Mol. Pharmacol.* 6, 13–23.

Schiffman, S. N., Libert, F., Vassart, G., and Vanderhagen, J. J. (1991) Distribution of adenosine A_2 receptor mRNA in the human brain. *Neurosci. Letts.* 130, 177–181.

Schiffman, S. N., Halleux, P., Menu, R., and Vanderhagen, J.-J. (1993) Adenosine A_2 receptors expressed in striatal neurons. Implications for basal ganglia pathophysiology. *Drug. Develop. Res.* 28, in press.

Schiffmann, S. N., Libert, F., Vassart, G., Dumont, J. E., and Vanderhaeghen, J. (1990) A cloned G protein-coupled protein with a distribution restricted to striatal medium-sized neurons: possible relationship with the D_1 dopamine receptor. *Brain Res.* 519, 333–337.

Stehle, J. H., Rivkees, S. A., Lee, J. J., Weaver, D. R., Deeds, J. D., and Reppert, S. M. (1992) Molecular cloning and expression of the cDNA for a novel A_2-adenosine receptor subtype, *Mol. Endocrinol.* 6, 384–393.

Suzuki, F., Shimada, J., Nonaka, H., Ishii, A., and Ichikawa, S. (1992) Potent and selective adenosine A_2 antagonist. *Int. J. Purine Pyrimidine Res.* 3, 107.

Tucker, A. L., Linden, L., Robevam, A. S., D'Angelo, D. D., and Lynch, K. R. (1992) Cloning and expression of a bovine adenosine A_1 receptor DNA. *FEBS Letts.* 297, 107–111.

Ueeda, M., Thompson, R. D., Arroyo, L. H., and Olsson, R. A. (1991) 2-Aralkoxyadenosines: potent and selective agonists at the coronary artery A_2 adenosine receptor. *J. Med. Chem.* 34, 1334–1344.

van Calker, D., Mueller, M., and Hamprecht, B. (1979) Adenosine regulates via two different types of receptors, the accumulation of cyclic AMP in cultured brain cells. *J. Neurochem.* 33, 999–1005.

van Galen, P. J. M., Stiles, G. L., Michaels, G., and Jacobson, K. A. (1992) Adenosine A_1 and A_2 receptors: structure-function relationships. *Med. Res. Rev.* 12, 423–471.

Varshavsky, A. (1983) Diadenosine 5', 5'''-P1, P4-tetraphosphate: a pleiotropically acting alarmone? *Cell* 34, 711–712.

Williams, M., Braunwalder, A., and Erickson, T. E. (1986) Evaluation of the binding of the A-1 selective adenosine radioligand, cyclopentyl-adenosine to rat brain tissue. *Naunyn-Schimedberg's Arch. Pharmacol.* 332, 179–183.

Williams, M., Francis, J. E., Ghai, G. R., Braunwalder, A., Psychoyos, S., Stone, G. A., and Cash, W. D. (1987) Biochemical characterization of the triazoloquinazoline, CGS 15943A, a novel, non-xanthine adenosine antagonist. *J. Pharmacol. Exp. Ther.* 241, 415–420.

Zhou, F. Q.-Y., Olah, M. E., Li, C., Johnson, R. A., Stiles, G. L., and Civelli, O. (1992) Molecular cloning and characterization of a novel adenosine receptor: the A_3 adenosine receptor. *Proc. Natl. Acad. Sci. U.S.A.* 89, 7432–7436.

Linden, J., H. E. Taylor, A. S. Robeva, A. L. Tucker, J. H. Stehle, S. A. Rivkees, J. S. Fink, and Reppert, S. M. (1993) Molecular cloning and functional expression of a sheep A_3 adenosine receptor with widespread tissue distribution. *Molecular Pharmacology,* in press.

Salvatore, C. A. M. A. Jacobson, H. E. Taylor, J. Linden, and Johnson, R. G. (1993) Molecular cloning and characterization of the human A_3 adenosine receptor. *Proc. Natl. Acad. Sci. U.S.A.* in press.

Lustig, K. D., A. K. Shiau, A. J. Brake, and Julius, D. (1993) Expression cloning of an ATP receptor from mouse neuroblastoma cells. *Proc. Natl. Acad. Sci. U.S.A.* 9-: 5113–5117.

Webb, T. E., J. Simon, B. J. Krishek, a. N. Bateson, T. G. Smart, B. F. King, G. Burnstock, and Barnard, E. A. (1993) Cloning and functional expression of a brain G-protein coupled ATP receptor. *FEBS Lett.* 324: 219–225.

3

Adrenergic Receptors

David J. Pepperl and John W. Regan

1.3.0 Introduction

The adrenergic receptors (ARs) comprise one of the largest and most extensively characterized families within the G-protein coupled receptor "superfamily". This superfamily includes not only adrenergic receptors, but also muscarinic cholinergic, dopaminergic, serotonergic, and histaminergic receptors. Numerous peptide receptors including glucagon, somatostatin, and vasopressin receptors, as well as sensory receptors for vision (rhodopsin), taste, and olfaction, also belong to this growing family. Despite the diversity of signaling molecules, G-protein coupled receptors all possess a similar overall primary structure, characterized by seven putative membrane-spanning alpha helices (Probst et al., 1992).

In the most basic sense, the adrenergic receptors are the physiological sites of action of the catecholamines epinephrine and norepinephrine. ARs were initially classified as either α or β by Ahlquist, who demonstrated that the order of potency for a series of agonists to evoke a physiological response was distinctly different at the two receptor subtypes (Ahlquist, 1948). Functionally, α adrenergic receptors were shown to control vasoconstriction, pupil dilation, and uterine inhibition, while β ARs were implicated in vasorelaxation, myocardial stimulation, and bronchodilation (Regan et al., 1990). Eventually, pharmacologists realized that these responses resulted from activation of several distinct AR subtypes. β-Receptors in the heart were defined as β_1, while those in the lung and vasculature were termed β_2 (Lands et al., 1967).

α-Adrenergic receptors, meanwhile, were first classified based on their anatomical location, as either pre- or postsynaptic (α_2 and α_1 ARs, respectively) (Langer et al., 1974). This classification scheme was confounded, however, by the presence of α_2 receptors in distinctly nonsynaptic locations, such as platelets (Berthelsen and Pettinger, 1977). With the development of radioligand binding techniques, α adrenergic receptors could be distinguished pharmacologically based on their affinities for the antagonists prazosin or yohimbine (Stark, 1981). Definitive evidence for adrenergic receptor subtypes, however, awaited purification and molecular cloning of AR subtypes. In 1986, the genes for the hamster β_2 (Dixon et al., 1986) and turkey β_1 ARs (Yarden et al., 1986) were cloned and sequenced. Hydropathy analysis revealed that these proteins contained seven hydrophobic domains similar to rhodopsin, the receptor for light. Since that time, the adrenergic receptor family has expanded to include three subtypes of β receptors (Emorine et al., 1989), three subtypes of α_1 receptors (Schwinn et al., 1990), and three distinct types of α_2 receptors (Lomasney et al., 1990).

The sequences for all known adrenergic receptor subtypes have been aligned (Figure 1), and compared by phylogenetic tree analysis (Feng and Doolittle, 1990) (Figure 2). Evolutionary relationships between receptors can be determined by comparing the length of each individual branch. Based on this analysis, the α_2 receptors appear to have diverged rather early from either the β or α_1 receptors. The α_2 receptors have been broken down into three molecularly distinct sutbtypes termed α_2C2, α_2C4, and α_2C10 based on their chromosomal location. These subtypes appear to correspond to the pharmacologically defined α_{2B}, α_{2C}, and α_{2A} subtypes, respectively (Bylund et al., 1992). By molecular genetic analysis, the α_2C2 appears more closely related to the α_2C10, while the α_2C4 formed a distinct branch at some earlier time. α_2C10 Receptors have been identified not only in human, but also pig, rat, and mouse, while both mouse and rat homologs for the α_2C2 and α_2C4 have been isolated. More recently, a receptor possessing the pharmacology of an α_2AR was cloned from fish pigment granule cells (Svensson et al., 1993) where it mediates pigment granule aggregation in response to epinephrine. This receptor, abbreviated α_2-F (fish-α_2) is pharmacologically similar to the previously cloned α_2C4 AR.

Phylogenetically, the α_1 receptors appear more closely related to the β receptors than the α_2 receptors. Presently, three distinct subtypes of α_1 receptors can be identified, termed α_{1a}, α_{1b}, and α_{1c}. Structurally, the α_{1b} receptors diverged much earlier than the α_{1a} or α_{1c} receptors. This divergence may account for the very different pharmacological properties exhibited by this receptor subtype. Despite their pharmacological similarities the α_{1a} and α_{1c} receptors are structurally distinct. The α_{1a} receptor has recently been cloned from both human and rat, while the α_{1c} receptor was similarly isolated from bovine.

The β adrenergic arm of the tree consists of 3 distinct subtypes, termed simply β_1, β_2, and β_3. Of these receptors, the β_3 or atypical β receptor is most evolutionarily divergent. Both the mouse and rat homologs to the human β_3 receptor have recently been identified. By far, the best characterized of the adrenergic receptors have been the β_2 receptors. This receptor was first characterized in hamster lung, and subsequently, the mouse, rat, and human β_2 adrenergic receptors have been cloned. These receptors can be genetically distinguished from the β_1 receptors, present primarily in the heart. The β_1 receptor was first identified in turkey erythrocytes; subsequently, both the human and rat homologs have been characterized.

This review offers a complete picture of the adrenergic receptor family as of early 1993. We hope to provide a structural basis for the distinct pharmacological and functional properties of each receptor species. We have systematically examined the primary sequence, tissue distribution, pharmacology, and second messenger systems for the various adrenergic receptors. This approach should clarify the extent of adrenergic receptor heterogeneity and provide a basis for any further classification or identification of adrenergic receptor subtypes.

1.3.11 α_{1a} Adrenergic Receptors

Until the early 1970s, the α adrenergic receptors were thought to be a single homogenous class. By this time, however, it became clear that the presynaptic α receptor involved in mediating norepinephrine release was distinct from the postsynaptic α receptor located directly on the effector organs. This postsynaptic receptor, termed the α_1, appeared to mediate stimulatory cellular responses via increases in cellular Ca^{+2} (McGrath, 1982). With the development of radioligand binding techniques, α_1 ARs were shown to possess high affinity for ligands [^{3}H]prazosin and

[³H]WB4101 (2,6-dimethoxyphenoxyethylaminomethyl-1,4-benzodioxane). Other ligands often used to bind α_1 ARs include [¹²⁵I]HEAT (2-{β-(4-hydroxy-3-[¹²⁵I]iodophenyl)-ethylaminomethyl}-tetralone), [³H]idazoxan and the nonselective catecholamine ligand [³H]norepinephrine (Table 2). α_1 AR subtypes have also been distinguished based on their sensitivity to inactivation by chlorethylclonidine (CEC). While the α_{1a} receptor appears insensitive to this drug, the α_{1b} AR can be totally inactivated by CEC pretreatment (Han et al., 1987).

```
alpha1A.human    1 .....MAAALRSVMMAGYLSEWRTPTYRST.....................
alpha1A.rat      1 .....M--TFRDILSVTFEGPRSSSSTGGSGAGGGAGTVGPEGGAVGGVP
alpha1B.hamster  1 .....MNPDLDTGHNTSAPAQW...........................
alpha1B.rat      1 .....MNPDLDTGHNTSAPAHW...........................
alpha1C.bovine   1 ...............MVF..............................
alpha2A.human    1 ......MGSLQPDAGNAS..............................
alpha2A.mouse    1 ......MGSLQPDAGNSS..............................
alpha2A.porcine  1 ......MGSLQPEAGNAS..............................
alpha2A.rat      1 ......MGSLQPDAGNSS..............................
alpha2B.human    1 MASPALAAALAVAAAAGPNAS............................
alpha2B.mouse    1 MASPALAAALAAAAAGPNGS............................
alpha2B.rat      1 MASPALAAALAAAAAEGPNGS...........................
beta1.human      1 .....MGAGVLVLGASEPGNL...........................
beta1.rat        1 .....MGAGALALGASEPCNL...........................
beta1.turkey     1 .....MGDGWLP-PDCGPHNR...........................
beta2.hamster    1 .....MGP.....PGNDSDFL...........................
beta2.human      1 .....MGQ.....PGNGSAFL...........................
beta2.mouse      1 .....MGP.....HGNDSDFL...........................
beta2.rat        1 .....MEP.....HGNDSDFL...........................
beta3.human      1 .....MAP.....WPHENSSL...........................
beta3.mouse      1 .....MAP.....WPHRNGSL...........................
beta3.rat        1 .....MAP.....WPHKNGSL...........................


alpha1A.human   26 ......EMVQRLRMEAVQHST...........STAAVGGLVVSAQGVGV
alpha1A.rat     44 GATGGGAVVGTGSGEDNQSSTGEPGAAASGEVNGSAAVGGLVVSAQGVGV
alpha1B.hamster 18 .....GELKDANFTGPNQTS...........SNSTLPQLDVTRAISV
alpha1B.rat     18 .....GELKDDNFTGPNQTS...........SNSTLPQLDVTRAISV
alpha1C.bovine   4 ...........LSGNASDSS...........NCTHPPPPVNISKAILL
alpha2A.human   13 .......W--NGTEAPGGGAR...........ATP---YSLQVTLTL
alpha2A.mouse   13 .......W--NGTEAPGGGTR...........ATP---YSLQVTLTL
alpha2A.porcine 13 .......W--NGTEAPGGGAR...........ATP---YSLQVTLTL
alpha2A.rat     13 .......W--NGTEAPGGGTR...........ATP---YSLQVTLTL
alpha2B.human   22 ...GAGERGSGGVANASGASW...........GPPRGQYSAGAVAGL
alpha2B.mouse   22 ...DAGEWGSGGGANASGTDW...........VPPPGQYSAGAVAGL
alpha2B.rat     22 ...DAGEWGSGGGANASGTDW...........APPPGQYSAGAVAGL
alpha2C.human    1 ...............MDH...........QDP---YSVQATAAI
alpha2C.mouse    1 ...............MVH...........QEP---YSVQATAAI
alpha2C.rat      1 ...........MSGPTMDH...........QEP---YSVQATAAI
beta1.human     77 .....SSAAPLPDGAATAARLLVPASPPASLLPPASESPEPLSQQWTAGM
beta1.rat       77 .....SSAAPLPDGAATAARLLVLASPPASLLPPASEGSAPLSQQWTAGM
beta1.turkey    16 .....SGGGG-ATAAPTGSRQV...........SAELLSQQWEAGM
beta2.hamster   12 .....LT----TNGSHVPDHDV...........TEERDEAWVVGM
beta2.human     12 .....LA----PNRSHAPDHDV...........TQQRDEVWVVGM
beta2.mouse     12 .....LA----PNGSRAPDHDV...........TQERDEAWVVGM
beta2.rat       12 .....LA----PNGSRAPGHDI...........TQERDEAWVVGM
beta3.human     12 .....APWPDLPTLAPNTANT...........SGLPGVPWEAAL
beta3.mouse     12 .....ALWSDAPTLDPSAANT...........SGLPGVPWAAAL
beta3.rat       12 .....AFWSDAPTLDPSAANT...........SGLPGVPWAAAL
```

FIGURE 1. Alignment of the amino acid sequences of adrenergic receptors. The sequences were obtained from commercial protein databases such as GenBank and EMBL. The references for each sequence are cited in the text.

```
alpha1A.human    58  - G V F L A A F I L M A V A G N L L V I L S V A C N R H L Q T V T N Y F I V N L A V A D L L L S A T
alpha1A.rat      94  - G V F L A A F I L T A V A G N L L V I L S V A C N R H L Q T V T N Y F I V N L A V A D L L L S A A
alpha1B.hamster  49  - G L V L G A F I L F A I V G N I L V I L S V A C N R H L R T P T N Y F I V N L A I A D L L L S F T
alpha1B.rat      49  - G L V L G A F I L F A I V G N I L V I L S V A C N R H L R T P T N Y F I V N L A I A D L L L S F T
alpha1C.bovine   30  - G V I L G G L I L F G V L G N I L V I L S V A C H R H L H S V T H Y Y I V N L A V A D L L L T S T
alpha2A.human    37  - V C L A G L L M L L T V F G N V L V I I A V F T S R A L K A P Q N L F L V S L A S A D I L V A T L
alpha2A.mouse    37  - V C L A G L L M L F T V F G N V L V I I A V F T S R A L K A P Q N L F L V S L A S A D I L V A T L
alpha2A.porcine  37  - V C L A G L L M L F T V F G N V L V I I A V F T S R A L K A P Q N L F L V S L A S A D I L V A T L
alpha2A.rat      37  - V C L A G L L M L F T V F G N V L V I I A V F T S R A L K A P Q N L F L V S L A S A D I L V A T L
alpha2B.human    55  - A A V V G F L I V F T V V G N V L V V I A V L T S R A L R A P Q N L F L V S L A S A D I L V A T L
alpha2B.mouse    55  - A A V V G F L I V F T V V G N V L V V I A V L T S R A L R A P Q N L F L V S L A S A D I L V A T L
alpha2B.rat      55  - A A V V G F L I V F T V V G T V L V V I A V L T S R A L R A P Q N L F L V S L A S A D I L V A T L
alpha2C.human    16  - A A A I T F L I L F T I F G N A L V I L A V L T S R S L R A P Q N L F L V S L A A A D I L V A T L
alpha2C.mouse    16  - A S A I T F L I L F T I F G N A L V I L A V L T S R S L R A P Q N L F L V S L A A A D I L V A T L
alpha2C.rat      21  - A S A I T F L I L F T I F G N A L V I L A V L T S R S L R A P Q N L F L V S L A A A D I L V A T L
beta1.human      62  - G L L M A L I V L L I V A G N V L V I V A I A K T P R L Q T L T N L F I M S L A S A D L V M G L L
beta1.rat        62  - G L L L A L I V L L I V V G N V L V I V A I A K T P R L Q T L T N L F I M S L A S A D L V M G L L
beta1.turkey     45  - S L L M A L V V L L I V A G N V L V I A A I G R T Q R L Q T L T N L F I T S L A C A D L V M G L L
beta2.hamster    37  - A I L M S V I V L A I V F G N V L V I T A I A K F E R L Q T V T N Y F I T S L A C A D L V M G L A
beta2.human      37  - G I V M S L I V L A I V F G N V L V I T A I A K F E R L Q T V T N Y F I T S L A C A D L V M G L A
beta2.mouse      37  - A I L M S V I V L A I V F G N V L V I T A I A K F E R L Q T V T N Y F I I S L A C A D L V M G L A
beta2.rat        37  - A I L M S V I V L A I V F G N V L V I T A I A K F E R L Q T V T N Y F I T S L A C A D L V M G L A
beta3.human      40  A G A L L A L A V L A T V G G N L L V I V A I A W T P R L Q T M T N V F V T S L A A A D L V M G L L
beta3.mouse      40  A G A L L A - - - L A T V G G N L L V I I A I A R T P R L Q T I T N V F V T S L A A A D L V V G L L
beta3.rat        40  A G A L L A - - - L A T V G G N L L V I T A I A R T P R L Q T I T N V F V T S L A T A D L V V G L L


alpha1A.human   107  V L P F S A T M E V L G F W A F G R A F C D V W A A V D V L C C T A S I L S L C T I S V D R Y V G V
alpha1A.rat     143  V L P F S A T M E V L G F W A F G R T F C D V W A A V D V L C C T A S I L S L C T I S V D R Y V G V
alpha1B.canine    1  V L P F S A A L E V L G Y W V L G R I F C D I W A A V D V L C C T A S I L S L C A I S I D R Y I G V
alpha1B.hamster  98  V L P F S A T L E V L G Y W V L G R I F C D I W A A V D V L C C T A S I L S L C A I S I D R Y I G V
alpha1B.rat      98  V L P F S A T L E V L G Y W V L G R I F C D I W A A V D V L C C T A S I L S L C A I S I D R Y I G V
alpha1C.bovine   79  V L P F S A I F E I L G Y W A F G R V F C N V W A A V D V L C C T A S I M G L C I I S I D R Y I G V
alpha2A.human    86  V I P F S L A N E V M G Y W Y F G K A W C E I Y L A L D V L F C T S S I V H L C A I S L D R Y W S I
alpha2A.mouse    86  V I P F S L A N E V M G Y W Y F G K V W C E I Y L A L D V L F C T S S I V H L C A I S L D R Y W S I
alpha2A.porcine  86  V I P F S L A N E V M G Y W Y F G K A W C E I Y L A L D V L F C T S S I V H L C A I S L D R Y W S I
alpha2A.rat      86  V I P F S L A N E V M G Y W Y F G K V W C E I Y L A L D V L F C T S S I V H L C A I S L D R Y W S I
alpha2B.human   104  V M P F S L A N E L M A Y W Y F G Q V W C G V Y L A L D V L F C T S S I V H L C A I S L D R Y W S V
alpha2B.mouse   104  V M P F S L A N E L M A Y W Y F G Q V W C G V Y L A L D V L F C T S S I V H L C A I S L D R Y W S V
alpha2B.rat     104  V M P F S L A N E L M A Y W Y F G Q V W C G V Y L A L D V L F C T S S I V H L C A I S L D R Y W S V
alpha2C.human    65  I I P F S L A N E L L G Y W Y F R R T W C E V Y L A L D V L F C T S S I V H L C A I S L D R Y W A V
alpha2C.mouse    65  I I P F S L A N E L L G Y W Y F W R A W C E V Y L A L D V L F C T S S I V H L C A I S L D R Y W A V
alpha2C.rat      70  I I P F S L A N E L L G Y W Y F W R A W C E V Y L A L D V L F C T S S I V H L C A I S L D R Y W A V
beta1.human     111  V V P F G A T I V V W G R W E Y G S F F C E L W T S V D V L C V T A S I E T L C V I A L D R Y L A I
beta1.rat       111  V V P F G A T I V V W G R W E Y G S F F C E L W T S V D V L C V T A S I E T L C V I A L D R Y L A I
beta1.turkey     94  V V P F G A T L V V R G T W L W G S F L C E C W T S L D V L C V T A S I E T L C V I A I D R Y L A I
beta2.hamster    86  V V P F G A S H I L M K M W N F G N F W C E F W T S I D V L C V T A S I E T L C V I A V D R Y I A I
beta2.human      86  V V P F G A A H I L M K M W T F G N F W C E F W T S I D V L C V T A S I E T L C V I A V D R Y F A I
beta2.mouse      86  V V P F G A S H I L M K M W N F G N F W C E F W T S I D V L C V T A S I E T L C V I A V D R Y V A I
beta2.rat        86  V V P F G A S H I L M K M W N F G N F W C E F W T S I D V L C V T A S I E T L C V I A V D R Y V A I
beta3.human      90  V V P P A A T L A L T G H W P L G A T G C E L W T S V D V L C V T A S I E T L C A L A V D R Y L A V
beta3.mouse      87  V M P P G A T L A L T G H W P L G E T G C E L W T S V D V L C V T A S I E T L C A L A V D R Y L A V
beta3.rat        87  V M P P G A T L A L T G H W P L G A T G C E L W T S V D V L C V T A S I E T L C A L A V D R Y L A V
```

FIGURE 1b.

Although α_1 ARs are typically associated with smooth muscle, where they mediate contraction, α_1 ARs are also located in the heart, liver, spleen, kidney, vas deferens, and brain. Within these tissues α_1 ARs couple to inositol phosphate (IP) release via a pertussis toxin (PTX) insensitive G protein. While this IP release leads

```
alpha1A.human    157  R H S L K Y P A I M T E R K A A A I L A L L W V V A L V V S V G P - L L G W - - - - - - K E P V P -
alpha1A.rat      193  R H S L K Y P A I M T E R K A A A I L A L L W A V A L V V S V G P - L L G W - - - - - - K E P V P -
alpha1B.canine    51  R Y S L Q Y P T L V T R R K A I L A L L G V W V L S T V I S I G P - L L G W - - - - - - K E P A P -
alpha1B.hamster  148  R Y S L Q Y P T L V T R R K A I L A L L S V W V L S T V I S I G P - L L G W - - - - - - K E P A P -
alpha1B.rat      148  R Y S L Q Y P T L V T R R K A I L A L L S V W V L S T V I S I G P - L L G W - - - - - - K E P A P -
alpha1C.bovine   129  S Y P L R Y P T I V T Q K R G L M A L L C V W A L S L V I S I G P - L F G W - - - - - - R Q P A P -
alpha2A.human    136  T Q A I E Y N L K R T P R R I K A I I I T V W V I S A V I S F P P L I S I E K K - - - - G G G G G P
alpha2A.mouse    136  T Q A I E Y N L K R T P R R I K A I I V T V W V I S A V I S F P P L I S I E K K - - - - G A G G G Q
alpha2A.porcine  136  T Q A I E Y N L K R T P R R I K A I I V T V W V I S A V I S F P P L I S I E K K - - - - A G G G G Q
alpha2A.rat      136  T Q A I E Y N L K R T R R R I K A I H C H C V V I S A V I S F P P L I S I E K K - - - - G A G G G Q
alpha2B.human    154  T Q A V E Y N L K R T P R R V K A T I V A V W L I S A V I S F P P L V S L Y R Q - - - - P D G - - -
alpha2B.mouse    154  T Q A V E Y N L K R T P R R V K A T I V A V W L I S A V I S F P P L V S F Y R R - - - - P D G - - -
alpha2B.rat      154  T Q A V E Y N L K R T P R R V K A T I V A V W L I S A V I S F P P L V S F Y R R - - - - P D G - - -
alpha2C.human    115  S R A L E Y N S K R T P R R I K C I I L T V W L I A A V I S L P P L I Y - - - K - - - - G D Q G P Q
alpha2C.mouse    115  S R A L E Y N S K R T P R R I K C I I L T V W L I A A V I S L P P L I Y - - - K - - - - G D Q R P E
alpha2C.rat      120  S R A L E Y N S K R T P C R I K C I I L T V W L I A A V I S L P P L I Y - - - K - - - - G D Q R P D
beta1.human      161  T S P F R Y Q S L L T R A R A R G L V C T V W A I S A L V S F L P I L M H W W R - A E S D E A R R C
beta1.rat        161  T L P F R Y Q S L L T R A R A R A L V C T V W A I S A L V S F L P I L M H W W R - A E S D E A R R C
beta1.turkey     144  T S P F R Y Q S L M T R A R A K V I I C T V W A I S A L V S F L P I M M H W W R - D E D P Q A L K C
beta2.hamster    136  T S P F K Y Q S L L T K N K A R M V I L M V W I V S G L T S F L P I Q M H W Y R - A T H Q K A I D C
beta2.human      136  T S P F K Y Q S L L T K N K A R V I I L M V W I V S G L T S F L P I Q M H W Y R - A T H Q E A I N C
beta2.mouse      136  T S P F K Y Q S L L T K N K A R V V I L M V W I V S G L T S F L P I Q M H W Y R - A T H K K A I D C
beta2.rat        136  T S P F K Y Q S L L T K N K A R V V I L M V W I V S G L T S F L P I Q M H W Y R - A T H K Q A I D C
beta3.human      140  T N P L R Y G A L V T K R C A R T A V V L V W V V S A A V S F A P I M S Q W W R V G A D A E A Q R C
beta3.mouse      137  T N P L R Y G T L V T K R R A R A A V V L V W I V S A A V S F A P I M S Q W W R V G A D A E A Q E C
beta3.rat        137  T N P L R Y G T L V T K R R A R A A V V L V W I V S A T V S F A P I M S Q W W R V G A D A E A Q E C


alpha1A.human    199  - P D E R F C G I T E E A G Y A V F S S V C S F Y L P M A V I V V M Y C R V Y V V A R S - T T R S L
alpha1A.rat      235  - P D E R F C G I T E E V G Y A I F S S V C S F Y L P M A V I V V M Y C R V Y V V A R S - T T R S L
alpha1B.canine    93  - N D D K E C G V T E E P F Y A L F S S L G S F Y I P L A V I L V M Y C R V Y I V A K R - T T K N L
alpha1B.hamster  190  - N D D K E C G V T E E P F Y A L F S S L G S F Y I P L A V I L V M Y C R V Y I V A K R - T T K N L
alpha1B.rat      190  - N D D K E C G V T E E P F C A L F C S L G S F Y I P L A V I L V M Y C R V Y I V A K R - T T K N L
alpha1C.bovine   771  - E D E T I C Q I N E E P G Y V L F S A L G S F Y V P L T I I L V M Y C R V Y V V A K R - E S R G L
alpha2A.human    182  Q P A E P R C E I N D Q K W Y V I S S C I G S F F A P C L I M I L V Y V R I Y Q I A K R - R T R V P
alpha2A.mouse    182  Q P A E P S C K I N D Q K W Y V I S S S I G S F F A P C L I M I L V Y V R I Y Q I A K R - R T R V P
alpha2A.porcine  182  Q P A E P R C E I N D Q K W Y V I S S C I G S F F A P C L I M I L V Y V R I Y Q I A K R - R T R V P
alpha2A.rat      182  Q P A E P S C K I N D Q K W Y V I S S S I G S F F A P C L I M I L V Y V R I Y Q I A K R - R T R V P
alpha2B.human    197  - A A Y P Q C G L N D E T W Y I L S S C I G S F F A P C L I M G L V Y A R I Y R V A K R - R T R T L
alpha2B.mouse    197  - A A Y P Q C G L N D E T W Y I L S S C I G S F F A P C L I M G L V Y A R I Y R V A K L - R T R T L
alpha2B.rat      197  - A A Y P Q C G L N D E T W Y I L S S C I G S F F A P C L I M G L V Y A R I Y R V A K L - R T R T L
alpha2C.human    158  P R G R P Q C K L N Q E A W Y I L A S S I G S F F A P C L I M I L V Y L R I Y L I A K R S N R R G P
alpha2C.mouse    158  P D G L P Q C E L N Q E A W Y I L A S S I G S F F A P C L I M I L V Y L R I Y V I A K R S H C R G L
alpha2C.rat      163  A R G L P Q C E L N Q E A W Y I L A S S I G S F F A P C L I M I L V Y L R I Y V I A K R S H C R G L
beta1.human      210  Y N D P K C C D F V T N R A Y A I A S S V V S F Y V P L C I M A F V Y L R V F R E A Q K - Q V K K I
beta1.rat        210  Y N D P K C C D F V T N R A Y A I A S S V V S F Y V P L C I M A F V Y L R V F R E A Q K - Q V K K I
beta1.turkey     193  Y Q D P G C C D F V T N R A Y A I A S S I I S F Y I P L L I M I F V Y L R V Y R E A K E - Q I R K I
beta2.hamster    185  Y H K E T C C D F F T N Q A Y A I A S S I V S F Y V P L V V M V F V Y S R V F Q V A K R - Q L Q K I
beta2.human      185  Y A N E T C C D F F T N Q A Y A I A S S I V S F Y V P L V I M V F V Y S R V F Q E A K R - Q L Q K I
beta2.mouse      185  Y T E E T C C D F F T N Q A Y A I A S S I V S F Y V P L C V M V F V Y S R V F Q V A K R - Q L Q K I
beta2.rat        185  Y A K E T C C D F F T N Q A Y A I A S S I V S F Y V P L V V M V F V Y S R V F Q V A K R - Q L Q K I
beta3.human      190  H S N P R C C A F A S N M P Y V L L S S S V S F Y L P L L V M L F V Y A R V F V V A T R - Q L R L L
beta3.mouse      187  H S N P R C C S F A S N M P Y A L L S S S V S F Y L P L L V M L F V Y A R V F V V A K R - Q R H L L
beta3.rat        187  H S N P R C C S F A S N M P Y A L L S S S V S F Y L P L L V M L F V Y A R V F V V A K R - Q R R F V
```

FIGURE 1c.

to intracellular Ca^{+2} release, some α_1ARs may couple directly to influx of Ca^{+2} by a dihydropyridine-sensitive Ca^{+2} channel (Lomasney et al., 1991b).

Molecular cloning of has revealed the existence of a third α_1 AR subtype, termed the α_{1c}. This receptor was cloned from bovine brain, and pharmacologically

```
alpha1A.human   247  E A G V K R E R G K A S E V V - - - - - - L R I H C R G A A T G A D G A H G M R S A K G H T F R S S
alpha1A.rat     283  E A G I K R E P G K A S E V V - - - - - - L R I H C R G A A T S A K G Y P G T Q S S K G H T L R S S
alpha1B.canine  141  E A G V M K E M S N S K E L T - - - - - - L R I H S K N F H E D T L S S T K A K - - - G H N P R S S
alpha1B.hamster 238  E A G V M K E M S N S K E L T - - - - - - L R I H S K N F H E D T L S S T K A K - - - G H N P R S S
alpha1B.rat     238  E A G V M K E M S N S K E L T - - - - - - L R I H S K N F H E D T L S S T K A K - - - G H N P R S S
alpha1C.bovine  219  K S G L K T D K S D S E Q V T - - - - - - L R I H R K N A Q V G G S G V T S A K N - - - - - - K T H
alpha2A.human   231  P S R R G - - - - - - - - P D - - - - - A V A A P P G G T E R R P N G L G P E R S A G P G G A E A
alpha2A.mouse   231  P S R R G - - - - - - - - P D - - - - - A C S A P P G G A D R R P N G L G P E R G A G P T G A E A
alpha2A.porcine 231  P S R R G - - - - - - - - P D - - - - - A A A A L P G G A E R R P N G L G P E R G V G R V G A E A
alpha2A.rat     231  P S R R G - - - - - - - - P D - - - - - A C S A P P G G A D R R P N G L G P E R G A G T A G G E A
alpha2B.human   245  S E K R A - - - - - - - - P - - - - - - - - V G P D G A S P T T E N G L G A A A G E A R T G T A R
alpha2B.mouse   245  S E K R G - - - - - - - - P - - - - - - - - A G P D G A S P T T E N G L G K A A G E N G H C A P P
alpha2B.rat     245  T E K R G - - - - - - - - P - - - - - - - - A G P D G A S P T T E N G L G K A A G E N G H C A P P
alpha2C.human   208  R A K G G P G Q G E S K Q P R - - - - - - P D H G G A L A S A K L P A L - A S V A S A R E V N G H S
alpha2C.mouse   208  G A K R G S G E G E S K K P R - - - - - - P A A G G V P A S A K V P T L V S P L S S V G E A N G H P
alpha2C.rat     213  G A K R G S G E G E S K K P Q - - - - - - P V A G G V P T S A K V P T L V S P L S S V G E A N G H P
beta1.human     259  D S C E R R F L G G P A R P P S P S P S P V P A P A P P P G P P R P A A A A A T A P L A N G R A G K
beta1.rat       259  D S C E R R F L T G P P R P P S - - - - - P A P S P S P G P P R P A - - - - - D S L A N G R S S K
beta1.turkey    242  D R C E G R F Y G S Q E Q P Q - - - - - - - - - P P P L P Q H Q P - - - - - I L G N G R A S K
beta2.hamster   234  D K S E G R F H S - - - - - - - - - - - - - P N L G Q V E Q D - - - - - - - G R S G H G L
beta2.human     234  D K S E G R F H V - - - - - - - - - - - - - Q N L S Q V E Q D - - - - - - - G R T G H G L
beta2.mouse     234  D K S E G R F H A - - - - - - - - - - - - - Q N L S Q V E Q D - - - - - - - G R S G H G L
beta2.rat       234  D K S E G R F H A - - - - - - - - - - - - - Q N L S Q V E Q D - - - - - - - G R S G H G L
beta3.human     239  R G E L G R F P P E E S P P A - - - - - - - - P S R S L A P A P V G - - - T C A P P E G V P A C G
beta3.mouse     236  R R E L G R F S P E E S P P S - - - - - - - - P S R S P S P A T G G - - - T P A A P D G V P P C G
beta3.rat       236  R R E L G R F P P E E S P R S - - - - - - - - P S R S P S P A T V G - - - T P T A S D G V P S C G


alpha1A.human   291  L S V R L L K F S R E K K A A K T L A I I V V G V F V L C W F P F F F V L P L G S L F P - - - - - - -
alpha1A.rat     327  L S V R L L K F S R E K K A A K T L A I I V V G V F V L C W F P F F F V L P L G S L F P - - - - - - -
alpha1B.canine  182  I A V K L F K F S R E K K A A K T L G I V V G M F I L C W L P F F I A L P L G S L F S - - - - - - -
alpha1B.hamster 279  I A V K L F K F S R E K K A A K T L G I V V G M F I I L C W L P F F I A L P L G S L F S - - - - - - -
alpha1B.rat     279  I A V K L F K F S R E K K A A K T L G I V V G M F I I L C W L P F F I A L P L G S L F S - - - - - - -
alpha1C.bovine  257  F S V R L L K F S R E K K A A K T L G I V V G C F V L C W L P F F L V M P I G S F F P - - - - - - -
alpha2A.human   267  E P L P T Q L N G A P G E P A P A G P R D T D A L D L E E S S S S D - H A E R P P G P - - - - - - R
alpha2A.mouse   267  E P L P T Q L N G A P G E P A P A G P R D G D A L D L E E S S S S E - H A E R P P G P - - - - - - R
alpha2A.porcine 267  E P L P V Q L N G A P G E P A P A G P R D A D G L D L E E S S S S E - H A E R P P G P - - - - - - R
alpha2A.rat     267  E P L P T Q L N G A P G E P A P T R P R D G D A L D L E E S S S S E - H A E R P Q G P - - - - - - G
alpha2B.human   278  P R P P T W S R T A A Q R P R G G A P G P L R R G G R R R A G A E G G A G G A D G Q - - - - - - G
alpha2B.mouse   278  R T E V E P D E S S A A E R R R - - R R G A L R R G G R R R E G A E G D T G S A D G P - - - - - - G
alpha2B.rat     278  R T E V E P D E S S A A E R R R - - R R G A L R R G G R R R E G A E G D T G S A D G P - - - - - - G
alpha2C.human   251  K S T G E K E E G E T P E D T G T R A L P P S W A A L P N S G Q G Q K E G V C G A S P E D E A E E
alpha2C.mouse   252  K P P R E K E E G E T P E D P E A R A L P P N W S A L P R S V Q D Q K K G T S G A T A E - - - K G A
alpha2C.rat     257  K P P R E K E E G E T P E D P E A R A L P P T W S A L P R S G Q G Q K K G T S G A T A E - - - E G D
beta1.human     309  R R P S R L V A L R E Q K A L K T L G I I M G V F T L C W L P F F L A N V V K A F - H - - - - - - -
beta1.rat       309  R R P S R L V A L R E Q K A L K T L G I I M G V F T L C W L P F F L A N V V K A F - H - - - - - - -
beta1.turkey    275  R K T S R V M A M R E H K A L K T L G I I M G V F T L C W L P F F L V N I V N V F - N - - - - - - -
beta2.hamster   259  R R S S K F - C L K E H K A L K T L G I I M G T F T L C W L P F F I V N I V H V I - Q - - - - - - -
beta2.human     259  R R S S K F - C L K E H K A L K T L G I I M G T F T L C W L P F F I V N I V H V I - Q - - - - - - -
beta2.mouse     259  R R S S K F - C L K E H K A L K T L G I I M G T F T L C W L P F F I V N I V H V I - R - - - - - - -
beta2.rat       259  R S S S K F - C L K E H K A L K T L G I I M G T F T L C W L P F F I V N I V H V I - R - - - - - - -
beta3.human     277  R R P A R L L P L R E H R A L C T L G L I M G T F T L C W L P F F L A N V L R A L G G - - - - - - -
beta3.mouse     274  R R P A R L L P L R E H R A L R T L G L I M G I F S L C W L P F F L A N V L R A L A G - - - - - - -
beta3.rat       274  R R P A R L L P L G E H R A L R T L G L I M G I F S L C W L P F F L A N V L R A L V G - - - - - - -
```

FIGURE 1d.

resembles the α_{1a}. Unlike the α_{1a}, however, this subtype is partially sensitive to inactivation by CEC, and is not expressed in tissues where the α_{1a} is prevalent (Schwinn et al., 1990).

Morrow and Creese were first to clearly demonstrate the existence of two distinct subtypes of α_1 receptors in rat brain. While these two subtypes, termed α_{1a} and α_{1b} both

```
alpha1A.human   334  - Q L K P S E G V F K V I F W L G Y F N S C V N P L I Y P C S S R E F K R A F L R L L R C Q C R - -
alpha1A.rat     370  - Q L K P S E G V F K V I F W L G Y F N S C V N P L I Y P C S S R E F K R A F L R L L R C Q C R - -
alpha1B.canine  225  - T L K P P D A V F K V V F W L G Y F N S C L N P I I I Y P C S S K E F K R A F V R I L G C Q C R G R
alpha1B.hamster 322  - T L K P P D A V F K V V F W L G Y F N S C L N P I I I Y P C S S K E F K R A F M R I I L G C Q C R S G
alpha1B.rat     322  - T L K P P D A V F K V V F W L G Y F N S C L N P I I I Y P C S S K E F K R A F M R I I L G C Q C R G G
alpha1C.bovine  300  - D F R P S E T V F K I A F W L G Y L N S C I N P I I I Y P C S S Q E F K K A F Q N V L R I Q C L - -
alpha2A.human   310  R P E R G P R G K G K A R A S Q V K P G D S L R G A G R G R R G S G R R L Q G R G R - - - S A S G L
alpha2A.mouse   310  R P D R G P R A K G K T R A S Q V K P G D S L P R R G P G A A G P G A S G S G H G E - - - E R G G G
alpha2A.porcine 310  R S E R G P R A K S K A R A S Q V K P G D S L P R R G P G A P G P G A P A T G A G E - - - E R G G V
alpha2A.rat     310  K P E R G P R A K G K T K A S Q V K P G D S L P R R G P G A A G P G A S G S G Q G E - - - E R A G G
alpha2B.human   322  A G P G A A Q S G A L T A S R S P G P G G R L S R - - A S S R S V E F F L S R R R R - - - A R S S V
alpha2B.mouse   320  P G L A A E Q - G A R T A S R S P G P G G R L S R - - A S S R S V E F F L S R R R R - - - A R S S V
alpha2B.rat     320  P G L A A E Q - G A R T A S R S P G P G G R L S R - - A S S R S V E F F L S R R R R - - - A R S S V
alpha2C.human   301  E E E E E E E E E C E P Q A V P V S P A S A C S P P L Q Q P Q G S R V L A T L R G Q V L L G R G V G
alpha2C.mouse   299  E E D E E E V E E C E P Q T L P A S P A S V F N P P L Q Q P Q T S R V L A T L R G Q V L L S K N V G
alpha2C.rat     304  E E D E E E V E E C E P Q T L P A S P A S V C N P P L Q Q P Q T S R V L A T L R G Q V L L G K N V G
beta1.human     351  - R E L V P D R L F V F F N W L G Y A N S A F N P I I Y - C R S P D F R K A F Q G L L - - - C C A R
beta1.rat       340  - R D L V P D R L F V F F N W L G Y A N S A F N P I I Y - C R S P D F R K A F Q R L L - - - C C A R
beta1.turkey    377  - R D L V P D W L F V F F N W L G Y A N S A F N P I I Y - C R S P D F R K A F K R L L - - - C F P R
beta2.hamster   300  - D N L I P K E V Y I L L N W L G Y V N S A F N P L I Y - C R S P D F R I A F Q E L L - - - C - L R
beta2.human     300  - D N L I R K E V Y I L L N W I G Y V N S G F N P L I Y - C R S P D F R I A F Q E L L - - - C - L R
beta2.mouse     300  - D N L I P K E V Y I L L N W L G Y V N S A F N P L I Y - C R S P D F R I A F Q E L L - - - C - L R
beta2.rat       300  - A N L I P K E V Y I L L N W L G Y V N S A F N P L I Y - C R S P D F R I A F Q E L L - - - C - L R
beta3.human     320  - P S L V P G P A F L A L N W L G Y A N S A F N P L I Y - C R S P D F R S A F R R L L - - - C - - -
beta3.mouse     377  - P S L V P S G V F I A L N W L G Y A N S A F N P V I Y - C R S P D F R D A F R R L L - - - C - - -
beta3.rat       377  - P S L V P S G V F I A L N W L G Y A N S A F N P L I Y - C R S P D F R D A F R R L L - - - C - - -


alpha1A.human   381  R R R R - - - - - R R P L - - - - - W R V Y - - - - G H H W R A S T S G L R Q D C A P S S G D A P P
alpha1A.rat     417  R R R R R L W S L R P P L A S L D R R R A F R L R P Q P S H R S P R G P S S P H C T P G C G L G R H
alpha1B.canine  274  R R R R R R R R L G G C A Y T Y R P W T R G G S L E R S Q S R K D S L D D S G S C L S G S Q R T L P
alpha1B.hamster 371  R R R R R R R R L G A C A Y T Y R P W T R G G S L E R S Q S R K D S L D D S G S C M S G S Q R T L P
alpha1B.rat     371  R R R R R R R R L G A C A Y T Y R P W T R G G S L E R S Q S R K D S L D D S G S C M S G Q K R T L P
alpha1C.bovine  347  R R K Q - - - S S K H T L - - - - - - - G Y - - - - - - T L H A P S H V L E G Q H K - - - D L V R I
alpha2A.human   357  P R R R - - - - - - - - - - - - - - - - - - A G A G G Q N R E K R F T F V L A V V I
alpha2A.mouse   357  A K A S - - - - - - - - - - - - - - - - - - - R W R G R Q N R E K R F T F V L A V V I
alpha2A.porcine 357  A K A S - - - - - - - - - - - - - - - - - - - R W R G R Q N R E K R F T F V L A V V I
alpha2A.rat     357  A K A S - - - - - - - - - - - - - - - - - - - R W R G R Q N R E K R F T F V L A V V I
alpha2B.human   367  - - - C - - - - - - - - - - - - - - - - - - - R R K V A Q A R E K R F T F V L A V V M
alpha2B.mouse   364  - - - C - - - - - - - - - - - - - - - - - - - R R K V A Q A R E K R F T F V L A V V M
alpha2B.rat     364  - - - C - - - - - - - - - - - - - - - - - - - R R K V A Q A R E K R F T F V L A V V M
alpha2C.human   351  A I G G Q W - - - - - - - - - - - - - W R R R A H V T R E K R F T F V L A V V I
alpha2C.mouse   349  V A S G Q W - - - - - - - - - - - - - W R R R T Q L S R E K R F T F V L A V V I
alpha2C.rat     354  V A S G Q W - - - - - - - - - - - - - W R R R T Q L S R E K R F T F V L A V V I
beta1.human     396  R A A R R - - - - - - - - - - - - - - - - - - R H A T H G D R P R A S G C L A R P G P
beta1.rat       385  R A A C R - - - - - - - - - - - - - - - - - - R R A A H G D R P R A S G C L A R A G P
beta1.turkey    362  K A D R R - - - - - - - - - - - - - - - - - - L H A G G Q P A P L P G G F I S T L G S
beta2.hamster   344  R S S S K - - - - - - - - - - - - - - - - - - A Y - G N G Y S S N S N G K T D Y M G E
beta2.human     344  R S S L K - - - - - - - - - - - - - - - - - - A Y - G N G Y S S N G N - - - - - T G E
beta2.mouse     344  R S S S K - - - - - - - - - - - - - - - - - - T Y - G N G Y S S N S N G R T D Y T G E
beta2.rat       344  R S S S K - - - - - - - - - - - - - - - - - - T Y - G N G Y S S N S N G R T D Y T G E
beta3.human     362  - - - - - - - - - - - - - - - - - - - - R C G R R L P P E P C A A - - - - - -
beta3.mouse     359  - - - - - - - - - - - - - - - - - - - - S Y G G R G P E E P - - - - - - - - - -
beta3.rat       359  - - - - - - - - - - - - - - - - - - - - S Y G G R G P E E P - - - - - - - - - -
```

FIGURE 1e.

possessed high affinity for [^{3}H]prazosin, only the α_{1a} subtype also bound [^{3}H]WB4101 and phentolamine with similar high affinity (Morrow and Creese, 1986). The α_{1a} receptor also appears insensitive to the effects of the alkylating agent CEC (Han et al.,1987). Other ligands exhibiting selectivity for the α_{1a} subtype include oxymetazoline, benoxathine, (+)niguldipine, amidephrine, and 5-methyl-urapidil (Lomasney et al., 1991b).

```
alpha1A.human   477  G A P L A L T A L P D P D P E P P G T P E M Q A P V A S R R S H - P A - P S A S G G C W G R S G D P
alpha1A.rat     467  A G D A G F G L Q Q S K A S L R L R E W R L L G P L Q R P T T Q L R A - K V S S L S H K I R S G A R
alpha1B.canine  324  S A S P S P G Y L - G R A A P P P V E L C A V P E W K A P G A L L S L P A P Q P P G R R G R R D S G
alpha1B.hamster 421  S A S P S P G Y L - G R G A Q P P L E L C A Y P E W K - S G A L L S L - - P E P P G R R G R L D S G
alpha1B.rat     421  S A S P S P G Y L - G R G T Q P P V E L C A F P E W K - P G A L L S L - - P E P P G R R G R L D S G
alpha1C.bovine  378  P V G S A E T F Y K I S K T D G V C E W K I F S S L P R G S A R M A V - A R D P S A C T T A R V R S
alpha2A.human   381  G V F V V C W F P F F F T Y T L T A V - - - - - G C S V P R T L F K F F F W F G Y C N S S L N P V I
alpha2A.mouse   381  G V F V V C W F P F F F T Y T L I A V - - - - - G C P V P S Q L F N F F F W F G Y C N S S L N P V I
alpha2A.porcine 381  G V F V V C W F P F F F T Y T L T A V - - - - - G C P V P P T L F K F F F W F G Y C N S S L N P V I
alpha2A.rat     381  G V F V V C W F P F F F T Y T L I A V - - - - - G C P V P Y Q L F N F F F W F G Y C N S S L N P V I
alpha2B.human   388  G V F V L C W F P F F F I Y S L Y G I - - C R E A C Q V P G P L F K F F F W I G Y C N S S L N P V I
alpha2B.mouse   385  G V F V L C W F P F F F S Y S L Y G I - - C R E A C Q L P E P L F K F F F W I G Y C N S S L N P V I
alpha2B.rat     385  G V F V L C W F P F F F S Y S L Y G I - - C R E A C Q L P E P L F K F F F W I G Y C N S S L N P V I
alpha2C.human   378  G V F V L C W F P F F F S Y S L G A I - - C P K H C K V P H G L F Q F F F W I G Y C N S S L N P V I
alpha2C.mouse   376  G V F V V C W F P F F F S Y S L G A I - - C P Q H C K V P H G L F Q F F F W I G Y C N S S L N P V I
alpha2C.rat     381  G V F V V C W F P F F F S Y S L G A I - - C P Q H C K V P H G L F Q F F F W I G Y C N S S L N P V I
beta1.human     421  P - P S P G A A S D D D D D V V G A - - - - - - T P P A R L L E - - - P W A G C N G G A A A - D
beta1.rat       410  P - P S P G A P S D D D D D - A G A - - - - - - T P P A R L L E - - - P W A G C N G G T T T V D
beta1.turkey    387  P E H S P G G T W S D C N G G T R G G - - - - - - S E S S - L E E - - - R H S K T S R S E S K M E
beta2.hamster   368  - - A S G C Q L G Q E K E S E R L C E - - - - - - D P P G - - T E - - - S F V N C Q G T V P S L S
beta2.human     363  - - Q S G Y H V E Q E K E N K L L C E - - - - - - D L P G - - T E - - - D F V G H Q G T V P S D N
beta2.mouse     368  - - P N T C Q L G Q E R E Q E L L C E - - - - - - D P P G - - M E - - - G F V N C Q G T V P S L S
beta2.rat       368  - - Q S A Y Q L G Q E K E N E L L C E - - - - - - E A P G - - M E - - - G F V N C Q G T V P S L S
beta3.human     375  - - A R P A L F P S G V P A A R - - S - - - - - - - S P A - - - Q P - - R L C Q R L D G - - - - -
beta3.mouse     369  - - - R A V T F P A S P V E A R - - Q - - - - - - - S P P - _ - L N - - - R - - - R
beta3.rat       369  - - - R V V T F P A S P V A S R - - Q - - - - - - - N S P - - - L N - - - R F D G Y E G E R P F P T


alpha1A.human   465  R - - P - - - - S C A P K S - - - P A C R T R S P P G A R S A Q R Q R A P S A Q R W R L C P - - - -
alpha1A.rat     516  R - - A - - E T A C A L R S E V E A V S L N V P Q D G A E A V I C Q A Y E P G D Y S N L R E T D I -
alpha1B.canine  373  P L F T F R L L A E R G S P A - - - - A G D G A C R P A P D A A N G Q P G F K T N M P L A P G Q F -
alpha1B.hamster 467  P L F T F K L L G E P E S P G T E G D A S N G G C D A T T D L A N G Q P G F K S N M P L A P G H F -
alpha1B.rat     467  P L F T F K L L G D P E S P G T E A T A S N G G C D T T T D L A N G Q P G F K S N M P L G P G H F -
alpha1C.bovine  427  K - - S F L Q V C C C L G P - - - S T P S H G E N H Q I P T I K I H T I S L S E N G E E V - - - - -
alpha2A.human   426  - - - - - - - - Y T I F N H D F R R A F K K I L C R G D R K R I V - - - - - - - - - - - - - - -
alpha2A.mouse   426  - - - - - - - - Y T I F N H D F R R A F K K I L C R G D R K R I V - - - - - - - - - - - - - - -
alpha2A.porcine 426  - - - - - - - - Y T I F N H D F R R A F K K I L C R G D R K R I V - - - - - - - - - - - - - - -
alpha2A.rat     426  - - - - - - - - Y T I F N H D F R R A F K K I L C R G N R K R I V - - - - - - - - - - - - - - -
alpha2B.human   436  - - - - - - - - Y T V F N Q D F R P S F K H I L F R R R R R G F R Q - - - - - - - - - - - - - - -
alpha2B.mouse   433  - - - - - - - Y T V F N Q D F R R S F K H I L F R R R R R G F R Q - - - - - - - - - - - - - -
alpha2B.rat     433  - - - - - - - - Y T V F N Q D F R R S F K H I L F R R R R R G F R Q - - - - - - - - - - - - - - -
alpha2C.human   426  - - - - - - - - Y T I F N Q D F R R A F R R I L C R P W T Q T A W - - - - - - - - - - - - - - -
alpha2C.mouse   434  - - - - - - - - Y T I F N Q D F R R A F R R I L C R Q W T Q T G W - - - - - - - - - - - - - - -
alpha2C.rat     429  - - - - - - - - Y T V F N Q D F R R A F R R I L C R P W T Q T G W - - - - - - - - - - - - - - -
beta1.human     459  S D S S - - - - - - - - - - L D E P C R P G F A S E S K V - - - - - - - - - - - - - - - - -
beta1.rat       448  S D S S - - - - - - - - - - L D E P G R Q G F S S E S K V - - - - - - - - - - - - - - - - -
beta1.turkey    426  R E K N I L A T T R F Y C T F L G N G D K A V F C T V L R I V K L F E D A T C T C P H T H K L K M K
beta2.hamster   404  - - - - - - - - - - - - - L D S Q G R N C S T N D S P L - - - - - - - - - - - - - - - -
beta2.human     399  - - - - - - - - - - - - - I D S Q G R N C S T N D S L L - - - - - - - - - - - - - - - -
beta2.mouse     404  - - - - - - - - - - - - - V D S Q G R N C S T N D S P L - - - - - - - - - - - - - - - -
beta2.rat       404  - - - - - - - - - - - - - I D S Q G R N C N T N D S P L - - - - - - - - - - - - - - - -
```

FIGURE 1f.

1.3.11.1 Human a₁ₐ Receptors

The human α_{1a} receptor was cloned from a hippocampal cDNA library using degenerate oligonucleotide primers directed toward conserved regions of previously cloned G-protein coupled receptors (Bruno et al., 1991). This clone encoded a protein of 501 amino acids possessing 78% overall homology to the rat α_{1a} AR. Between the first and seventh transmembrane domains these receptors are 95% homologous; whereas, the amino and

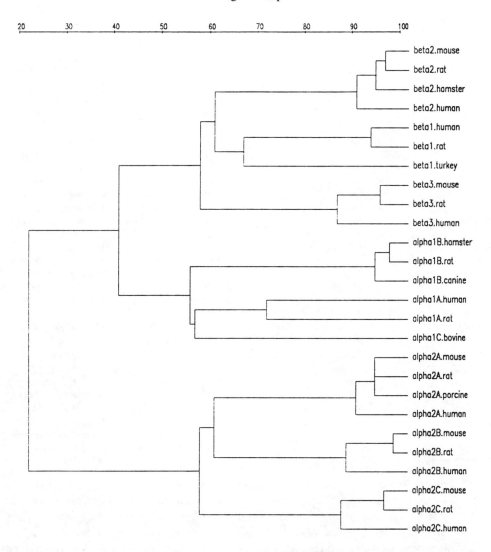

FIGURE 2. Phylogenetic tree of adrenergic receptors. The phylogenetic tree was constructed according to the method of Feng and Doolittle (1990). The length of each branch of the tree correlates with the evolutionary distance between receptor subpopulations.

carboxyl termini are far less homologous, possessing 50 and 47% homology, respectively. Furthermore, in contrast to most other adrenergic receptors, this human α_{1a} receptor clone does not contain sites for N-linked glycosylation. The human α_{1a} receptor has yet to be expressed and, subsequently, its pharmacological properties have not been assessed. However, its high degree of sequence homology with the rat α_{1a} receptor within the first and seventh transmembrane segments (95%) suggests that these receptors are pharmacologically similar.

1.3.11.10   Rat a_{1A} Receptors

A gene encoding the rat α_{1a} AR was cloned from a rat cerebral cortex cDNA library using a DNA probe from the hamster α_{1b} AR (Lomasney et al., 1991a). This gene encodes a protein of 560 amino acids (Table 1), with an overall topology similar to the

Table 1.   Molecular Biological Properties of Cloned Adrenergic Receptors

Receptor	Second messenger	Species cloned	Chromosomal location	a.a. Sequence	Accession number	Primary reference
Alpha 1A	+PI	Human	5q23-32	501	P25100	Bruno et al., 1991
		Rat		560	P23944	Lomasney et al., 1991
Alpha 1B	+PI	Canine		417	P11615	Liebert et al., 1989
	+AC	Rat		515	P15823	Voigt et al., 1990
		Hamster		515	P18841	Cotecchia et al., 1988
Alpha 1C	+PI	Bovine		466	P18130	Schwinn et al., 1990
Alpha 2A	–AC	Human	10q24-26	450	P08913	Kobilka et al., 1987
		Porcine		450	P18871	Guyer et al., 1990
		Rat		450	P22909	Chalberg et al., 1990
		Mouse		450	M99377	Link et al., 1992
Alpha 2B	–AC	Human	2	450	P18089	Lomasney et al., 1990
		Rat		453	P19328	Zeng et al., 1990
		Mouse		448	L00979	Chruscinski et al., 1992
Alpha 2C	–AC	Human	4	461	P18825	Regan et al., 1988
		Rat		458	P22086	Voigt et al., 1991
		Mouse		458	M99376	Link et al., 1992
		Fish		432		Svensson et al., 1993
Alpha 2D						
Beta 1	+AC	Human		477	P08588	Frielle et al., 1987
		Rat		466	P18090	Machida et al., 1990
		Turkey		483	M14379	Yarden et al., 1986
Beta 2	+AC	Human	5q31-32	413	P07550	Kobilka et al., 1987
		Hamster		418	P04274	Dixon et al., 1986
		Rat		418	P10608	Gocayne et al., 1987
		Mouse		418	P18762	Allen et al., 1988
Beta 3	+AC	Human		402	P13945	Emorine et al., 1989
		Rat		400	P26255	Granneman et al., 1991
		Mouse		388	P25962	Nahmias et al., 1991

other members of the adrenergic receptor family. The rat α_{1a} receptor possesses a third intracellular loop which is intermediate in size between the β ARs (short) and the α_2 ARs (long). The intracellular carboxyl terminus however, is among the longest of any adrenergic receptor identified to date (~161 amino acids). This α_1 receptor is 73% homologous with the rat α_{1b} AR within the transmembrane regions. Moreover, the rat α_{1a} receptor possesses conserved elements implicated in ligand binding to other ARs. These include asp-170, thought to interact with the catecholamine nitrogen moeity

Table 2.   Radioligands Used to Bind Adrenergic Receptors

α_1ARs	α_2ARs	βARs
[³H] Prazosin	[³H] Clonidine	[¹²⁵I] Iodocyanopindolol
[¹²⁵I] HEAT[1]	[³H] p-Aminoclonidine	[³H] CGP12177A[2]
[³H] Idazoxan	[³H] Idazoxan	[³H] Dihydroalprenalol
[³H] Norepinephrine	[¹²⁵I] Iodoclonidine	[³H] Norepinephrine
[³H] WB4101[3]	[³H] Norepinephrine	[³H] Pindolol
	[³H] Rauwolscine	[³H] Propranolol
	[³H] UK-14304[4]	[¹²⁵I] Iodopindolol
	[³H] Guanfacine	

[1]   2-{b-(4-hydroxy-3-iodophenyl)-ethylaminomethyl}-tetralone
[2]   (–)-4-3-t-butylamino-2-hydroxypropoxy)-benzimidazol-2-one
[3]   2,6-dimethoxyphenoxyethylaminomethyl-1,4-benzodioxane
[4]   5-bromo-6-N-(2-4,5-dihydroimidazolyl)-quinoxaline

(Dixon et al., 1988) and serine residues in the fifth transmembrane region which may form hydrogen bonds with the catecholamine hydroxyl goups (Strader et al., 1989). The rat α_{1a} receptor also exhibits two potential glycosylation sites in the amino terminus (asn-60 and asn-76), and a number of putative phosphorylation sites for protein kinase C in the intracellular loops and carboxyl terminus. Unlike either the α_{1b} or α_{1c} receptors, this clone lacks potential sites for phosphorylation by cAMP-dependent protein kinase in the third intracellular loop.

Somatic cell hybridization analysis has localized the gene for the α_{1a} AR to human chromosome 5. Genes encoding the human β_2 and hamster α_{1b} receptors have also been localized to chromosome 5 (Table1), suggesting that these related genes may have been duplicated from a single precursor. Northern blot analysis performed in various tissues using a cDNA probe from the rat α_{1a} receptor revealed ample expression in the vas deferens, hippocampus, and cerebral cortex (Lomasney et al., 1991a).

Expression of the rat α_{1a} subtype in COS-7 cells resulted in saturable, high-affinity binding of the α-antagonist [125I]HEAT, (Kd = 70 to 100 pM). Competition binding studies with various adrenergic compounds revealed a distinct pharmacology for the α_{1a} AR subtype. This subtype has traditionally been distinguished from the α_{1b} receptor by its higher affinity for antagonists WB4101 and phentolamine. When expressed in COS-7 cells however, the α_{1a} subtype possessed a 10-fold higher affinity for both WB4101 and the agonist methoxamine (Table 3). In contrast to earlier studies in which the α_{1a} receptor bound the agonist oxymetazoline with high affinity (Han et al., 1987), the expressed α_{1a} receptor binds this compound much more poorly compared to either the α_{1b} or α_{1c} receptor subtypes, whereas endogenous catecholamines epinephrine and norepinephrine bind with slightly higher affinity to this subtype. The α_{1a} receptor expressed in these cells also exhibited insensitivity to the alkylating agent CEC. Transfected cells expressing either the α_{1a} or α_{1b} subtype were treated for 10 min with 10 μM CEC and washed extensively. While over 95% of the expressed α_{1b} receptors were inactivated, only 15% of the α_{1a} receptors were similarly affected, again suggesting that this receptor clone encoded the pharmacological α_{1a} receptor subtype.

1.3.12 α_{1b} Adrenergic Receptors

Like the α_{1a} receptor, the α_{1b} ARs were initially characterized in rat brain (Morrow and Creese, 1986). This subtype possesses lower affinity for antagonists WB4101 and phentolamine than the α_{1a} subtype, and unlike the α_{1a}, the α_{1b} AR is highly sensitive to inactivation by CEC pretreatment. Moreover, α_{1b} ARs are predominantly expressed in liver, kidney, and spleen; whereas, α_{1a} ARs are localized to hippocampus and vas deferens (Han et al., 1987).

1.3.12.6 Canine α_{1b} Receptors

A partial clone encoding a putative canine α_{1b} AR was obtained by screening a dog thyroid cDNA library (Liebert et al., 1989). A partial clone was isolated which exhibited 91% identity to the hamster α_{1b} AR. Northern blot analysis of RNA from various dog tissues revealed abundant expression in a number of tissues including liver, kidney, spleen, and stomach. This clone was not detected in thyroid tissues, however, arguing that this α_{1b} receptor is particularly rare, and this rarity may explain the inability to obtain a full-length clone.

Table 3. α_1 Adrenergic Receptor Pharmacology

K_i (nM) for Competition of [^{125}I] HEAT in COS-7 Cell Membranes

Drug	α_{1a}	α_{1b}	α_{1c}
Prazosin	0.33	2.5	0.27
WB4101	2.1	10	0.55
Coryanthine	253	517	142
Phentolamine	111	300	4.8
Oxymetazoline	2,140	140	27
(–)Epinephrine	546	2,000	6,550
(–)Norepinephrine	100	10,500	9,730
Phenylephrine	1,400	6,500	15,000

α_{1a} = Rat α_{1a} Receptor
α_{1b} = Hamster α_{1b} Receptor
α_{1c} = Bovine α_{1c} Receptor

1.3.12.10   Rat α_{1b} Receptors

Screening of a rat brain cDNA library with oligonucleotides corresponding to the hamster α_{1b} receptor yielded the rat α_{1b} AR (Voigt et al., 1990). This clone encoded a protein of 515 amino acids, and displayed 97% overall homology with the hamster α_{1b} receptor. This receptor was simultaneously cloned by another group (Lomasney et al., 1991b), who showed it to possess high affinity for prazosin (Ki = 0.56 n*M)*. This receptor exhibits lower affinity for WB4101 and fairly low affinity for most other adrenergic antagonists, similar to the hamster α_{1b} AR. Northern blot analysis using rat tissues revealed expression in liver, heart, cerebral cortex, brain stem, lung, kidney, and spleen. This pattern of expression closely corresponds to the pharmacological localization of the α_{1b} receptor in rat tissues.

1.3.12.12   Hamster α_{1b} Receptors

The first α_1 AR cloned, the hamster α_{1b} AR, was initially purified from a hamster smooth muscle cell line DDT$_1$MF-2 (Cotecchia et al., 1988). Screening of a DDT$_1$MF-2 cDNA library yielded a 2-kb clone encoding a polypeptide of 515 amino acids. In contrast to most other adrenergic receptors, the gene encoding this receptor subtype was shown to contain an intron. Moreover, somatic cell hybridization analysis localized the gene for the hamster α_{1b} receptor to human chromosome 5, similar to the α_{1a} receptor subtype. When compared to other α_1 adrenergic receptors, the hamster α_{1b} AR exhibits 73% homology with the rat α_{1a} receptor and 65% homology with the bovine α_{1C} receptor in the transmembrane regions. Furthermore, the hamster α_{1c} AR is intermediate in size between the rat α_{1a} and bovine α_{1c}. The size of the hamster α_{1b} has also been determined biochemically. Following purification and deglycosylation, the α_{1b} exhibited a molecular weight of 50,000 to 55,000, which correlates closely with the predicted weight for the cloned protein (Sawurz et al., 1987). Like the rat α_{1a} receptor, the hamster α_{1b} possesses multiple sites for N-linked glycosylation (asn-10, asn-23, asn-28, asn-33) near its amino terminus. Other potential sites for modification include a consensus recognition site for cAMP-dependent protein kinase (residues 230 to 234), and several potential protein kinase C recognition sites in the second and third intracellular loops.

Following expression in COS-7 cells, the hamster α_{1b} exhibited high affinity for the α_1 antagonists [^{125}I]HEAT (Kd = 50 to 100 p*M)* and prazosin (Ki = 2.5 nM). The antagonist WB4101 bound the α_{1b} receptor with only slightly lower affinity (Table 3), while most other α_1 receptor ligands bound rather poorly to this receptor subtype. Only the α_2 receptor agonist oxymetazoline possessed submicromolar affinity (Ki = 140 nM) for the α_{1b} receptor. These data essentially agree with previous ligand binding studies which suggested that the α_{1b} AR possesses lower affinity for most α_1-selective ligands compared with the α_{1a} AR (Cotecchia et al., 1988).

Functionally, the hamster α_{1b} AR couples to inositol phosphate (IP) hydrolysis. The cloned α_{1b} AR was expressed in HeLa cells and shown to stimulate IP hydrolysis when challenged with agonist. The major cellular messenger produced following α_{1b} AR activation was inositol 1,4,5 triphosphate, $(IP_3)_{1,4,5}$. Epineprine, norepinephrine, and phenylepinephrine were all full agonists at the α_{1b}, whereas methoxamine and oxymetazoline were only weak partial agonists. Epinephrine was the most potent in stimulating IP release with an EC_{50} of 2 μM, compared to nearly 20 μM for phenlylephrine and over 80 μM for methoxamine. Previous studies comparing the α_{1a} and α_{1b} ARs demonstrated that the α_{1b} AR could couple to IP hydrolysis in the spleen, while α_{1a} receptors could stimulate IP hydrolysis in the vas deferens. α_1-Mediated IP hydrolysis is insensitive to the effects of the Ca^{+2} channel blocker nifedipine, whereas IP hydrolysis in the vas deferens could be completely attenuated by prior nifedipine treatment. These results suggest that these α_1 AR subtypes increase intracellular Ca^{+2} by distinct mechanisms (Schwinn et al., 1991). While the α_{1a} AR receptor couples to a dihydropyridine (DHP)-sensitive Ca^{+2} channel to permit the influx of Ca^{+2}, activation of the α_{1b} AR promotes IP hydrolysis and Ca^{+2} release from intracellular stores.

α_1-Receptor coupling to IP release appears to involve amino acid residues in the third intracellular loop. Cotecchia et al. demonstrated that substitution of residues 252 to 259 near the amino terminus of the third intracellular loop abolished α_{1B} receptor coupling to IP release (Cotecchia et al., 1990). A subsequent study revealed that replacement of amino acids in the corresponding region of the β_2 AR with residues 233 to 259 from the α_{1b} receptor yielded a β_2 receptor which coupled to inositol phosphate release (Cotecchia et al., 1992). The same group demonstrated that substitution of ala-293 near the carboxyl terminus of the third intracellular loop with any other amino acid produced a constitutively active receptor (Kjelsberg et al., 1991). This replacement also increased the receptor affinity for agonists, suggesting that this region, and especially Ala-293, is critical for maintaining the receptor in an inactive conformation. Even minor alterations in this region appear to substantially enhance receptor-G protein coupling.

1.3.13 α_{1c} Adrenergic Receptors

1.3.13.4 Bovine α_{1c} Receptors

A third distinct subtype of α_1 adrenengic receptor was cloned from a bovine brain cDNA library (Schwinn et al., 1990). In contrast to the α_{1a} and α_{1b} receptors, the gene for the α_{1c} receptor was localized to human chromosome 8. This receptor consists of 466 amino acids and is most closely related to the hamster α_{1b} AR (72% homology in membrane-spanning domains). Homology with other members of the adrenergic receptor family in similar regions was typically about 40%. Other regions of this receptor displayed much lower homology with the hamster α_1 receptor. While the third intracellular loops were 50% identical, the amino and carboxyl termini exhibited only 27 and 12% identity, respectively. The α_{1c} receptor possesses several potential glycosylation sites near the amino terminus (asn-7, asn-13, and asn-22), as well as a number of potential sites for PKC phosphorylation in the second and third cytoplasmic loops. Similarly, a potential protein kinase A phosphorylation site is present between amino acids 211 and 215. Since the α_{1b} receptor can be regulated by phosphorylation (Leeb-Lundberg et al., 1987; Bouvier et al., 1987), these sites may represent a potential means of regulating the α_{1c} AR.

Despite its isolation from a bovine cDNA library, the α_{1c} AR does not appear to be expressed in bovine tissues. Northern blot analysis of both rat and bovine tissues did not reveal expression of the α_{1c} messenger RNA. α_{1c} Expression was detected only in rabbit liver and human dentate gyrus, suggesting that the bovine α_{1c} is expressed either at very low levels or only in very isolated cell populations.

The pharmacology of the bovine α_{1c} AR resembles that of the α_{1a} AR. The α_{1c} receptor possesses high affinity for [^{125}I]HEAT (Kd = 60 to 70 pM), and prazosin (Table 3). A number of ligands however bind the α_{1c} receptor with much higher affinity than the α_{1a} AR. WB4101 was 4-fold more selective for the α_{1c}, while phentolamine binds this subtype with 23-fold higher affinity compared to the α_{1a} receptor. Agonists epinephrine and norepinephrine, however, bind the α_{1c} with substantially lower affinities relative to the α_{1a} receptor. The bovine α_{1c} receptor may also be be distinguished based on its sensitivity to chlorethylclonidine. CEC pretreatment inactivated only a portion of the α_{1c} binding sites, suggesting that the bovine α_{1c} is partially sensitive to CEC relative to the α_{1a} (insensitive) or the α_{1b} receptor (sensitive).

Like the α_{1a} and α_{1b} ARs, the α_{1c} AR activates a phospholipase C, leading to IP production. This response appears to involve a pertussis toxin-insensitive G-protein. Activation of the α_{1c} receptor stably expressed in HeLa cells produced a 2- to 3-fold greater increase in inositol phosphate release as compared with the α_{1b} AR. Moreover, this IP release could be enhanced by addition of extracellular Ca^{+2} (Han et al., 1987). Since this enhancement was not blocked by the Ca^{+2} channel antagonist nifedipine, this effect does not appear to involve a DHP-sensitive Ca^{+2} channel. Extracellular calcium, then, may indirectly facilitate α_{1c}-mediated IP release. Epinephrine and norepinephrine were both full agonists at the α_{1c} receptor, whereas oxymetazoline was only a weak agonist. Methoxamine, meanwhile was nearly a full agonist at the α_{1c} compared to only a weak partial agonist at the α_{1b} receptor subtype. Oxymetazoline and epinephrine were the most potent stimulators of IP release with EC_{50} values of 7.4 and 5.8 μM, respectively. Norepinephrine was only slightly less potent (EC_{50} = 22 μM), followed by phenylephrine (64 μM), and finally methoxamine with an EC_{50} of 180 μM. This order of potency was identical to that for stimulation of IP release by the hamster α_{1b} receptor subtype (Schwinn et al., 1991).

1.3.21 α_{2A} Adrenergic Receptors

α_2 Adrenergic receptors were first distinguished from α_1 receptors based on their synaptic location. Postsynaptic α receptors located on the effector organ were termed α_1, while presynaptic α receptors which mediated neurotransmitter release were classified as α_2 (Langer, 1974). Unfortunately, this classification scheme could not account for the existence of postsynaptic α_2 ARs or α_2 ARs in distinctly nonsynaptic locations. α_2 ARs have also been distinguished from α_1 ARs based on their function. α_1 Receptors were thought to mediate excititory functions, while α_2 AR activation would evoke an inhibitory response (Berthelsen and Pettinger, 1977). Similarly, α_1 receptors were shown to couple to the release of intracellular Ca^{+2}, while α_2 receptors mediated responses subsequent to inhibition of adenylyl cyclase (Fain and Garcia-Sainz, 1980). With the development of radioligand binding techniques and ligands

[³H]prazosin and [³H]yohimbine (Stark et al., 1981), α adrenergic receptors could be classified based on their pharmacology. While [³H]prazosin has been successfully used to label α_1 ARs, α_2ARs may be labeled by agonists [³H]clonidine, [³H]p-aminoclonidine, [³H]UK-14304, and the antagonists [³H]idazoxan and [³H]rauwolscine (α-yohimbine) (Table 2).

Whereas α_1 receptors mainly couple to PI turnover, the α_2ARs exhibit numerous potential signaling interactions. Traditionally, α_2ARs have been characterized as coupling to inhibition of adenylyl cyclase. However, recent studies have demonstrated that α_2ARs may couple to a number of intracellular effector systems including phosphatidylinositol hydrolysis, phospholipase A_2, and Na^+/H^+ exchange (Limbird, 1988)

By the early 1980s, radioligand binding studies were pointing to the existence of multiple α_2AR subtypes. Cheung et al. showed that prazosin labeled α_2 receptors in rat brain with much higher affinity compared to human platelet α_2ARs (Cheung et al., 1982). In another study, neonatal rat lung was shown to express a single α_2AR population with high affinity for prazosin, while human platelets expressed another distinct population with much lower affinity for prazosin. The human platelet was subsequently termed the α_{2A} receptor, while the neonatal rat lung receptor with high affinity for prazosin was designated the α_{2B} subtype (Bylund, 1985).

1.3.21.1 Human α_{2A} Receptors

The first of the α adrenergic receptors to be cloned was the human platelet α_2 adrenergic receptor (Kobilka et al., 1987). The lack of any α adrenergic receptor sequence information prompted purification of native receptor. An α_2 AR was purified from human platelets using a combination of lectin-agarose, heparin-agarose, and yohimbinic acid affinity chromatography (Regan et al., 1986). Peptides from the purified receptor subsequently yielded suitable primary sequence information for cloning the platelet α_2AR gene. A clone encoding a 450 amino acid protein was isolated from a human genomic library, and shown to possess an overall structural homology to the previously cloned human β_2 AR. The gene coding for this receptor was localized to human chromosome 10 and is often referred to as α_2C10 (human chromosome 10).

Like the β_2 receptor, this α_2AR clone exhibits seven hydrophobic regions thought to span the plasma membrane, with an extracellular amino terminus and intracellular carboxyl terminus. However, whereas the β ARs have long carboxyl termini and fairly short third intracellular loops, this α_2AR exhibits a short carboxyl terminus (21 residues) and a large third cytoplasmic loop (150 residues). The amino terminal region possesses several potential glycosylation sites at residues asn-10 and asn-14. This was not surprising, as the purified receptor possesses a molecular weight of 64,000 Da, whereas the predicted molecular weight of the cloned protein is only 49,000 Da, suggesting that the receptor undergoes posttranslational modification. Numerous potential phosphorylation sites were also evident in the second and third intracellular loops. Since the purified α_2AR can be phosphorylated stoichiometrically by the β-adrenergic receptor kinase, these sites may prove critical for regulation of α_2AR function (Benovic et al., 1987). Other potential sites for posttranslational modification include a cysteine residue at position 442 which may be covalently modified by fatty acid acylation. Both rhodopsin (Ovchinnikov et al., 1988) and the β AR (O'Dowd et al., 1989) may be palmitoylated at cysteine residues near their carboxyl termini. In a similar fashion, addition of a palmitic acid residue to this cysteine near the carboxyl terminus may permit formation of an additional intracellular loop which may be involved in coupling the receptor to its intracellular effectors.

The pharmacology of the cloned α_2C10 AR has been extensively studied following transient expression in COS-7 cells. Initially defined as the pharmacologic α_{2A} receptor subtype (Bylund, 1985), the α_2C10 AR possesses high affinity for the antagonists rauwolscine, yohimbine, and phentolamine (Table 4). Like the native α_2AR, the α_2C10 exhibits rather low affinity for the α_1 antagonist prazosin, and suprisingly low affinities for the endogenous catecholamines epinephrine and norepinephrine. The agonist oxymetazoline, however, binds much more selectively to the α_2C10 subtype compared to other human α_2AR subtypes.

Since both α and β ARs bind the endogenous catecholamines epinephrine and norepinephrine, much of our information concerning ligand binding to α receptors is based on results with the β ARs. Like the β_2 AR, the α_2C10 contains an aspartic acid residue at position 113 which is thought to interact with the charged amine group of the catecholamines. Mutation of this asp-113 residue to asn eliminated high-affinity agonist binding, and substantially attenuated agonist-induced inhibition of forskolin-stimulated cAMP production (Wang et al., 1991). Similarly, mutation of two conserved serine residues in the fifth transmembrane region (positions 200 and 204) suggest that these residues are required for hydrogen bonding interactions with the catecholamine hydroxyl groups. While these studies implicated the fifth transmembrane domain in receptor-ligand interactions, the fourth membrane-spanning domain may also be involved. Matsui et al. found that photoaffinity ligands [^{3}H]SKF102229 and p-azido[^{3}H]clonidine covalently bound to specific residues located in the fourth transmembrane domain, pointing to involvement of this region in receptor-ligand interactions (Matsui et al., 1989). More recently, residues in the seventh transmembrane domain have been shown to confer specificity for the α_2AR. Construction of a chimeric receptor with the first six transmembrane domains from the β_2AR and the seventh from the α_2AR produced a receptor which selectively bound the α_2 antagonist [^{3}H]yohimbine, but not the β receptor antagonist [^{125}I]cyanopindolol (Kobilka et al., 1988). Moreover, replacement of a single residue, phe-412, in the seventh membrane-spanning segment conferred βAR ligand binding specificity to the α_2AR. These results imply that the seventh transmembrane segment contains elements necessary for maintaining α_2AR selectivity (Suryanarayana et al., 1991).

While ligand binding involves the hydrophobic core of the receptor, α_2 receptor-effector coupling is thought to involve the hydrophilic intracellular loops. Using chimeric receptor analysis, Kobilka et al. demonstrated that an α_2AR containing the third intracellular loop from the β_2AR could stimulate adenylyl cyclase with α_2 receptor pharmacology (Kobilka et al., 1988). These results suggested that effector coupling specificity resides within the third intracellular loop. More recent studies have specifically implicated the second cytoplasmic loop, the C-terminal portion of the third loop (Dalman and Neubig, 1991) and the amino terminal region of the third loop (Okamoto and Nishimoto, 1992) in α_2 receptor-G protein coupling.

Typically, the α_2C10 AR couples to the inhibition of adenylyl cyclase via a pertussis toxin (PTX) sensitive G protein, Gi. This was first demonstrated in human platelet membranes (Jakobs et al., 1976), and has since been observed virtually wherever α_2ARs are located. In addition to inhibition of adenylyl cyclase, recent evidence argues for a number of distinct α_2AR-effector interactions. In the platelet, α_2AR activation also stimulates a Na^+/H^+ antiporter leading to increased arachidonic acid release (Sweatt et al., 1985). This effect appears to involve activation of phospholipase A_2 (PLA_2) following increased Na^+/H^+ exchange, independent of any effect on adenylyl cyclase. However, in HT29 human colonic adenocarcinoma cells, which endogenously express the α_2C10 AR, activation of PLA_2 was observed without prior Na^+/H^+ exchange (Canitello et al., 1989), suggesting a direct activation of this pathway by the α_2C10 AR. Similar to inhibition of adenylyl cyclase, this α_2AR-mediated

Table 4. α_2 Adrenergic Receptor Pharmacology

K_i for Competition of [³H] Yohimbine in
COS-7 Cell Membranes

Drug	K_i (nM)		
	α_2C2	α_2C4	α_2C10
(–)Epinephrine	851	318	1671
(–)Norepinephrine	1265	606	3677
Oxymetazoline	1506	125	13.2
p-Aminoclonidine	120	97	31
Coryanthine	1002	182	1188
Rauwolscine	11	2.1	7.1
Prazosin	293	67.7	2237
Phentolamine	9.2	14.4	6.2
SKF104078	105	41	97
WB4101	132	13	47

stimulation of PLA_2 could be blocked by pretreatment of the cells with PTX, suggesting involvement of an inhibitory G protein in this process (Jones et al., 1991).

Depending on the cell type, the α_2C10 may activate a variety of distinct signaling pathways. In hamster lung fibroblasts activation of the α_2C10 both inhibits adenylyl cyclase and stimulates IP turnover. These effects were both attenuated by PTX pretreatment, implicating a single class of G proteins in these responses (Cotecchia et al., 1990). In transfected CHO cells however, α_2AR activation decreased cAMP levels but did not affect IP turnover (Fraser et al., 1989). α_2ARs have also been shown to stimulate intracellular Ca^{+2} mobilization in human erythroleukemia cells (Michel et al., 1989), as well as activate phospholipase D in rat-1 fibroblasts (Macnutty et al., 1992). These studies demonstrate that the ultimate response to α_2AR activation is cell-type specific.

The majority of these responses appear to involve pertussis toxin-sensitive, inhibitory G proteins (G_i). G proteins are comprised of three distinct subunits, an α subunit, which exhibits GTPase activity, and a $\beta\gamma$ complex, which is thought to anchor the protein in the plasma membrane. Although only a few $\beta\gamma$ subunits have been identified, a number of unique G protein α subunits have been cloned and sequenced (Jackson, 1991). Reconstitution of purified α_2ARs and G protein α subunits in phospholipid vesicles demonstrated that α_2ARs can potentially interact with several distinct G proteins. In this study, the α_2C10 coupled most strongly to $G\alpha_{i3}$, followed by $G\alpha_{i1}$, $G\alpha_{i2}$, and finally Go, with little observable coupling to the stimulatory G protein Gs (Kurose et al., 1991). More recent studies, however, suggest that the α_2C10 may indeed couple to other functionally distinct G proteins. Although α_2ARs typically inhibit adenylyl cyclase, Fraser et al. demonstrated that a stably expressed α_2C10 AR could potentiate forskolin-stimulated cAMP production at high doses of epinephrine (Fraser et al., 1989). In a related study, Jones et al. found that α_2AR-mediated stimulation of cAMP production in transfected CHO cells was insensitive to pertussis toxin (Jones et al., 1991). These results suggested involvement of a different class of G protein. Subsequently, physical α_2AR-G_s coupling was demonstrated in membranes from stably transfected CHO cells (Eason et al., 1992). The physiological implications of this α_2AR-Gs coupling or α_2AR-mediated stimulation of cAMP production, however, have yet to be elucidated.

1.3.21.7   Porcine α_{2A} Receptors

The gene encoding the porcine analogue of the human α_2C10 AR was cloned from a porcine genomic library (Guyer et al., 1990). This receptor is identical in size (450 amino acids) and possesses 94% overall sequence homology with the human receptor.

Like the human α_2C10 AR, the porcine α_2AR appears to be glycosylated near the amino terminus (asn-10 and asn-14). Furthermore, a single potential protein kinase A (PKA) phosphorylation site is present at thr-373 in the third intracellular loop. Phosphorylation of the porcine α_{2A} receptor has been observed using the catalytic subunit of cAMP-dependent protein kinase (Wilson et al., 1990). Other similarities with the human α_2C10 AR include conserved aspartic acid residues in transmembrane domains 2 and 3 and a potential palmitoylation site at cys-442.

Expression of the cloned porcine α_2AR reveals an "α_{2A}-like" pharmacology. This receptor displays highest affinity for yohimbine (Kd = 5.8 nM), followed by agonist oxymetazoline, and antagonists idazoxan and phentolamine. Similar to the human α_2C10, the porcine α_2AR binds the antagonist prazosin rather poorly (Ki = 7 μM), and exhibits a similar low affinity for the catecholamines epinephrine and norepinephrine. The porcine α_{2A} receptor appears to functionally couple to inhibition of adenylyl cyclase. Following stable expression in LLC-PK1 cells, which lack endogenous α_2 receptors, the α_2 receptor agonist UK-14304 inhibited both forskolin and arginine vasopression-induced cAMP accumulation, suggesting that this receptor is functionally similar to the human α_2C10 AR.

Early studies with the porcine α_{2A} receptor demonstrated that ligand binding can be allosterically modulated by Na$^+$, H$^+$, and 5-amino substituted amiloride analogs (Nunnari et al., 1987) In the presence of Na$^+$ ions, the porcine α_2AR exhibits a noticeably higher affinity for [^{3}H]yohimbine. The increase in antagonist binding affinity appears to result from an overall increase in the rate of ligand-receptor association. Modulation of antagonist binding was also observed in preparations from COS cells transfected with the porcine α_{2A} receptor, suggesting that this effect is a direct allosteric modulation of the α_2 receptor itself (Guyer et al., 1990).

1.3.21.10   Rat α_{2A} Receptors

Screening of a rat genomic library with a probe from the human α_2C4 AR yielded two unique rat α_2AR clones (Lanier et al., 1991). One of these, termed RG20, coded for a protein of 450 amino acids, possessing 89% overall homology with the human α_2C10 AR. Like the human and porcine α_2 receptors, this rat α_2 clone exhibits potential glycosylation sites near the amino terminus and the two conserved aspartic acid residues implicated in catecholamine binding. mRNA Expression of this clone was detected in numerous rat tissues, including brain, kidney, and salivary gland.

Pharmacologically, this receptor appeared to be unique. Following expression in COS-1 cells, the RG20 clone exhibited a nearly 10-fold lower affinity for the α_2AR antagonists rauwolscine and yohimbine compared to the human and porcine α_2ARs. Moreover, the rat α_2 clone possessed a different rank order of potency for inhibition of [^{3}H] rauwolscine binding than the human α_2C10. For the human receptor this order was rauwolscine > yohimbine > idazoxan > phentolamine > prazosin, while for the rat receptor the order was phentolamine > idazoxan > yohimbine > rauwolscine > prazosin. This apparent low affinity for the α_2-selective ligands rauwolscine and yohimbine prompted classification of this receptor as an α_{2D} subtype. Subsequently, work with the mouse homolog to the α_2C10 receptor suggested that the RG20 clone indeed corresponded to the human α_2C10 receptor. Link et al., identified a single amino acid variation present in both the rat and mouse receptors responsible for a decreased affinity for yohimbine. Mutation of residue ser-201 in the mouse α_2C10 homolog to cysteine increased the affinity for yohimbine 5-fold, suggesting that this single variation in the rodent receptor was responsible for the lower affinity binding. While binding of antagonists rauwolscine and yohimbine appeared to be affected, this variation does not alter the apparent affinity for other compounds. Antagonists phentolamine and idazoxan, and agonists epinephrine, norepinephrine, and oxymetazoline all bound with similar affinity to the rat vs. human receptors. These data and the high degree of sequence homology between the human $\alpha_2 C10$ and the RG20 clone (89%), suggest that this rat clone indeed codes for an α_{2A} receptor subtype (Link et al., 1992).

This single amino acid variation, while having distinct effects on ligand binding, does not appear to alter the functional properties of the rat receptor. Upon stable expression in NIH3T3 fibroblasts, the rat α_{2A} receptor inhibited both forskolin and PGE_1-stimulated cAMP accumulation. Like both the human and porcine α_{2A} receptors, this rat α_2 receptor also couples to the inhibition of adenylyl cyclase in transfected cells.

1.3.21.11 Mouse α_{2A} Receptors

The mouse homolog to the human $\alpha_2 C10$ AR has been cloned from a murine genomic library (Link et al., 1992). This receptor is composed of 450 amino acids, and possesses 92 and 96% overall homology to the human and rat α_{2A} receptors, respectively. Like the human $\alpha_2 C10$ AR, this mouse $\alpha_2 AR$ exhibits potential glycosylation sites at positions 10 and 14, as well as consensus PKA phosphorylation sites at threonines 227 and 373. Aspartic acid residues 79 and 113, implicated in catecholamine binding, are also present in the second and third transmembrane domains, respectively.

The mouse α_{2A} receptor exhibits a pharmacological profile similar to the aforementioned rat α_{2A} receptor. Like the human and rat receptors, the murine receptor displays a characteristically low affinity for prazosin, but somewhat lower affinity for WB4101 compared to the human $\alpha_2 C10$. Antagonists idazoxan and antipamezole bind to the mouse α_{2A} receptor with fairly high affinity (Ki < 10nM), wereas most agonists including oxymetazoline, epinephrine, and norepinephrine bind this mouse receptor with affinities similar to the human $\alpha_2 C10$. Like the rat α_{2A} receptor, however, the mouse α_{2A} receptor exhibits a fairly low affinity for rauwolscine and yohimbine (K_i = 50 to 60 nM). This diminished affinity appears to be due to a single amino acid variation in the fifth transmembrane domain. Both the rat and mouse α_{2A} receptors possess a serine at position 201, whereas the human and porcine receptors exhibit a cysteine at this location. Mutation of this residue in the mouse receptor to cysteine substantially increased the affinity for yohimbine, pointing to its importance in ligand-receptor interactions. While a cysteine to serine change is fairly conservative, this single difference may eliminate disulfide bind formation with other transmembrane domains, or perhaps lead to formation of an obstructive hydrogen bond. In any event, the importance of such specific residues in ligand-receptor interactions is only beginning to be understood.

1.3.22 α_{2B} Adrenergic Receptors

While the human platelet α_{2A} receptor exhibits a rather low affinity for the α_1 antagonist prazosin, radioligand binding studies pointed to the existence of an α_2 AR with much higher affinity for prazosin. A homogenous population of α_2 receptors located in neonatal rat lung was shown to possess nearly 50-fold higher affinity for prazosin relative to the human platelet $\alpha_2 AR$ (Latifpour et al., 1982). Kobilka et al. first provided evidence for human $\alpha_2 AR$ heterogeneity. Southern blot analysis of human genomic DNA using a probe from the $\alpha_2 C10$ AR revealed the existence of two additional $\alpha_2 AR$ genes (Kobilka et al., 1987). One of these genes, located on human chromosome 4 was cloned and termed the α_2C4 AR. While this receptor exhibited many of the properties of the pharmacologically defined α_{2B} receptor, including high affinity for prazosin, this receptor has more recently been classified as an α_{2C} AR (Bylund et al., 1992).

1.3.22.1   Human α_{2B} Receptors

A third human α_2AR was cloned from a genomic library using oligonucleotide probes corresponding to conserved regions of the two previously cloned α_2AR subtypes (Lomasney et al., 1990). The gene encoding this α_2AR was localized to chromosome 2 and was subsequently termed the α_2C2 AR. The α_2C2 is comprised of 450 amino acids and displays 75 and 74% amino acid homology with the α_2C4 and α_2C10 ARs in the transmembrane domains, respectively. While the α_2C2 AR exhibits an overall structural similarity to the α_2C4 and C10, this receptor possesses a much shorter amino terminus (12 residues), and lacks potential sites for N-linked glycosylation present in most other adrenergic receptor subtypes. Multiple potential phosphorylation sites are evident in the intracellular loops, as well as a potential palmitoylation site at cys-432. Aspartic acid residues implicated in ligand binding are also evident in the second and third transmembrane loops. Northern blot analysis of rat tissues revealed that this clone is expressed predominantly in the liver and kidney.

The α_2C2 AR expressed in COS cells binds antagonists rauwolscine and phentolamine with high affinity (Table 4). This receptor also displayed intermediate affinity for prazosin (Ki = 293 nM) and 10-fold lower affinity for antagonist WB4101 than the α_2C4 AR. Moreover, this subtype exhibited a much lower affinity for oxymetazoline (Ki = 1500 nM) compared to either the α_2C4 or α_2C10 ARs. Most other agonists texted bound the α_2C2 with low affinities similar to oxymetazoline, except for p-aminoclonidine, which was over 10-fold more potent (Ki = 120 nM). At present, little is known about the functional properties of the α_2C2, although the α_2C2 has been shown to inhibit forskolin-stimulated adenylyl cyclase in transfected fibroblast cells (Eason et al., 1992). Coupling of this receptor subtype to other second messenger pathways has not been extensively examined.

1.3.22.10   Rat α_{2B} Receptors

Zeng et al., have cloned the rat homolog for the human α_2C2 AR from a rat kidney cDNA library (Zeng et al., 1990). This clone, termed RNGα_2, encodes a protein of 453 amino acids which displays 84% overall homology to the human α_2C2 AR. Like the human α_2C2, the rat α_{2B} receptor lacks potential glycosylation sites near its amino terminus and shares other structural elements with its human counterpart, including serine residues at positions 181 and 185, as well as asp-97, which corresponds to asp-113 in the human α_2C10 AR. Expression of this clone has been detected in neonatal rat lung, the prototypical tissue for the α_{2B} AR subtype.

The rat RNGα_2 receptor exhibits an "α_{2B}-like" pharmacology. Prazosin displays fairly high affinity (Ki = 27 nM), while oxymetazoline binds rather poorly (Ki = 613 nM). The overall order of potency for inhibition of [^{3}H]rauwolscine binding was yohimbine > chlorpromazine > prazosin > clonidine > norepinephrine > oxymetazoline. Interestingly, a related rat α_{2B} clone possessing a phenylalanine substitution at position 169 in the third extracellular loop did not specifically bind [^{3}H]rauwolscine. All other adrenergic receptors cloned to date exhibit a cysteine at the homologous position, suggesting that this residue, albeit located in an extracellular loop is intimately involved in ligand-receptor interactions.

1.3.22.11 Mouse α_{2B} Receptors

The murine homolog to the human α_2C2 AR has been cloned from a mouse genomic library (Chruscinski et al., 1992). This receptor (Mα_2-2H) is 448 amino acids in size and possesses 82 and 96% identity to the human α_2C2 and rat RNGα_2 ARs, respectively. Within the transmembrane regions, virtually complete identity is observed, whereas the third intracellular loop is only 64% homologous to the human α_2C2 AR. Like other α_{2B} receptors, the Mα_2-2H possesses a short amino terminus lacking potential glycosylation sites. Potential phosphorylation sites are evident in the third intracellular loop (ser-231 and thr-336) as well as a carboxyl-terminal cysteine residue (cys-440).

The Mα_2-2H binds [^{3}H]yohimbine with high affinity (Kd = 12 nM) and possesses a pharmacological profile characteristic of an α_{2B} AR, including a relatively high affinity for prazosin (Ki = 59 nM) and low affinity for oxymetazoline (Ki = 1200 nM).

1.3.23 α_{2C} Adrenergic Receptors

Extensive radioligand binding analysis revealed the existence of a third pharmacologically distinct α_2AR subtype in opossum kidney (OK) cells (Murphy and Bylund, 1988). This receptor exhibited many of the characteristics of the α_{2B} receptor present in neonatal rat lung, including a fairly high affinity for prazosin. Other drugs such as oxymetazoline, spiroxatrine, and chlorpromazine also bound the OK α_2 receptor with affinities similar to the α_{2B}. However, a number of compounds including the α_1 antagonist WB4101 bound the the opossum kidney receptor with substantially higher affinity compared to the α_{2B} AR. Moreover, antagonists rauwolscine and yohimbine possessed higher affinity for the OK cell AR compared to either the α_{2A} or α_{2B} receptor subtypes. These results led to the classification of this opossum kidney α_2AR as the pharmacologic α_{2C} AR subtype (Blaxall et al., 1991).

1.3.23.1 Human α_{2C} Receptors

The gene for the human platelet α_2C10 AR was used as a probe to screen a human kidney cDNA library for additional α_2AR genes. A clone encoding a 462-amino acid protein was obtained which exhibited 75% homology with the α_2C10 AR within the membrane-spanning domains (Regan et al., 1988). Somatic cell hybridization analysis revealed that this gene resided on human chromosome 4 and subsequently this clone was termed the α_2C4. Like the α_2C10, the α_2C4 possesses a very long third intracellular loop (~150 amino acids) and a short carboxyl terminus of only 21 residues. The α_2C4 however, possesses a somewhat longer putative extracellular amino terminus (51 residues). Despite the extensive amino acid homology with the α_2C10 within the membrane-spanning regions, the intracellular and extracellular domains exhibit little similarity. The amino terminus and third intracellular loop possess only 14 and 21% amino acid homology with the α_2C10, respectively. Potential glycosylation sites are present near the amino terminus (asn-19 and asn-33); however, this receptor lacks the conserved cysteine residue near the carboxyl terminus thought to undergo palmitoylation. Elements typically implicated in ligand binding, including asp-131 and serines 214 and 218 are present in the second and third transmembrane domains, respectively.

The α_2C4 AR expressed in COS-7 cells possesses 5-fold higher affinity for [^{3}H]rauwolscine and substantially higher affinity for prazosin than the α_2C10 AR (Table 4). Likewise, catecholamine agonists epinephrine and norepinephrine were nearly 10-fold more potent at the α_2C4 subtype. The overall order of potency for inhibition of [^{3}H]rauwolscine binding to the α_2C4 was yohimbine > WB4101 > idazoxan > phentolamine > prazosin > oxymetazoline > p-aminoclonidine > epinephrine > norepinephrine. The high affinity for prazosin demonstrated by the α_2C4 subtype initially prompted its classification as an α_{2B} subtype. Comparison of the radioligand binding properties of the α_2C4 receptor subtype with the neonatal rat lung α_{2B} subtype and the opossum kidney α_{2C} subtype revealed much better correlation to the α_{2C} subtype. Both the α_2C4 and the OK cell α_{2C} receptor possess much higher affinity for [^{3}H]rauwolscine than either the α_{2A} or α_{2B} subtypes. Moreover, the agonist oxymetazoline possessed identical affinity for the α_2C4 AR and the OK cell α_2 receptor. This affinity appears intermediate to those for the α_{2A} (high) and α_{2B} subtypes (low), suggesting that the α_2C4 encodes the pharmacologic α_{2C} AR subtype (Bylund et al., 1992).

Although the α_2C4 was isolated from a kidney cDNA library, northern blots of rat kidney using the coding sequence of the α_2C4 as probe did not reveal expression of this clone. α_2C4 Expression was detected, however, in various brain regions including cerebral cortex, cerebellum, brainstem, and hippocampus (Lorenz et al., 1990). Subsequently, a rat homolog to the α_2C4 has been isolated, and appears to be strongly expressed in kidney (Voigt et al., 1991). Despite the overall homology between the rat and human receptors, the human α_2C4 probe simply does not appear to hybridize rat α_{2C} receptor mRNA.

Like the α_2C10, the α_2C4 appears to couple primarily to inhibition of adenylyl cyclase. Cotecchia et al. have stably expressed the α_2C4 in chinese hamster ovary cells and demonstrated that agonist activation of the α_2C4 could inhibit forskolin-stimulated cAMP accumulation (Cotecchia et al., 1990). Although the EC$_{50}$ for inhibition of cAMP production was similar for both receptors (5 μM), the α_2C10 inhibited basal cAMP levels 68%, while the α_2C4 elicited only 27% inhibition. This inhibition appears to involve a pertussis toxin (PTX)-sensitive G protein, as treatment of α_2C4 transfected cells with the toxin completely attenuated this effect. Reconstitution of purified α_2C4 ARs and G proteins in phospholipid vesicles revealed that the α_2C4 coupled most effectively to G_{i3}, followed by G_{i1}, G_{i2}, and finally, G_o. Neither the α_2C4 nor the α_2C10 coupled significantly to the stimulatory G protein, Gs (Kurose et al., 1991). The α_2C4 has also been shown to couple weakly to IP production. Stimulation of COS-7 cells expressing the α_2C10 or α_2C4 subtypes produced small but significant increases in IP release. This IP release was not attenuated by prior PTX treatment, implying that both subtypes can couple to independent signaling pathways involving distinct G protein effectors (Cotecchia et al., 1990).

1.3.23.10 Rat α_{2C} Receptors

A rodent homolog to the human α_2C4 AR has been cloned from a rat genomic library (Voigt et al., 1991). This clone codes for a protein of 458 amino acids which displays over 90% homology to the human α_2C4 AR subtype. The rat α_{2C} receptor appears to be expressed most strongly in brain, with lower levels found in the kidney and heart.

The rat α_{2C} receptor appears pharmacologically similar to the human α_2C4 subtype. [^{3}H]Rauwolscine binds the rat α_{2C} receptor with high affinity (Kd = 0.37 nM), as do WB4101 (Ki = 1.6 nM), and prazosin (Ki = 20 nM). Oxymetazoline possesses intermediate affinity for this rat receptor (Ki = 34 nM) compared to the α_{2A} and α_{2B} receptor subtypes. Functionally, the rat α_{2C} receptor has been shown to inhibit forskolin-stimulated cAMP production in response to agonists clonidine and norepinephrine.

1.3.23.11 Mouse α_{2C} Receptors

Link et al. have cloned the mouse homolog to the human α_2C4 AR. This 458-amino acid protein is 99% homologous to the aforementioned rat α_{2C} and 92% homologous to the human α_2C4 receptors (Link et al., 1992). The mouse α_{2C} receptor possesses a pharmacology similar to the human α_2C4 receptor subtype, with good affinity for prazosin (Ki = 97 n*M*), WB4101 (Ki = 7.8 n*M*), and idazoxan (Ki = 9.8 n*M*), as well as intermediate affinity for p-aminoclonidine and oxymetazoline with Ki values of 102 and 109 n*M*, respectively.

1.3.24 Other α_2 Receptors

α_2 ARs are found in non-mammalian species as well. Receptors with an α_2 AR pharmacology have been characterized in the skin of lower vertebrates, where they appear to mediate pigment cell (melanophore) aggregation, leading to a color change (Berthelsen and Pettinger, 1977). Subsequently, a fish-skin α_2 AR has been cloned from a cuckoo wrasse (*Labrus ossifagus*) genomic library (Svensson et al., 1993b). This fish receptor, termed α_2-F, is the first non-mammalian α_2 AR to be cloned and sequenced. This receptor consists of 432 amino acids and possesses 57% overall homology to the human α_2C4 AR subtype. Homology with the human α_2C10 and α_2C2 subtypes were 50 and 47%, respectively. Within the transmembrane regions, this receptor was 82% homologous to the α_2C4 AR. Sequence homology with α_1 and β ARs in these regions were typically about 40%. Like most adrenergic receptors, this fish clone does not contain introns and possesses many of the conserved structural features involved in ligand binding and effector coupling. Potential glycosylation sites are evident near the putative amino terminus, as well as phosphorylation sites for cAMP-dependent protein kinase and the β-AR kinase. Consensus sequences for regulatory transcription factors are also present in the 5'-untranslated region, suggesting that expression of this fish α_2 gene may be regulated by both cAMP and steroid hormones. One notable exception, however, is a phenylalanine residue located in the seventh transmembrane region. This residue, homologous to phe-412 in the human α_2C10 AR, has recently been implicated in antagonist binding specificity (Suryanarayana et al., 1991). While this residue is conserved among all previously cloned α ARs, the corresponding amino acid is absent in this fish α_2 AR.

The expressed fish α_2 clone exhibits a pharmacology characteristic of an α_2 AR. The FPC AR binds [³H]rauwolscine with high affinity (Kd = 0.8 n*M*), and possesses high affinity for both oxymetazoline and yohimbine. Prazosin binds this receptor with fairly low affinity (Ki = 469 n*M*), similar to the α_2C2 AR. The overall antagonist order of potency was rauwolscine > atipamezole > yohimbine > phentolamine > prazosin, while for agonists, this order was medetomidine > clonidine > BHT 920 > guanfacine > UK14304 > norepinephrine. These results imply that this FPC clone indeed codes for an α_2 AR. While UK14304 is an agonist at most α_2ARs, this compound possesses no intrinsic activity at the α_2-F AR. Svensson et al., recently demonstrated that UK14304 is without activity at fish melanophore α_2ARs, suggesting UK14304 is an antagonist at this α_2AR subtype (Svensson et al., 1993a)

1.3.31 β_1 Adrenergic Receptors

The β adrenergic receptors are perhaps the most extensively characterized class of any G protein coupled receptors. Ahlquist first classified adrenergic receptors as either α

or β based on pharmacological criteria. He demonstrated that at α ARs, the order of potency for agonists to evoke a physiological response was epinephrine > norepineph-rine > isoproterenol, while at β ARs, this order was isoproterenol > epinephrine > norepinephrine (Ahlquist, 1948). Subsequently, β ARs were subclassified as either β_1 or β_2 based on agonist potency and tissue localization. Lands et al. examined the relative potency of sympathomimetic amines on fatty acid mobilization (lipolysis), cardiac stimulation, vasodepression, and bronchodilation. They noted a marked simi-larity between the relative potencies for lipolysis and cardiac stimulation, and a similar relation between the potencies for inducing bronchodilation and vasodepression (Lands et al., 1967). Receptors mediating lipolysis and cardiac stimulation were designated β_1, whereas those linked to bronchodilation and vasodilation were termed β_2. Radioligand binding studies and Northern blot analysis confirm that β_1 receptors are indeed localized primarily in the heart while β_2 receptors are concentrated in the lung and vasculature.

β ARs are typically characterized as possessing high affinity for [^{125}I]cyanopindolol. Other radioligands including [^{3}H]dihydroalprenolol, [^{3}H]pindolol and [^{3}H]propranolol have also been used to characterize β ARs. Compounds exhibiting selectivity for the various subtypes of β receptors include CGP-20712A for β_1 receptors and ICI 118551 for β_2 receptor subtypes. Epinephrine typically does not distinguish between these subtypes, whereas norepinephrine has somewhat higher affinity for β_1 ARs.

The β ARs were among the first of the G protein coupled receptors to be cloned and sequenced. In 1986 genes encoding the hamster β_2 (Dixon et al., 1986) and turkey β_1 ARs (Yarden et al., 1986) were isolated and shown to possess an overall structure similar to bovine rhodopsin. Hydropathy analysis revealed the presence of seven hydrophobic, membrane-spanning α helices, with extracellular amino termini and intracellular carboxyl termini. Connecting these domains were hydrophilic extracellu-lar and intracellular loops. In contrast to α ARs, β receptors typically possess fairly short third intracellular loops and long cytoplasmic carboxyl termini. More recently, a third subtype of β AR termed the β_3 receptor has been cloned. This receptor is localized primarily in liver, adipocytes, and skeletal muscle, where it regulates a number of functions including intestinal mobility, lipolysis, and glycogen synthesis (Emorine et al., 1989).

β_1 ARs are located primarily in the atrial and ventricular myocardium, where they mediate increases in both cardiac rate and contractility. Depending on the species however, β_1 receptors have been localized to brain regions including cortex, hippo-campus, and thalamus, as well as juxtaglomerular cells of the kidney, where they mediate renin release (Summers and McMartin, 1993). Like the β_2 ARs, the β_1 ARs stimulate adenylyl cyclase by activating a stimulatory G protein, G_s.

1.3.31.1 Human β_1 Receptors

The human β_1 AR was cloned in 1987 from a placental cDNA library. This receptor consists of 477 amino acids and possesses 54% overall homology with the human β_2 AR subtype. A consensus glycosylation site is present at asn-15, as well as two sites for phosphorylation by cAMP-dependent protein kinase (ser 312 and ser 412). Serine residues located in the carboxyl terminus may also serve as potential regulatory sites for phosphorylation by the β AR kinase (Benovic et al., 1986). The third intracellular loop of this receptor is the longest of any β receptor at 75 residues, and this loop was shown to contain an unusually large number of proline residues (13). Northern blot analysis on total cellular RNA from various rat tissues revealed the highest levels of β_1 AR mRNA in the heart, pineal gland, and cerebral cortex.

Table 5. β Adrenergic Receptor Pharmacology

K_i (n*M*) for Competition of [^{125}I] Cyanopindolol in
CHO Cell Membranes

Drug	β_1	β_2	β_3
[^{125}I]$_{ICYP}$ (K$_D$)	0.017	0.031	0.23
Isoproterenol	50	140	620
(–)Epinephrine	900	370	20,650
(–)Norepinephrine	250	740	475
ICI118551	150	0.93	257
CGP20712A	1.5	1,800	2,300
Alprenolol	5.0	0.27	ND
Betaxolol	15	100	ND
Oxprenolol	ND	ND	70
Pindolol	ND	ND	11
BRL37344	ND	ND	183

Note: ND = not determined.

The β_1 AR expressed in CHO cells displays high affinity for [^{125}I]CYP (Kd = 0.017 n*M*), as well as high affinity for the β_1 antagonist CGP20171A (Ki = 1.5 n*M*) (Table 5). Other compounds exhibiting high affinity for the β_1 receptor include alprenolol (Ki = 5 n*M*), betaxolol (15 n*M*), and isoproterenol (50 n*M*). Agonist potency for inhibition of [^{125}I]CYP binding was identical to that for stimulation of adenylyl cyclase in transfected *Xenopus* oocyte membranes (isoproterenol > norepinephrine > epinephrine).

1.3.31.10 Rat β_1 Receptors

A gene encoding a β_1 AR was cloned from a rat genomic library (Machida et al., 1990). The cloned protein contains 466 residues and possesses a consensus glycosylation site at asn-15. Potential sites for phosphorylation by cAMP-dependent protein kinase are also evident at serines 296, 301, and 401. The rat receptor is 91% identical to the human β_1 receptor overall, and 98% homologous within the membrane-spanning domains. The third intracellular loop of the rat β_1 receptor, however, is 12 amino acids shorter than the corresponding loop in the human receptor. Since a consensus 3' mRNA splice site has been identified in this region, this deletion suggests the presence of an intron in the rat β_1 receptor. As expected, β_1 message was detected in the heart, as well as in the brain, particularly in the pineal gland, thalamus, amygdala, hippocampus, and anterior basal ganglia.

Following stable expression in mouse L cells, the rat β_1 AR bound [^{125}I]CYP with high affinity (Kd = 0.011 n*M*). The β_1-selective antagonist ICI89,406 also bound with high affinity (Ki = 1.0 n*M*), while the β_2 antagonist ICI118,506 possessed much lower affinity (Ki = 126 n*M*). When challenged with agonist isoproterenol, the rat β_1 AR stimulated adenylyl cyclase activity 4- to 5-fold. Although epinephrine and norepinephrine stimulated adenylyl cyclase to a similar extent, they were somewhat less potent. EC$_{50}$ values for stimulation of adenylyl cyclase by isoproterenol, norepinephrine, and epinephrine were 10, 35, and 160 n*M*, respectively.

1.3.31.14 Turkey β_1 Receptors

Only the second adrenergic receptor to be cloned, the turkey β_1 AR was shown to be structurally homologous to both rhodopsin and the hamster β_2 AR (Yarden et al.,

1986). Oligonucleotide probes synthesized based on primary sequence information from the purified receptor were used to screen a fetal turkey cDNA library. The deduced amino acid sequence codes for a protein of 483 amino acids which exhibits 69% overall homology to the human β_1 AR. Within the transmembrane domains, these receptors were 84% identical. A single potential glycosylation site is present near the amino terminus (asn-14), as well as numerous potential phosphorylation sites near the carboxyl terminus.

1.3.32 β_2 Adrenergic Receptors

The β_2 ARs are among the most extensively characterized members of the G protein coupled receptor superfamily. The β_2 AR was the first AR to be purified, cloned, and sequenced (Dixon et al., 1986). Furthermore, biochemical studies have revealed that the β_2 AR undergoes extensive posttranslational modification, including glycosylation, phosphorylation, and palmitoylation. The receptor has been mapped structurally through mutational analysis, and β_2 AR expression has been shown to be regulated by both cAMP and glucocorticoids (Collins et al., 1990).

Although β_2 ARs have typically been characterized in the lung and vasculature, these receptors can also be found in the brain, particularly in the cerebellum and pia matter. β_2 ARs have also been identified in the kidney and uterus (Summers and McMartin, 1993). Compounds selective for β_2 ARs include procaterol, alprenolol, and ICI 118,551. Other β_2-selective agonists such as albuterol and pirbuterol have been widely used to treat bronchial asthma (Hoffman and Lefkowitz, 1990).

1.3.32.1   Human β_2 Receptors

The human β_2 AR was cloned in 1987 from a placental cDNA library, and shown to be 87% identical to the previously cloned hamster β_2 AR (Kobilka et al., 1987). This clone encodes a protein of 413 amino acids with a deduced molecular weight of 46,000. Somatic cell hybridization has localized this gene to human chromosome 5q31-32. Two glycosylation sites are present at residues 6 and 15, as well as two potential sites for phosphorylation by cAMP-dependent protein kinase at positions 262 and 345. A number of serine and threonine residues located within the intracellular domains also serve as phosphorylation sites for the β AR kinase (Benovic et al., 1986).

Many of these domains appear to be involved in β_2 AR function. β ARs were shown to be extensively glycosylated, and while removal of the extracellular glycosylation sites does not impare ligand-receptor interactions or receptor-G protein coupling, it does dramatically reduce the level of β_2 receptor expressed at the cell surface (Rands et al., 1990). Phosphorylation, however, plays a much more significant role in β_2 AR function. Short-term exposure of the β_2 AR to agonist produces a rapid desensitization and uncoupling of the receptor from Gs (Clark et al., 1988). This initial response appears to involve phosphorylation of ser-262 by cAMP-dependent protein kinase, as replacement of this residue by alanine, prevents rapid desensitization (Clark et al., 1989). Further exposure to agonist promotes phosphorylation by the β AR kinase, leading to receptor sequestration (Benovic et al., 1986). The carboxyl tail of the β_2 receptor contains a number of serine residues, elimination of which slows both phosphorylation and receptor sequestration, suggesting that these residues are critical for β AR function (Bouvier et al., 1988).

Structural domains involved in ligand binding to adrenergic receptors were initially defined in the β_2 AR. These studies implicated a conserved aspartic acid residue (asp-113) as the counterion for catecholamine binding. Mutation of this residue eliminated both high affinity agonist and antagonist binding (Dixon et al., 1988). Subsequently, Strader et al. demonstrated that two serine residues at positions 204 and 207 in the fifth transmembrane domain were required for agonist binding and receptor activation (Strader et al., 1989). They proposed that the hydroxyl groups of these serine residues served as hydrogen bond donors for the catecholamine ligands epinephrine and norepinephrine.

Extensive work has also focused on defining the effector coupling domains for the β_2 AR. Kobilka et al. first demonstrated that the third intracellular loop of the β_2 AR was required for effector coupling specificity. Construction of chimeric α_2 ARs with the third intracellular loop from the β_2 receptor stimulated adenylyl cyclase similar to a β_2 receptor, with an α_2 AR pharmacology. (Kobilka et al., 1988). Similarly, using substitution and deletion mutagenesis, O'dowd et al. found that the carboxyl terminal portion of the third intracellular loop and amino-terminal region of the carboxyl tail were both involved in receptor-effector coupling (O'dowd et al., 1988). More recently, it was shown that β_2 ARs containing α_2 receptor substitutions from the carboxyl terminal region of the third intracellular loop or amino terminal portion of the carboxyl tail were significantly uncoupled from G_s (Liggett et al., 1991). Following pertussis toxin treatment, basal adenylyl cyclase activity was increased over 500% in one of these mutants, containing three individual α_2 AR substitutions. These results implicate multiple cytoplasmic domains in the receptor-G protein coupling process, and particularly the amino terminal region of the third intracellular loop in conferring G protein specificity.

Palmitoylation of the β_2 AR is also required for proper receptor-effector coupling. Mutation of cysteine 341 in the carboxyl tail to glycine prevented not only palmitoylation, but also functional coupling of the β_2 receptor to G_s (O'dowd et al., 1989). Palmitoylation of this cysteine is thought to form an additional intracellular loop which may interact with G proteins. Although this cysteine is conserved in nearly every adrenergic receptor, palmitoylation has been demonstrated only for the β_2 AR and rhodopsin (Ovchinnikov et al., 1988).

The β AR expressed in CHO cells appears pharmacologically similar to the native receptor (Tate et al., 1991). This receptor exhibited high affinity for [^{125}I]CYP (Kd = 0.031 nM) and ICI118551 (Ki = 0.93 nM), whereas agonists possess much lower affinity for this receptor. Norepinephrine in particular was threefold less potent at this subtype compared to the β_1 AR. The human β_2 AR has also been successfully expressed in mouse B-82 cells (Fraser et al., 1987). In these cells, the β_2 AR possesses similar high affinity not only for ICYP and ICI118551, but also propranolol (Ki = 2.5 nM). Norepinephrine however, was far less potent in these cells (Ki = 4.5 μM). Agonist activation of the β_2 receptor expressed in these cells stimulated adenylyl cyclase over fourfold.

1.3.32.10   Rat β_2 Receptors

The rat homolog to the human β_2 AR was cloned from a genomic library. This receptor is identical in size and possesses 90% overall homology with the human gene. This protein has not yet been expressed (Buckland et al., 1990).

1.3.32.11   Mouse β_2 Receptors

The murine β_2 AR consists of 413 amino acids and possesses 93% homology with the corresponding hamster gene (Allen et al., 1988). This clone has been stably expressed in Y1 cells, and was shown to bind [^{125}I]CYP with high affinity (Kd = 0.026 nM). Isoproterenol also binds this receptor with fairly high affinity (Ki = 5.3 nM), while epinephrine was nearly 25-fold less potent, and norepinephrine over 500-fold less potent. Activation of the β_2 receptor in Y1 cells promoted cAMP-dependent steroid secretion, indirectly suggesting that this receptor functionally couples to adenylyl cyclase in these cells.

1.3.32.12   Hamster β_2 Receptors

The hamster β_2 AR was purified from lung and cloned from a cDNA library (Dixon et al., 1986). This protein consists of 418 amino acids, five more than any other cloned β_2 receptor. The hamster β_2 possesses two consensus glycosylation sites near the amino terminus (asn-6 and asn-15), as well as consensus phosphorylation sites for cAMP-dependent protein kinase and the β AR kinase. Within the transmembrane domains, this protein is 95% identical to the human β_2 receptor. All other residues implicated in ligand binding and effector coupling for the human receptor are evident in this protein.

1.3.33 β_3 Adrenergic Receptors

A third subtype of β AR has recently been identified in the digestive tract (Bond and Clark, 1987), skeletal muscle (Challis et al., 1988), and adipose tissue (Arch et al., 1984) which may be responsible for some of the metabolic effects of the catecholamines. In adipose tissue, activation of this receptor stimulates lipolysis, whereas in muscle it appears to inhibit glycogen synthesis. Strangely, this β receptor possesses an unusually low affinity for the typical β-blockers including propranolol and, subsequently, these receptors have been termed atypical β ARs. Compounds exhibiting affinity for this receptor include the β_1 receptor antagonist CGP20712A and the β_2 antagonist ICI118551. One drug which appears selective for this subtype is BRL37344. While this compound only weakly activates β_1 and β_2 receptors, it potently stimulates lipolysis in rat adipocytes (Arch et al., 1984) and inhibits contractions of guinea pig ileum (Bond and Clark, 1988).

1.3.33.1   Human β_3 Receptors

The human β_3 AR was cloned from a genomic library and shown to possess 51 and 46% homology to the human β_1 and β_2 ARs, respectively (Emorine et al., 1989). Somatic cell hybridization localized the gene for this receptor to human chromosome 8p11.1-8p12. The protein is composed of 402 amino acids with a predicted molecular weight of 42,800. Glycosylation sites are present near the amino terminus, while the third intracellular loop and carboxyl terminus exhibit a number of potential phosphorylation domains. Like both the β_1 and β_2 ARs, this receptor possesses the conserved residues implicated in ligand binding and effector coupling. As expected, Northern blot of various tissues revealed expression in the ileum, soleus muscle, liver, and both brown and white adipose tissues. When expressed in CHO cells lacking endogenous β ARs, the β_3 AR stimulated cAMP accumulation in response to epinephrine, norepinephrine, oxprenolol, pindolol, and isoproterenol. Other compounds displaying functional activity at the β_3 AR include fenoterol, salbutamol, prenaterol, and BRL28410. The β_3 AR bound [^{125}I]CYP with much lower affinity than either the β_1 or β_2 AR subtypes (Kd = 0.49 nM), while antagonists ICI118551 and CGP20712A bound this receptor with Ki values of 257 nM and 2.3 μM, respectively.

1.3.33.10   Rat β₃ Receptors

The rat β_3 receptor consists of 400 amino acids and possesses 79% homology to the human β_3 receptor (Granneman et al., 1991). Within the transmembrane domains however, these proteins are 98% homologous. The majority of the amino acid differences reside in the carboxyl-terminal region, except for a three-amino-acid deletion found within the first transmembrane domain (residues 48 to 50 in the human receptor). Other elements are highly conserved within the membrane-spanning regions and intracellular loops. Like the human β_3 receptor, this receptor does not contain consensus sites for phosphorylation by cAMP dependent protein kinase.

The rat β_3 receptor expressed in CHO cells exhibited rather low affinity for the common β-receptor ligands, and subsequently radioligand binding was not performed. However, agonists epinephrine, norepinephrine, isoproterenol, and BRL37344 all stimulated adenylyl cyclase in transfected CHO cell membranes. The overall order of potency for this stimulation was BRL37344 > isoproterenenol > norepinephrine > epinephrine. Compounds possessing partial agonist activity at this β_3 receptor include CGP12177, alprenolol, and pindolol.

Expression of this clone was detected most strongly in adipose tissue, with smaller amounts found in the ileum.

1.3.33.11   Mouse β₃ Receptors

The cloned mouse β_3 AR consists of 388 amino acids and possesses 82% homology with its human counterpart (Nahmias et al., 1991). This mouse receptor is the shortest adrenergic receptor identifed to date, with 11 fewer amino acids than the human β_3 AR. The carboxyl tail is clearly the most divergent from the human receptor, although a conserved cysteine is potentially available for palmitoylation at position 358. The mouse β_3 AR expressed in CHO cells possesses high affinity for [^{125}I]CYP (Kd = 0.88 nM), similar to the human β_3 receptor. Extensive radioligand binding has not been performed, but this receptor does appear functionally similar to the human β_3 AR. Agonist potency for stimulation of cAMP accumulation was BRL37344 > isoproterenol > epinephrine > norepinephrine. CGP 12177, oxprenolol, and pindolol were all partial agonists, whereas propranolol, ICI118551, and CGP 20172A had little effect on isoproterenol-induced cAMP accumulation.

REFERENCES

Ahlquist, R. P. (1948) A study of adrenotropic receptors. *Am. J. Physiol.* 153, 586–600.

Allen, J. M., Baetge, E. E., Abrass, I. B., and Palmiter, R. D. (1988) Isoproterenol response following transfection of the mouse β_2-adrenergic receptor gene into Y1 cells. *EMBO J.* 7, 133–138.

Arch, J. R. S., Ainsworth, A. T., Cawthorne, M. A., Piercy, V., Sennitt, M. V., Thody, V. E., Wilson, C., and Wilson, S. (1984) Atypical b-adrenoceptor on brown adipocytes as target for anti-obesity drugs. *Nature* 309, 163–165.

Battaglia, G., Shannon, M., Borgundvaag, B., and Titeler, M. (1983) Properties of [^{3}H]prazosin-labelled α_1-adrenergic receptors in rat brain and porcine neurointermediate lobe tissue. *J. Neurochem.* 41, 538–542.

Benovic, J. L., Regan, J. W., Matsui, H., Mayor, F., Jr., Cotecchia, S., Leeb-Lundberg, L. M. F., Caron, M. G., and Lefkowitz, R. J. (1987) Agonist dependent phosphorylation of the α_2-adrenergic receptor by the β-adrenergic receptor kinase. *J. Biol. Chem.* 262, 17251–17253.

Benovic, J. L., Strasser, R. H., Benovic, J. L., Daniel, and Lefkowitz, R. J. (1986) Beta adrenergic receptor kinase: identification of a novel protein kinase that phosphorylates the agonist-occupied form of the receptor. *Proc. Natl. Acad. Sci. U.S.A.* 83, 2797–2801.

Berthelsen, S. and Pettinger, W. A. (1977) A functional basis for classification of a-adrenergic receptors. *Life Sci.* 21, 595–606.

Blaxall, H. S., Murphy, T. J., Baker, J. C., Ray, C., and Bylund, D. B. (1991) Characterization of the alpha-2C adrenergic receptor subtype in the opossum kidney and in the OK cell line. *J. Pharm. Exp. Ther.* 259, 323–329.

Boer, R., Brassegger, A., Schudt, C., and Glossman, H. (1989) (+) Niguldipine binds with very high affinity to Ca^{+2} channels and to a subtype of a_1 adrenoceptors. *Eur. J. Pharmacol.* 172, 131–145.

Bond, R. A. and Clark, D. E. (1988) Agonist and antagonist characterization of a putative adrenoceptor with distinct pharmacological properties from the a- and b-subtypes. *Br. J. Pharmacol.* 95, 723–734.

Bouvier, M. W., Hausdorff, A., DeBlasi, A., O'Doud, B. F., Kobilka, B. K., Caron, M. G., and Lefkowitz, R. J. (1988) Removal of phosphorylation sites from the b-adrenergic receptor delays the onset of agonist-promoted desensitization. *Nature* 333, 370–373.

Bouvier, M., Leeb-Lundberg, L. M. F., Benovic, J. L., Caron, M. G., and Lefkowitz, R. J. (1987) Regulation of adrenergic receptor function by phosphorylation. II. Effects of agonist occupancy on phosphorylation of a_1 and b_2 adrenergic receptors by protein kinase C and the cyclic AMP-dependent protein kinase. *J. Biol. Chem.* 262, 3106–3113.

Bruno, J. F., Whittaker, J., Song, J., and Berelowitz, M. (1991) Molecular cloning and sequencing of a cDNA encoding a human a_{1A} adrenergic receptor. *Biochem. Biophys. Res. Commun.* 179, 1485–1490.

Buckland, P. R., Hill, R. M., Tidmarsh, S. F., and McGuffin, P. (1990) Primary structure of the rat beta-2 adrenergic receptor gene. *Nucl. Acids Res.* 18, 682.

Bylund, D. B. (1985) Heterogeneity of alpha-2 adrenergic receptors. *Pharmacol. Biochem. Behav.* 22, 835–843.

Bylund, D. B., Blakall, H. J., Iversen, L. J., Caron, M. G., Lefkowitz, R. J., and Lomasney, J. W. (1992) Pharmacological characteristics of a_2-adrenergic receptors: comparison of pharmacologically defined subtypes with subtypes identified by molecular cloning. *Mol. Pharmacol.* 42, 1–5.

Bylund, D. B. and U'Prichard, D. C. (1983) Characterization of alpha-1 and alpha-2 adrenergic receptors. *Int. Rev. Neurobiol.* 24, 343–431.

Canitello, H. F. and Lanier, S. M. (1989) Alpha-2 adrenergic receptors and the Na^+/H^+ exchanger in the intestinal epithelial cell line HT-29. *J. Biol. Chem.* 264, 16000–16007.

Chalberg, S. C., Duda, T., Rhine, J. A., and Sharma, R. K. (1990) Molecular cloning, sequencing and expression of an a_2-adrenergic receptor complementary DNA from rat brain. *Mol. Cell. Biol.* 97, 161–172.

Challiss, R. A. J., Leighton, B., Wilson, S., Thurlby, P. L., and Arch, J. R. S. (1988) An investigation of the b-adrenoceptor that mediates metabolic responses to the novel agonist BRL28410 in rat soleus muscle. *Biochem. Pharmacol.* 37, 947–950.

Cheung, Y., Barnett, D. B., and Nahorski, S. R. (1982) [³H]Rauwolscine and [³H]yohimbine binding to rat cerebral and human platelet membranes: possible heterogeneity of a_2-adrenoceptors. *Eur. J. Pharmacol.* 84, 79–85.

Chruscinski, A. J., Link, R. E., Daunt, D. A., Barsh, G. S., and Kobilka, B. K. (1992) Cloning and expression of the mouse homolog of the human a_2-C2 adrenergic receptor. *Biochem. Biophys. Res. Commun.* 186, 1280–1287.

Clark, R. B., Friedman, J., Dixon, R. A. F., and Strader, C. D. (1989) Identification of a specific site required for rapid heterologous desensitization of the b-adrenergic receptor by cAMP-dependent protein kinase. *Mol. Pharmacol.* 36, 348–356.

Clark, R. B., Kunkel, M. W., Friedman, J., Goka, T. J., and Hohnson, J. A. (1988) Activation of cAMP-dependent protein kinase is required for heterologous desensitization of adenylyl cyclase in S49 wild-type lymphoma cells. *Proc. Natl. Acad. Sci. U.S.A.* 85, 1442–1446.

Collins, S., Altschmied, J., Herbsman, O., Caron, M. G., Mellon, P. L., and Lefkowitz, R. J. (1990) A cAMP response element in the b_2-adrenergic receptor gene confers transcriptional autoregulation by cAMP. *J. Biol. Chem.* 265, 19330–19335.

Cotecchia, S., Exum, S., Caron, M. G., and Lefkowitz, R. J. (1990) Regions of the a1-adrenergic receptor involved in coupling to phosphatidylinositol hydrolysis and enhanced sensitivity of biological function. *Proc. Natl. Acad. Sci. U.S.A.* 87, 2896–2900.

Cotecchia, S., Kobilka, B. K., Daniel, K. W., Nolan, R. D., Lapetina, E. Y., Caron, M. G., Lefkowitz, R. J., and Regan, J. W. (1990) Multiple second messenger pathways of a-adrenergic receptor subtypes expressed in eukaryotic cells. *J. Biol. Chem.* 265, 63–69.

Cotecchia, S., Ostrowski, J., Kjelsberg, M. A., Caron, M. G., and Lefkowitz, R. J. (1992) Discrete amino acid sequences of the a_1-adrenergic receptor determine the selectivity of coupling to phosphatidylinositol hydrolysis. *J. Biol. Chem.* 267, 1633–1639.

Cotecchia, S., Schwinn, D. A., Randall, R. R., Lefkowitz, R. J., Caron, M. G., and Kobilka, B. K. (1988) Molecular cloning and expression of the cDNA for the hamster a_1-adrenergic receptor. *Proc. Natl. Acad. Sci. U.S.A.* 85, 7159–7163.

Dalman, H. M. and Neubig, R. R. (1991) Two peptides from the a_{2A} adrenergic receptor alter receptor G protein coupling by distinct mechanisms. *J. Biol. Chem.* 266, 11025–11029.

Dixon, R. A. F., Kobilka, B. K., Strader, D. J., Benovic, J. L., Dohlman, H. G., Frielle, T., Bolanowski, M. A., Bennett, C. D., Rands, E., Diehl, R. E., Mumford, R. A., Slater, E. E., Sigal, I. S., Caron, M. G., Lefkowitz, R. J., and Strader, C. D. (1986) Cloning of the gene and cDNA for mammalian b-adrenergic receptor and homology with rhodopsin. *Nature* 321, 75–79.

Dixon, R. A. F., Sigal, I. S., and Strader, C. D. (1988) Structure-function analysis of the b-adrenergic receptor. *Cold Spring Harbor Symp. Quant. Biol.* 53, 487–497.

Eason, M. G., Kurose, H., Holt, B. D., Raymond, J. R., and Liggett, S. B. (1992) Simultaneous coupling of a_2-adrenergic receptors to two G-proteins with opposing effects. Subtype-selective coupling of a_2C10, a_2C4 and a_2C2 adrenergic receptors to Gi and Gs. *J. Biol. Chem.* 267, 15795–15801.

Emorine, L. J., Marullo, S., Briend-Sutrem, M. M., Patey, G., Tate, K., Delavier-Klutchko, C., and Strosberg, A. D. (1989) Molecular characterization of the human b_3 adrenergic receptor. *Science* 245, 1118–1121.

Fain, J. N. and Gracia-Sainz, J. A. (1980) Role of phosphatidylinositol turnover in $alpha_1$ and of adenylate cyclase inhibition in $alpha_2$ effects of catecholamines. *Life Sci.* 26, 1183–1194.

Feng, D. F. and Doolittle, R. F. (1990) Progressive alignment and phylogenetic tree construction of protein sequences. *Meth. Enzymol.* 183, 375–387.

Fraser, C. M., Arakawa, S., McCombie, W. R., and Venter, J. C. (1989) Cloning, sequence analysis, and permanent expression of a human a_2-adrenergic receptor in chinese hamster ovary cells. Evidence for independent pathways of receptor coupling to adenylate cyclase attenuation and activation. *J. Biol. Chem.* 264, 11754–11761.

Fraser, C. M., Chung, F. Z., and Venter, J. C. (1987) Continuous high density expression of human b_2 adrenergic receptors in a mouse cell line previously lacking b-receptors. *J. Biol. Chem.* 262, 14843–14846.

Frielle, T., Collins, S., Daniel, K. W., Caron, M. G., Lefkowitz, R. J., and Kobilka, B. K. (1987) Cloning of the cDNA for the human b_1 adrenergic receptor. *Proc. Natl. Acad. Sci. U.S.A.* 84, 7920–7924.

Granneman, J. G., Lahners, K. N., and Chaudhry, A. (1991) Molecular cloning and expression of the rat b_3-adrenergic receptor. *Mol. Pharmacol.* 40, 889–895.

Guyer, C. A., Horstman, D. A., Wilson, A. L., Clark, J. D., Cragoe, E. J., Jr., and Limbird, L. E. (1990) Cloning, sequencing and expression of the gene encoding the porcine a_2-adrenergic receptor. Allosteric modulation by Na^+, H^+ and amiloride analogs. *J. Biol. Chem.* 265, 17307–17317.

Han, C., Abel, P. W., and Minneman, K. P. (1987a) Heterogeneity of a_1-adrenergic receptors revealed by chlorethylclonidine. *Mol. Pharmacol.* 32, 505–510.

Han, C., Abel, P. W., and Minneman, K. P. (1987b) a_1 Adrenoceptor subtypes linked to different mechanisms for increasing intracellular Ca^{+2} in smooth muscle. *Nature* 329, 333–335.

Hoffman, B. B. and Lefkowitz, R. J. (1990) Catecholamines and sympathomimetic drugs. In *Goodman and Gilman's The Pharmacological Basis of Theraputics,* Gilman, A. G., Rall, R. W., Nies, A. S., and Taylor, P., Eds, Pergamon Press, New York, pp. 187–220.

Jackson, T. (1991) Structure and function of G protein coupled receptors. *Pharmacol. Ther.* 50, 425–442.

Jakobs, K. H., Saur, W., and Schultz, G. (1976) Reduction of adenylate cyclase activity in lysates of human platelets by the alpha-adrenergic component of epinephrine. *J. Cyclic Nucl. Res.* 2, 381–392.

Jones, S. B., Halenda, S. P., and Bylund, D. B. (1991) a_2-Adrenergic receptor stimulation of phospholipase A_2 and of adenylate cyclase in transfected chinese hamster ovary cells is mediated by different mechanisms. *Mol. Pharmacol.* 39, 239–245.

Kjelsberg, M. A., Cotecchia, S., Ostrowski, J., Caron, M. G., and Lefkowitz, R. J. (1992) Constitutive activation of the a_{1b} adrenergic receptor by all amino acid substitutions at a single site. *J. Biol. Chem.* 267, 1430–1433.

Kobilka, B. K., Dixon, R. A. F., Frielle, T., Dohlman, H. G., Bolanowski, M. A., Sigal, I. S., Yang-Feng, T. L., Francke, U., Caron, M. G., and Lefkowitz, R. J. (1987) cDNA for the human b_2 adrenergic receptor: a protein with multiple membrane-spanning domains and encoded by a gene whose chromosomal location is shared with that of the receptor for platelet-derived growth factor. *Proc. Natl. Acad. Sci. U.S.A.* 84, 46–50.

Kobilka, B. K., Kobilka, T. S., Daniel, K., Regan, J. W., Caron, M. G., and Lefkowitz, R. J. (1988) Chimeric a_2-, b_2-adrenergic receptors: delineation of domains involved in effector coupling and ligand binding specificity. *Science* 240, 1310–1316.

Kobilka, B. K., Matsui, H., Kobilka, T. S., Yang-Feng, T. L., Francke, U., Caron, M. G., Lefkowitz, R. J., and Regan, J. W. (1987) Cloning, sequencing and expression of the gene coding for the human platelet a_2-adrenergic receptor. *Science* 238, 650–656.

Kurose, H., Regan, J. W., Caron, M. G., and Lefkowitz, R. J. (1991) Functional interactions of recombinant a_2 adrenergic receptor subtypes and G proteins in reconstituted phospholipid vesicles. *Biochemistry* 30, 3335–3341.

Lands, A. M., Arnold, A., McAuliff, J. P., Ludena, F. P., and Brown, T. G. (1967) Differentiation of receptor systems activated by sympathomimetic amines. *Nature* 214, 597–598.

Langer, S. Z. (1974) Presynaptic regulation of catecholamine release. *Biochem. Pharmacol.* 23, 1793–1800.

Lanier, S. M., Downing, S., Duzic, E., and Homey, C. J. (1991) Isolation of rat genomic clones encoding subtypes of the a_2-adrenergic receptor. *J. Biol. Chem.* 266, 10470–10478.

Latifpour, J., Jones, S. B., and Bylund, D. B. (1982) Characterization of [³H]yohimbine binding to putative alpha-2 adrenergic receptors in neonatal rat lung. *J. Pharmacol. Exp. Ther.* 223, 606–611.

Leeb-Lundberg, L. M. F., Cotecchia, S., Deblasi, A., Caron, M. G., and Lefkowitz, R. J. (1987) Regulation of adrenergic receptor function by phosphorylation. I. Agonist-promoted desensitization and phosphorylation of a_1-adrenergic receptors coupled to inositol phospholipid metabolism in DDT$_1$MF-2 smooth muscle cells. *J. Biol. Chem.* 262, 3098–3105.

Liebert, F., Parmentier, M., Lefort, A., Dinsart, C., Van Sande, J., Maenhaut, C., Simons, M.-J., Dumont, J. E., and Vasssart, G. (1989) Selective amplification and cloning of four new members of the G-protein-coupled receptor family. *Science* 244, 569–572.

Liggett, S. B., Caron, M. G., Lefkowitz, R. J., and Hnatowich, M. (1991) Coupling of a mutated form of the human b_2-adrenergic receptor to G_i and G_s. Requirement for multiple cytoplasmic domains in the coupling process. *J. Biol. Chem.* 266, 4816–4821.

Link, R., Daunt, D., Barsh, G., Chruscinski, A., and Kobilka, B. (1992) Cloning of two mouse genes encoding a_2-adrenergic receptor subtypes and identification of a single amino acid in the mouse a_2-C10 homolog responsible for an interspecies variation in antagonist binding. *Mol. Pharmacol.* 42, 16–27.

Limbird, L. E. (1988) Receptors linked to adenylate cyclase: additional signalling mechanisms. *FASEB J.* 2, 2686–2695.

Lomasney, J. W., Cotecchia, S., Lorenz, W., Leung, W-Y., Schwinn, D. A., Yang-Feng, T. L., Brownstein, M., Lefkowitz, R. J., and Caron, M. G. (1991a) Molecular cloning and expression of the cDNA for the a_{1a} adrenergic receptor: the gene for which is located on human chromosome 5. *J. Biol. Chem.* 266, 6365–6369.

Lomasney, J. W., Cotecchia, S., Lefkowitz, R. J., and Caron, M. G. (1991b) Molecular biology of a-adrenergic receptors: implications for receptor classification and for structure-function relationships. *Biochem. Biophys. Acta.* 1095, 127–139.

Lomasney, J. W., Lorenz, W., Allen, L. F., King, K., Regan, J. W., Yang-Feng, T. L., Caron, M. G., and Lefkowitz, R. J. (1990) Expansion of the a_2-adrenergic receptor family: cloning and characterization of a human a_2-adrenergic receptor subtype, the gene for which is located on chromosome 2. *Proc. Natl. Acad. Sci. U.S.A..* 87, 5094–5098.

Machida, C. A., Bunzow, J. R., Searles, R. P., Van Tol, H., Tester, B., Neve, K. A., Teal, P., Nipper, V., and Civelli, O. (1990) Molecular cloning and expression of the rat b_1-adrenergic receptor gene. *J. Biol. Chem.* 265, 12960–12965.

Macnutty, E. E., McClue, S. J., Carr, I. C., Jess, T., Wakelam, M. J. O., and Milligan, G. (1992) a_2-C10 adrenergic receptors expressed in rat 1 fibroblasts can regulate both adenylylcyclase and phospholipase D-mediated hydrolysis of phosphatidylcholine by interacting with pertussis toxin-sensitive guanine nucleotide-binding proteins. *J. Biol. Chem.* 267, 2149–2156.

Marrow, A. L. and Creese, I. (1986) Characterization of a_1 adrenergic receptor subtypes in rat brain: a reevaluation of [^{3}H]WB4101 and [^{3}H]prazosin binding. *Mol. Pharmacol.* 29, 321–330.

Matsui, H., Lefkowitz, R. J., Caron, M. G., and Regan, J. W. (1989) Localization of the fourth membrane spanning domain as a ligand binding site in the human platelet a_2-adrenergic receptor. *Biochemistry* 28, 4125–4130.

McGrath, J. C. (1982) Evidence for more than one type of postjunctional a-adrenoceptor. *Biochem. Pharmacol.* 31, 467–484.

Michel, A. D., Lowry, D. N., and Whiting, R. L. (1989) Identification of a single a_1 adrenoceptor corresponding to the a_{1a} subtype in rat submaxillary gland. *Br. J. Pharmacol.* 98, 883–889.

Michel, M. C., Brass, L. F., Williams, A., Bokoch, G. M., Lamorte, V. J., and Motulsky, H. J. (1989) a_2-adrenergic receptor stimulation mobilizes intracellular Ca^{+2} in human erythroleukemia cells. *J. Biol. Chem.* 264, 4986–4991.

Murphy, T. J. and Bylund, D. B. (1988) Characterization of alpha-2 adrenergic receptors in the OK cell, an opossum kidney cell line. *J. Pharm. Exp. Ther.* 244, 571–578.

Nahmias, C., Blin, N., Elalouf, J. M., Mattei, M. G., Strosberg, A. D., and Emorine, L. J. (1991) Molecular characterization of the mouse b_3-adrenergic receptor: relationship with the atypical receptor of adipocytes. *EMBO J.* 10, 3721–3727.

Nunnari, J. M., Repaske, M. G., Brandon, S., Cragoe, E. J., Jr., and Limbird, L. E. (1987) Regulation of porcine brain a_2-adrenergic receptors by Na^+, H^+ and inhibitors of Na^+/H^+ exchange. *J. Biol. Chem.* 262, 12387–12392.

O'dowd, B. F., Hnatowich, M., Caron, M. G., Lefkowtz, R. J., and Bouvier, M. (1989) Palmitoylation of the human b2-adrenergic receptor. *J. Biol. Chem.* 264, 7564–7569.

O'dowd, B. F., Hnatowich, M., Regan, J. W., Leader, W. M., Caron, M. G., and Lefkowitz, R. J. (1988) Site-directed mutagenesis of the cytoplasmic domains of the human b2 adrenergic receptor. Localization of regions involved in G protein-receptor coupling. *J. Biol. Chem.* 263, 15985–15992.

Okamoto, T. and Nishimoto, I. (1992) Detection of G protein-activator regions in M4 subtype muscrinic cholinergic, and a_2-adrenergic receptors based upon characteristics in primary structure. *J. Biol. Chem.* 267, 8342–8346.

Ovchinnikov, Y. A., Abdulaev, N. G., and Bogachuk, A. S. (1988) Two adjacent cysteine residues in the C-terminal cytoplasmic fragment of bovine rhodopsin are palmitylated. *FEBS Lett.* 230, 1–5.

Probst, W. C., Snyder, L. A., Schuster, D. I., Brosius, J., and Sealfon, S. C. (1992) Sequence alignment of the G-protein coupled receptor superfamily. *DNA Cell Biol.* 11, 1–20.

Rands, E., Candelore, M. R., Cheung, A. H., Hill, W. S., Strader, C. D., and Dixon, R. A. F. (1990) Mutational analysis of b-adrenergic receptor glycosylation. *J. Biol. Chem.* 265, 10759–10764.

Regan J. W., Caron, M. G., and Lefkowitz, R. J. (1990) Structural determinants of ligand binding to alpha- and beta-adrenoceptors. In *Transmembrane Signalling, Intracellular Messengers and Implications for Drug Development*, Nahorski, S. R., Ed, John Wiley & Sons, New York, pp. 1–9.

Regan, J. W., Kobilka, T. S., Yang-Feng, T. L., Caron, M. G., Lefkowitz, R. J., and Kobilka, B. K. (1988) Cloning and expression of a human kidney cDNA for an a_2-adrenergic receptor subtype. *Proc. Natl. Acad. Sci. U.S.A.* 85, 6301–6305.

Regan, J. W., Nakata, H., Demarinis, R. M., Caron, M. G., and Lefkowitz, R. J. (1986) Purification and characterization of the human platelet a_2 adrenergic receptor. *J. Biol. Chem.* 261, 3894–3900.

Sawurz, D. G., Lanier, S. M., Warren, C. D., and Grahm, R. M. (1987) Glycosylation of the mammalian a_1-adrenergic receptor by complex type N-linked oligosaccharides. *Mol. Pharmacol.* 32, 565–571.

Schwinn, D. A., Lomasney, J. W., Lorenz, W., Szklut, P. J., Fremeau, R. T., Jr., Yang-Feng, T. L., Caron, M. G., Lefkowitz, R. J., and Cotecchia, S. (1990) Molecular cloning and expression of the cDNA for a novel a_1 adrenergic receptor subtype. *J. Biol. Chem.* 265, 8183–8189.,

Schwinn, D. A., Page, S. O., Middleton, J. P., Lorenz, W., Liggett, S. B., Yamamoto, K., Lapetina, E. G., Caron, M. G., Lefkowitz, R. J., and Cotecchia, S. (1991) The a_{1c} adrenergic receptor: characterization of signal transduction pathways and mammalian tissue heterogeneity. *Mol. Pharmacol.* 40, 619–626.

Starke, K. (1981) Alpha adrenoceptor subclassification. *Rev. Physiol. Biochem. Pharmacol.* 88, 199–236.

Strader, C. D., Candelore, M. R., Hill, W. S., Sigal, I. S., and Dixon, R. A. F. (1989) Identification of two serine residues involved in agonist activation of the b-adrenergic receptor. *J. Biol. Chem.* 264, 13572–13578.

Summers, R. J. and McMartin, L. R. (1993) Adrenoceptors and their second messenger systems. *J. Neurochem.* 60, 10–23.

Suryanarayana, S., Daunt, D. A., Von Zastrow, M., and Kobilka, B. K. (1991) A point mutation in the seventh hydrophobic domain of the a_2 adrenergic receptor increases its affinity for a family of b receptor antagonists. *J. Biol. Chem.* 266, 15488–15492.

Svensson, S. P. S., Andersson, R. G. G., Grundstrom, N., Regan, J. W., and Karlsson, J. O. G. (1993a) UK14,304 acts as an a_2-adrenoceptor antagonist in fish melanophores, *Br. J. Pharmacol.,* submitted.

Svensson, S. P. S., Bailey T. J., Pepperl D. J., Grundstrom N., Ala-Uotila, S., Scheinin, M., Karlsson, J. O. G., and Regan, J. W. (1993b) Cloning and expression of a fish a_2-adrenoceptor, *Br. J. Pharmacol.,* 110, 54260–54266.

Sweatt, J. D., Johnson, S. L., Cragoe, E. J., and Limbird, L. E. (1985) Inhibitors of Na^+/H^+ exchange block stimulus-provoked arachidonic acid release in human platelets. *J. Biol. Chem.* 260, 12910–12919.

Tate, K. M., Briend-Sutren, M. M., Emorine, L. J., Delavier-Klutchko, C., Marullo, S., and Strosberg, A. D. (1991) Expression of three human b-adrenergic-receptor subtypes in transfected chinese hamster ovary cells. *Eur. J. Biochem.* 196, 357–361.

Voigt, M. M., Kispert, J., and Chin, H. (1990) Sequence of a rat brain cDNA encoding an a_{1b} adrenergic receptor. *Nucl. Acids Res.* 18, 1053.

Voigt, M. M., McCune, S. K., Kanterman, R. Y., and Felder, C. C. (1991) The rat a_2-C4 adrenergic receptor gene encodes a novel pharmacological subtype. *FEBS Lett.* 278, 45–50.

Wang, C. D., Buck, M. A., and Fraser, C. M. (1991) Site-directed mutagenesis of a_{2A}-adrenergic receptors: identification of amino acids involved in ligand binding and receptor activation by agonists. *Mol. Pharmacol.* 40, 168–179.

Wilson, A. L., Guyer, C. A., Cragoe, E. J., Jr., and Limbird, L. E. (1990) The hydrophobic tryptic core of the porcine a_2-adrenergic receptor retains allosteric modulation of binding by Na^+, H^+, and 5-amino-substituted amiloride analogs. *J. Biol. Chem.* 265, 17318–17322.

Yarden, Y., Rodriquez, H., Wong, S. K. F., Brandt, D. R., May, D. C., Burnier, J., Harkins, R. N., Chen, E. Y., Ramachandran, J., Ullrich, A., and Ross, E. M. (1986) The avian b-adrenergic receptor: primary structure and membrane topology. *Proc. Natl. Acad. Sci. U.S.A.* 83, 6795–6799.

Zeng, D. W., Harrison, J. K., D'Angelo, D. D., Barber, C. M., Tucker, A. L., Lu, Z. H., and Lynch, K. R. (1990) Molecular characterization of a rat a_2B-adrenergic receptor. *Proc. Natl. Acad. Sci. U.S.A.* 87, 3102–3106.

6

Cannabinoid Receptors

Tom I. Bonner and Lisa A. Matsuda

1.6.0   Introduction

Cannabinoid receptors are the receptors which mediate the psychoactive effects of marijuana (*Cannabis sativa*) (see Table 1). Since the major psychoactive component of marijuana, Δ^9-tetrahydrocannibinol (Δ^9-THC) (Gaoni and Mechoulam, 1964), is very hydrophobic, it was long thought that its cellular effects were due to membrane perturbation rather than activation of a specific receptor. In addition, a pharmacological antagonist for cannabinoid-induced effects was not available. However, within the last 5 to 10 years it has become clear that there are receptors in rat brain and in neuroblastoma cell lines that respond specifically to psychoactive cannabinoids (*Cannabis*-derived compounds and synthetic analogs). Characterization of the effects of cannabinoid receptor activation was first accomplished using both naturally occurring compounds (Δ^9-THC and Δ^8-THC) and a synthetic analog, desacetyllevonantradol. In a plasma membrane fraction of the N18TG-2 neuroblastoma cell line, cannabinoids inhibited basal and forskolin-stimulated adenylate cyclase in a reversible, dose-dependent, and stereoselective manner and did not act on receptors known to be present on the cells (Howlett and Fleming, 1984; Howlett, 1985; Howlett et al., 1986). Furthermore, the ability of pertussis toxin to block the cannabinoid-induced effect on adenylate cyclase activity confirmed that a guanine-nucleotide binding protein (G protein) was involved, thus establishing the cannabinoid receptor as a G protein-coupled receptor (Howlett, et al., 1986). This sensitivity to pertussis toxin has also been observed in NG-108-15 neuroblastoma-glioma hybrid cells in which cannabinoid receptor activation inhibits calcium channels (Mackie and Hille, 1992) as well as as reduces adenylate cyclase activity (Howlett et al., 1986).

The expression of the cannabinoid receptor in brain tissues was initially demonstrated *in vitro* by a tissue homogenate ligand binding assay and by the effects of cannabinoids on cAMP accumulation in tissue slices. Binding studies performed with synaptosomes or P2 membranes of rat brain demonstrated reversible, specific, and saturable binding of ^{3}H-CP 55,940 (Devane et al., 1988). In cortical membranes binding occurred at a single class of sites with a K_d of 133 p*M* and a B_{max} of 1.9 pmol/mg protein. The rank order of potency for displacement of ^{3}H-CP 55,940 binding by natural and synthetic cannabinoids paralleled the effects of these compounds on adenylate cyclase activity in neural cell lines. The binding properties of the cannabinoid receptor have since been demonstrated using a variety of ligands which have been used to localize the receptor in a variety of tissues and species (Table 2). The high

Table 1.

Receptor	Second messenger	Species cloned	Chromosomal location	a.a. Sequence	GenBank accession	Primary reference
Cannabinoid	–AC	Human	bq14-15	472	X54937	Gerard et al., 1990
		Rat		473	X55812	Matsuda et al., 1990

correlation between the affinities of numerous cannabinoids and their potencies in a series of *in vivo* cannabinoid responses implicates the receptor in many of the behavioral and physiological effects that typify the cannabinoid response in animals (Abood and Martin, 1992).

The ability of cannabinoids to reduce cAMP accumulation in brain tissue slices was evident in rat cortex, striatum, hippocampus, and cerebellum (Bidaut-Russell et al., 1990). Subsequently, decreased activity of adenylate cyclase was observed in membrane preparations of rat cerebellum or striatum treated with the novel cannabinoid receptor agonist, WIN 55212-2 (Pacheco et al., 1991). In addition to adenylate cyclase activity, the cannabinoids have been reported to alter numerous enzymes, second messengers, and ion channels in the brain (Pertwee, 1988).

The effects of cannabinoids have also been examined in a variety of other tissues and *in vivo* but it is difficult to directly link some of these effects with the cannabinoid receptor because stereospecificity was not always established and no antagonist was available. Furthermore, many early studies tested cannabinoids (Δ^9-THC, Δ^8-THC) that demonstrate only moderate potency at the receptor and therefore were often used at relatively high concentrations. More recent studies benefit from the availability of several agonists which not only have substantially higher affinity for the receptor but also differ substantially from the classical chemical structure of Δ^9-THC (Johnson and Melvin, 1986; D'Ambra et al., 1992). An antagonist would still be desirable to clearly define the role of the cannabinoid receptor in a variety of systems.

The direct proof that the cannabinoid receptor is a distinct receptor was provided by the cloning and expression of a rat cannabinoid receptor cDNA (Matsuda et al.,

Table 2. Radioligands (Agonists) Used to Label G Protein-Coupled Cannabinoid Receptors In Vitro

Ligand	Tissue preparation	Kd	Bmas (pmol/mg protein)	Reference
3H-CP 55940[a,g]	Brain membranes (rat cortex)	33 pM	1.85 ± 0.26	Devane et al., 1988
	Tissue sections (rat brain sausage)	19 nM	3.2 ± 0.4	Herkrenham et al., 1990
	Intact cells (mouse spleen)	860 pM	n.d.[e]	Kaminski et al., 1992
	Intact cells (sea urchin sperm)	5.16 nM	2.44 ± 0.42f	Chang et al., in preparation
	CHO-K1 membranes (rat cDNA)	~4.0 nM	~0.34	Felder et al., 1992
	COS-7 membranes (human cDNA)	2.9 nM	6.4	Felder et al., 1992
3H-WIN 55212-2[b,g]	Brain membranes (rat, cerebellum)	2 nM	1.2	D'Ambra et al., 1992
	Tissue sections (rat, cerebellum)	15 nM	n.d.	Jansen et al., 1992
3H-HU-243[c]	Brain synaptosomes (rat)	45 pM	4.41 ± 0.08	Devane et al., 1992
3H-11-OH-Δ^9-TCH-DMH[d]	Tissue sections (rat, brain sausage)	29 nM	4.3 ± 0.6 27	

[a] [1a,2b(R)5a]-(–)-5(1,1-dimethylheptyl)-2-[5-hydroxy-2-(3-hydroxypropyl)cyclohexyl]-phenol

[b] R-(+)-(2,3-dihydro-5-methyl-3-[{4-morphonolinyl}methyl]pyrrolo[1,2,3-de]-1,4-benzoxazin-6-yl)(1-napthalenyl)methanone monomethanesulfonate

[c] 11-hydroxy-hexahydrocannabinol-dimethyl heptyl

[d] 11-hydroxy-Δ^9-tetrahydrocannabinol-dimethyl heptyl

[e] receptor density estimated to be 800–1000 receptors per cell

[f] receptor density estimated to be 712 ± 122 receptors per cell

[g] commercially available

1990). The existence of this receptor suggests that there must be an endogenously synthesized agonist in mammals. Arachidonylethanolamide has recently been identified as the endogenous agonist (Deware et al., 1992). Although the pharmacology of cannabinoid receptors is in its infancy compared to many G protein-coupled receptors, there is currently no pharmacological evidence for subtypes. Furthermore, Southern blot analysis of genomic DNA provides no evidence for closely related genes that could represent subtypes (Bonner, unpublished). Thus, for the present, we can consider there to be only a single cannabinoid receptor.

1.6.1   Cannabinoid Receptors

1.6.1.1   Human Cannabinoid Receptors

The first report of a human cannabinoid receptor (Gerard et al., 1990) was a cDNA isolated from a human brain stem cDNA library by polymerase chain reaction using consensus primers based on known G protein-coupled receptors and identified by its high sequence similarity to the published rat cDNA (Matsuda et al., 1990). The 472 amino acid protein encoded by the cDNA has 98% identity to the rat sequence and the corresponding DNA sequences have 90% identity (Figure 1). Cosmids spanning 56 kb of the human gene have also been cloned (Bonner et al., unpublished). Sequence analysis of the gene indicates that the entire coding sequence is contained within a single exon which also contains 63 bases of 5′ untranslated sequence and the entire 3′ untranslated sequence of about 3.8 kb. In addition, there are two exons containing 5′ untranslated sequence. One contains a sequence homologous to the 5′ end of the rat cerebral cortex cDNA but which is absent, due to alternative splicing, in the human brain stem cDNA while the other exon contains sequence corresponding to the 5′ end of the human brain stem cDNA. However, the first 44 bases of the brain stem cDNA are derived from the noncoding strand presumably through formation of a hairpin which primed the synthesis of the second strand of the cDNA. The gene has been mapped by chromosomal *in situ* hybridization to 6q14-15 (Hoehe et al., 1991).

Both the brain stem cDNA and the coding sequence of the gene have been inserted into expression vectors and expressed in mammalian cells for pharmacological evaluation (Gerard et al., 1991; Felder et al., 1992). The cDNA was stably expressed in CHO-K1 cells and the effect of several cannabinoids on cAMP accumulation was determined (Table 3). The response of these transfected cells mirrored that previously described for the neural cell lines. The same cDNA was transiently expressed in COS-7 cells and saturable, specific binding of ^{3}H-CP 55,940 was demonstrated. The coding region of the human gene was stably expressed in mouse L cells using an expression system designed to create cell lines containing very high numbers of receptors. While inhibition of cAMP accumulation could not be observed in this high expression cell line, the binding properties of the human gene product were readily determined. The ability of various cannabinoids to compete with ^{3}H-CP 55,940 binding were measured in membranes prepared from the transfected L cells (Table 4).

The stably expressed human receptor has been used to determine whether cannabinoid induced effects on arachidonic acid release and uptake and on intracellular calcium concentrations are receptor-mediated (Felder et al., 1992). This study indicated that although cannabinoids altered these biochemical measures of cell activity, the effects did not appear to be receptor mediated.

```
cannabinoid.human   1  MKSILDGLADTTFRTITTDLLYVGSNDIQYEDIKGDMASKLGYFPQKFPL
cannabinoid.rat     1  MKSILDGLADTTFRTITTDLLYVGSNDIQYEDIKGDMASKLGYFPQKFPL

cannabinoid.human  51  TSFRGSPFQEKMTAGDNPQLVPA-DQVNITEFYNKSLSSFKENEENIQCG
cannabinoid.rat    51  TSFRGSPFQEKMTAGDNSPLVPAGDTTNITEFYNKSLSSFKENEENIQCG

cannabinoid.human 100  ENFMDIECFMVLNPSQQLAIAVLSLTLGTFTVLENLLVLCVILHSRSLRC
cannabinoid.rat   101  ENFMDMECFMILNPSQQLAIAVLSLTLGTFTVLENLLVLCVILHSRSLRC

cannabinoid.human 150  RPSYHFIGSLAVADLLGSVIFVYSFIDFHVFHRKDSRNVFLFKLGGVTAS
cannabinoid.rat   151  RPSYHFIGSLAVADLLGSVIFVYSFVDFHVFHRKDSPNVFLFKLGGVTAS

cannabinoid.human 200  FTASVGSLFLTAIDRYISIHRPLAYKRIVTRPKAVVAFCLMWTIAIVIAV
cannabinoid.rat   201  FTASVGSLFLTAIDRYISIHRPLAYKRIVTRPKAVVAFCLMWTIAIVIAV

cannabinoid.human 250  LPLLGWNCEKLQSVCSDIFPHIDETYLMFWIGVTSVLLLFIVYAYMYILW
cannabinoid.rat   251  LPLLGWNCKKLQSVCSDIFPLIDETYLMFWIGVTSVLLLFIVYAYMYILW

cannabinoid.human 300  KAHSHAVRMIQRGTQKSIIIHTSEDGKVQVTRPDQARMDIRLAKTLVLIL
cannabinoid.rat   301  KAHSHAVRMIQRGTQKSIIIHTSEDGKVQVTRPDQARMDIRLAKTLVLIL

cannabinoid.human 350  VVLIICWGPLLAIMVYDVFGKMNKLIKTVFAFCSMLCLLNSTVNPIIYAL
cannabinoid.rat   351  VVLIICWGPLLAIMVYDVFGKMNKLIKTVFAFCSMLCLLNSTVNPIIYAL

cannabinoid.human 400  RSKDLRHAFRSMFPSCEGTAQPLDNSMGDSDCLHKHANNAASVHRAAESC
cannabinoid.rat   401  RSKDLRHAFRSMFPSCEGTAQPLDNSMGDSDCLHKHANNTASMHRAAESC

cannabinoid.human 450  IKSTVKIAKVTMSVSTDTSAEAL
cannabinoid.rat   451  IKSTVKIAKVTMSVSTDTSAEAL
```

FIGURE 1. Alignment of the amino acid sequences of cannabinoid receptors. The sequences were obtained from commercial protein databases such as GenBank and EMBL. The references for each sequence are cited in the text.

1.6.1.10   Rat Cannabinoid Receptors

The first clone of the cannabinoid receptor was a 5.5-kb rat cerebral cortex cDNA (Matsuda et al., 1990) which was isolated by hybridization to a 56-base oligonucleotide based on the second transmembrane domain sequence of the substance K receptor. The clone isolated by virtue of 68% identity to the probe encoded a 473 amino acid protein containing seven potential membrane spanning domains and many of the 30 to 40 residues which are highly conserved in G protein-coupled receptors (Figure 1). Notable among the highly conserved residues which were not present are the cysteine just before the third transmembrane domain, the proline in the fifth transmembrane domain and the glycine in the first transmembrane domain which is replaced by a glutamic acid. These three features as well as a modest degree of similarity, primarily in transmembrane domains three and four, are shared by two other orphan receptors, edg-1 (Hla and Maciag, 1990) and R334 (Eidne et al., 1991; Bonner

Table 3   Cannabinoid-Induced Effects on cAMP Accumulation** in CHO-K1 Cells Expressing Rat and Human Cannabinoid Receptor Clones

Ligand	ED_{50} (nM) Human	ED_{50} (nM) Rat	Reference
(–)Δ⁹THC*	13	55 ± 1.2	Gerard et al. (1990), Felder et al. (1992)
(+)Δ⁹THC*	—	13.5 ± 2.7	Matsuda et al. (1990)
		>900	Felder et al. (1992)
		773 ± 187	Matsuda et al. (1990)
(–)Δ⁸THC	82	27.4 ± 8.4	Matsuda et al. (1990), Gerard et al. (1990)
11-OH-Δ⁹THC	—	8.9 ± 1.8	Matsuda et al. (1990)
Nabilone	—	27.4 ± 8.4	Matsuda et al. (1990)
CP 55,940[†]	0.99	1.8 ± 0.2	Gerard et al. (1990), Felder et al. (1992)
		0.87 ± 0.20	Matsuda et al. (1990)
CP 56,667[†]	346	96.3 ± 7.1	Matsuda et al. (1990), Gerard et al. (1990)
CP 50,556	—	81.4 ± 0.9	Felder et al. (1992)
CP 55,244[§]	0.128	0.2 ± 0.01	Gerard et al. (1990), Felder et al. (1992)
CP 55,243[§]	>10000	—	Gerard et al. (1990)
WIN 55-212-2[¥]	—	24 ± 3.7	Felder et al. (1992)
WIN 55,212-3[¥]	—	>2000	Felder et al. (1992)
HU-210[□]	—	0.02 ± 0.001	Felder et al. (1992)
HU-211[□]	—	191 ± 2.6	Felder et al. (1992)
RM-8ç	—	8.5 ± 0.2	Felder et al. (1992)
RM-7ç	—	2060 ± 24	Felder et al. (1992)
Propano-Δ⁸THC	—	78.3 ± 4.2	Felder et al. (1992)
2-Iodo-Δ⁸THC-DMH	—	36.8 ± 1.4	Felder et al. (1992)

** forskolin-stimulated cAMP production

*, †, §, □, ç enantiomeric pair

and Brownstein, unpublished). However, the protein sequence was not similar enough to any known receptor sequence to suggest its identity.

To identify the receptor a variety of ligands were tested by binding to CHO-K1 cells transfected with the clone, by cyclic AMP assays on these cells, or by electrophysiological responses in *Xenopus* oocytes injected with RNA transcribed from the clone. The ligands tested in these assays and found to be without effect were neurokinin B, substance P, neurokinin A, neuropeptide Y, peptide YY, PHI-27, secretin, gastrin, vasoactive intestinal peptide, prostaglandins E2 and F2, α-melanocyte stimulating hormone, ACTH, α-MSH inhibiting factor, calcitonin gene related peptide, pituitary adenylate cyclase activating polypeptide, D-Ala D-Leu enkaphalin, α-neo endorphin, dynorphin B, atrial naturetic factor, leutinizing hormone releasing hormone, growth hormone releasing hormone, thyrotropin-releasing hormone, corticotropin-releasing hormone, GABA, glycine, aspartate, glutamate kainate, dopamine, serotonin, acetylcholine, histamine, bombesin, neurotensin, somatostatin, adenosine receptor agonists (NECA and CPA), oxytocin, cholecystokinin, vasopressin, angiotensin II, galanin, and bradykinin. Receptors for most of these agonists have been cloned subsequently and found to be distinct from the cannabinoid receptor. Ultimately, the clue which pointed to the identity of the receptor was the expression of related mRNA in NG108-15 and its parent N18TG-2, cell lines in which cannabinoid receptor had been identified. The distribution of mRNA in rat brain also was similar to, at the time unpublished, data on the distribution of cannabinoid binding in brain (Herkrenham et al., 1990).

The identification of the clone as a cannabinoid receptor was established by demonstrating that, when expressed in CHO-K1 cells, it responds to cannabinoids by inhibiting adenylate cyclase in a dose-dependent, stereoselective, and pertussis toxinsensitive manner similar to that observed the neuroblastomas (Howlett et al., 1986). In

Table 4. Competetion of ³H-CP 55,940 Binding to Cloned Human Receptor Expressed in L Cells by Cannabinoid Agonists

Agonist	K_i (nM)
(–) Δ-9-THC	53 ± 0.2
(+) Δ-9-THC	>900
Nabilone	79.1 ± 2.5
CP 55,940	3.7 ± 0.1
CP 50,556	81.4 ± 0.9
CP 55,244	0.51 ± 0.02
WIN-55212-2	564 ± 12.5
WIN-55212-3	>3,500
HU-210	0.06 ± 0.0002
HU-211	1364 ± 14
RM-8	32.7 ± 2.5
RM-7	3,994 ± 53
2-Iodo-Δ-8-THC (DMH)	164 ± 21
O,2-Propano-Δ-8-THC	262 ± 2.7
TMA	>2000
THC-7-oic Acid	>2000

this cell line, relative high concentrations of cannabinoids increased arachidonic acid release (Felder et al., 1992). Although this effect was concentration dependent, it was not stereoselective and was indistinguishable from the effect observed in nontransfected CHO cells. A nonreceptor mediated cannabinoid induced inhibition of arachidonate uptake probably contributes to this apparent increase in arachidonic acid release. Similarly, the cannabinoid receptor did not appear to be involved in the cannabinoid-induced, dose-dependent increase in internal calcium concentrations (Felder et al., 1992). This increase appeared to occur intracellularly and was not stereoselective nor found exclusively in the transfected cell line. The binding properties of the rat receptor expressed in the CHO were also determined (Felder et al., 1992) and while receptor densities were relatively low, the affinity of the receptor for the radioactive ligand was similar to that observed for the receptor in rat brain tissues (Table 2).

1.6.1.11   Mouse Cannabinoid Receptors

The cloning of the mouse cannabinoid receptor has not been reported but most (443 amino acids) of the coding sequence has been obtained by PCR amplification of mouse genomic DNA (Bonner, unpublished) and is quite similar to the rat sequence (95% DNA sequence identity, 99.5% amino acid identity).

1.6.99   Other Cannabinoid Receptors

Although the existence of subtypes of the cannabinoid receptor would offer some hope of disassociating the psychoactive responses of cannabinoid drugs from therapeutically useful analgetic or antiemetic responses, there is currently no evidence to suggest their existence. Low stringency Southern blot analysis of rat and human genomic DNAs provides evidence for a few weakly hybridizing genes but low stringency screening of genomic libraries has resulted in the isolation of clones with insignificant homology to the cannabinoid receptor (Lautens and Bonner, unpublished). Although they do not appear to be similar enough to the cannabinoid

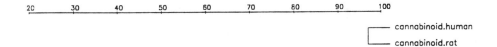

FIGURE 2. Phylogenetic tree of cannabinoid receptors. The phylogenetic tree was constructed according to the method of Feng and Doolittle (1990). The length of each "branch" of the tree correlates with the evolutionary distance between receptor subpopulations.

receptor to represent subtypes, the distantly related clones edg-1 and R334 might conceivably represent receptors for natural agonists which are related to the endogenous cannabinoid agonist.

REFERENCES

Abood, M. E. and Martin, B. R. (1992) Neurobiology of marijuana abuse. *Trends Pharmacol. Sci.* 13, 201–206.

Bidaut-Russell, M., Devane, W. A., and Howlett, A. C. (1990) Cannabinoid receptors and modulation of cyclic AMP accumulation in the rat brain. *J. Neurochem.* 55, 21–26.

Caulfield, M. P. and Brown, D. A. (1992) Cannabinoid receptor agonists inhibit Ca current in NG108–15 neuroblastoma cells via a Pertussis toxin-sensitive mechanism. *Brit. J. Pharmacol.* 106, 231–232.

Chang, M. C., Berkery, D., Schuel, R., Laychock, S. G., Zimmerman, A. M., Zimmerman, S., and Schuel, H. (1993) Evidence for a cannabinoid receptor in sea urchin sperm and its role in blockade of the acrosome reaction *Mol. Reprod. Dev,* in press.

D'Ambra, T. E., Estep, K. G., Bell, M. R. et al. (1992) Conformationally restrained analogues of Pravadoline: nanomolar potent, enantioselective (aminoalkyl)indole agonists of the cannabinoid receptor. *J. Med. Chem.* 35, 124–135.

Devane, W. A., Dysarz, F. A. I., Johnson, M. R., Melvin, L. S., and Howlett, A. C. (1988) Determination and characterization of a cannabinoid receptor in rat brain. *Mol. Pharmacol.* 34, 605–613.

Devane, W. A., Breuer, A., Sheskin, T., Jarbe, T. U. C., Eisen, M. S., and Mechoulam, R. (1992) A novel probe for the cannabinoid receptor. *J. Med. Chem.* 35, 2065–2069.

Devane, W. A., Hanuš, L., Brener, A., Pertwee, R. G., Stevenson, L. A., Grittin, G., Gibson, D., Mandelbaum, A., Etinger, A., and Mechonlam, R. (1992) Isolation and structure of a brain constituent that binds to the Cannabinoid receptor, *Science,* 258, 1946–1949.

Eidne, K. A., Zabavnik, J., Peters, T., Yoshida, S., Anderson, L., and Taylor, P. L. (1991) Cloning, sequencing and tissue distribution of a candidate G protein-coupled receptor from rat pituitary gland. *FEBS Lett.* 292, 243–248

Felder, C. C., Veluz, J. S., Williams, H. L., Briley, E. M., and Matsuda, L. A. (1992) Cannabinoid Agonists Stimulate both Receptor- and Non-Receptor-Mediated Signal Transduction Pathways in Cells Transfected with and Expressing Cannabinoid Receptor Clones. *Mol. Pharmacol.,* 42, 838–845.

Gaoni, Y. and Mechoulam, R. (1964) Isolation, structure, and partial synthesis of an active constituent of hashish. *J. Am. Chem. Soc.* 86, 1646–1647.

Gerard, C., Mollereau, C., Vassart, G., and Parmentier, M. (1990) Nucleotide sequence of a human cannabinoid receptor cDNA. *Nucl. Acids Res.* 18, 7142

Gerard, C. M., Mollereau C., Vassart, G., and Parmentier, M. (1991) Molecular cloning of a human cannabinoid receptor which is also expressed in testis. *Biochem. J.* 279, 129–134

Herkrenham, M., Lynn, A. B., Little, M. D., Johnson, M. R., Melvin, L. S., deCosta, B. R., and Rice, K. C. (1990) Cannabinoid receptor localization in brain. *Proc. Natl. Acad. Sci. U.S.A.* 87, 1932–1936.

Hla, T. and Maciag, T. (1990) An abundant transcript induced in differentiating human endothelial cells encodes a polypeptide with structural similarities to G-protein-coupled receptors. *J. Biol. Chem.* 265, 9308–9313

Hoehe, M. R., Caenazzo, L., Martinez, M. M., Hsieh, W.-T., Modi, W. S., Gershon, E. S., and Bonner, T. I. (1991) Genetic and Physical Mapping of the Human Cannabinoid Receptor Gene to Chromosome 6q14–15. *New Biologist* 3, 880–885

Howlett, A. C. (1985) Cannabinoid inhibition of adenylate cyclase. Biochemistry of the response in neuroblastoma cell membranes. *Mol. Pharmacol.* 27, 429–436.

Howlett, A. C. and Fleming, R. M. (1984) Cannabinoid inhibition of adenylate cyclase. Pharmacology of the response in neuroblastoma membranes. *Mol. Pharmacol.* 26, 532–538.

Howlett, A. C., Qualy, J. M., and Khachatrian, L. L. (1986) Involvement of G_i in the inhibition of adenylate cyclase by cannabimimetic drugs. *Mol. Pharmacol.* 29, 307–313.

Kaminski, N. E., Abood, M. E., Kessler, F. K., Martin, B. R., and Schatz, A. R. (1992) Identification of a functionally relevant cannabinoid receptors on mouse spleen cells that is involved in Cannabinoid-mediated immune modulation. *Mol. Pharmacol.*, 42, 736–742.

Jansen, E. M., Haycock, D. A., Ward, S. J., and Seybold, V. S. (1992) Distribution of cannabinoid receptors in rat brain determined with aminoalkylindoles. *Brain Res.* 575, 93–102.

Johnson, M. R. and Melvin, L. S. (1986) The discovery of nonclassical cannabinoid analgetics. In *Cannabinoids as Therapeutic Agents*, Mechoulam, R., Ed., CRC Press, Boca Raton, FL, pp. 121–145.

Mackie, K. and Hille, B. (1992) Cannabinoids inhibit N-type calcium channels in neuroblastoma-glioma cells. *Proc. Natl. Acad. Sci. U.S.A.* 89, 3825–3829.

Matsuda, L. A., Lolait, S. J., Brownstein, M. J., Young, A. C., and Bonner, T. I. (1990) Structure of a cannabinoid receptor and functional expression of the cloned cDNA. *Nature* 346, 561–564

Pacheco, M., Childers, S. R., Arnold, R., Casiano, F., and Ward, S. J. (1991) Aminoalkylindoles: actions on specific G-protein-linked receptors. *J. Pharmacol. Exp. Therap.* 257, 170–183.

Pertwee, R. G. (1988) The central neuropharmacology of psychotropic cannabinoids. *Pharmac. Ther.* 36, 189–261.

Thomas, B. F., Wei, X., and Martin, B. R. (1992) Characterization and autoradiographic localization of the cannabinoid binding site in rat brain using [3H] 11-OH-Δ^9-THC-DMH. *J. Pharmacol. Exp. Therap.*, 263, 1383–1390.

7

Cytokine Receptors

Richard Horuk, Ph.D.

1.7.0   Introduction

Interleukin-8 (IL-8) is a member of a family of related proinflammatory cytokines or chemokines that have a variety of biological properties including leukocyte chemotaxis and activation (Oppenheim et al., 1991; Yoshimura et al., 1987; Moser et al., 1990; Clark-Lewis et al., 1991; Moser et al., 1991). The family has been divided into two separate classes dependent on whether the first two conserved cysteine residues are separated by an intervening amino acid (C-X-C) or whether they are adjacent (C-C) (Schall, 1991). The C-X-C class members include Interleukin-8 (IL-8), melanocyte growth-stimulatory activity (MGSA), and platelet factor 4, while the C-C class includes RANTES and monocyte chemoattractant protein-1 (MCP-1) (Schall, 1991; Oppenheim et al., 1991). These chemokine superfamily members have been postulated to play a major role in both acute and chronic inflammation.

In vitro IL-8 is produced by a variety of cells including monocytes, macrophages, endothelial cells, and fibroblasts (Oppenheim et al., 1991). IL-8 is induced in response to a variety of stimuli, including cytokines (Oppenheim et al., 1991). IL-8 is a powerful chemoattractant and has been shown to be required for the transvenule migration of neutrophils at sites of inflammation (Huber et al., 1991). In addition IL-8 inhibits the adhesion of neutrophils to endothelial cells although in some circumstances enhancement of adhesion occurs (Oppenheim et al., 1991). Thus, *in vivo,* IL-8 appears to be an important regulator of neutrophil activation and migration.

The increasing availability of recombinant IL-8 in unlimited amounts has enabled rapid progress to be made both in solving the tertiary structure of the molecule and in the identification and molecular cloning of some of its receptors. Two recent studies, for example, employing alanine scanning mutagenesis and the synthesis of N-terminal truncated IL-8 variants, have identified residues of IL-8 that are important for receptor binding and the expression of biologic activity (Hébert et al., 1991; Clark-Lewis et al., 1991). Both studies suggest that the ELR region of amino acids near the N-terminus are essential for receptor binding and biological activity. In contrast, modification at the C-terminus end of the molecule appears to have little impact on binding or activity but there is evidence to suggest that it may play a role in maintaining the overall structural integrity of the molecule (Clark-Lewis et al., 1991).

The biology and physiology of IL-8 have been adequately reviewed elsewhere (Oppenheim et al., 1991) and the focus of this chapter will be to review the current information on the identification, molecular cloning, and biochemical characterization of the IL-8 receptors and their intracellular signalling mechanisms.

1.7.1   IL-8R-A Receptors

IL-8 elicits its biological effects by binding to high affinity cell surface receptors. A number of cell lines and tissues have been surveyed and IL-8 binding sites appear to be restricted to neutrophils and some monocytic-like cell lines including, HL-60, U937, and THP-1 cells (Grob et al., 1990). Based on chemical crosslinking experiments with IL-8 the molecular mass of the receptor has been reported as 58 to 69 kDa (Moser et al., 1991; Grob et al., 1990; Samanta et al., 1989). Of the several cell lines examined so far neutrophils appear to express the highest density of IL-8 binding sites: 20,000-75,000 sites per cell with a K_D of 0.18-4 nM (Moser et al., 1991; Grob et al., 1990; Samanta et al., 1989). Competition experiments have demonstrated that the related C-X-C ligands MGSA, NAP-2, and ENA-78 can compete for IL-8 binding to neutrophils suggesting a common receptor or receptors for these ligands on these cells (Moser et al., 1991; Samanta et al., 1989; Walz et al., 1991; Derynck et al., 1990). In contrast no displacement of IL-8 binding is observed with any of the C-C ligands, or the chemotactic agents fMLP, or C5a (Samanta et al., 1989).

It is now clear from studies on neutrophils that more than one type of IL-8 receptor exists (Moser et al., 1991). Several groups have shown that there is a single class of high affinity IL-8 binding sites on human neutrophils (Moser et al., 1991; Grob et al., 1990; Samanta et al., 1990; Besemer et al., 1989). However, competition studies with ^{125}I-IL-8 and unlabeled MGSA have shown that the IL-8 binding sites on neutrophils can be resolved into two classes, one that binds MGSA with high affinity and another that binds MGSA with low affinity (Moser et al., 1991). Furthermore, cross-desensitization studies of the intracellular Ca^{2+} response in neutrophils by IL-8 and MGSA corroborates the binding data discussed above. Initial stimulation by IL-8 blocks any subsequent Ca^{2+} response by MGSA, but initial stimulation by MGSA does not block a subsequent Ca^{2+} response to IL-8 (Moser et al., 1991).

1.7.1.1   Human IL-8R-A Receptors

cDNA clones encoding two human IL-8 receptors have been isolated and characterized (Holmes et al., 1991; Murphy and Tiffany, 1991). One receptor, termed the IL-8R-A, was isolated by Holmes et al. (Holmes et al., 1991) from a human neutrophil cDNA library by expression cloning and binding to ^{125}I-IL-8. Characterization of mammalian cells transfected with IL-8R-A shows that they encode a 350 amino acid receptor that binds IL-8 with high (i.e., nanomolar) affinity (Holmes et al., 1991). The receptor binds MGSA with a 200-fold reduced affinity (K_D = 450 nM). The transfected receptor is also able to signal in response to ligand binding. A transient increase in the intracellular Ca^{2+} concentration is found in response to its high affinity ligand, IL-8, but not in response to MGSA. This receptor has been localized to human chromosome 2q35 (Morris et al., 1992).

1.7.1.6   Rabbit IL-8R-A Receptors

A protein that was initially identified as the rabbit fMLP receptor has been shown by two separate groups to be the rabbit homolog of the human IL-8 receptor (Beckmann et al., 1991; Lee et al., 1992). The predicted amino acid sequence of the rabbit IL-8 receptor shows 84 and 73% homology to the human types A and B IL-8 receptors, respectively (Figure 1). Cells transfected with the rabbit IL-8 receptor bind human IL-

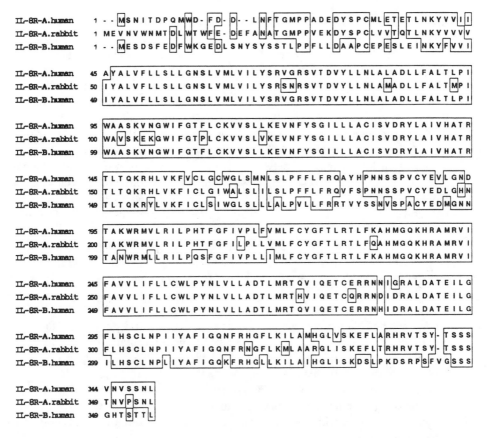

FIGURE 1. Alignment of the amino acid sequences of cytokine receptors. The sequences were obtained from commercial protein databases such as GenBank and EMBL. The references for each sequence are cited in the text.

8 with high affinity (K_D 1.9 to 3.6 n*M*) and were able to induce a rapid increase in the mobilization of intracellular free Ca^{2+} (Lee et al., 1992). Interestingly, cells transfected with the rabbit IL-8 receptor did not bind radiolabeled MGSA nor did the unlabeled MGSA cross compete for IL-8 binding (Beckmann et al., 1991). It is unlikely that the failure of MGSA to bind to the rabbit IL-8 receptor reflects species differences since experiments with rabbit neutrophils demonstrate that these cells respond chemotactically to MGSA (Beckmann et al., 1991). The failure of the rabbit IL-8 receptor to bind to MGSA suggests that it more closely resembles the human IL-8-RA receptor than the IL-8-RB receptor. Indeed, based on sequence homologies (Figure 1) the rabbit IL-8 receptor is more closely related to the type A receptor than to the type B receptor.

1.7.2 IL-8R-B Receptors

1.7.2.1 Human IL-8-B Receptors

A second, "low affinity" IL-8 receptor, termed the IL-8R-B, was isolated by Murphy and Tiffany (1991) from a neutrophil-like cell line, HL-60 (stimulated with dibutyryl cAMP). This receptor was reported to have a low affinity for IL-8 and for the ligands MGSA and NAP-2 (Murphy and Tiffany, 1991). No high affinity ligand for this receptor was identified. Blot hybridization with cDNA probes for the two cloned IL-8 receptors have shown that mRNA for both receptors is present in neutrophils (Holmes et al., 1991; Murphy and Tiffany, 1991). Since the sequences of the two receptors are quite similar (77% identical), it was suggested that the second receptor might be a high affinity receptor for one of the several cytokines that are related to IL-8 (Holmes et al., 1991; Murphy and Tiffany, 1991). Both receptors are members of the superfamily of seven transmembrane domain-containing proteins that bind to guanine nucleotide binding proteins (G proteins). An alignment of the encoded amino acid sequence of these two receptors is shown in Figure 1. Although the two receptors share a sequence identity of 77% overall, the sequence similarity is low in some regions (28% for the amino terminus) and high in others with two blocks of sequence identity of 105 and 64 amino acids. The DNA sequence encoding these blocks is highly conserved as well with only 8 of the 105 codons and 5 of the 64 codons differing between the two sequences (data not shown). This receptor has been localized to human chromosome 2q35 (Morris et al., 1992).

We have recently expressed both types of IL-8 receptors in mammalian cells, and performed binding assays with ^{125}I- IL-8 and ^{125}I-MGSA (Lee et al., 1992). In direct contrast to the studies of Murphy and Tiffany (Murphy and Tiffany, 1991), which suggested that the IL-8RB, expressed in *Xenopus* oocytes, is a low affinity receptor for IL-8 and MGSA, analysis of the binding properties and intracellular Ca^{2+} response of the two human IL-8 receptors have demonstrated that the affinity of IL-8 for the A and B IL-8 receptors is similar ($K_D = 1.2$ to 3.6 nM). These receptors differ considerably, however, in their affinities for the related ligand MGSA. IL-8-RB binds both IL-8 and MGSA with similar affinities ($K_D 2$ nM), while IL-8-RA binds MGSA with a 200-fold reduced affinity ($K_D = 450$ nM). The transfected receptors are also able to signal in response to ligand binding. A transient increase in the intracellular Ca^{2+} concentration is found for IL-8RB with both IL-8 and MGSA, while IL-8RA responds only to its high affinity ligand, IL-8.

1.7.3   IL-8R Pseudogenes

1.7.3.1   Human IL-8R-B Pseudogene

An apparent human IL-8 pseudogene has been identified and localized to chromosome 2q35 (Morris et al., 1992).

1.7.99   Other Cytokine Receptors

Recent work suggests that the IL-8 receptor in U937 cells, a human monocytic-like cell line, might be different to the cloned IL-8 receptors (Horuk et al., 1992). Several lines of evidence suggest that these cells possess a distinct type of IL-8 receptor. First, IL-8 competition studies yielded linear Scatchard plots for both IL-8 and MGSA binding in these cells suggesting binding to a single class of sites (Horuk et al., 1992). Second, although both MGSA and IL-8 were able to induce a rapid increase in the mobilization of intracellular free Ca^{2+} in U937 cells in a dose-responsive manner, cross-tolerization studies with MGSA and IL-8 indicate that pretreatment of cells with either ligand at maximally stimulating doses renders them unresponsive to further doses of either MGSA or IL-8 (Horuk et al., 1992). These data are consistent with the binding data discussed above and different from what is observed in neutrophils. Third, in contrast to the heterogeneity of crosslinked IL-8 labeling observed in other cells U937 cells labeled only one protein of molecular mass 69 kDa (Horuk et al., 1992). Finally, the cDNA for the cloned IL-8 receptor failed to hybridize to mRNA from U937 cells under

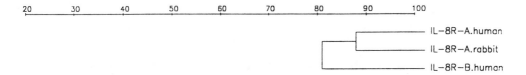

FIGURE 2. Phylogenetic tree of cytokine receptors. The phylogenetic tree was constructed according to the method of Feng and Doolittle (1990). The length of each "branch" of the tree correlates with the evolutionary distance between receptor subpopulations.

conditions where it hybridized to mRNA from neutrophils (Holmes et al., 1991; Lee et al., 1992). This observation together with the receptor binding, Ca^{2+} flux, and chemical crosslinking data raise the intriguing possibility that the IL-8 receptor detected in U937 cells is different from the cloned neutrophil IL-8 receptor.

In contrast to the cloned receptors described above, which appear to be specific for IL-8, an IL-8 binding protein which binds both IL-8 and the C-C chemokine MCP-1 with high affinity was recently described in human erythrocytes (Darbonne et al., 1991). Since the erythrocyte IL-8 receptor has been shown to efficiently remove IL-8 from whole blood (Darbonne et al., 1991), it has been postulated that erythrocytes might play a major role in inflammation by limiting these, and other, chemokines in the circulation.

In a recent communication, we characterized the ligand binding profile of the erythrocyte IL-8 receptor and found that unlike specific chemokine receptors on monocytes or neutrophils it binds nondiscriminately to a whole array of inflammatory peptides of the IL-8 superfamily (Neote et al., 1993). Based on its broad ligand specificity we have renamed the erythrocyte IL-8 receptor the multispecific chemokine (CK) receptor. So far no sequence information is available for the CK receptor, but initial studies demonstrate that it is a glycoprotein of molecular mass 39 kDa which is considerably smaller than the cloned IL-8 receptors (Horuk et al., 1993).

1.7.99.8 Mouse IL-8 Receptors

Screening of a mouse library with a cDNA probe for the type A human IL-8 receptor has identified a single nucleotide sequence that has around 70% homology with both type A and type B human IL-8 receptors (Lee, J., Wood, W. I., and Moore, M., unpublished observations). Whether this sequence codes for a putative mouse IL-8 receptor remains to be determined. Nevertheless, the failure to detect mouse versions of the human type A and type B receptors is somewhat surprising. These preliminary findings suggest that mice may have a different type of IL-8 receptor than those cloned from human cells. This is especially interesting in light of a recent communication claiming that the IL-8 gene is absent in mice and rats (Yoshimura, 1992).

REFERENCES

Beckmann, M. P., Munger, W. E., Koslosky, C., VandenBos, T., Price, V., Lyman, S., Gerard, N. P., Gerard, C., and Cerretti, D. P. (1991) Molecular characterization of the Interleukin-8 receptor. *Biochem. Biophys. Res. Commun.* 179, 784–789.

Besemer, J., Hujber, A., and Kuhn, B. (1989) Specific binding, internalization, and degradation of human neutrophil activating factor by human polymorphonuclear leukocytes. *J. Biol. Chem.* 264, 17409–17415.

Clark-Lewis, I., Moser, B., Walz, A., Baggiolini, M., Scott, G. J., and Aebersold, R. (1991) Chemical synthesis, purification, and characterization of two inflammatory proteins, neutrophil activating peptide 1 (interleukin-8) and neutrophil activating peptide 2. *Biochemistry* 30, 3128–3135.

Clark-Lewis, I., Schumacher, C., Baggiolini, M., and Moser, B. (1991) Structure-activity relationships of interleukin-8 determined using chemically synthesized analogs. Critical role of amino terminal residues and evidence for uncoupling of neutrophil chemotaxis, exocytosis, and receptor binding activities. *J. Biol. Chem.* 266, 23128–23134.

Darbonne, W. C., Rice, G. C., Mohler, M. A., Apple, T., Hébert, C. A., Valente, A. J., and Baker, J. B. (1991) Red blood cells are a sink for interleukin-8, a leukocyte chemotaxin. *J. Clin. Invest.* 88, 1362–1369.

Derynck, R., Han, J. H., Thomas, H. G., Wen, D., Samantha, A. J., Zachariae, C. O., Griffin, P. R., Brachmann, R., Balentien, E., Wong, W. L., Matsushima, K., and Richmond, A. (1990) Recombinant expression, biochemical characterization, and biological activities of the human MGSA/gro protein. *Biochemistry* 29, 10225–10233.

Grob, P. M., David, E., Warren, T. C., DeLeon, P., Farina, P. R., and Homon, C. A. (1990) Characterization of a receptor for human monocyte-derived neutrophil chemotactic factor/interleukin-8. *J. Biol. Chem.* 265, 8311–8316.

Hébert, C. A., Vitangcol, R. V., and Baker, J. B. (1991) Scanning mutagenesis of interleukin-8 identifies a cluster of residues required for receptor binding. *J. Biol. Chem.* 266, 18989–18994.

Holmes, W. E., Lee, J., W.-J., Kuang, Rice, G., and Wood, W. I. (1991) Structure and functional expression of a human interleukin-8 receptor. *Science* 253, 1278–1280.

Horuk, R., Colby, T. J., Darbonne, W. C., Schall, T. J., and Neote, K. (1993) The human erythrocyte inflammatory peptide (chemokine) receptor. Biochemical characterization, solubilization and development of a binding assay for the soluble receptor. *Biochemistry,* 32, 5733–5738.

Horuk, R., Yansura, D. G., Reilly, D., Spencer, S., Bourell, J., Henzel, W., Rice, G., and Unemori, E. (1993) Purification, receptor binding analysis and biological characterization of human melanoma growth stimulating activity (evidence for a novel MGSA receptor). *J. Biol. Chem.* 268, 541–546.

Huber, A. R., Kunkel, S. R., Todd, R. F., and Weiss, S. J. (1991) Regulation of transendothelial neutrophil migration by endogenous interleukin-8. *Science* 254, 99–102.

Lee, J., Horuk, R., Rice, G. C., Bennett, G. L., Camerato, T., and Wood, W. I. (1993) Characterization of two high affinity human interleukin-8 receptors. *J. Biol. Chem.* 267, 16283–16287.

Lee, J., Kuang, W.-J., Rice, G. C., and Wood, W. (1992) Characterization of complementary DNA clones encoding the rabbit IL-8 receptor. *J. Immunol.* 148, 1261–1264.

Morris, S. W., Nelson, N., Valentine, M. B., Shapiro, D. N., Look, A. T., Kozlowsky, C. J., Beckmann, M. P., and Cerretti, D. P. (1992) Assignment of the genes encoding human interleukin-8 receptor types 1 and 2 an an interleukin-8 receptor pseudogene to chromosome 2q35. *Genetics* 14, 685–691.

Moser, B., Clark-Lewis, I., Zwahlen, R., and Baggiolini, M. (1990) Neutrophil-activating properties of the melanoma growth-stimulatory activity. *J. Exp. Med.* 171, 1797–1802.

Moser, B., Schumacher, C., Tscharner, V., Clark-Lewis, I., and Baggiolini, M. (1991) Neutrophil-activating peptide 2 and gro/melanoma growth stimulatory activity interact with neutrophil-activating peptide 1/interleukin 8 receptors on human neutrophils. *J. Biol. Chem.* 266, 10666–10671.

Murphy, P. M. and Tiffany, H. L. (1991) Cloning of complementary DNA encoding a functional human interleukin-8 receptor. *Science* 253, 1280–1283.

Neote, K., Darbonne, W. C., Ogez, J., Horuk, R., and Schall, T. J. (1993) Identification of a promiscous inflammatory peptide receptor on the surface of red blood cells. *J. Biol. Chem.,* 268, 12247–12249.

Oppenheim, J. J., Zachariae, C. O. C., Mukaida, N., and Matsushima, K. (1991) Properties of the novel proinflammatory supergene "intercrine" family. *Annu. Rev. Immunol.* 9, 617–648.

Samanta, A. K., Oppenheim, J. J., and Matsushima, K. (1989) Identification and characterization of specific receptors for monocyte-derived neutrophil chemotactic factor (MDNCF) on human neutrophils. *J. Exp. Med.* 169, 1185–1189.

Samanta, A. J., Oppenheim, J. J., and Matsushima, K. (1990) Interleukin-8 (MDNCF) dynamically regulates its own receptor expression on human neutrophils. *J. Biol. Chem.* 265, 183–189.

Schall, T. J. (1991) Biology of the RANTES/SIS cytokine family. *Cytokine* 3, 165–183.

Walz, A., Burgener, R., Car, B., Baggiolini, M., Kunkel, S. L., and Strieter, R. M. (1991) Structure and neutrophil-activating properties of a novel inflammatory peptide (ENA-78) with homology to interleukin-8. *J. Exp. Med.* 174, 1355–1362.

Yoshimura, T. (1992) Neutrophil attractant protein1 (NAP1) is highly conserved in guinea pig but not in mouse or rat. *FASEB J.* 6, A1340.

Yoshimura, T. K., Matsushima, K., Tanaka, S., Robinson, E. A., Apella, E., Oppenheim, J. J., and Leonard, E. J. (1987) Purification of a human monocyte-derived neutrophil chemotactic factor that has peptide sequence similarity to other host defense cytokines. *Proc. Natl. Acad. Sci. U.S.A.* 84, 9233–9237.

8

Dopamine Receptors

Brian F. O'Dowd, Philip Seeman, and Susan R. George

1.8.0   Introduction

The cloning of the G protein-coupled receptor genes, including the adrenergic (AR), serotonergic, and dopamine receptors genes, has revealed that the proteins encoded by these genes are remarkably similar (reviewed by Dohlman et al., 1991, and O'Dowd, 1991). Characteristic in the structure of the proteins are regions of hydrophobic amino acid residues, which span the plasma membrane seven times. This conserved topography of seven transmembrane (TM) regions suggests a common tertiary structure which endows these proteins with an ability to interact with the ligand and transmit a signal through the cell membrane. The similarity of the shared structural features have been exploited to isolate other receptors (reviewed by Probst et al., 1992).

Until recently, only two dopamine receptors were defined on the basis of pharmacological and biochemical criteria: D1 dopamine receptors (D1 receptor), mediating the activation of adenylyl cyclase (Kebabian and Greengard, 1971) through the G protein, Gs, and D2 dopamine receptors (D2 receptor), inhibiting the activation of adenylyl cyclase through the G protein, G_i (Onali et al., 1985). Cloning studies have now determined that the physiological effects of dopamine are mediated by the products of at least five genes representing three more than had been identified by radioligand binding and functional studies alone. The primary structures of the dopamine receptors, D1, D2, D3, D4, and D5, encoded by these genes are aligned in Figure 1. It has also been revealed that some of these newly discovered dopamine receptors have G protein coupling specificities different than either the D1 receptor or D2 receptor, thus this larger number of receptors ensures a greater repertoire of effects within the cell.

The molecular biology strategies that have resulted in the identification and cloning of the five dopamine receptor proteins and their isoforms, have revealed the limitations of the available pharmacological tools for the characterization of these receptors. On the basis of the defining pharmacological characteristics, the receptors fall into two broad categories, "D1-like" and "D2-like". Among the subtypes, certain individual pharmacological properties have been identified, e.g., the higher affinity of D5 receptor for dopamine compared to D1 receptor, and the higher affinity of D4 receptor for the antagonist clozapine compared to D2 receptor. The development of dopamine receptor subtype-specific ligands has seriously lagged behind the rapidly evolving molecular biology, and there are few pharmacological strategies available to definitively and selectively label a single subtype of dopamine receptor within a tissue

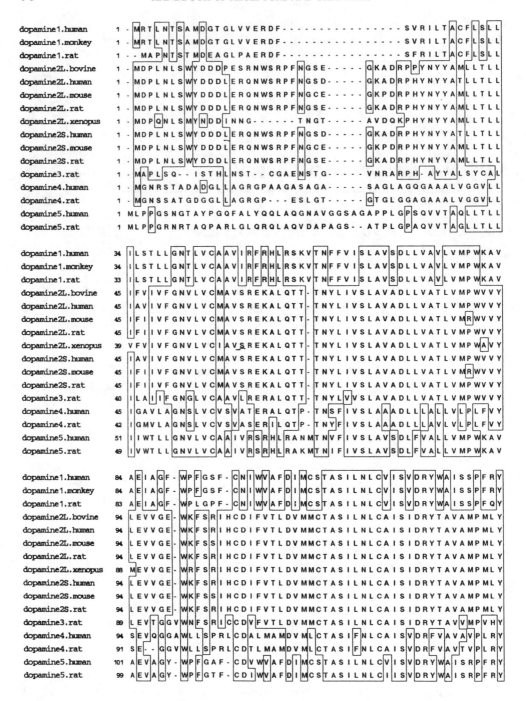

FIGURE 1. Alignment of the amino acid sequences of dopamine receptors. The sequences were obtained from commercial protein databases such as GenBank and EMBL. The references for each sequence are cited in the text.

expressing multiple subtypes. The precise mapping of the distribution of the dopamine receptors in brain, therefore, awaits the development of selective and specific pharmacological or immunological probes. There have been some preliminary reports of receptor protein localization using immunocytochemistry, as will be described in the following sections; however, the definitive mapping depends on the specificity and

```
dopamine1.human     132  E - - - R K M T P K A A F I L I S V A W T L S V L I S F I P V Q L S W H K A K P T S P S D - - - - -
dopamine1.monkey    132  E - - - R K M T P K A A F I L I S V A W T L S V L I S F I P V Q L S W H K A K P T S P S D - - - - -
dopamine1.rat       131  E - - - R K M T P K A A F I L I S V A W T L S V L I S F I P V Q L S W H K A K P T W P L D - - - - -
dopamine2L.bovine   143  N - - T R Y S S K R R V T V M I A I V W V L S F T I S - C P M L F G L N N - - T - D Q - - - - - -
dopamine2L.human    143  N - - T R Y S S K R R V T V M I S I V W V L S F T I S - C P L L F G L N N - - A - D Q - - - - - -
dopamine2L.mouse    143  N - - T R Y S S K R R V T V M I A I V W V L S F T I S - C P L L F G L N N - - T - D Q - - - - - -
dopamine2L.rat      143  N - - T R Y S S K R R V T V M I A I V W V L S F T I S - C P L L F G L N N - - T - D Q - - - - - -
dopamine2L.xenopus  137  N - - T R Y S S K R R V T V M I S V V W V L S F A I S - C P L L F G L N N - - T - G S - - - - - -
dopamine2S.human    143  N - - T R Y S S K R R V T V M I S I V W V L S F T I S - C P L L F G L N N - - A - D Q - - - - - -
dopamine2S.mouse    143  N - - T R Y S S K R R V T V M I A I V W V L S F T I S - C P L L F G L N N - - T - D Q - - - - - -
dopamine2S.rat      143  N - - T R Y S S K R R V T V M I A I V W V L S F T I S - C P L L F G L N N - - T - D Q - - - - - -
dopamine3.rat       139  Q H G T G Q S S C R R V A L M I T A V W V L A F A V S - C P L L F G F N T - - T G D P - - - - - -
dopamine4.human     144  N - - - R Q G G S R R Q L L L I G A T W L L S A A V A - A P V L C G L N D V R G R D P - - - - - - -
dopamine4.rat       139  N - - - Q Q G Q C - - Q L L L I A A T W L L S A A V A - A P V V C G L N D V P G R D P - - - - - - -
dopamine5.human     149  K - - - R K M T Q R M A L V M V G L A W T L S I L I S F I P V Q L N W H R D Q A A S W G G L D L P N
dopamine5.rat       147  E - - - R K M T Q R V A L V M V G L A W T L S I L I S F I P V Q L N W H R D K A G S Q G Q E G L - -

dopamine1.human     174  - - G N A T S L A E - - - - - - T - I D N C D S S L S R T Y A I S S S V I S F Y I P V A I M I V T Y
dopamine1.monkey    174  - - G N A T S L A E - - - - - - T - I D N C D S S L S R T Y A I S S S V I S F Y I P V A I M I V T Y
dopamine1.rat       173  - - G N F T S L E D - - - - - - T E D D N C D T R L S R T Y A I S S S L I S F Y I P V A I M I V T Y
dopamine2L.bovine   180  - - - - - - - - - - - - - - - - N E C I I A - N P A F V V Y S S I V S F Y V P F I V T L L V Y
dopamine2L.human    180  - - - - - - - - - - - - - - - - N E C I I A - N P A F V V Y S S I V S F Y V P F I V T L L V Y
dopamine2L.mouse    180  - - - - - - - - - - - - - - - - N E C I I A - N P A F V V Y S S I V S F Y V P F I V T L L V Y
dopamine2L.rat      180  - - - - - - - - - - - - - - - - N E C I I A - N P A F V V Y S S I V S F Y V P F I V T L L V Y
dopamine2L.xenopus  174  - - - - - - - - - - - - - - - - K V C I I D - N P A F V I Y S S I V S F Y V P F I V T L L V Y
dopamine2S.human    180  - - - - - - - - - - - - - - - - N E C I I A - N P A F V V Y S S I V S F Y V P F I V T L L V Y
dopamine2S.mouse    180  - - - - - - - - - - - - - - - - N E C I I A - N P A F V V Y S S I V S F Y V P F I V T L L V Y
dopamine2S.rat      180  - - - - - - - - - - - - - - - - N E C I I A - N P A F V V Y S S I V S F Y V P F I V T L L V Y
dopamine3.rat       179  - - - - - - - - - - - - - - - S I C S I S - N P D F V I Y S S V V S F Y V P F G V T V L V Y
dopamine4.human     183  - - - - - - - - - - - - - - A V C R L E - D R D Y V V Y S S V C S F F L P C P L M L L L Y
dopamine4.rat       176  - - - - - - - - - - - - T V C C L E - D R D Y V V Y S S I C S F F L P C P L M L L L Y
dopamine5.human     196  N L A N W T P W E E D F W E P D V N A E N C D S S L N R T Y A I S S S L I S F Y I P V A I M I V T Y
dopamine5.rat       192  - L S N G T P W E E G - W E L E G R T E N C D S S L N R T Y A I S S S L I S F Y I P V A I M I V T Y

dopamine1.human     215  T R I Y R I A Q K Q I R R I A A L E R A A V H A K N C Q T T T G N G K P V E - - - - - - - - - C S Q
dopamine1.monkey    215  T R I Y R I A Q K Q I R R I A A L E R A A V H A K N C Q T T T G N G K P V E - - - - - - - - - C S Q
dopamine1.rat       215  T S I Y R I A Q K Q I R R I S A L E R A A V H A K N C Q T T A G N G N P V E - - - - - - - - - C A Q
dopamine2L.bovine   210  I K I Y I V L R R R - R K R V N T K R S S R A F R A N L K A P L K G N C T H P E D M K L C T V I M K
dopamine2L.human    210  I K I Y I V L R R R - R K R V N T K R S S R A F R A H L R A P L K G N C T H P E D M K L C T V I M K
dopamine2L.mouse    210  I K I Y I V L R K R - R K R V N T K R S S R A F R A N L K T P L K G N C T H P E D M K L C T V I M K
dopamine2L.rat      210  I K I Y I V L R K R - R K R V N T K R S S R A F R A N L K T P L K G N C T H P E D M K L C T V I M K
dopamine2L.xenopus  204  V Q I Y I V L R K R - R K R V N T K R N S R G V A V D - - - A H K D K C T H P E D V K L C S V F V K
dopamine2S.human    210  I K I Y I V L R R R - R K R V N T K R S S R A F R A H L R A P L K - - - - - - - - - - - - - - - -
dopamine2S.mouse    210  I K I Y I V L R K R - R K R V N T K R S S R A F R A N L K T P L K - - - - - - - - - - - - - - - -
dopamine2S.rat      210  I K I Y I V L R K R - R K R V N T K R S S R A F R A N L K T P L K - - - - - - - - - - - - - - - -
dopamine3.rat       209  A R I Y I V L R Q R Q R K R I L T R Q N S Q C I S I R P G F P Q Q S S C L R L H P I R Q F S I R A R
dopamine4.human     213  W A T F R G L Q R W E V A R - R A K L H G R - - - - - - - - - - - - - - - - - - - - - - - - - -
dopamine4.rat       206  W A T F R G L R R W E A A R - H T K L H S R - - - - - - - - - - - - - - - - - - - - - - - - - -
dopamine5.human     246  T R I Y R I A Q V Q I R R I S S L E R A A E H A Q S C R S S A A - - - - - - - - - - - - - - - C A
dopamine5.rat       240  T R I Y R I A Q V Q I R R I S S L E R A A E H A Q S C R S R G A - - - - - - - - - - - - - - - Y E
```

FIGURE 1b.

lack of crossreactivity of the antisera used for the very closely related receptor
subtypes and isoforms. The distribution of the mRNA for the various receptors is quite
heterogenous, mapped by *in situ* hybridization histochemistry. Since this technology
relies on the design of specific oligonucleotide, cDNA or cRNA probes, the results are
more reliable; however, the considerable degree of homology between the DNA
sequences of the closely related receptor proteins make the validation of the specificity
and lack of crossreactivity of the hybridization probes also critical and equally essential for accurate interpretation of results.

```
dopamine1.human     256  PESSFKMSFKRE----TKVLKTLSVIMGVFVCCWLPFFILNCILPFCGSG
dopamine1.monkey    256  PESSFKMSFKRE----TKVLKTLSVIMGVFVCCWLPFFILNCILPFCGSG
dopamine1.rat       256  SESSFKMSFKRE----TKVLKTLSVIMGVFVCCWLPFFISNCMVPFCGSE
dopamine2L.bovine   259  SNGSFPVNRRRV----EAARRAQELEMEMLSSTSPPERTRYSPIP---PS
dopamine2L.human    259  SNGSFPVNRRRV----EAARRAQELEMEMLSSTSPPERTRYSPIP---PS
dopamine2L.mouse    259  SNGSFPVNRRRM----DPARRAQELEMEMLSSTSPPERTRYSPIP---PS
dopamine2L.rat      259  SNGSFPVNRRRM----DAARRAQELEMEMLSSTSPPERTRYSPIP---PS
dopamine2L.xenopus  250  SNGSFPADKKKVILVQEAGKHPEDMEMEMMSSTSPPEKTKHKSAS---PD
dopamine2S.human    242  ----------------EAARRAQELEMEMLSSTSPPERTRYSPIP---PS
dopamine2S.mouse    242  ----------------DPARRAQELEMEMLSSTSPPERTRYSPIP---PS
dopamine2S.rat      242  ----------------DAARRAQELEMEMLSSTSPPERTRYSPIP---PS
dopamine3.rat       259  FLSDATGQMEHI----EDKQYPQKCQDPLLSHLQPPSPGQTHGGL---KR
dopamine4.human     234  ----------APRRP--------SGPGPPSPT--PPAP---RL
dopamine4.rat       227  ----------APRRP--------SGPGPPVSD--PTQG---PL
dopamine5.human     280  PDTSLRASIKKE----TKVLKTLSVIMGVFVCCWLPFFILNCMVPFC-SG
dopamine5.rat       274  PDPSLRASIKKE----TKVFKTLSMIMGVFVCCWLPFFILNCMVPFCSSG

dopamine1.human     302  ETQ----PF-CIDSNTFD---VFVWFGWANSSLNPIIYAFNADFRKAFST
dopamine1.monkey    302  ETQ----PF-CIDSITFD---VFVWFGWANSSLNPIIYAFNADFRKAFST
dopamine1.rat       302  ETQ----PF-CIDSITFD---VFVWFGWANSSLNPIIYAFNADFQKAFST
dopamine2L.bovine   302  HHQ-----LTLPDPSHH----GLHSTPD-SPAKPEKNGHAKTVNPKIAK
dopamine2L.human    302  HHQ-----LTLPDPSHH----GLHSTPD-SPAKPEKNGHAKD-HPKIAK
dopamine2L.mouse    302  HHQ-----LTLPDPSHH----GLHSNPD-SPAKPEKNGHAKIVNPRIAK
dopamine2L.rat      302  HHQ-----LTLPDPSHH----GLHSNPD-SPAKPEKNGHAKIVNPRIAK
dopamine2L.xenopus  297  HNQ-----LAVPATSNQCKNASLTSPVE-SPYKAEKNGHPK-DSTKPAK
dopamine2S.human    273  HHQ-----LTLPDPSHH----GLHSTPD-SPAKPEKNGHAK-DHPKIAK
dopamine2S.mouse    273  HHQ-----LTLPDPSHH----GLHSNPD-SPAKPEKNGHAKIVNPRIAK
dopamine2S.rat      273  HHQ-----LTLPDPSHH----GLHSNPD-SPAKPEKNGHAKIVNPRIAK
dopamine3.rat       302  YYS------ICQDTALRH---PSLEGGAGMSPVERTRNSLSPTMAPKLSL
dopamine4.human     254  PQD------PCGPDCA------P--PAPG-LPPDP----CGSNCAP--PD
dopamine4.rat       247  FSD------CPPPSP------SLRTSPT-VSSRPESDLSQSPCSP--GC
dopamine5.human     325  HPEGPPAGFPCVSETTFD---VFVWFGWANSSLNPVIYAFNADFQKVFAQ
dopamine5.rat       320  DAEGPKTGFPCVSETTFD---IFVWFGWANSSLNPIIYAFNADFRKVFAQ

dopamine1.human     344  LLGCYRLC-PATNNAIETVSINNNGAAMFSSHHEPRGSISKECNLVYLIP
dopamine1.monkey    344  LLGCYRLC-PATNNAIETVSINNNGAAMFSSHHEPRGSISKECNLVYLIP
dopamine1.rat       344  LLGCYRLC-PTTNNAIETVSINNNGAVVFSSHHEPRGSISKDCNLVYLIP
dopamine2L.bovine   341  IFEIQSMPNGKTRTSLKTMS-RRKLSQQKEKKATQMLAIVLGVFIICWLP
dopamine2L.human    340  IFEIQTMPNGKTRTSLKTMS-RRKLSQQKEKKATQMLAIVLGVFIICWLP
dopamine2L.mouse    341  FFEIQTMPNGKTRTSLKTMS-RRKLSQQKEKKATQMLAIVLGVFIICWLP
dopamine2L.rat      341  FFEIQTMPNGKTRTSLKTMS-RRKLSQQKEKKATQMLAIVLGVFIICWLP
dopamine2L.xenopus  339  VFEIQSMPNGKTRTSIKTMS-KKKLSQHKEKKATQMLAIVLGVFIICWLP
dopamine2S.human    311  IFEIQTMPNGKTRTSLKTMS-RRKLSQQKEKKATQMLAIVLGVFIICWLP
dopamine2S.mouse    312  FFEIQTMPNGKTRTSLKTMS-RRKLSQQKEKKATQMLAIVLGVFIICWLP
dopamine2S.rat      312  FFEIQTMPNGKTRTSLKTMS-RRKLSQQKEKKATQMLAIVLGVFIICWLP
dopamine3.rat       343  --EVRKLSNGRLSTSLRLGPLQPRGVPLREKKATQMVVIVLGAFIVCWLP
dopamine4.human     283  AVRAAALPPQTPPQTRRR---RRAKITGRERKAMRVLPVVVGAFLLCWTP
dopamine4.rat       281  LLPDAALAQPPAPSSRRK---RGAKITGRERKAMRVLPVVVGAFLMCWTP
dopamine5.human     372  LLGCSHFC---SRTPVETVNISNE---LISYNQDIVFHKEIAAAYIHMMP
dopamine5.rat       367  LLGCSHFC---FRTPVQTVNISNE---LISYNQDTVFHKEIATAYVHMIP
```

FIGURE 1c.

The question of whether these dopamine receptor subtypes are colocalized within the same neurons has generated interest, because of the possibility of direct interactions between receptor types involving intracellular mechanisms. In a significant fraction of striatonigral neurons, there is a colocalization of functional D1 receptor, D2 receptor, and D3 receptor (Surmeier et al., 1992). In primate and human motor cortex, there is dense localization of the mRNA for D1 receptor, D2 receptor, and D5 receptor

```
dopamine1.human    393  HAVGSSE-DLKKEEAAGIARPLEKLSP--------ALSV-ILDYDTDVSL
dopamine1.monkey   393  HAVGSSE-DLKKEEAAGIARPLEKLSP--------ALSV-ILDYDTDVSL
dopamine1.rat      393  HAVGSSE-DLKKEEAGGIAKPLEKLSP--------ALSV-ILDYDTDVSL
dopamine2L.bovine  390  FFITHI--LNIHCD-CNIPPVLYSAFTWL-------------------
dopamine2L.human   389  FFITHI--LNIHCD-CNIPPVLYSAFTWL-------------------
dopamine2L.mouse   390  FFITHI--LNIHCD-CNIPPVLYSAFTWL-------------------
dopamine2L.rat     390  FFITHI--LNIHCD-CNIPPVLYSAFTWL-------------------
dopamine2L.xenopus 388  FFIIHI--LNMHCN-CNIPQALYSAFTWLGYVNSAVNPIIYTTFNVEFRK
dopamine2S.human   360  FFITHI--LNIHCD-CNIPPVLYSAFTWLGYVNSAVNPIIYTTFNIEFRK
dopamine2S.mouse   361  FFITHI--LNIHCD-CNIPPVLYSAFTWLGYVNSAVNPIIYTTFNIEFRK
dopamine2S.rat     361  FFITHI--LNIHCD-CNIPPVLYSAFTWLGYVNSAVNPIIYTTFNIEFRK
dopamine3.rat      391  FFLTHV--LNTHCQACHVSPELYRATTWLGYVNSALNPVIYTTFNVEFRK
dopamine4.human    330  FFVVHI--TQALCPACSVPPRLVSAVTWLGYVNSALNPVIYTVFNAEFRN
dopamine4.rat      328  FFVVHI--TRALCPACFVSPRLVSAVTWLGYVNSALNPIIYTIFNAEFRS
dopamine5.human    416  NAVTPGNREVDNDEEEGPFDRMFQIYQTSPDGDPVAESVWELDCEGEISL
dopamine5.rat      411  NAVSSGDREVGEEEEEGPFDHMSQISPTTPDGDLAAESVWELDCEEEVSL


dopamine1.human    433  EKIQPITQNGQHPT-
dopamine1.monkey   433  EKIQPITQNGQHPT-
dopamine1.rat      433  EKIQPVTHSGQHST-
dopamine2L.xenopus 435  AFIKILH--C-----
dopamine2S.human   407  AFLKILH--C-----
dopamine2S.mouse   408  AFMKILH--C-----
dopamine2S.rat     408  AFMKILH--C-----
dopamine3.rat      439  AFLKILS--C-----
dopamine4.human    378  VFRKALRACC-----
dopamine4.rat      376  VFRKTLRLRC-----
dopamine5.human    466  DKITPFTPNGFH---
dopamine5.rat      461  GKISPLTPNCFDKTA
```

FIGURE 1d.

in all layers (Huntley et al., 1992). Although the definitive double- or triple-labeling experiments were not performed, the intense localization of each of the mRNA types in virtually all the Betz cells of the cortex suggests that the RNA species are coexpressed within the same neurons. Although not yet available, such studies also require the confirmation that the translated proteins are present within the cells. The colocalization or the single expression of the various dopamine receptor RNA and protein species and their interaction at an as yet undefined locus (loci) may help to explain the often paradoxical ability of D1 and D2 receptor-like agonists to interact in an opposing or synergistic manner in different brain regions.

Little is currently known about the tertiary structures of the G-protein coupled receptors. However, evidence is available from electron diffraction studies in the bacteriorhodopsin system, a protein with a similar topography, which indicate the TM regions are in an α-helical formation (Henderson et al., 1990). The three-dimensional modeling of the G-protein receptors has been reviewed elsewhere (Findlay and Eliopoulos, 1992). More recently a mutagenesis approach has been utilized to identify which TM regions are aligned (Suryanarayama et al., 1992). The results indicate that amino acids in the seventh TM region of the α_2-AR (adrenergic) and β_2-AR lie next to the first TM region, indicating the arrangement shown in Figure 2. Many of the receptors in this family have cysteine residues in the cytoplasmic loops, and Strader and Dixon (as reviewed, 1992) suggested in reference to the β-AR that a disulfide bond exists joining two of these loops. The effect of a disulfide bridge would be to constrain the receptor into a conformation whereby TM3 would be in close proximity to TM5. A model of the ligand binding site of the adrenergic receptor (also based on mutagenesis studies, reviewed by Strader and Dixon, 1992) requires the TM3 and TM5 to be in close contact,

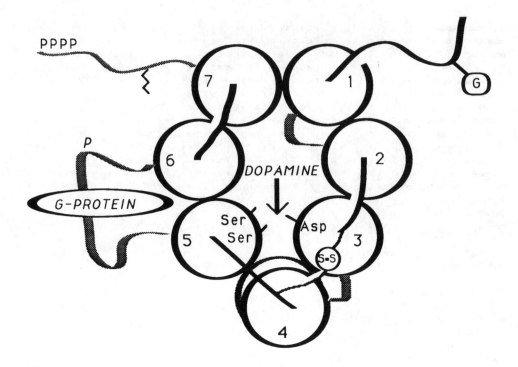

FIGURE 2. A model proposed for the interaction of the TM regions of the D1 receptor. The TM regions (1 to 7) are shown as viewed from the top of the receptor outside of the cell, in accordance with the model described by Suryanarayama et al., 1992. Shaded connecting loops are inside the cell, and a G protein is shown interacting with the loop formed between TM5 and TM6. Amino acids postulated to be directly involved in ligand binding, Asp in TM3, and the two serine residues in TM5 are shown. A disulfide bond may exist between TM3 and TM4. TM4 is shown offset from the vertical, as the presence of proline residues is postulated to put a tilt in some TM regions. The receptor is also shown as glycosylated, palmitoylated, and phosphorylated.

also indicated in Figure 2. The single polypeptide chain of each dopamine receptor has the seven hydrophobic regions, and a comparison of the deduced amino acid sequences of each of the five dopamine receptors reveals that 75 amino acids are identical among them (Niznik et al., 1992), as demonstrated in Figure 1.

A comparison of the deduced amino acid sequence of the human D1 receptor and D5 receptor with that of the D2, D3, and D4 receptors reveals that the main structural features that vary between these two receptor types are the sizes of the third cytoplasmic loop and the carboxy tail. The D1 receptor and D5 receptor more closely resemble the β_2-AR (with a shorter third loop and a longer tail) while the D2D, D3D, or D4D receptors resemble the α_2-AR (with the large third cytoplasmic loop and the short carboxy tail). A model of the D1 receptor is shown inserted in the membrane in Figure 3.

Although the functional significance of these differences in structure are poorly understood, the size of the loop may affect the specific functional interactions of receptors with the cytoplasmic transducing molecules, such as G proteins (O'Dowd et al., 1991). Inspection of the primary structures of all the dopamine receptors, aligned

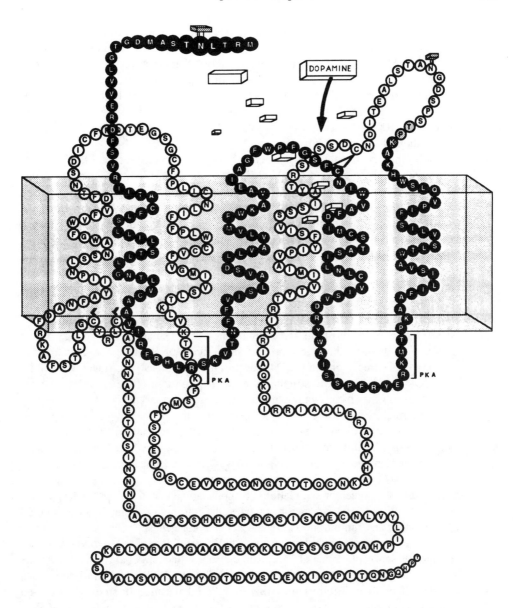

FIGURE 3. A three-dimensional representation of the human D1 receptor inserted in the plasma membrane. The TM regions are arranged in accordance with the model described by Suryanarayama et al., 1992. At the top of the figure dopamine molecules are represented (by boxes) entering the ligand binding pocket, formed by TM3 and TM5, in the receptor. The receptor is shown glycosylated (extracellular) and palmitoylated (intracellular) and with a disulfide bond between the top of TM3 and the top of TM4.

in Figure 1, reveals that putative sites of phosphorylation by PKA are common to all. These PKA sites are found both in the second and third cytoplasmic loops of the D1 receptor (Figure 3) and D5 receptor. Also, a putative site for palmitoylation at a cysteine residue in the cytoplasmic tail is present in all dopamine receptors. Important functional roles for phosphorylation and palmitoylation at the β_2-AR have been shown in coupling with Gs, and thus activation of adenylyl cylase (reviewed by O'Dowd et al., 1992a). Although no evidence exists to suggest that either the D1 or D5 receptor are substrates for the β_2-AR kinase (β-ARK), many serine and threonine residues also exist in the carboxy terminus of the D1 receptor and D5 receptor. In the β_2-AR, serine and threonine residues found in the equivalent positions represent sites of phosphorylation by β-ARK. β-ARK contributes to the desensitization of β_2-AR but its role in the densitization of D1 receptor is unknown.

In common with many of the genes in the G protein-linked receptor family, genes encoding the D1 receptor and the D5 receptor and the transcribed D5 receptor-related pseudogenes (Nguyen et al., 1991a) also lack introns in their coding regions. However, the D1 receptor does contain a small intron in the 5' untranslated region (Minowa et al., 1991). The coding sequence of the D2, D3, and D4 receptor genes are separated by introns, and analysis of the gene structure of the D2 receptor has revealed that the coding region is divided into 7 exons, with introns found following TM2, 3, 4, and before TM6 (Grandy et al., 1989., and Gandelman et al., 1991). Exon 1 of the human D2 receptor gene appears to be separated from exon 2 by a large intron of at least 38 kb, and exons 2 to 8 are clustered in approx 14 kb, but because of the size of intron 1, the entire gene is estimated to span a region in excess of 50 kb (O'Malley et al., 1991). One set of introns flank the coding region of the small 87 bp exon (29 amino acids) which distinguishes the long and short forms of D2 receptor. In the genes encoding the D3 receptor and the D4 receptor, introns are also positioned after TM2, TM3, TM4, and before TM6 strictly corresponding to those of the D2 receptor gene. Thus, all three genes have a common phylogenetic origin, and the positions of several of these introns are also shared with the opsin genes, in agreement with the suggestion that a common primordial ancestral gene also contained these intron positions (Grandy et al., 1989). The aligned sequences of the five dopamine receptors genes were compared and a phylogenetic tree (Figure 4) was constructed using the method of Feng and Doolittle (1990). The length of each branch correlates with the evolutionary distance between receptor populations, demonstrating that the dopamine receptors have differentiated into two major branches, the intron-containing D2, D3, and D4 receptor genes and the intronless D1 and D5 receptor genes.

Dopamine receptors have been suggested as candidate genes involved in disorders such as schizophrenia and Parkinson's disease (Seeman et al., 1987), Gilles de la Tourette syndrome (reviewed by Leckman et al., 1988), and drug addictions including alcoholism (Blum et al., 1990). Family linkage analysis studies have failed to implicate either the D2 or D4 receptor in the etiology of schizophrenia (Moises et al., 1991, Kennedy et al., 1992), and D1 and D2 receptors in Gilles de la Tourette syndrome (Gelernter et al., 1990., and Gelernter et al., 1992), or bipolar affective disorder (Nothen et al., 1992). An association study has reported a link between the D2 receptor gene and alcoholism. Using a TaqI polymorphism associated with the D2 receptor gene, Blum and colleagues examined DNA from alcoholic subjects. The alcoholics showed a higher frequency of the A1 allele (50%) than nonalcoholics (21%) (Blum et al., 1991). These researchers have also examined the binding characteristics of the D2 receptor in alcoholics and they found that among individuals with the A1 allele, the number of receptor binding sites was reduced in alcoholics compared to nonalcoholics (Noble et al., 1991). They have suggested that the D2 receptor is involved in conferring susceptibility to at least one type of alcoholism. Subsequently, Cloninger's group reported that they have replicated this association between the A1 allele and severe alcoholism and have suggested that the A1 allele may affect the severity of alcoholism (Parsian et al., 1991). Comings et al. (1990)

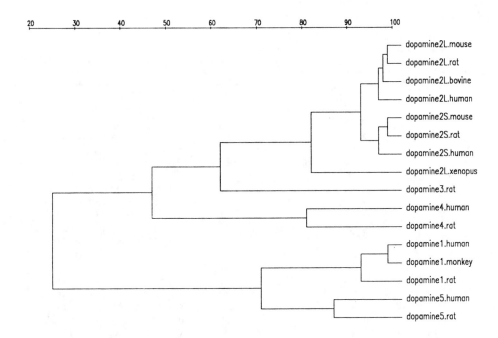

FIGURE 4. Phylogenetic tree of dopamine receptors. The phylogenetic tree was constructed according to the method of Feng and Doolittle (1990). The length of each "branch" of the tree correlates with the evolutionary distance between receptor subpopulations.

also suggested that the A1 allele is associated with a number of behavioral disorders including alcoholism, but that the D2 receptor acts as a modifying gene rather than the major gene causing the disorder. Two other reports have failed to support this allelic association between the A1 allele and alcoholism (Bolos et al., 1991, Gelernter et al., 1991), so at the time of writing the role of the D2 receptor gene in the various illnesses still needs further investigation.

In summary, the genes for five human dopamine receptors and the two pseudogenes have been cloned, and from the gene structures the protein structures of the dopamine receptors and the products of the pseudogenes have been deduced. Various characteristics of these dopamine receptors are summarized in Table 1. The distribution of each dopamine receptor has been mapped in rat and human brain, demonstrating that the mRNA encoding each dopamine receptor has a different pattern of distribution. The expression of these receptors in transfected cell lines is allowing a greater understanding of the pharmacology and function of the individual receptors. Regulation of each dopamine subtype is probably differentiated at the level of the individual neurons as well as in gross regions of the brain. Much remains to be understood concerning the physiological significance of the D3, D4, and D5 receptors revealed by cloning.

1.8.1   Dopamine D1 Receptors

The highest density of D1 receptor mRNA signal is seen in the caudate-putamen, nucleus accumbens, olfactory tubercle, and the islands of Calleja in rat brain (Fremeau et al., 1991; Mengod et al., 1991). In the human brain, too, high density D1 receptor

Table 1.

Receptor	Second messenger	Species cloned	Chromosomal location	a.a. Sequence	Accession number	Primary reference
D1	+AC	Human	5q31-34	446	P21728	Sunahara et al., 1990
		Monkey		446	—	Machida et al., 1992
		Rat		446	P18901	Zhou et al., 1990
D2 (Short)	−AC	Human	11q22-23	414	R11800	Brann and Stormann, 1991
		Rat		415	R05539	Bunzow et al., 1988
		Mouse		415	—	Mountayeur et al., 1991
D2 (Long)	−AC	Human	11q22-23	443	P14416	Selbie et al., 1989
		Bovine		444	X51657	Chio et al., 1990
		Rat		444	P13953	Selbie et al., 1989
		Mouse		444	—	Mountayeur et al., 1991
		Xenopus		442	P24628	Martens et al., 1991
D3		Human	3q13.3	—	—	Giros et al., 1990
		Rat		455	M69194	Sokoloff et al., 1990
D3 (TM3 del)		Rat		446	P19020	Giros et al., 1991
D3 (O2 del)		Rat		—		Giros et al., 1991
D4	−AC	Human	11q22-23	387	P21917	van Tol et al., 1991
		Rat		385	M84009	O'Malley et al., 1991
D5	+AC	Human	4q15.3	477	P21918	Sunahara et al., 1991
		Rat		475	P25115	Tiberi et al., 1991
D (pseudogene 1)		Human	1		M77185	Nguyen et al., 1991a
		Gorilla			—	Marchese et al., 1992
		Monkey			—	Marchese et al., 1992
D (pseudogene 2)		Human	2		M77186	Nguyen et al., 1991a
		Gorilla			—	Marchese et al., 1992
		Monkey			—	Marchese et al., 1992

mRNA is localized to caudate, putamen, and accumbens nuclei (Mengod et al., 1992). Within the caudate-putamen, there is a medial-to-lateral gradient of expression, with the D1 receptor mRNA observed in ~50% of neurons, the majority of which are medium-sized neurons (Weiner et al., 1991). Most of these neurons have been characterized as substance P neurons, and distinct from those containing D2 receptor mRNA (Le Moine et al., 1991). This latter finding is controversial, as the debate of whether D1 receptor and D2 receptor mRNA (or that of other receptor subtypes) is colocalized within the same neurons is not yet completely resolved. The distribution of D1 receptor in striatum is highly concentrated (Boyson et al., 1986; Ariano, 1988), but the cellular resolution achieved by the use of fluorescently derivatized ligands has shown a pattern of receptor binding analogous to the striatal mosaic patch compartment, and very similar to the dopamine terminal histofluorescence (Ariano et al., 1989).

Lesser amounts of D1 receptor mRNA are visualized in cells throughout the cerebral cortex, with the most prominent labeling occurring in the deeper laminae of the frontal cortex and other neocortical areas (Weiner et al., 1991; Fremeau et al., 1991). The colocalization of D1 receptor protein and mRNA is in keeping with previous studies supporting a primarily postsynaptic location of these receptors in cells of the cerebral cortex (Tassin et al., 1982). The hippocampus contains several discrete areas of neurons containing D1 receptor mRNA. D1 receptor mRNA is also detected in the claustrum, some amygdaloid and thalamic nuclei, and certain anterior lobules of the cerebellum. It is significant that in midbrain no D1 receptor mRNA hybridization signal is detectable in the substantia nigra or in the ventral tegmentum. The distribution of D1 receptor binding in rat and human brain corresponds exceedingly well with the reported mRNA distribution in many of the regions studied (Mengod et al., 1992; Wamsley et al., 1991), with the notable exceptions of the globus pallidus, substantia nigra pars reticularis and entopeduncular nucleus, where high levels of D1 receptor density are seen together with a complete lack of D1 receptor mRNA hybridization signal. In these regions, particularly in the substantia nigra, it appears that D1 receptors are localized to afferent nondopaminergic neurons projecting to the area, rather than on the dopamine neurons with cell bodies originating there. Cells expressing D1 receptor mRNA are present in a number of hypothalamic nuclei, that are implicated in a variety of endocrine, autonomic, and behavioral processes. Immunohistochemical localization of the D1 receptor in rat brain using a purified polyclonal antibody shows dense labeling throughout the basal ganglia, and heavily concentrated receptor density in cortical and subcortical neurons of limbic forebrain, including the hippocampal formation, amygdala, septum, thalamic, and hypothalamic nuclei, and the neurohypophysis (Huang et al., 1992). The extensive occurrence of the receptors, together with the widespread distribution of D1 receptor mRNA in the above cortical, limbic, and thalamic regions favor a critical role for the D1 receptor in cognitive, mnemonic, affective, motor, and neuroendocrine functions.

1.8.1.1 Human D1 Receptors

The rat cDNA clone was used to screen a human genomic library and a clone with an open reading frame on a single exon was obtained, which encoded a protein of 446 amino acids, the sequence of which is shown in Figures 1 and 3 (Sunahara et al., 1990). The selective D1 receptor antagonist, [^{3}H]SCH-23390 bound to transfected cell membranes with high affinity. These results were confirmed by other laboratories (Dearry et al., 1990.; Zhou et al., 1990; Monsma et al., 1990). The gene encoding the D1 receptor is located on chromosome 5, at q31-34, near the structurally homologous genes for the β_2-AR and the α_{1B}-AR genes (Sunahara et al., 1990).

Although the properties of the D1 receptor have been extensively characterized from radioligand and photoaffinity-labeling studies using selective ligands (Niznik et al.,1992), virtually nothing was known about regulatory mechanisms such as desensitization or the posttranslational modifications of the receptor. To address these important issues, a baculovirus expression-Sf9 cell system has been used to study human D1 receptor regulation and signal transduction. The initial step was to determine whether D1 receptor expressed in Sf9 cells was similar to the neuronal D1 receptor. Also, the strategy chosen was to express the human D1 receptor and a ten amino acid c-myc epitope-tagged D1 receptor (mD1 receptor) so that specific antibodies directed against the c-myc epitope in the amino terminal domain of mD1 receptor enabled receptor identification by immunoblotting, purification by immunoprecipitation, and whole cell mapping by immunofluorescence. The level of expression of D1 receptor in Sf9 cells showed selective, reversible, and saturable binding of [3H]SCH-23390 (Ng et al., 1993). Evidence of functional coupling of D1 receptor with endogenous Gs in Sf9 cells was also obtained, via the presence of agonist detected high-affinity states sensitive to Gpp(NH)p, and the presence of dopamine-stimulated adenylyl cylase.

The baculovirus-Sf9 cell expression system was used to study the posttranslational modifications of phosphorylation and palmitoylation in the D1 receptor (Ng et al., 1993). Following metabolic labeling of the Sf9 cells with ^{32}P, mD1 receptor was immunoprecipated. As shown in Figure 5, autoradiography of an SDS-PAGE of the immunoprecipated material identified a major phosphorylated species of 48 kDa. Following metabolic labeling of the cells with [^{3}H]palmitic acid, mD1 receptor was immunoprecipitated (Ng et al., 1993) and fluorograph of an SDS-PAGE of the immu-

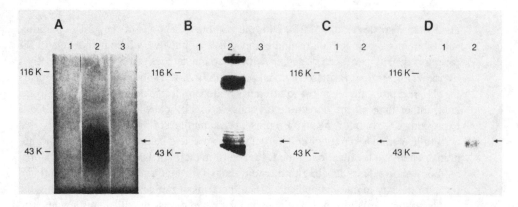

FIGURE 5. Biochemical characterization of the mD1 receptor. (A) Identification of photolabeled mD1 in Sf9 membranes. Membranes derived from mD1 receptor baculovirus infected cells at 48 h postinfection were photolabeled with [I^{125}] MAB and visualized by autoradiography. Lane 1, membranes from cells infected with wild-type baculovirus. Lane 2, photoincorporation of [I^{125}] MAB upon photolysis in absence, and Lane 3 in presence of 1mM (+) butaclamol. (B) Identification of immunoreactive mD1 receptor in Sf9 cell membranes. Membranes were prepared at 48 h postinfection from cells infected with the wild-type baculovirus (lane 1) or from cells expressing mD1 receptor (lane 2) or D1 receptor (lane 3). (C) Phosphorylation of mD1 receptor. The figure shows an autoradiogram of immunoprecipitated mD1 receptor prepared from Sf9 cells infected with wild-type baculovirus (lane 1) or mD1 receptor (lane 2). (D) Palmitoylation of mD1 receptor. The figure shows a fluograph of immunoprecipitated sample from cells prelabeled with [^{3}H] palmitic acid infected with the wild-type baculovirus (lane 1) or mD1 receptor (lane 2) (Ng et al., 1993).

noprecipitated material identified the presence of a single radiolabeled band (Figure 5). Palmitoylation of the D1 receptor may involve the cysteine residues (Cys 348, and Cys 352) in the carboxy tail, Figure 3, since it has previously been shown that Cys 341 located in an equivalent region of the β_2-AR is the site for palmitoylation (O'Dowd et al., 1989).

In Sf9 cells expressing mD1 receptor, preincubation with 1 mM dopamine for 15 min led to a 25% reduction of dopamine-stimulated adenylyl cylase, and immunofluorescent labeling showed that agonist treatment of the D1 receptor-expressing Sf9 cells induced a rapid redistribution of the surface D1 receptor and subsequent decrease of fluorescence. This response is consistent with rapid agonist-induced sequestration (Ng et al., 1993). The desensitization of the D1 receptor stimulated adenylyl cyclase was accompanied by an increase in D1 receptor phosphorylation. Thus, as was previously observed for the β_2-AR, an increase in phosphorylation of the D1 receptor appears to be involved in the process of rapid agonist-promoted desensitization.

The effect of mutations at serine residues in the TM 5 of the human D1 receptor have been studied (Pollock et al., 1992). Mutations in the D1 receptor, at serines 198, 199, and 202, all had an effect on the affinity of dopamine binding. Conversion of Ser 202 to an alanine in the TM 5 of the D1 receptor caused a 48-fold decrease in dopamine binding, but had little effect on the binding of other agonists. Mutations at Ser 198 and Ser 199 affected the binding of several agonists and antagonists, with agonists SKF-38393 and SKF-82958 decreasing in potency by 13- and 5-fold, respectively. Each of these mutated receptors was less efficient at stimulating adenylyl cyclase activity, confirming that these residues in TM5 of D1 receptor affect both ligand binding and signal transduction.

1.8.1.2 Monkey D1 Receptors

By using the rat D1 receptor as a probe, the cloning of the rhesus macaque monkey D1 receptor was achieved (Machida et al., 1992). The dopamine receptor in the monkey is 99.6% homologous to the human receptor.

1.8.1.10 Rat D1 Receptors

The cloning and sequence of the rat D1 receptor has been reported by Zhou et al. (1990).

1.8.2 Dopamine D2 Receptors

The D2 receptor exists as two isoforms, which are generated through alternative splicing of mRNA. The two forms of the receptor are identical except for a 29 amino acid insert in the large third cytoplasmic loop. Some differences between the two forms have been described in both their biological activities (Dal Toso et al., 1989) and tissue distributions (Neve et al., 1991), but not in their pharmacological profiles (Giros et al., 1989).

The levels of expression of the D2 receptor as revealed by receptor binding and autoradiography, closely matches the distribution of D2 receptor mRNA. The D2 receptor has been shown by numerous previous studies to be located predominantly in the basal ganglia, mesolimbic system, and the pituitary gland (Altar and Marien, 1987; Boyson et al., 1986; Dawson et al., 1986). The distribution of the D2 receptor and mRNA shows significant overlap with that of the D1 receptor in the traditional well-characterized dopamine neuronal systems in brain, yet has a distinct pattern of expression. The highest concentrations of D2 receptor mRNA has been localized to caudate-putamen, nucleus accumbens, and olfactory tubercle, with lower levels present in lateral septum, olfactory bulb, hypothalamus, and cortex (Meador-Woodruff et al., 1991; Chen et al., 1991; Weiner et al., 1991; Mengod et al., 1992). In contrast to D1 receptor mRNA, D2 receptor mRNA is present in high concentrations in the midbrain dopamine cell groups, confirming the D2 receptor identity of the autoreceptors in substantia nigra and ventral tegmentum area. Further differences in the distribution of D2 receptor mRNA compared to D1 receptor mRNA are evident in the regional localizations in hippocampus, amygdala, and pituitary. In the caudate-putamen, it has been estimated that approximately 33% of cells contain both D1 receptor and D2 receptor mRNA, using a technique of ultrathin sections (2.5 μm) to visualize adjacent serial sections through the same cells (Meador-Woodruff et al., 1991). However, although both types of RNA are seen in the medium-sized neurons (10 to 15 μm), only D2 receptor mRNA is present in the larger neurons (>20 μm) (Weiner et al., 1991). Using fluorescently labeled D2 receptor-selective ligands, the anatomical localization of D2 receptor in striatum is defined in the medium spiny neurons and the large-diameter cholinergic interneurons (Ariano et al., 1989).

The D2 receptor is far more abundant and expressed more widely than the D3 receptor. Overall, the abundance of D3 receptor by membrane binding and autoradiographic studies has been estimated to be 100-fold lower than D2 receptor (Levesque et al., 1992). Although mRNA levels may not necessarily reflect the level of protein expression, it appears that D3 receptor mRNA is also much less abundant than D2 receptor mRNA. D2 receptors are present in lower concentrations (two- to threefold) than D1 receptors as shown by the above radioligand and fluorescent binding studies.

D5 receptor density also appears to be far less than D1 receptor. The results of many of the studies that have used probes that fail to sufficiently discriminate between D1 and D5 receptors, and among D2, D3, and D4 receptors will need to be reinterpreted when more specific and selective methods are available.

The two isoforms of the D2 receptor that arise by alternative RNA splicing of the D2 receptor gene transcript are expressed together (Montmayeur et al., 1991; Rao et al., 1990; Chio et al., 1990). The relative amounts vary in all the regions where the D2 receptor is located, as indicated by the presence of the mRNA encoding for the two receptor forms (Dal Toso et al., 1989; Giros et al., 1989; Monsma et al., 1989). Using oligonucleotide probes that would specifically hybridize each of the two D2 receptor transcripts, it is seen that the distribution of the splice variants are heterogenous, suggesting that the regulation of the processing of the primary transcript varies among tissues and is tissue specific (Neve et al., 1991). The ratio of D2 receptor (long): D2 receptor (short) is 6 in pituitary, but 2 in striatum, and approximately equivalent in areas such as substantia nigra and midbrain, hippocampus, brainstem and parts of cerebral cortex. In fact, the relative regional proportions of D2 receptor (long): D2 receptor (short) mRNA remains invariant, even though total D2 receptor mRNA levels increase following denervation (Neve et al., 1991). In addition to these regional differences in tissue-specific RNA processing, it appears there may be regional differences also in the relative trancription rates of the D2 receptor mRNA, as indicated by the mapping of heteronuclear RNA by *in situ* hybridization, using an intronic sequence probe (Fox et al., 1992). Since intronic sequences are found only in the very short-lived heteronuclear RNA species, its distribution and quantitation may be representative of RNA turnover. The D2 receptor intronic probe was localized to caudate-putamen, nucleus accumbens, olfactory tubercle, substantia nigra, ventral tegmental area, and zona incerta, but not found in regions of lower D2 receptor mRNA expression, such as the globus pallidus, cerebral cortex, septum, and amygdala (Fox et al., 1992). Whether these differences relate to the limitations of the sensitivity of the method or are reflective of D2 receptor gene transcription remain to be confirmed.

The structurally similar genes shared by D2, D3, and D4 receptors suggested that the protein functions of these receptors would also be similar. The primary structures of each of these 3 receptors demonstrate that they contain 125 identical amino acids, equivalent to an overall homology of about 30%, (Figures 1 and 6). The common elements of their structures include a similarly sized amino terminus, a large third cytoplasmic loop and the very short carboxy tail with each receptor terminating at the same cysteine residue, Figure 1. Cysteine residues at equivalent positions in the β_2-AR and rhodopsin have been shown to be palmitolyated (O'Dowd et al., 1989., Ovchinikov et al., 1989). Within the transmembrane regions, the D2 receptor has a close homology to the α_2-AR (48%), and like this receptor it has a large third cytoplasmic loop (157 amino acids) and a very short cytoplasmic tail (16 amino acids). This type of structure appears to be important for those receptors, (including the m2 muscarinic receptor), which are required to couple to Gi proteins.

The site directed mutagenesis experiments of the β_2-AR, α_2-AR, and muscarinic receptors demonstrated that an aspartic acid residue in TM3 and several serine residues in TM5 are involved in ligand binding. Hibert et al., (1991) prepared a model of the D2 receptor suggesting that dopamine would also interact with Asp 114 in TM3 and serine residues 194 and 197 in TM5. Mutation analysis of the Asp 114 in the D2 receptor to either a Gly or Asn did result in a loss of binding of both agonists and antagonists (Mansour et al., 1992). Mutation of either serines 194 and 197 in TM 5 of the D2 receptor suggested that Ser 197 may be more important for dopaminergic binding. In the D2 receptor, mutation of the Ser 197 to Ala produced a greater decrease in agonist affinities, compared to the mutations of Ser 194. This result contrasts with the β_2-AR where the two serines in TM 5 are equally important for agonist binding. Interestingly, binding of one of the D2 receptor agonists N-0437, was unaffected by single mutations at either of the serines in TM 5 of D2 receptor. However, mutation of the dual serines produced a complete loss of binding suggesting that ligand N-0437 can interact equally well with either serine residue, and only in the case of both mutations is the binding of this agonist affected (Mansour et al., 1992).

1.8.2.1   Human D2 Receptors

The human D2 receptor (Stormann et al., 1990) gene is located on chromosome 11 at q22-23 (Grandy et al., 1989).

1.8.2.4   Bovine D2 Receptors

The cloning and sequence of the bovine D2 receptor has been reported by Chio et al. (1990).

1.8.2.10   Rat D2 Receptors

The first member of the dopamine receptor family to be cloned and characterized was the D2 receptor. Bunzow et al. (1988) screened a rat genomic library under low stringency conditions using the β_2-AR gene as a probe. One of the clones isolated shared homology with the β_2-AR. However, unlike most of the previously cloned members of this G protein-coupled receptor family, this genomic clone contained exons interrupted by introns in the coding sequence. A full-length clone was obtained from a rat cDNA library, and the longest open reading frame of this putative receptor cDNA clone was found to encode a 415 amino acid receptor protein. The characteristics of the receptor encoded by the cDNA were revealed by expression in two different cell lines, pituitary GH4C1 cells and Ltk-fibroblasts, where it has been shown to induce inhibition of adenylyl cyclase; membranes prepared from these cells specifically bound D2 receptor agonists and antagonists (Bunzow et al., 1988). Valler et al. (1990) showed that the role of the D2 receptor in transmembrane signaling is determined not only by the receptor itself but also by the membrane components of the cell. Effector systems coupled to the D2 receptor in GH4C1 cells and Ltk-fibroblasts were different. In the Ltk-cells the D2 receptor induced a rapid stimulation of inositol (1,4,5)-triphosphate and an increase of $[Ca^{2+}]$. However, in GH4Cl cells D2 receptor failed to affect phosphoinositide hydrolysis and induced a decrease of $[Ca^{2+}]$.

In an attempt to obtain this rat D2 receptor cDNA clone (and at the same time to investigate whether there were other genes encoding D2 receptor subtypes), PCR and the use of only one oligonucleotide with cDNA from a rat striatal library as template were used (O'Dowd et al., 1990). In addition to finding the clone described by Bunzow et al. (1988), another clone was isolated and identified as a D2 receptor variant because, compared to the first D2 receptor gene, it contained an additional 87-bp insert in the coding region. This result was confirmed by other research groups (Selbie et al., 1989, Grandy et al., 1989, Monsma et al., 1989, Dal Toso et al., 1989, Giros et al., 1989, Chio et al., 1990, O'Malley et al., 1990, Rao et al., 1990). When expressed in cultured mammalian cells, the receptor isoforms displayed similar pharmacology with selective D2 receptor ligands (Giros et al., 1989), and, when tested for their ability to inhibit adenylyl cyclase activity, the shorter form inhibits to a greater extent than the longer isoform (Dal Toso et al., 1989).

1.8.2.11 Mouse D2 Receptors

The cloning and sequence of the mouse D2 receptor has been reported by Mountayeur et al. (1991).

1.8.2.20 *Xenopus* D2 Receptors

The cloning and sequence of the *Xenopus* D2 receptor has been reported by Martens et al. (1991).

1.8.3 Dopamine D3 Receptors

1.8.3.1 Human D3 Receptors

The D3 receptor has been localized to chromosome 3 at q13.3 (Giros et al., 1991).

1.8.3.10 Rat D3 Receptors

A cDNA clone encoding the rat D3 dopamine receptor was obtained using a probe derived from the D2 receptor sequence (Sokoloff et al., 1990). When expressed in COS cells the D3 receptor was clearly distinguished from the D2 receptor by the binding patterns of several dopamine receptor agonists and antagonists. For instance, dopamine was 20 times more potent at the D3 receptor than the D2 receptor. Furthermore, the D3 receptor did not inhibit cAMP formation induced by forskolin and does not appear to interact with G_i (unlike the D2 receptor). Castro (University of Kent) has reported that the rat D3 receptor expressed in CHO cells exhibits high- and low-affinity sites for dopamine and does appear to be functionally coupled to a G-protein (Mills, 1992). Seabrook et al. (1992) have prepared stable transfection cell lines of GH4C1 expressing D3 receptors. These cells are known to contain 8 types of G proteins including Gi and Go subtypes. Their results indicated that D3 receptors do not couple to second messenger systems in these cell lines. Thus, it appears that the transduction mechanism involved with the D3 receptor is different from the D2 receptor.

Schwartz and co-workers have shown, using the PCR amplification of mRNAs from several areas of rat brain, the occurrence of two shorter transcripts of the D3 receptor receptor gene (Giros et al., 1991). Thus, as demonstrated previously for the D2 receptor gene which gave rise to the long and short forms of the D2 receptor, multiple transcripts also originate through alternative splicing mechanisms of the D3 receptor gene. One transcript, D3 (TM3 del) corresponded to a clone in which 113 bp (mainly corresponding to TM3) are deleted and a frameshift in the coding sequence gives rise to a stop codon shortly after the deletion. Thus, the encoded protein is a truncated receptor only 100 amino acids long. Snyder and co-workers (1991) also described this variant. However, their nucleotide sequence predicts a transcript of 109 amino acids compared to the 100 predicted by Giros et al., (1991). A second transcript called D3 (O2 del) revealed a clone with a 54-bp deletion which corresponds to the last 10 amino acids of the second extracellular loop and the first 8 amino acids of TM5 of the D3 receptor. The abundances of D3 receptor, D3 (O2 del), and D3 (TM3 del) mRNAs were evaluated in various rat brain regions, including cortex, substantia nigra, and striatum and found to have relative ratios of 5:1:4, respectively (Giros et al., 1991). The structure of D3 (TM3 del) makes it unlikely that this protein can function as a receptor. However, D3 (O2 del) may be compatible with the occurrence of seven transmembrane domains, although cell lines transfected with D3 (O2 del) failed to show any dopamine-binding activity (Giros et al., 1991).

The amino acid sequence of the rat D3 receptor is the most similar to that of the D2 receptor and together they share 200 identical amino acids, exhibiting an overall homology of 52% (Sokoloff et al., 1990). As expected, most of this sequence homology is found in the seven TM regions, but these receptors also share extensive homology in the first cytoplasmic loop and the extracytoplasmic loop joining TM regions 2 and 3 (Figures 1 and 6). The region reported to be important in determining G protein coupling such as the amino terminal and cytoplasmic regions of the large third loop region, and the short cytoplasmic tail, are also highly conserved in the D2 receptor and D3 receptor. However, the D3 receptor appears not to be involved in cAMP formation and thus far the evidence indicates that it does not interact with the same G protein(s) as the D2 receptor (Sokoloff et al., 1991, Seabrook et al., 1992). However, outside of these terminal regions important for G protein coupling, the third loops of the D2 receptor and D3 receptor are distinctly different. The rat D3 receptor structure contains a large third loop of similar size (162 amino acids compared to 156 amino acids) to the long form of the D2 receptor, and surprisingly the third loop of the human D3 receptor is 46 amino acids shorter than the equivalent loop in the rat (Giros et al., 1990).

The only information available regarding the pharmacological profile and expression of the D3 receptor was from transfected cells, since there was no specific pharmacological probe that could detect D3 receptor in brain membranes, until the recent identification of 7-OH-DPAT as a selective ligand for this receptor (Levesque et al., 1992). Among the D2 receptor-like receptors, 7-OH-DPAT was shown to have 100-fold lower affinity for D2 receptor, and 1000-fold lower affinity for D4 receptor receptors than for D3 receptor, to which it binds with subnanomolar affinity. These studies have also shed light on a very important characteristic of D3 receptor, namely the inability of guanine nucleotides to modulate dopamine binding to this receptor, unlike the effect of guanine nucleotides on agonist binding to D1 receptor and D2 receptor, extensively characterized previously (George et al., 1985; Grigoriadis et al., 1986). The demonstration of this property in neuronal membranes is critical to the understanding of D3 receptor function, since it verifies the lack of modulation by guanine nucleotides as physiological in brain, rather than the consequence of a deficiency of the cellular components of the particular expression system used.

The distribution of the D3 receptor in brain appears confined largely to the phylogenetically old regions, with highest expression in the ventral striatal complex, including the olfactory tubercle, islands of Calleja, and the anterior part of the nucleus accumbens, and in lobules 9 and 10 of cerebellum (Levesque et al., 1992). Interestingly, these very same regions transcribe D3 receptor mRNA, which is mainly detected in the areas receiving input from the A10 cell group, such as the nucleus accumbens, islands of Calleja, bed nucleus of the stria terminalis and other limbic areas, e.g., hippocampus and the mammillary nuclei (Bouthenet et al., 1991; Sokoloff et al., 1992; Levesque et al., 1992). The correlation of distribution of the D3 receptor and mRNA suggest localization to perikarya, dendrites, or short neurons in these regions, and correspond to some of the minor dopamine projection fields described (Swanson, 1982).

1.8.4   Dopamine D4 Receptors

1.8.4.1   Human D4 Receptors

Investigators used a probe from the D2 receptor gene to screen a cDNA library prepared from the human neuroblastoma cell line SK-N-MC (an mRNA signal distinct from D2 receptor was detected in this cell line). One isolated clone had a high degree of homology to the D2 receptor. A 5-kb clone obtained by screening a human genomic library, contained the full coding sequence (Van Tol et al., 1991). The sequence encoded by this gene, the D4 receptor, is shown in Figures 1 and 6. Initially, only low binding of [³H]spiperone was detected when the intron-containing human genomic clone encoding human D4 receptor was used to transfect COS cells. The original report stated that the low binding occurred because of a novel splice mechanism requirement of the third intron of the D4 receptor. It was reported that the COS cells were unable to process this type of splice junction, and a hybrid gene was constructed in which the problematic genomic sequence was replaced by cDNA sequence. Cells transfected with this construct expressed the receptor at 200 to 300 fmol/mg protein and with the tested agonists and antagonists displayed affinities similar to the D2 receptor. Clozapine, however, has an affinity for this receptor that is ten times higher than for either the D2 receptor or the D3 receptor.

More recently it has been reported that a 48-bp sequence in the putative third cytoplasmic loop exists (Van Tol et al., 1992; O'Dowd, 1992b) as a 2-, 3-, 4-, 5-, 6-, or 7-fold repeat, resulting in the formation of at least 6 different forms of the human D4 receptor gene (Van Tol et al., 1992). This loop, between transmembrane regions 5 and 6, can vary in size in this receptor from 106 to 186 amino acids long. The receptor with the sevenfold repeat in the third cytoplasmic loop is shown in Figure 6. The receptor with the fewest number of repeats is quite short when compared to the equivalent loop in D2 receptor and D3 receptor, each with 156 and 162 amino acids, respectively, while the receptor with the larger number of repeats has a loop larger than either D2 receptor or D3 receptor. These polymorphic variants are not found in the rat D4 receptor (O'Malley et al., 1991), suggesting that the emergence of these repeat sequences probably constitute a recent internal duplication events in the D4 receptor gene (O'Dowd, 1992b). The functional significance of these recent changes in the D4 receptor gene size are not understood, but the size of the third loop may affect interactions with cytoplasmic transducing molecules, such as G proteins, although the signal transduction pathway associated with the human D4 receptor has not been identified. It will be interesting to determine whether the recent changes evident in the structure of human D4 receptor have altered its coupling properties relative to the rat D4 receptor. The implications of these changes in the size of the loop may have resulted in a series of seven D4 receptors each with different properties in the human CNS. The frequencies of these various D4 receptor alleles in the population were calculated. In 150 individuals the frequencies were as follows: 0.01, 0.02, 0.04, 0.11, 0.14, and 0.68 for 5, 6, 3, 2, 7, and 4 repeats, respectively (Petronis et al., 1992). The identification of these various forms of the D4 receptor in the human population is the first example of such a polymorphic variation for the G protein-coupled receptors. Expression of three of these receptor variants showed different properties for the long form and the shorter forms with respect to clozapine and spiperone binding. The longest seven-repeat form of the D4 receptor appears to be less sensitive to changes in agonist or antagonist affinity induced by the addition of sodium chloride.

The deduced amino acid sequence of human D4 receptor reveals that this receptor shares almost equal identity with D2 receptor and D3 receptor (see Figure 1). The D2 receptor has 137 amino acids, and the D3 receptor 125 amino acids, identical with the shortest form of the D4 receptor (the D2 receptor and D3 receptor share 200 identical amino acids), demonstrating that D4 receptor is the most distantly related of this subgroup of the dopamine receptors. Again, the identical amino acids in D4 receptor that are shared with the D2 receptor and D3 receptor are concentrated in the seven TM regions, with significant identity also present in the short connecting loops and the cytoplasmic tail.

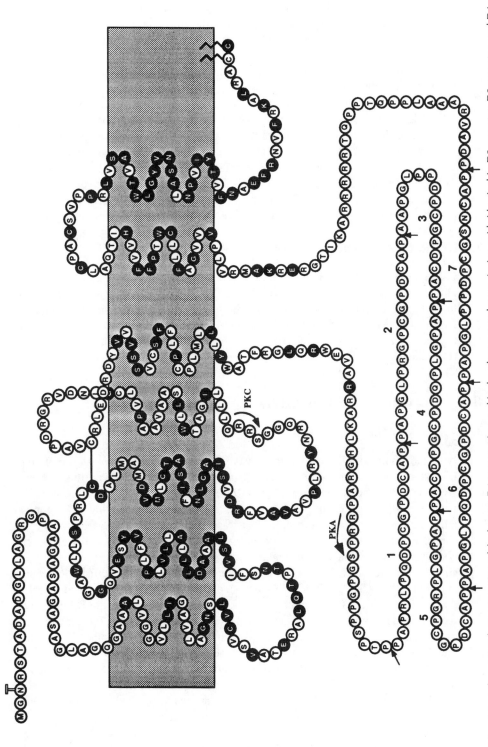

FIGURE 6. A two dimensional representation of the human D4 receptor inserted in the plasma membrane. Amino acids identical in D2 receptor, D3 receptor, and D4 receptor are indicated by highlighting with black circles. The seven repeat sequences in the third cytoplasmic loop of the D4 receptor are indicated by small arrows (1 to 7). The two larger arrows indicate the presence of the PKA and PKC sites. This receptor is also shown with a disulfide bond joining TM3 and the top of TM5.

The distribution of the D4 receptor RNA in human and rat brain is localized to only a few regions and is transcribed at low levels only (Van Tol et al., 1991; O'Malley et al., 1992). In rat, D4 receptor mRNA is found primarily in hypothalamic and thalamic areas, overlapping but not restricted to the dopaminergic A11, A12, A13, and A14 cell groups (O'Malley et al., 1992). Some positive cells are present in hippocampus, basal ganglia, and cerebral cortex as well. However, the transcription of the D4 receptor mRNA is at least 20 times more abundant in the cardiovascular system than in the CNS, with predominant expression in the vascular smooth muscle cells and myocytes of the proximal aorta, atria, and the outflow tract of the left ventricle (O'Malley et al., 1992). The mapping of the D4 receptor has not been satisfactorily accomplished because of the lack of D4 receptor selective pharmacological or immunological probes. Although D4 receptor have been shown to have relatively higher affinity for the drug clozapine than D2 receptor and D3 receptor by a factor of 10 (Van Tol et al., 1991), this difference is not sufficient to discriminate definitively among D2 receptor, D3 receptor, and D4 receptor binding. The D4 receptor gene is located on chromosome 11 the tip of the short arm near the teleomere (Gelernter et al., 1992).

1.8.4.10   Rat D4 Receptors

O'Malley et al. (1992) screened a rat library, also with a part of the D2 receptor cDNA clone, and sequence comparison revealed that one of the clones isolated was analogous to the human D4 receptor. For expression studies with the rat D4 receptor, the rat gene was transfected into fibroblast cell lines. Similar to the results reported for the human D4 receptor, the rat D4 receptor has a two- to threefold higher affinity for clozapine (O'Malley et al., 1992).

1.8.5   Dopamine D5 Receptors

1.8.5.1   Human D5 Receptors

After the cloning of D1 receptor, there was evidence to suggest the existence of other D1-like receptors expressed in the central nervous system and kidney (this evidence we reviewed previously in Niznik et al., 1992). The sequence of a clone (obtained by screening a human genomic library with a genomic clone encoding D1 receptor) indicated a gene encoding a receptor that was structurally similar to the D1 receptor (Sunahara et al., 1991). To identify the pharmacological profile of this clone, the coding sequence was again subcloned into a mammalian expression vector and transiently expressed in COS cells. The gene product bound the specific D1 receptor antagonist [^{3}H]SCH-23390 in a dose-dependent manner with an affinity identical to that observed for the D1 receptor. In general, this receptor, named the D5 dopamine receptor, binds drugs with a pharmacological profile similar to that of D1 receptor, but displays a 10-fold higher affinity for the endogenous agonist dopamine. To examine the ability of the D5 receptor to couple to adenylyl cyclase, COS cells transiently expressing D5 receptor were shown to exhibit an increase in cAMP formation in response to dopamine. Two other later reports described similar results (Weinshank et al., 1991; Grandy et al., 1991). Of note is the fact that Weinshank et al. (1991) have chosen to name the receptor D_{1b} rather than D5 receptor.

When the D5 receptor probe was used to assign the position of the D5 receptor gene, signals were observed on three different chromosomes: proximal 1q immediately adjacent to the centromere; 2p immediately adjacent to the centromere; and near the telomere in cytogenic band 4p 15.3 (Nguyen et al., 1991a) consistent with the fact that in addition to the functional gene D5 receptor there are also the two homologous pseudogenes in the human genome. Investigators (Eubanks et al., 1992; Polymeropoulos et al., 1991; Grandy et al., 1992) have ascertained that the signal on chromosome 4 is due to the presence of the D5 receptor gene.

1.8.5.10 Rat D5 Receptors

The rat D5 receptor (initially named the D_{1B} receptor) does not appear to display higher affinity for dopamine although it is structurally similar to the human D1 receptor (Tiberi et al., 1991). A comparison of the deduced amino acid structures of the rat D1 receptor and D5 receptor reveals an overall homology of 50% (represented by 250 identical amino acids), and the homology in the seven TM regions is particularly high at over 80% (129 identical amino acids) (Figure 1). Although the size of the third cytoplasmic loops and the carboxy tails are similar in the D1 receptor and D5 receptor, the sequence homologies in these areas are only 50 and 24%, respectively, any differences in the biology of these receptors may be as a result of sequence differences in these regions. The extra cytoplasmic loop between TM 4 and 5 differs considerably in size in the two receptors, with a larger loop of 41 amino acids in the D5 receptor compared to only 27 amino acids in the D1 receptor. This size difference may be related to the differences observed in the affinity for dopamine, although the rat D5 receptor with the lower affinity for dopamine, also has the larger sized loop (Tiberi et al., 1991). However, the amino acid sequence in this loop of the D5 receptor in human and rat is quite different.

The distribution of the dopamine D5 receptor mRNA is distinctly different from that of D1 receptor mRNA, with greatest concentrations in hippocampus, hypothalamus, and midbrain although there is significant overlap with the D1 mRNA pattern (Sunahara et al., 1991; Tiberi et al., 1991; Laurier and George, unpublished). Lesser concentrations are also evident in cerebral cortex, striatum, and spinal cord. Preliminary studies examining the localization of the D5 receptor reveals a pattern consistent with the above mRNA localization, and quite distinct fron D1 receptor and D2 receptor mRNA mapping (Ciliax et al., 1992). The D5 receptor was most dense in the olfactory tubercle, islands of Calleja, and outer molecular layer of the piriform cortex, with lesser concentrations present in the superficial layers of the other cortical areas, and in pyramidal neurons in layers 2, 3, and 5. The receptor was also present in hippocampus, and throughout the dorsal and ventral striatum. Receptor positive neurons are also present in the substantia nigra, ventral tegmental area, hypothalamus, raphe, and certain brainstem nuclei (Ciliax et al., 1992).

1.8.6 Dopamine Receptor Pseudogenes

1.8.6.1 Human Dopamine Receptor Pseudogene 1

In the search for other dopamine receptor genes in the human genome, oligonucleotides based on the D5 receptor sequence were used to amplify human genomic DNA in the PCR, and the resulting fragments were subcloned. The nucleotide sequence of one of these PCR-amplified clones shared 95% homology with the D5 receptor gene. This PCR-generated clone was used to screen a human genomic library. Sequence analysis of the probe-binding clones revealed a population of three different genes: D5 receptor and two genes initially named PG-1 and PG-2 (Nguyen et al., 1991a). PG-1 shares 98% homology with PG-2 and 94% with the D5 receptor gene. Each of the genes, PG-1 and PG-2, contained defects in the sequence that render these genes incapable of encoding functional receptors. Thus, the human genome contains two pseudogenes that share a high degree of homology with D5 receptor. Similar results have also been reported by Grandy et al. (1991). Other human pseudogenes reported in this family include a 5-HT 1D analog (Nguyen et al., 1993) and an interleukin-8 receptor analog gene (Ahuja et al., 1992).

The dopamine receptor pseudogene sequences, when compared with the coding region of the D5 receptor gene, shows a difference of 5.7 and 6.4%, respectively, at the nucleotide level, indicating a divergence from D5 receptor of about 20 million years ago. We have also found the D5 receptor gene and the two pseudogenes in both chimpanzee and gorilla (Marchese et al., 1992). The two pseudogenes share the deleterious mutations of stop codons and frame shifts and, therefore, were presumably derived from a common defective sequence. The 2% nucleotide difference between these pseudogenes suggests a divergence from a common precursor about 7 million years ago.

The extensive homology in the 5' and 3' untranslated regions of D5 receptor, 1D5 receptor analog, and 2D5 receptor analog indicated that these three genes were probably the most recent duplications in the G protein-receptor family. In our studies thus far we have found that the 3 genes are very homologous over a region of at least 6 kb. In the 5' untranslated region of these genes, following 1.9 kb of homologous sequences, a transition to nonhomologous sequences occurs in the D5 receptor gene. The transition site to nonhomologous sequences occurs at an Alu element that is present in both of the pseudogenes, but is not found in the D5 receptor gene (Marchese et al., 1992). It is possible that, given the position of the Alu element in the pseudogenes, that this element was involved in the duplication of the D5 receptor gene which produced the first pseudogene (Marchese et al., 1992). Gene duplications have played a major role in the evolution of this gene family. However, nothing was previously known about the molecular mechanism of their formation, and these three genes provide the opportunity to obtain information about the original sequences of the duplication junctions and to determine the method of gene duplication.

At issue is whether these pseudogenes are transcriptionally competent. By surveying for the presence of pseudogene cDNAs in human cDNA libraries, including the putamen, substantia nigra, hippocampus, and the nucleus accumbens (Nguyen et al., 1991b; Weinshank et al., 1991), transcription of the pseudogenes was detected in many areas of the human brain. Weinshank et al. (1991) reported that the pseudogene has a similar, but more restricted, pattern of distribution compared to the D5 receptor mRNA. High levels of pseudogene mRNA were detected in the frontal, cerebellar, and temporal cortex, choroid plexus, hippocampus, and brain stem. Lower levels of the pseudogene mRNA were found in temporal cortex and choroid plexus compared to the D5 receptor mRNA. Each group predicted that this mRNA is capable of forming a polypeptide of 154 amino acids, a receptor truncated as a result of a frame shift in the loop between TM3 and TM4. Transcription of the gorilla pseudogenes would produce truncated receptors of only 95 amino acids (Marchese et al., 1992). These peptides may have a biological role, although with only two or three TM regions they cannot function as G protein-coupled receptors.

1.8.7   Dopamine Receptor Pseudogene 2

1.8.7.1   Human Dopamine Receptor Pseudogene 2

The cloning and sequence of the human dopamine receptor pseudogene 2 has been reported (Nguyen et al., 1991; Grandy et al., 1991).

Table 2.   Characteristics of Dopamine Receptors

	D1 Group		D2 Group		
	D1	**D5**	**D2**	**D3**	**D4**
Found alone in:	Parathyroid	—	Ant. pit	—	—
Adenylate cyclase	Stimulates	Stimulates	Inhibits		Inhibits?
P-inositol metab.	—	—	May inhibit or stimulate		
Amino acids	446	475–477	414–444	400–446	387–467
Introns in gene	No	No	Yes	Yes	Yes
Chromosome (C)	5q34.5	4p15.2	11q22.5	3q	11p,ter
Pseudogenes		On C1,C2			
Polymorphism	EcoR1	None	Taq I		HinCII

Agonist K at high-affinity state (NaCl absent)

	nM	**nM**	**nM**	**nM**	**nM**
Apomorphine-(–)	0.7	?	0.66	?	~2
at low state	~450	363	127	73	
Bromocriptine*	~700	454	4.8	7.4	340
Dopamine	0.8	?	7.5	3.9	28
at low state	~2,000	228	4,300	73	450
Fenoldopam-R	1.6	?	2.8	?	?
at low state	~39	15	1,000	?	321
Pergolide	0.8	?	0.75	?	?
at low state	~800	918	60	0.6 to 2	?
(+)PHNO	75	?	0.98	?	~13
at low state	5,000	?	645	?	79
Quinpirole	1,900	?	3.9	?	~12
at low state	42,000	50,000	3,680	5 to 39	46
SKF 38393	1 to 6	?	157	?	?
at low state	~200	~100	8,800	5,000	1,800
SKF 81297	3.6	?	320	?	?
at low state	~60	?	8,000	?	?

Antagonist K (with or without Na)

	nM	**nM**	**nM**	**nM**	**nM**
Chlorpromazine	96	133	8.5	~6	37
Clozapine (no Na)	—	—	86	—	4.2
Clozapine (with Na)	172	250	182	479	9
Haloperidol	60	48	1.3	3 to 10	5.1
Raclopride	18,000	?	2.9	3.5	237
Remoxipride	?	?	447	2,300	3,685
SCH 23390	0.37	0.3	1,430	?	3,560
Spiperone	258	4,500	0.08	0.6	0.08
Sulpiride-S-(–)	34,500	77,270	14.7	~23	52
Sulpiride-R-(+)	25,800	28,636	868	422	?

* Bromocriptine has same affinity for high and low states.

Note: K values for D1 from Sunahara et al. (1990) and this lab.
K values for D3 from Sokoloff et al. (1990, 1992).
K values for D4 from Van Tol et al. (1991).
K values for D5 from Sunahara et al. (1991).

Acknowledgments

This work was supported by grants from the Medical Research Council of Canada, the National Institute for Drug Abuse (U.S.) in the form of grant 1 R01-DA07223-01, and the Addiction Research Foundation of Ontario.

REFERENCES

Ahuja, S. K., Ozcelik, T., Milatovitch, A., Francke, U., and Murphy, P. M. (1992) Molecular evolution of the human interleukin-8 receptor gene cluster. *Nature Genetics* 2, 33–36.

Altar, C. A. and Marien, M. R. J. (1987) Picomolar affinity of 125I-SCH 23982 for D1 receptors in brain demonstrated with digital substraction autoradiography. *Neuroscience* 7, 213–222.

Ariano, M. A. (1987) Comparison of dopamine binding sites in the rat superior cervical ganglion and caudate nucleus. *Brain Res.* 421, 245–254.

Ariano, M. A., Monsma, F. J., Barton, A. C., Kang, H. C., Haugland, R. P., and Sibley, D. R. (1989) Direct visualization and cellular localization of D1 and D2 dopamine recptors in rat forebrain by use of fluorescent ligands. *Proc. Natl. Acad. Sci. U.S.A.* 86, 8570–8574.

Blum, K., Noble, E. P., Sheridan, P. J., Montgomery, A., Ritchie, T., Jagadeeswaran, P., Nogami, H., Briggs, A. H., and Cohn, J. B. (1990) Allelic association of human dopamine D2 receptor genome in alcoholism. *JAMA* 263, 2055–2096.

Blum, K., Noble, E. P., Sheridan, P. J., Finley, O., Montgomery, A., Ritchie, T., Ozkaragoz, T., Fitch, R. J., Sadlack, F., Sheffield, D., Dahlmann, T., Halbardier, S., and Nogami, H. (1991) Association of the A1 allele of the D2 dopamine receptor gene with severe alcoholism. *Alcohol* 8, 409–416.

Bolos, A. M., Dean, M., Lucas-Derse, S., Ramsburg, M., Brown, G. L., and Goldman, D. (1990) Population and pedigree studies reveal a lack of association between the dopamine D2 receptor gene and alcoholism. *JAMA* 264, 3156–3160.

Boyson, S. J., McGonigle, P., and Molinoff, P. (1986) Quantitative autoradiographic localization of the D1 and D2 subtypes of dopamine receptors in rat brain. *J. Neurosci.* 6, 3177–3188.

Bouthenet, M. L., Souil, E., Martres, M. P., Sokoloff, P., Giros, B., and Schwartz, J. C. (1991) Localization of dopamine D3 receptor mRNA in the rat brain using in situ hybridization histochemistry: comparison with dopamine D2 receptor mRNA. *Brain Research* 564, 203–219.

Bunzow, J. R., Van Tol, H. M., Grandy, D. K., Albert, P., Salon, J., Chrisre, M., Machida, C. A., Neve, K. A., and Civelli, O. (1988) Cloning and expression of a rat D2 dopamine receptor cDNA. *Nature* 336, 783–787.

Ciliax, B. J., Nash, N., Heilman, C., and Levey, A. (1992) Immunocytochemical localization of dopamine D5 receptor in rat brain. *Soc. Neuroscience Abstr.* 18, #124.3.

Chen, J. F., Qin, Z. H., Szele, F., Bai, G., and Weiss, B. (1991) Neuronal localization and modulation of the D2 dopamine receptor mRNA in brain of normal mice and mice lesioned with 6-hydroxydopamine. *Neuropharmacology* 30, 927–941.

Chio, C. L., Hess, G. F., Graham, R. S., and Huff, R. M. (1990) A second form of D2-dopamine in rat and bovine caudate nucleus. *Nature* 343, 266–269.

Comings, D. E., Comings, B., Muhleman, D., Dietz, G., Shahbahrami, B., Tast, D., Knell, E., Kocsis, P., Baumgarten, R., Kovacs, B. W., Levy, D. L., Smith, M., Borison, R. L., Evans, D., Klein, D. N., MacMurray, J., Tosk, J. M., Sverd, J., Gysin, R., and Flanagan, S. D. (1991) The dopamine D2 receptor locus as a modifying gene in neuropsychiatric disorders. *JAMA* 266, 1793–1799.

Dal Toso, R., Sommer, B., Ewert, M., Herb, A., Pritchett, D. B., Bach, A., Shivers, B. D., and Seeburg, P. H. (1989) The dopamine D2 receptor: two molecular forms generated by alternative splicing. *EMBO J.* 8, 4025–4034.

Dawson, T. M., Barone, P., Sidhu, A., Wamsley, J. K., and Chase, T. N. (1986) Quantitative autoradiographic localization of D1 dopamine receptors in the rat brain: use of the iodinated ligand [125I]SCH 23982. *Neurosci. Lett.* 68, 261–266.

Dearry, A., Gingrich, J. A., Falardeau, P., Fremeau, R. T., Bates, M. D., and Caron, M. G. (1990) Molecular cloning and expression of the gene for a human D1 dopamine receptor. *Nature* 347, 72–75.

Dohlman, H. G., Thorner, J., Caron, M. G., and Lefkowitz, R. L. (1991) Model systems for the study of seven-transmembrane segment receptors. *Annu. Rev. Biochem.* 60, 653–688.

Eubanks, J. H., Altherr, M., Wagner-McPherson, C., Mc Pherson, J., Wasmuth, J. L., and Evans, G. A. (1992) Localization of the D5 Dopamine receptor gene to human chromosome 4p15.1–15.33 centromeric to the Huntingtons disease locus. *Genomics* 12, 510–516.

Feng, D. F. and Doolittle, R. F. (1990) Progressive alignment and phylogenetic tree construction of protein sequences. *Methods Enzymol.* 183, 375–387.

Findlay, J. and Eliopoulos, E. (1990). Three-dimensional modelling of G protein-linked receptors. *Trends Pharmacol. Sci.* 11, 492–499.

Fox, C. A., Thompson, R. C., Bunzow, J., Civelli, O., and Watson, S. J. (1992) The distribution of dopamine D2 receptor heteronuclear RNA by intronic in situ hybridization of the rat brain. *Soc. Neurosci. Abstr.* 18, #281.5.

Fremeau, R. T., Duncan, G. E., Fornaretto, M. G., Dearry, A., Gingrich, J. A., Breese, G. R., and Caron, M. G. (1991) Localization of D1 dopamine receptors mRNA in brain supports a role in cognitive, affective, and neuroendocrine aspects of dopaminergic transmission. *Proc. Natl. Acad. Sci. U.S.A.* 88, 3772–3776.

Gale, K., Guidotti, A., and Costa, E. (1977) Dopamine sensitive adenylate cyclase location in substantia nigra. *Science* 195, 503–505.

Gandelman, K.-Y., Harmon, S., Todd, R. D., and O'Malley, K. L. (1992) Analysis of the structure and expression of the human dopamine D2A receptor gene. *J. Neurochem.* 56, 1024–1029.

Gelernter, J., Van Tol, H. M., Civelli, O., and Kidd, K. (1992) The D4 dopamine receptor (DRD4) maps to distal 11p close to HRAS. *Genomics* 13, 208–210.

Gelernter, J., Kennedy, J. L., Grandy, D. K., Zhou, Q.-Y., Civelli, O., Pauls, D. L., Pakstis, A., Kurlan, R., Sunahara, R. K., Niznik, H., O'Dowd, B. F., Seeman, P., and Kidd, K. K. (1993) Exclusion of close linkage of Giles de la Tourette Syndrome to D1 dopamine receptor. *Arch. Gen. Psychiatry,*150, 449–453.

Gelernter, J., O'Malley, S., Risch, N., Kranzler, H. R., Krystal, J., Merikangas, K., Kennedy, J. L., and Kidd, K. (1991) No association between an allele at the D2 dopamine receptor gene (D2 receptor) and alcoholism. *JAMA* 266, 1801–1808.

Giros, B., Martres, M.-P., Pilon, C., Sokoloff, P., and Schwartz. (1991) Shorter variants of the dopamine receptor produced through various patterns of alternative splicing. *Biochem. Biophy. Res. Commun.* 176, 1584–1592.

Giros, B., Martres, M.-P., Sokoloff, P., and Schartz, J.-C. (1990) cDNA cloning of the human dopaminergic D3 receptor and chromosome identification. *C. R. Acad. Sci.* 311, 501–508.

Grandy, D. K., Marchionni, M. A., Makam, H., Stofko, R. E., Alfano, M., Frothingham, L., Fischer, J. B., Burke-Howie, K. J., Bunzow, J. B., Server, A. C., and Civelli, O. (1989a) Cloning of the cDNA and gene for a human D2 dopamine receptor. *Proc. Natl. Acad. Sci. U.S.A.* 86, 9762–9766.

Grandy, D. K., Zhou, Q.-Y., Allen, L., Litt, R., Magenis, E., Civelli, O., and Litt, M. (1990) A human D1 receptor gene is located on chromosome 5 at q35.1 and identifies an EcoR1 RFLP. *Am. J. Hum. Genet.* 47, 828–834.

Grandy, D. K., Litt, M., Allen, L., Bunzow, J. R., Marchionni, M., Makam, H., Reed, L., Magenis, R. E., and Civelli, O. (1989) The human dopamine D2 receptor gene is located on chromosome 11 at q22-q23 and identifies a TaqI RFLP. *Am. J. Hum. Genet.* 45, 778–785.

Grandy, D. K., Zhang, Y., Bouvier, C., Zhou, Q.-Y., Johnson, R. A., Allen, L., Buck, K., Bunzow, J. R., Salon, J., and Civelli, O. (1991) Multiple human D5 dopamine receptor genes: A functional receptor and two pseudogenes. *Proc. Natl. Acad. Sci. U.S.A.* 88, 9175–9179.

Hauge, X. Y., Grandy, D. K., Eubanks, J. H., Evans, G. A., Civelli, O., and Litt, M. (1991) Detection and characterization of additional DNA polymorphisms in the dopamine D2 receptor gene. *Genomics* 10, 527–530.

Henderson, R., Baldwin, J. M., Ceska, T. A., Zemlin, F., Beckmann, E., and Downing, K. H. (1990) Model for the structure of bacteriorhodopsin based on high-resolution electron cryumicroscopy. *J. Mol. Biol.* 213, 899–929.

Hibert, M. F., Trumpp-Kallmeyer, A., Bruinvels, A., and Hoflack, J. (1991) Three dimensional models of neurotransmitter G-binding protein-coupling receptors. *Mol. Pharmacol.* 40, 8.

Huntley, G. W., Morrison, J. H., Prikhozhan, A., and Sealfon, S. C. (1992) Localization of multiple dopamine receptor subtype mRNAs in human and monkey motor cortex and striatum. *Molec. Brain Res.* 15, 181–188.

Huang, O., Zhou, D., Chase, K., Gusella, J. F., Aronin, N., and DiFiglia, M. (1992) Immunohistochemical localization of the D1 dopamine receptor in the rat brain. *Soc. Neurosci. Abstr.* 18, 124.

Kebabian, J. W. and Greengard, P. (1971) Dopamine-sensitive adenylyl cyclase: possible role in synaptic transmission. *Science* 174, 1346–1349.

Kennedy, J. L., Sidenberg, D. G., Van Tol, H. M., and Kidd, K. K. (1991) A HincII RFLP in the human D4 dopamine receptor locus (D4 receptor). *Nucleic Acid Res.* 19, 5801.

Leckman, J. F., Riddle, M. A., and Cohen, D. J. (1988) Pathobiology of Tourettes syndrome. In *Tourettes Syndrome and Tic Disorders: Clinical Understanding and Treatment.* Cohen, D. J., Brunn, R. D., and Leckman, J. F., Eds., John Wiley & Sons, 103–118.

Le Moine, C., Normand, E., and Bloch, B. (1991) Phenotypical characterization of the rat striatal neurons expressing the D1 dopamine receptor gene. *Proc. Natl. Acad. Sci. U.S.A.* 88, 4205–4209.

Levesque, D., Diaz, J., Pilon, C., Martres, M.-P., Giros, B., Souil, E., Schott, D., Morgat, J.-L., Schwartz, J.-C., and Sokoloff, P. (1992) Identification, characterization and localization of the dopamine D3 receptor in rat brain using 7-[3H] hydroxy-N, N-di-n-propyl-2 aminotetralin. *Proc. Natl. Acad. Sci. U.S.A.* 89, 8155–8159.

Machida, C. A., Searles, R. P., Nipper, V., Brown, J. A., Kozell, L. B., and Neve, K. (1992) Molecular cloning and expression of the rhesus macaque D1 dopamine receptor gene. *Mol. Pharmacol.* 41, 652–659.

Mack, K. J., Todd, R. D., and O'Malley, K. (1991) The mouse dopamine D2A receptor gene:sequence homology with the rat and human genes and expression of alternative transcripts. *J. Neurochem.* 57, 795–801.

Mansour, A., Meador-Woodruff, J., Burke, S., Bunzow, J., Akil, H., Van Tol, H. M., Civelli, O., and Watson, S. J. Differential distribution of D2 and D4 dopamine receptor mRNAs in the rat brain: an in situ hybridization study. Abstract # 238.7. Society for Neuroscience (New Orleans, 1991).

Mansour, M., Meng, F., Meador-Woodruff, J. H., Taylor, L. P., Civelli, O., and Akil, H. (1992) Site-directed mutagenesis of the human dopamine D2 receptor. *Eur. J. Pharmacol.* 227, 205–214.

Martens, G. J. M., Molhuizen, H. O. F., Groneveld, D., and Roubos, E. W. (1991) Cloning and sequence analysis of brain cDNA encoding a *Xenopus* D2 dopamine receptor. *FEBS Lett.* 281, 85–89.

Marchese, A., Beischlag, T., Nguyen, T., Niznik, H., Grupp, L., Seeman, P., Seeman, P., and O'Dowd, B. F. Alu repeat involved in the duplication of the human D5 dopamine receptor gene. Abstract # 124.10. Society for Neuroscience (New Orleans, 1991).

Meador-Woodruff, J. H., Mansour, A., Healy, D. J., Kuehn, R., Zhou, Q. Y., Bunzow, J. R., Akil, H., Civelli, O., and Watson, S. J. (1991) Comparison of the distribution of D_1 and D_2 dopamine receptor mRNAs in rat brain. *Neuropsychopharmacology* 5, 231–242.

Meador-Woodruff, J. H., Mansour, A., Bunzow, J. R., Van Tol, H. M., Watson, S. J., and Civelli, O. (1989). Distribution of D2 dopamine receptor mRNA in rat brain. *Proc. Natl. Acad. Sci. U.S.A.* 86, 7625–7628.

Mengod, G., Martinez-Mir, M. I., Vilaro, M. T., and Palacios, J. M. (1989) Localization of the mRNA for the dopamine D2 receptor in the rat brain by in situ hybridization histochemistry. *Proc. Natl. Acad. Sci. U.S.A.* 86, 8560–8564.

Mengod, G., Villaro, M. T., Landwehrmeyer, G. B., Martinez-Mir, M. I., Niznik, H. B., Sunahara, R. K., Seeman, P., O'Dowd, B. F., Probst, A., and Palacios, J. M. (1992) Visualization of a dopamine D1, D2 and D3 receptor mRNA's in human and rat brain. *Neurochem. Int.* 20, 33S–43S.

Mills, A. (1992) Dopamine: from cinderella to holy grail. *Trends Pharmacol. Sci.* 13, 399–400.

Minowa, M. T., Minowa, T., Monsma, F. J., Sibley, D., and Mouradian, M. M. (1992) Characterization of the 5' flanking region of the human D-1 dopamine receptor gene. *Proc. Natl. Acad. Sci. U.S.A.* 89, 3045–3049.

Moises, H. W., Gelernter, J., Giuffra, L. A., Zarcone, V., Wetterberg, L., Civelli, O., Kidd, K., and Cavalli-Sforza, L. (1991) No linkage between D2 dopamine receptor gene region and schizophrenia. *Arch. Gen. Psychiatry* 48, 643–647.

Montmayeur, J. P., Bausero, P., Amlaiky, N., Maroteaux, L., Hen, R., and Borrelli, E. (1991) Differential expression of the mouse D2 dopamine receptor isoforms. *FEBS Lett.* 278, 239–243, 1991.

Monsma, F. J., Mahan, L. C., McVittie, L. D., Gerfen, C. R., and Sibley, D. R. (1990) Molecular cloning and expression of a dopamine receptor linked to adenyl cyclase activation. *Proc. Natl. Acad. Sci. U.S.A.* 87, 6723–6727.

Monsma, F. J., McVittie, L. D., Gerfen, C. R., Mahan, L. C., and Sibley, D. R. (1989) Multiple D2 dopamine receptors produced by alternative RNA splicing. *Nature* 342, 926–929.

Neve, K. A., Cox, B. A., Henningsen, R. A., Spanoyannis, A., and Neve, R. L. (1991) Pivotal role for aspartate-80 in the regulation of dopamine D2 receptor affinity for drugs and inhibition of adenylylcylase. *Mol. Pharmacol.* 39, 733–739.

Ng, G. Y. K., Mouillac, B., George, S., Caron, M., Dennis, M., Bouvier, M., and O'Dowd, B. F. (1993) Desensitization, phosphorylation and palmitoylation of the human D1 dopamine receptor. *Eur. J. Pharmacol.*, (submitted).

Nguyen, T., Marchese, A., Kennedy, J. L., Petronis, A., Peroutka, S. J., Wu, P. H., and O'Dowd, B. F. (1993) An Alu sequence interrupts a human 5-Hydroxytryptamine 1D receptor pseudogene. *Gene* 124, 295–230.

Nguyen, T., Bard, J., Jin, H., Taruscio, D., Ward, D. C., Kennedy, J. L., Weinshank, R., Seeman, P., and O'Dowd, B. F. (1991a) Human dopamine pseudogenes. *Gene* 109, 211–219.

Nguyen, T., Sunahara, R., Marchese, A., Van Tol, H. M., Seeman, P., and O'Dowd, B. F. (1991b) Transcription of a human dopamine D5 pseudogene. *Biochem. Biophys. Res. Commun.* 181, 16–21.

Niznik, H., Sunahara, R. K., Van Tol, H. M., Seeman, P., Weiner, D. M., Stormann, T. M., Brann, M. R., and O'Dowd, B. F. (1992) The D1 Dopamine receptor. In *Molecular Biology of Receptors Which Couple to G-Proteins*, Brann, M., Eds., Birkhauser, Boston.

Noble, E. P., Blum, K., Ritchie, T., Montgomery, A., and Sherida, P. J. (1991) Allelic association of the D_2 dopamine receptor gene with receptor-biding characteristics in alcoholism. *Arch. Gen. Psychiatry* 48, 648–654.

Nothen, M. M., Erdmann, J., Korner, J., Lanczik, M., Fritze, J., Fimmers, R., Grandy, D. K., O'Dowd, B. F., and Propping, P. (1992) Lack of association between dopamine D1 and D2 receptor genes and bipolar affective disorder. *Am. J. Psychiatry* 149, 199–201.

O'Dowd, B. F., Hnatowich, M., Caron, M., Lefkowitz, R. J., and Bouvier, M. (1989). Palmitoylation of the human β_2-adrenergic receptor. *J. Biol. Chem.* 264, 7564–7569.

O'Dowd, B. F., Lefkowitz, R. J., and Caron, M. (1989) Adrenergic and related G protein-coupled receptors, structure and function. *Annu. Rev. Neurosci.* 12, 67–83.

O'Dowd, B. F., Nguyen, T., Tirpak, A., Jarvie, K., Israel, Y., Seeman, P., and Niznik, H. (1990) Cloning of two additional catecholamine receptors from rat brain. *FEBS Lett.* 262, 8–12.

O'Dowd, B. F., Hnatowich, M., and Lefkowitz, R. J. (1991) Adrenergic and related G protein-coupled receptors, structure and function. In *Encyclopedia of Human Biology*. Volume 1, Academic Press, San Diego, 81–92.

O'Dowd, B. F., Collins, S., Bouvier, M., Caron, M. G., and Lefkowitz, R. L. (1992a) Structural, functional, and genetic aspects of receptors coupled to G-proteins. In *Molecular Biology of Receptors Which Couple to G-Proteins*, Brann, M. Eds., Birkhauser, Boston.

O'Dowd, B. F. (1992b). Repeat sequences in the gene encoding the human D4 dopamine receptor. *Gene* 118, 301–302.

O'Dowd, B. F. (1993) Structures of dopamine receptors. *J. Neurochem.*, 60, 804–816.

O'Malley, K. L., Mack, K. L., Gandelman, K., and Todd, R. (1990) Organization and expression of the rat D2a receptor gene: Identification of alternative transcripts and a variant donor splice site. *Biochemistry* 29, 1367–1371.

O'Malley, K. L., Harmon, S., Tang, L., and Todd, R. D. (1992) The rat D4 dopamine receptor. *New Biol.* 4, 137–146.

Onali, P., Olianas, M. C., and Gessa, G. L. (1985). Characterization of dopamine receptors mediating inhibition of adenylate cyclase activity in rat striatum. *Mol. Pharmacol.* 28, 138–145.

Ovchinikov, Y. A., Abdulaev, N. G., and Bogachuk, A. S. (1989) Two adjacent cysteine residues in the C-terminal cytoplasmic fragment of bovine rhodopsin are palmitoylated. *FEBS Lett.* 230, 1–5.

Parmentier, M., Libert, F., Schurmans, S., Schiffmann, S., Lefort, A., Eggerickx, D., Ledent, C., Mollereau, C., Gerard, C., Perret, J., Grootegoed, A., and Vassart, G. (1992) Expression of members of the putative olfactory receptor gene family in mammalian germ cells. *Nature* 355, 453–455.

Parsian, A., Todd, R. D., Devor, E. J., O'Malley, K. L., Suarez, B. K., Reich, T., and Cloninger, C. L. (1991) Alcoholism and alleles of the human D2 dopamine receptor locus. *Arch. Gen. Psychiatry* 48, 655–663.

Petronis, A., Van Tol, H., Livak, K. J., Sidenberg, D. G., Macciardi, F. M., and Kennedy, J. L. (1992) Genetic analysis of variable repeat sequence in DRD4 gene exon. Abstract # 779. Human Genetics meeting in San Francisco, 1992.

Pollock, N. J., Manelli, A. M., Hutchins, C. W., Steffey, M. E., Mac Kenzie, R. G., and Frail, D. (1992) Serine mutations in transmembrane V of the dopamine D1 receptor affect ligand interactions and receptor activation. *J. Biol. Chem.* 267, 17780–17786.

Polymeropoulos, M. H., Xiao, H., and Merril, C. R. (1991) The human D5 dopamine receptor (D5 receptor) maps on chromosome 4. *Genomics* 11, 777–778.

Probst, W. C., Snyder, L. A., Schuster, D. I., Brosius, J., and Sealfon, S. C. (1992) Sequence alignment of the G-protein coupled receptor superfamily. *DNA Cell Biol.* 11, 1–20.

Rao, D. D., Mc Kelvy, J., Kebabian, J., and Mac Kenzie, R. G. (1990) Two forms of the rat dopamine receptor as revealed by the polymerase chain reaction. *FEBS Lett.* 263, 18–22.

Sarkar, G., Kapelner, S., Grandy, D. K., Marchionni, M., Civelli, O., Sobell, J., Heston, L., and Sommer, S. S. (1991) Direct sequencing of the dopamine receptor (D2 receptor) in schizophrenics reveals three polymorphisms but no structural change in the receptor. *Genomics* 11, 8–14.

Seabrook, G. R., Patel, S., Marwood, R., Emms, F., Knowles, M. R., Freedman, S. B., and Mc Allister, G. (1992) Stable expression of human D3 dopamine receptors in GH4C1 pituitary cells. *FEBS Lett.* 312, 123–126.

Selbie, L. A., Hayes, G., and Shine, J. (1989) The major dopamine D2 receptor: molecular analysis of the human D2$_A$ subtype. *DNA* 8, 683–689.

Senogles, S. E., Amlaiky, N., Falardeau, P., and Caron, M. G. (1988) Purification and characterization of the D2-dopamine receptor from bovine anterior pituitary. *J. Biol. Chem.* 263, 18996–19002.

Seeman, P. (1987) Dopamine receptors and the dopamine hypothesis of schizophrenia. *Synapse* 1, 133–152.

Seeman, P., Niznik, H. B., Guan, H.-C., Booth, G., and Ulpian, C. (1989) Link between D1 and D2 dopamine receptors is reduced in schizophrenia and Huntington diseased brain. *Proc. Natl. Acad. Sci. U.S.A.* 86, 10156–10160.

Sibley, D. R. (1991) Cloning of a D3 receptor subtype expands dopamine receptor family. *Trends Pharmacol. Sci.* 12, 7–9.

Snyder, L. A., Roberts, J. L., and Sealfon, S. C. (1991) Alternative transcripts of the rat and human dopamine D3 receptor. *Biochem. Biophys. Res. Commun.* 180, 1031–1035.

Sokoloff, P., Giros, B., Martres, M.-P., Andrieux, M., Besancon, A., Pilon, C., Bouthenet, M. L., Souil, E., and Schwartz, J.-C. (1992) Localization and function of the D3 dopamine receptor. *Arnzeimittel Forschung* 42, 224–230.

Sokoloff, P., Giros, B., Martres, M.-P., Bouthenet, M. L., and Schwartz, J.-C. (1990). Molecular cloning and characterization of a novel dopamine receptor (D3) as a target for neuroleptics. *Nature* 347, 146–151.

Strader, C. D. and Dixon, R. A. (1992) Genetic analysis of the β_2-AR receptor. In *Molecular Biology of Receptors Which Couple to G-Proteins*, Brann, M., Ed., Birkhauser, Boston.

Stormann, T. M., Gdula, D. C., Weiner, D. M., and Brann, M. (1990) Molecular cloning and expression of a dopamine D2 receptor from human retina. *Mol. Pharm.,* 37, 1–6.

Sunahara, R. K., Niznik, H., Weiner, D. M., Stormann, T., Brann, M. R., Kennedy, J. L., Gelernter, J. E., Rozmahel, R., Yang, Y., Israel, Y., Seeman, P., and O'Dowd, B. F. (1990). Human dopamine D1 receptor encoded by an intronless gene on chromosome 5. *Nature* 347, 80–83.

Sunahara, R. K., Guan, H.-C., O'Dowd, B. F., Seeman, P., Laurier, L. G., George, S. R., Torchia, J., Van Tol, H., and Niznik, H. (1991) Cloning of a human dopamine receptor gene (D5) with higher affinity for dopamine than D1. *Nature* 350, 614–619.

Surmeier, D. J., Eberwine, J., Wilson, C. J., Cao, Y., Stefani, A., and Kitai, S. T. (1992) Dopamine receptor subtypes colocalize in rat striatonigral neurons. *Proc. Natl. Acad. Sci. U.S.A.* 89, 10178–10182.

Swanson, L. W. (1982) The projections of the ventral tegmental area and adjacent regions: a combined fluorescent retrograde tracer and immunofluorescence study in the rat. *Brain Res. Bull.* 9, 321–353.

Suryanarayama, S., Van Zastrow, M., and Kobilka, B. K. (1992) Identification of intramolecular interactions in adrenergic receptors. *J. Biol. Chem.* 267, 21991–21994.

Tassin, J. P., Simon, D., Herve, G., Blanc, M., Le Moal, M., Glowinski, J., and Bockaert, J. (1982) Non dopaminergic fibres may regulate dopamine-sensitive adenylate cyclase in prefrontal cortex and nucleus accumbens. *Nature* 295, 696–698.

Tiberi, M., Jarvie, K. R., Silva, C., Falardeau, P., Gingrich, J. A., Godinot, N., Bertrand, L., Yang-Feng, T. L., Fremeau, R. T., and Caron, M. G. (1991) Cloning, molecular characterization, and chromosomal assignment of a gene encoding a second D1 dopamine receptor subtype: Differential expression pattern in rat brain compared with the D1A receptor. *Proc. Natl. Acad. Sci. U.S.A.* 88, 7491–7495.

Weiner, D. M., Levey, A. I., Sunahara, R. K., Niznik, H., Seeman, P., O'Dowd, B. F., and Brann, M. R. (1991) Dopamine D1 and D2 receptor mRNA expression in rat brain. *Proc. Natl. Acad. Sci. U.S.A.* 88, 1859–1863.

Weinshank, R. L., Adham, N., Macchi, M., Olsen, M. A., Branchek, T. A., and Hartig, P. R. (1991) Molecular cloning and characterization of a high affinity dopamine receptor (D1b) and its pseudogene. *J. Biol. Chem.* 266, 22427–22435.

Valler, L., Muca, C., Magni, M., Albert, P., Bunzow, J., Meldolesi, J., and Civelli, O. (1992) Differential coupling of dopaminergic D2 receptors expressed in different cell lines. *J. Biol. Chem.* 265, 10320–10326.

Van Tol, H. M., Bunzow, J. R., Guan, H.-G., Sunahara, R. K., Seeman, P., Niznik, H., and Civelli, O. (1991) Cloning of the gene for a human dopamine D4 receptor with high affinity for the antipsychotic clozapine. *Nature* 350, 610–614.

Van Tol, H. M., Wu, C. M., Guan, H.-G., Ohara, K., Bunzow, J., Civelli, O., Kennedy, J., Seeman, P., Niznik, H., and Jovanovic, V. (1992) Multiple dopamine D4 variants in the human population. *Nature* 358, 149–152.

Yang-Feng, T. L., Xue, F., Zhong, W., Cotecchia, S., Frielle, T., Caron, M. G., and Lefkowitz, R. L. (1990) Chromosomal organization of adrenergic receptor genes. *Proc. Natl. Acad. Sci. U.S.A.* 87, 1516–1520.

Zhou, Q.-Y., Grandy, D. K., Thambi, L., Kushner, J. A., Van Tol, H. M., Cone, R., Pribnow, D., Salon, J., Bunzow, J. R., and Civelli, O. (1990) Cloning and expression of human and rat D1 dopamine receptors. *Nature* 347, 76–79.

9

Endothelin Receptors

Stephen J. Peroutka, M.D., Ph.D.

1.9.0   Introduction

The endothelins (ETs) are a family of peptides with potent vasoconstrictor effects. The molecules, designated ET-1, ET-2, and ET-3, were identified initially in endothelial cells but are now known to be present in many nonvascular tissues. Each molecule is 21 amino acids in length. However, the molecules display a varied and diverse set of pharmacological properties which have been reviewed in detail by Sokolovsky (1992).

1.9.1   Endothelin$_A$ Receptors

1.9.1.1   Human ET$_A$ Receptors

A cDNA coding for the human ET$_A$ receptor was cloned from a human placenta cDNA library by Adachi et al. (1991). The receptor is a 427 amino acid protein with > 90% identity to the bovine and rat ET$_A$ receptors but only 64% homology to the human ET$_B$ receptor. The receptor mRNA is distributed widely and can be identified in the placenta, uterus, testis, heart and adrenal glands of monkeys.

1.9.1.4   Bovine ET$_A$ Receptors

The bovine ET$_A$ receptor was the first ET receptor cloned (Arai, 1990). The receptor mRNA was found in multiple tissues including the central nervous system, heart, and lung. The receptor has a relatively long N terminus preceding transmembrane segment I. This portion of the molecule was hypothesized to be involved in the binding of the ETs to the receptor.

1.9.2   Endothelin$_B$ Receptors

1.9.2.1   Human ET$_B$ Receptors

The human ET$_B$ receptor was cloned by Sakamoto et al. (1991). The receptor is a 442 amino acid protein which shares approximately 90% homology to the other ET$_B$ receptors. Activation of the receptor leads to a transient increase in intracellular

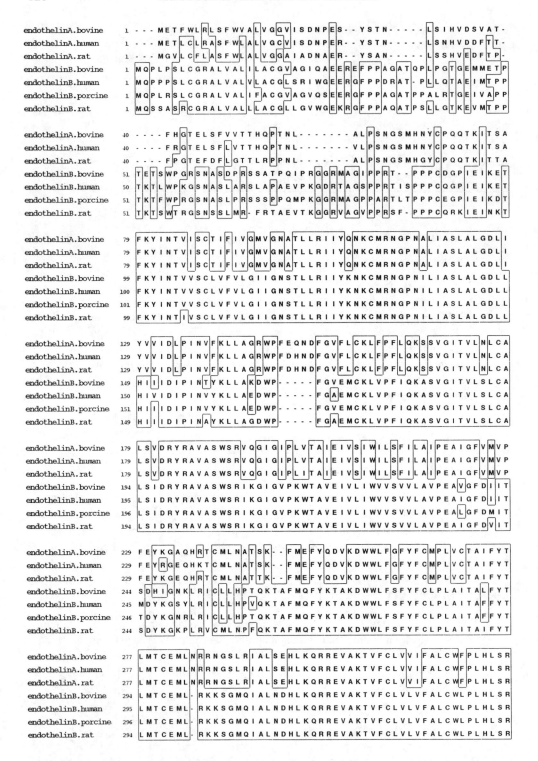

FIGURE 1. Alignment of the amino acid sequences of endothelin receptors. The sequences were obtained from commercial protein databases such as GenBank and EMBL. The references for each sequence are cited in the text.

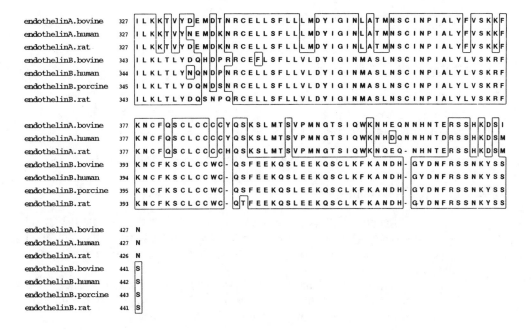

FIGURE 1b.

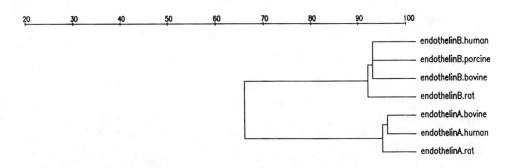

FIGURE 2. Phylogenetic tree of endothelin receptors. The phylogenetic tree was constructed according to the method of Feng and Doolittle (1990). The length of each "branch" of the tree correlates with the evolutionary distance between receptor subpopulations.

Table 1.

Receptor	Second messenger	Species cloned	Chromosomal location	a.a. Sequence	Accession number	Primary reference
Endothelin A	+PI	Human		427	P25101	Adachi et al., 1991
		Bovine		427	P21450	Arai et al., 1990
		Rat		426	P26684	Lin et al., 1991
Endothelin B	+PI	Human		422	P24530	Nakamuta et al., 19??
		Bovine		441	P28088	Saito et al., 1991
		Porcine		443		Elshourbagy et al., 1992
		Rat		441	P21451	Sakurai et al., 1990

calcium concentrations. The receptor mRNA was detected in brain, kidney, and placenta.

1.9.2.4   Bovine ET$_B$ Receptors

The bovine ET$_B$ receptor was cloned by Saito et al. (1991).

1.9.2.7   Porcine ET$_B$ Receptors

The porcine ET$_B$ receptor was cloned by Elshourbagy et al. (1992).

1.9.2.10   Rat ET$_B$ Receptors

The rat ET$_B$ receptor was cloned by Sakurai et al. (1990). The receptor is a 441 amino acid protein which shares significant homology to the other ET$_B$ receptors. Activation of the receptor leads to increases in the production of inositol phosphates and a transient increase in intracellular calcium concentrations. The receptor mRNA was detected in brain, kidney, and lung but not in vascular smooth muscle cells. The protein was designated the ET$_B$ receptor since the amino acid sequence differed significantly from the bovine ET$_A$ receptor (which had been identified concurrently).

Acknowledgments

I thank Jean M. Peroutka for excellent editorial assistance. This work was supported in part by the Kleiner Family Foundation and NIH Grant NS 25360-06.

REFERENCES

Adachi, M., Yang, Y. Y., Furuichi, Y., and Miyamoto, C. (1991) Cloning and characterization of cDNA encoding human A-type endothelin receptor. *Biochem. Biophys. Res. Commun.* 180, 1265–1272.

Arai, H., Hori, S., Aramori, I., Ohkubo, and Nakanishi, S. (1990) Cloning and expression of a cDNA encoding an endothelin receptor. *Nature* 348, 730–732.

Elshourgaby, N. A., Lee, J. A., Korman, D. R., Nuthalaganti, P., Sylvester, D. R., Dilella, A. G., Sutiphong, J. R., and Kumar, C. S. (1992) Molecular cloning and characterization of the major endothelin receptor subtype in the porcine cerebellum. *Mol. Pharmacol.* 41, 465–473.

Sakamoto, M., Yanagisawa, M., Sakurai, T., Takuwa, Y., Yanagisawa, H., and Masaki, T. (1991) Cloning and functional expression of human cDNA for the ET$_B$ endothelin receptor. *Biochem. Biophys. Res. Commun.* 178, 656–663.

Saito, Y., Mizuno, T., Itakura, M., Suzuki, Y., Ito, Y., Hagiwara, H., and Hirose, S. (1991) *J. Biol. Chem.* 266, 23433–23437.

Sakurai, T., Yanagisawa, M., Takuwa, Y., Miyazaki, H., Kimura, S., Goto, K., and Masaki, T. (1990) Cloning of a cDNA encoding a non-isopeptide-selective subtype of the endothelin receptor. *Nature* 348, 732–735.

Sokolovsky, M. (1992) Structure-function relationship of endothelins, sarafotoxins, and their receptors. *J. Neurochem.* 59, 809–821.

13

Glutamate Receptors

Darryle D. Schoepp, Ph.D.

1.13.0   Introduction

The existence of a G-protein coupled glutamate receptor that was linked to phosphoinositide hydrolysis was initially reported by Sladeczek et al. (1985). Their study showed that L-glutamate and quisqualate potently increased phosphoinositide hydrolysis in cultured murine striatal neurons. Phosphoinositide hydrolysis induced by these excitatory amino acid agonists was not mimicked by N-methyl-D-aspartate (NMDA) and kainate, selective ligand-gated ion channel (ionotropic) glutamate receptor agonists. At that time, it was not clear if this effect was directly linked to activation of phospholipase C (PLC) or was an indirect consequence of ion (i.e., calcium) entry through ionotropic glutamate receptors. Other early studies by Nicoletti et al. (1986a,b) showed that stimulation of phosphoinositide hydrolysis in rat brain slices had a unique pharmacology compared to other glutamate receptors known at that time, and a unique developmental profile compared to other phosphoinositide linked receptor systems. Many subsequent studies in brain slices, cultured neurons, synaptoneurosomes, and glial cells clearly distinguished glutamate agonist-induced phosphoinositide hydrolysis from agonist effects at ionotropic glutamate receptors (see Schoepp et al., 1990). For example, the NMDA receptor agonist ibotenate greatly stimulates phosphoinositide hydrolysis, but this effect is not blocked by 2-amino-5-phosphonopentanoic acid (AP5), an NMDA receptor antagonist. Furthermore, although the AMPA receptor agonist quisqualate is a highly potent stimulant of phosphoinositide hydrolysis, this effect was not mimicked by AMPA or blocked by quinoxalinediones such as 6-cyano-7-nitroquinoxaine-2,3-dione (CNQX) which blocks the ionotropic effects of AMPA and quisqualate. The term "metabotropic" glutamate receptor was used to describe this new glutamate receptor type that was linked to phosphoinositide hydrolysis (Sladeczek et al., 1988; Schoepp et al., 1990). This led to the reclassification of quisqualate receptors into (1) "AMPA" receptors: to reflect the greater selectivity of this agent vs. quisqualate for ligand-gated ion channel glutamate receptors and (2) "metabotropic glutamate receptors" (mGluRs) (or Q_p receptors): to reflect quisqualate-activated receptors which are coupled to phosphoinositide hydrolysis.

The role of guanine nucleotide binding proteins (G-proteins) in mediating coupling to metabotropic glutamate receptors was initially described by Sugiyama et al. (1987). In *Xenopus* oocytes injected with rat brain mRNA, phosphoinositide-coupled receptors are distinguished by the activation of an oscillating chloride current that is due to inositol trisphosphate (IP_3)-mediated mobilization of intracellular calcium. In this system, quisqualate and L-glutamate stimulate chloride currents in the absence of extracellular calcium and this is attenuated by pertussis toxin. This data clearly showed that quisqualate could produce G-protein mediated "metabotropic" effects in addition to its well-characterized ionotropic actions.

The availability of selective compounds in which to study mGluRs greatly hampered characterization of these receptors. Recently, a highly useful agonist compound with selectivity for mGluRs as a class has emerged. Using rat brain slices Palmer et al. (1989) demonstrated that the rigid glutamate analog (±)trans-1-aminocyclopentane-1,3-dicarboxylic acid (trans-ACPD) stimulated phosphoinositide hydrolysis at concentrations which do not activate ionotropic glutamate receptors. The use of this compound, or its active stereoisomer 1S,3R-ACPD (Irving et al., 1990; Schoepp et al., 1991), has provided a way to further study the *in situ* cellular effectors and functional consequences linked to mGluR activation (see Schoepp and Conn, 1992). Use of 1S,3R-ACPD has established that mGluRs are linked to modulation of a number of cellular effectors in addition to PLC. These include adenylate cyclase, phospholipase D, phosphoplipase A_2, and the modulation of ion channels (K^+ and Ca^{++}) (see Schoepp and Conn, 1993). 1S,3R-ACPD has also been useful to show the effects of mGluR activation on synaptic function in a number of brain regions including hippocampus, striatum, nucleus tractus solitarius, thalmus, amygdala, and olfactory bulb (see Schoepp and Conn, 1993).

Establishment of mGluRs as a novel receptor family at the molecular level has evolved from the use of the *Xenopus* oocyte system to expression clone a phosphoinositide-linked mGluR. Two groups independently reported the cloning and expression of rat mGluR1 (Masu et al., 1991) (or Glu_GR, Houamed et al., 1991) from a rat cerebellar library. Upon expression into *Xenopous* oocytes, the rat mGluR1 protein can be activated by agonists such as L-glutamate, quisqualate, ibotenate, and (±)trans-ACPD to induce an oscillating chloride current. The amino acid sequence for rat mGluR1 was used to probe rat brain libraries for related clones. In this manner, a family of mGluRs has been established (see Table 1; Nakanishi, 1992). These mGluRs are differentially distributed throughout rat brain, and show coupling to multiple G-protein modulated effector systems. Members of the mGluR family each possess seven putative transmembrane regions, but have no sequence homology to other G-protein linked receptors. Compared to other G-protein linked receptors, mGluRs are quite large (871 to 1199 amino acids) and have a large N-terminal extracellular domain of about 570 amino acids. Consistent with the N-terminus being extracellular, a number of N-glycosylation sites are found in this region of the protein (Masu et al., 1991; Houamed et al., 1991). Little is known yet about the site at which glutamate binds to activate mGluRs, or the structural motifs responsible for G-protein interaction and receptor regulation. A number of cysteine residues are highly conserved in the extracellular NH_2-terminus regions and the loops between the transmembrane regions of all mGluRs (Nakanishi, 1992), suggesting their importance in mGluR secondary structure. Also, phosphorylation sites are found at the putative intracellular region near the carboxy-terminus and thus may play a role in mGluR regulation.

Rat mGluRs cloned to date can be placed into three groups that have 60 to 70% sequence homology to each other, but only 40 to 50% homology to other mGluR groups (Tables 2 and 3). Within each of these groups, mGluRs share similar agonist pharmacology and second messenger responses (Table 1 and 2; Nakanishi, 1992). Group one includes rat mGluR1 and mGluR5, which are each coupled to phosphoinositide hydrolysis and are potently activated by quisqualate, but show low trans-ACPD sensitivity. Group 2 includes rat mGluR2 and mGluR3 which are nega-

Table 1. Molecular and Pharmacological Characteristics of Cloned Rat mGluRs

Clone	#amino acids	2nd messenger	Agonist potency
mGluR1α	1199	+PI, +cAMP, +AA	quis > glut > ibo > t-ACPD
mGluR1β	906	+PI	quis > glut > ibo > t-ACPD
mGluR1c	897	+PI	quis > glut > ibo > t-ACPD
mGluR2	872	−cAMP	glut = t-ACPD > ibo > quis
mGluR3	879	−cAMP	glut = t-ACPD > ibo > quis
mGluR4	912	−cAMP	AP4 > glut » t-ACPD » quis
mGluR5	1171	−cAMP	quis > glut > ibo > t-ACPD

Table 2. % Homology Among Rat mGluRs

	mGlu1	mGluR2	mGluR3	mGluR4	mGluR5
mGLuR1	100	46	44	43	60
mGluR2	—	100	70	43	46
mGLuR3	—	—	100	47	44
mGluR4	—	—	—	100	42
mGluR5	—	—	—	—	100

Table 3.

Receptor	Second messenger	Species cloned	Chromosomal location	a.a. Sequence	GenBank accession	Primary reference
mGlu1α	+PI	Rat		1199	P23385	Masu et al., 1991
mGlu1β	+PI	Rat		906		Tanabe et al., 1992a
mGlu1c	+PI	Rat		897		Pin et al., 1992
mGlu2	−cAMP	Rat		872	JH0561	Tanabe et al., 1992a
mGlu3	−cAMP	Rat		879	JH0562	Tanabe et al., 1992a
mGlu4	−cAMP	Rat		912	JH0563	Tanabe et al., 1992a
mGlu5	+PI	Rat		1171	S39116	Abe et al., 1992

tively coupled to cAMP formation, and are potently activated by trans-ACPD, but are relatively insensitive to quisqualate. Although mGluR6 has not been fuctionally expressed, it is included in group 3 along with mGluR4 based on their high sequence homology. Interestingly, mGluR4 (unlike mGluRs in the other groups) is potently activated by L-2-amino-4-phosphonobutyrate (L-AP4) (Tanabe et al., 1993; Thomsen et al., 1992b). Thus, mGluR4 (and possibly mGLuR6) may represent the pharmacologically characterized L-AP4 presynaptic receptors which control glutamate release (Monaghan et al., 1989).

The pharmacology of mGluRs is clearly in its infancy, and selective radioligands for the different mGluRs are not yet known. Although 1S,3R-ACPD is useful to selectively activate mGluRs at concentrations that have no effect on ionotropic glutamate receptors, this requires μM concentrations and prohibits the use of this compound as a selective radioligand. *In situ* mGluRs can be labeled with ^{3}H-glutamate in the presence of blockers for ionotropic sites (Cha et al., 1990; Schoepp and True, 1992). This method likely reflects ^{3}H-glutamate binding to multiple mGluRs. Using ^{3}H-glutamate binding in brain membranes under these conditions, multiple mGluR populations can be distinguished by high and low affinity displacement sites for quisqualate (Schoepp and True, 1992). Likewise, cloned rat mGluRs also differ in their sensitivity to activation by quisqualate (see Table 1). The binding of ^{3}H-glutamate to expressed mGluRs was attempted but was unsuccessful (Tanabe et al., 1992). Thus, radioligands for specific cloned and *in situ* mGluRs must await the discovery of more potent and selective compounds than those currently known.

The discovery of these mGluR clones and their expression in nonneuronal cells will be highly useful for future progress in this area. For example, they can be used for the development and characterization of subtype selective agonists/antagonists and mGluR selective antibody probes, which can then be used to sort out the role of each specific mGluR in brain function and pathology. Future discovery of human mGluRs could be used to clearly define mGluR pharmacology in the species for which drugs agents would be ultimately targeted.

1.13.1   Glutamate mGluR1 Receptors

1.13.1.10   Rat mGluR1 Receptors

Rat mGluR1α was the first mGluR to be cloned (see above). This receptor protein was expression cloned from a rat cerebellar library using the *Xenopus* oocyte expression system (Masu et al., 1991; Houamed et al., 1991). mGluR1α is thought to be phophoinositide coupled *in situ* based on glutamate-induced oscillating chloride currents following expression in the *Xenopus* oocyte, and glutamate-induced phosphoinositide hydrolysis in transfected nonneuronal (chinese hamster ovary, CHO) cells (Aramori and Nakanishi, 1992). In CHO cells transfected with rat mGluR1α, glutamate and other mGluR agonists not only increases phosphoinositide hydrolysis, but also increase cyclic AMP formation and arachadonic acid release (Aramori and Nakanishi, 1992). These three second messenger responses all show an identical agonist potency order, but are differentially sensitive to pertussis toxin. This suggests that mGluR1α interacts with multiple G-protein complexes to activate multiple cellular effectors such as phospholipase C, adenylate cyclase, and phospholipase A_2. The relevance of mGluR1α coupling to to these multiple effectors in the CHO cell to mGluR1α responses is not clear (see Schoepp and Conn, 1992).

Message for mGluR1α and the receptor protein itself is highly expressed in the rat cerebellum, primarily within Purkinje cells of this region (Shigemoto et al., 1992; Martin et al., 1992). mGluR1α message is also found in primary cultures of rat cerebellar granule cells, with its RNA expression dependent on the KCl concentration of the culture media (Favaron et al., 1992). In addition to the cerebellum, mGluR1α is highly expressed in neurons of the olfactory bulb, hippocampus, thalmus, septum, globus pallidus, and superior colliculus (Shigemoto et al., 1992; Martin et al., 1992).

The limited agonist pharmacology known to date coupled with knowledge of regional expression suggests that mGluR1α contributes to phosphoinositide hydrolysis elicited from rat cerebellar slices. Cloned mGluR1α expressing cells and rat cerebellar slices each show an agonist potency order of quisqualate > ibotenate > trans-ACPD (Aramori and Nakanishi, 1992; Schoepp et al., 1990; Schoepp and Conn, 1993). However, mGluR1a does not appear to contribute to phosphoinositide hydrolysis elicited in the rat hippocampus. Trans-ACPD stimulated phosphoinositide hydrolysis in the rat hippocampus is blocked by L-2-amino-3-phosphonopropionate (L-AP3) and L-AP4 (see Schoepp et al., 1990), but these compounds are weak to inactive agents in mGluR1α expressing cells (Masu et al., 1991; Houamed et al., 1991; Aramori and Nakanishi, 1992).

The expression of message for mGluR1a is developmentally regulated in the rat brain. In the olfactory bulb, hypothalamus, and cerebral cortex mGluR1α mRNA greatly increases from 1 to 8 d of age, then remains constant at later ages (Condorelli et al., 1992; Shigemoto et al., 1992). In the rat cerebellum and hippocampus mGluR1α mRNA increases progressively with age all the way into adulthood (Condorelli et al., 1992; Shigemoto et al., 1992). These changes in mGluR expression with age do not correlate with the mGluR1α functional response (phosphoinositide hydrolysis) in these tissues (Condorelli et al., 1992). This suggests that other factors such as G-protein and effector coupling or the expression of other phosphoinositide linked mGluRs is responsible for the greatly increased glutamate agonist-induced phosphoinositide hydrolysis found in various neonatal brain regions compared to the adult (see Schoepp et al., 1990). How these changes in mGluR1α mRNA expression relate to actual receptor expression as measured by mGluR selective antibodies or selective ligand binding is not known.

In addition to rat mGluR1α, two other forms of mGluR1 have been described that are likely alternative splice variants of mGluR1α. These are mGluR1β (Tanabe et al., 1992a) and mGluR1c (Pin et al., 1992). All three forms of mGluR1 exhibit a similar pharmacological profile of agonist potency (quisqualate > glutamate > ibotenate > trans-ACPD) and are linked to increased phosphoinositide hydrolysis when expressed in *Xenopus* oocytes (Tanabe et al., 1992a; Thomsen et al., 1992; Pin et al., 1992). Structurally, these forms of mGluR1 differ after the seventh transmembrane region corresponding to the C-terminal intracellular domain. Rat mGluR1α possesses about 360 amino acid in this region of the protein. In mGluR1β, 312 amino acids at the C-terminus are alternatively spliced with 20 different amino acids resulting in a much shorter (about 28 amino acid) C-terminal intracellular region (Tanabe et al., 1992a). Likewise, mGluR1c has a 10 amino acid insert in the same region resulting in a very short (about 18 amino acid) C-terminus (Pin et al., 1992). Since these truncations at this part of the mGluR protein do not greatly affect receptor transduction and agonist pharmacology, the C-terminus region is apparently not responsible for agonist binding and G-protein interaction. However, compared to mGluR1α and mGluR1β, mGluR1c generates a more slowly elicited and longer-lasting chloride current when expressed in the *Xenopus* oocyte, and is expressed in lower amounts in the rat brain (Pin et al., 1992). Therefore, the selective expression of different mGluR1 forms *in situ* may have functional relevance in certain qualitative aspects of cell signaling.

1.13.2   Glutamate mGluR2 Receptors

1.13.2.10   Rat mGluR2 Receptors

Rat mGluR2 was initially described by Tanabe et al. (1992a) as a 3294 base-pair (872 amino acid) clone with a calculated molecular weight of 95,770 and 46% sequence homology to mGluR1α. When compared to rat mGluR1α, rat mGluR2 has a relatively short C-terminal intracellular domains (similar to mGluR1β and mGluR1c). Differences in amino acid sequence between mGluR2 and other mGluRs is found throughout all domains of the protein. Among members of this family mGluR2 is most closely related to mGluR3 (70% sequence homology).

Unlike mGluR1 and mGluR5, injection of mGluR2 mRNA into *Xenopus* oocytes does not lead to oscillating chloride currents, suggesting that this receptor is not coupled to phosphoinositide hydrolysis *in situ*. Likewise, CHO cells transfected with mGluR2 exhibit only a very weak phosphoinositide response to glutamate (Tanabe et al., 1992). However, these cells when activated by glutamate show negative coupling to adenylate cyclase. The agonist potency order for decreasing forskolin-stimulated cyclic AMP formation is glutamate = trans-ACPD > ibotenate $\gg$ quisqualate. This effect in these cells is fully reversed by pertussis toxin. (Tanabe et al., 1992). This pharmacology is similar to mGluR-mediated inhibition of cyclic AMP formation that is found in the rat hippocampus (Schoepp and Johnson, 1992), and suggests that mGluR2 contributes to this mGluR rsponse *in situ*.

Recently, the glutamate analog compound L-2-(carboxycyclopropyl)glycine-I isomer (L-CCG -I) has been shown to potently and selectively activate mGluR2 with an EC_{50} of 0.3 μM (Hayashi et al., 1992). The EC_{50} values of L-CCG-I activation of mGluR1 and mGluR4 was much higher (about 50 μM). Thus L-CCG-I is an mGluR2 selective compound which may be useful to identify *in situ* mGluR2 responses, and as a potential mGluR selective radioligand.

In situ hybridization showed mGluR2 to be widely distributed to neurons throughout the rat brain. Prominent expression of mGluR2 was observed in golgi cells of the cerebellum, granule cells of the dentate gyrus, many neurons of the cerebral cortex, and small neurons within the olfactory bulbs (Tanabe et al., 1992).

1.13.3   Glutamate mGluR3 Receptors

1.13.3.10   Rat mGluR3 Receptors

Rat mGluR3 was initially described by Tanabe et al. (1992) as a 3215 base-pair (879 amino acid, mw = 98,960) clone with 47% sequence identity to mGluR1, but relatively higher homology to mGluR2 (70%). Similar to mGluR2, mGluR3 is negatively coupled to adenylate cyclase when expressed in CHO cells, shows an agonist potency order of L-glutamate > trans-ACPD > ibotenate > quisqualate, and can be fully uncoupled with pertussis toxin (Tanabe et al., 1992b).

High expression of mGluR3 mRNA is found in neurons of the cerebral cortex, thalmic reticular neuclesus, and granule cells of the dentate gyrus. Unlike mGluR2, mGluR3 is also expressed in glial cells throughout different brain regions (Tanabe et al., 1992a,b). mGluRs have been shown to be present in glial cells since agonists such as quisqualate, glutamate, and trans-ACPD stimulate phosphoinositide hydrolysis in cultured astrocytes (Pearce et al., 1986; Nicoletti et al., 1990). Presently, mGluR3 is the only cloned mGluR with expression found in glial cells. However, unlike *in situ* mGluR responses in glia, mGluR3 is not phosphoinositide linked (at least when expressed in nonneuronal cells), and is not potently activated by quisqualate. *In situ* mGluR-mediated changes in cyclic AMP formation have not been reported in glial cells, but might be present based on what is known about rat mGluR3.

1.13.4   Glutamate mGluR4 Receptors

1.13.4.10   Rat mGluR4 Receptors

The cDNA sequence for rat mGluR4 consists of 3704 base pairs coding for a 912 amino acid protein with a molecular weight of 101,810 Da (Tanabe et al., 1992a). This protein shows 40 to 50% sequence homology with rat mGluRs 1, 2, 3, and 5. Like mGluR2 and mGluR3, mGluR4 is negatively coupled to adenylate cyclase when expressed in CHO cells and is highly sensitive to pertussis toxin inhibition (Tanabe et al., 1992b). However, mGluR4 has a pharmacology unique from all other functionally expressed mGluRs. The compound L-2-amino-4-phosphonobutyrate (L-AP4) is a relatively potent (EC_{50} 0.5 μM) and selective agonist for mGluR4. This effect is also mimicked by L-glutamate (EC_{50} 5 μM) and L-serine-O-phosphate (EC_{50} 4 μM), but trans-ACPD and quisqualate are relatively inactive (Tanabe et al., 1992b). In contrast to the other negatively coupled cyclic AMP linked mGluRs which fully inhibit cyclic AMP formation, all these agonists only partially (about 50%) reduce forskolin-stimulated cyclic AMP formation at maximally effective concentrations in the same cell.

The distribution of mGluR4 is restricted to neurons in the rat brain. Relatively high mGluR4 expression is found in the granular layer of the main olfactory bulb and neurons of the thalamus, lateral septum, and pontine nucleus and granule cells of the cerebellum. Unlike mGluR1, mGluR2, mGluR3, and mGluR5, only weak expression of mGluR4 is found in the dentate gyrus (Tanabe et al., 1993).

1.13.5   Glutamate mGluR5 Receptors

1.13.5.10   Rat mGluR5 Receptors

The cDNA encoding rat mGluR5 was isolated from a rat brain library using degenerate probes to mGluRs 1 to 4 (Abe et al., 1992). Similar to the other mGluRs, this protein has eight hydrophobic regions corresponding to on N-terminus signal peptide and the seven putative transmembrane regions. Among the rat mGluRs, mGluR5 has highest homology to mGluR1 (60% absolute and 87% if conservative substitutions are included) and consists of 1171 amino acids with a molecular weight of 128,289. Like mGluR1, mGluR5 is coupled to phosphoinositide hydrolysis when expressed in *Xenopus* oocytes or transfected CHO cells (Abe et al., 1992). The agonist profile for mGluR5 stimulated phosphoinositide hydrolysis is also similar to mGluR1 (quisqualate > ibotenate > trans-ACPD). However, the compounds L-AP3 and L-AP4 are inactive as either agonists or antagonists at the mGluR5 receptor. This contrasts mGluR5 from mGluR1, where AP3 is a weak antagonist (Houamed et al., 1991), and mGluR4 where L-AP4 is a potent agonist (Tanabe et al., 1993). Also unlike mGluR1a, mGluR5 expressing CHO cells are not coupled to increased cyclic AMP formation or arachidonic acid release. mGluR5 and mGluR1a share the structural property of a relatively large (about 350 amino acid) intracellular region at the carboxy-terminus, in addition to the large (about 560 amino acid) N-terminus extracellular region.

In adult rats, mGluR5 is expressed in many brain regions. Unlike mGluR1, its expression is highest in the striatum and relatively low in the cerebellum and pons/medulla. In the hippocampus, mGluR5 message is localized to pyramidal cells throughout the CA1-CA4 regions, and is found in granule cells of the dentate gyrus (Abe et al., 1992). Rat mGluR5 is highly expressed within the telencephalon of 6-d old rats, and thus might contribute to the relatively robust phosphoinositide mGluR responses found in the neonatal forebrain (Nicoletti et al., 1986b; Schoepp and Hillman, 1990).

1.13.99   Other G-Protein Linked Glutamate Receptors

A G-protein coupled glutamate receptor with high affinity for kainate has been recently described (Willard and Oswald, 1992). This protein was cloned from *Rana pipiens* and has about 35% sequence homology with mammalian non-NMDA ionotropic glutamate receptors. In common with ionotropic glutamate receptors, this protein binds ^{3}H-kainate with high afinity (K_D 2 nM) and has four hydrophobic putative membrane spanning regions. However, when expressed in CHO cells, the binding of ^{3}H-kainate to this protein

```
glutamate1alpha.rat   1  MVRLLLIFFPMIFLEMSILPRMPDRKVLLAGASSQ----RSVARMDGDVI
glutamate2.rat        1  -----------MESLLGFLALLLLWGAVAEGPA---KKVLTLEGDLV
glutamate3.rat        1  --------MKMLTRLQILMLALFSKGFLLSLGDHNF--MRREIKIEGDLV
glutamate4.rat        1  MSGKGGWAWWWARLPLCLLLSLYAPWVPSSLGKPKGHPHMNSIRIDGDIT
glutamate5.rat        1  ----------MVLLLILSVLLLKEDVRGSAQSSER----RVVAHMPGDII

glutamate1alpha.rat  47  IGALFSVHHQPPAEKVPERKCGEIREQYGIQRVEAMFHTLDKINADPVLL
glutamate2.rat       34  LGGLFPVHQKGG----PAEECGPVNEHRGIQRLEAMLFALDRINRDPHLL
glutamate3.rat       41  LGGLFPINEKGT----GTEECGRINEDRGIQRLEAMLFAIDEINKDNYLL
glutamate4.rat       51  LGGLFPVHGRGS----EGKACGELKKEKGIHRLEAMLFALDRINNDPDLL
glutamate5.rat       37  IGALFSVHHQPTVDKVHERKCGAVREQYGIQRVEAMLHTLERINSDPTLL

glutamate1alpha.rat  97  PNITLGSEIRDSCWHSSVALEQSIEFIRDSLISIRDEKDGLNRCLPDGQT
glutamate2.rat       80  PGVRLGAHILDSCSKDTHALEQALDFVRASLSRGADGSRHICP---DGSY
glutamate3.rat       87  PGVKLGVHILDTCSRDTYALEQSLEFVRASLTK-VDEAEYMCP---DGSY
glutamate4.rat       97  PNITLGARILDTCSRDTHALEQSLTFVRALIEK--DGTEVRCG---SGG-
glutamate5.rat       87  PNITLGCEIRDSCWHSAVALEQSIEFIRDSLIS-SEEEGLVRCV-DGSS

glutamate1alpha.rat 147  LPPGRTKKP--IAGVIGPGSSSVAIQVQNLLQLFDIPQIAYSATSIDLSD
glutamate2.rat      127  --ATHSDAPTAVTGVIGGSYSDVSIQVANLLRLFQIPQISYASTSAKLSD
glutamate3.rat      133  --AIQENIPLLIAGVIGGSYSSVSIQVANLLRLFQIPQISYASTSAKLSD
glutamate4.rat      141  --PPIITKPERVVGVIGASGSSVSIMVANILRLFKIPQISYASTAPDLSD
glutamate5.rat      135  --SFRSKKP--IVGVIGPGSSSVAIQVQNLLQLFNIPQIAYSATSMDLSD

glutamate1alpha.rat 195  KTLYKYFLRVVPSDTLQARAMLDIVKRYNWTYVSAVHTEGNYGESGMDAF
glutamate2.rat      175  KSRYDYFARTVPPDFFQAKAMAEILRFFNWTYVSTVASEGDYGETGIEAF
glutamate3.rat      181  KSRYDYFARTVPPDFYQAKAMAEILRFFNWTYVSTVASEGDYGETGIEAF
glutamate4.rat      189  NSRYDFFSRVVPSDTYQAQAMVDIVRALKWNYVSTLASEGSYGESGVEAF
glutamate5.rat      181  KTLFKYFMRVVPSDAQQARAMVDIVKRYNWTYVSAVHTEGNYGESGMEAF

glutamate1alpha.rat 245  KELAAQE-GLCIAHSDKIYSNAGEKSFDRLLRKLRERLPKARVVCFCEG
glutamate2.rat      225  ELEARAR-NICVATSEKVGRAMSRAAFEGVVRALLQK-PSARVAVLFTRS
glutamate3.rat      231  EQEARLR-NICIATAEKVGRSNIRKSYDSVIRELLQK-PNARVVVLFMRS
glutamate4.rat      239  IQKSRENGGVCIAQSVKIPREPKTGEFDKIIKRLLET-SNARGIIIFANE
glutamate5.rat      231  KDMSAKE-GICIAHSYKIYSNAGEQSFDKLLKKLRSHLPKARVVACFCEG

glutamate1alpha.rat 294  MTVRGLLSAMRRLGVVGEFSLIGSDGWADRDEVIEGYEVEANGGITIKLQ
glutamate2.rat      273  EDARELLAATQR--LNASFTWVASDGWGALESVVAGSERAAEGAITIELA
glutamate3.rat      279  DDSRELIAAANR--VNASFTWVASDGWGAQESIVKGSEHVAYGAITLELA
glutamate4.rat      288  DDIRRVLEAARRANQTGHFFWMGSDSWGSKSAPVLRLEEVAEGAVTILPK
glutamate5.rat      280  MTVRGLLMAMRRLGLAGEFLLLGSDGWADRYDVTDGYQREAVGGITIKLQ

glutamate1alpha.rat 344  SPEVRSFDDYFLKLRLDTNTRNPWFPEFWQHRFQCRLPGHLLENPNFKKV
glutamate2.rat      321  SYPISDFASYFQSLDPWNNSRNPWFREFWEERFHCSFRQRDCAAHSLRAV
glutamate3.rat      327  SHPVRQFDRYFQSLNPYNNHRNPWFRDFWEQKFQCSLQNKRNHRQVCDKH
glutamate4.rat      338  RMSVRGFDRYFSSRTLDNNRRNIWFAEFWEDNFHCKLSRHALKKGSHIKK
glutamate5.rat      330  SPDVKWFDDYYLKLRPETNLRNPWFQEFWQHRFQCRLEGFAQENSKYNKT

glutamate1alpha.rat 394  C--TGNESLEENYVQDSKMGFVINAIYAMAHGLQNMHHALCPGHVGLCDA
glutamate2.rat      371  ----------PFEQESKIMFVVNAVYAMAHALHNMHRALCPNTTHLCDA
glutamate3.rat      377  L--AIDSS---NYEQESKIMFVVNAVYAMAHALHKMQRTLCPNTTKLCDA
glutamate4.rat      388  CTNRERIGQDSAYEQEGKVQFVIDAVYAMGHALHAMHRDLCPGRVGLCPR
glutamate5.rat      380  C--NSSLTLRTHHVQDSKMGFVINAIYSMAYGLHNMQMSLCPGYAGLCDA

glutamate1alpha.rat 442  MKPIDGRKLL-DFLIKSSFVG-----VSGEEVWFDEKGDAPGRYDIMNL
glutamate2.rat      410  MRPVNGRRLYKDFVLNVKFDA-PFRPADTDDEVRFDRFGDGIGRYNIFTY
glutamate3.rat      422  MKILDGKKLYKEYLLKINFTAPFNPNKGADSIVKFDTFGDGMGRYNVFNL
glutamate4.rat      438  MDPVDGTQLL-KYIRNVNFSG-----IAGNPVTFNENGDAPGRYDIYQY
glutamate5.rat      428  MKPIDGRKLL-DSLMKTNFTG-----VSGDMILFDENGDSPGRYEIMNF
```

FIGURE 1. Alignment of the amino acid sequences of glutamate receptors. The sequences were obtained from commercial protein databases such as GenBank and EMBL. The references for each sequence are cited in the text.

```
glutamate1alpha.rat  485  QYTEANRYDYVHVGTWHEGVLNIDDYKI---QMNKSGMVRSVCSEPCLKG
glutamate2.rat       459  LRAGSGRYRYQKVGYWAEG-LTLDTSFIPWASPSAGPLPASRCSEPCLQN
glutamate3.rat       472  QQTGGK-YSYLKVGHWAET-LSLDVDSI---HWSRNSVPTSQCSDPCAPN
glutamate4.rat       481  QLRNGS-AEYKVIGSWTDH-LHLRIERM-QWPGSGQQLPRSICSLPCQPG
glutamate5.rat       471  KEMGKDYFDYINVGSWDNGELKMDDDEV---WSKKNNIIRSVCSEPCEKG

glutamate1alpha.rat  532  QIKVIRKGEVSCCWICTACKENEFVQDEFTCRACDLGWWPNAELTGCEPI
glutamate2.rat       508  EVKSVQPGEV-CCWLCIPCQPYEYRLDEFTCADCGLGYWPNASLTGCFEL
glutamate3.rat       517  EMKNMQPGDV-CCWICIPCEPYEYLVDEFTCMDCGPGQWPTADLSGCYNL
glutamate4.rat       528  ERKKTVKGMA-CCWHCEPCTGYQYQVDRYTCKTCPYDMRPTENRTSCQPI
glutamate5.rat       518  QIKVIRKGEVSCCWTCTPCKENEYVFDEYTCKACQLGSWPTDDLTGCDLI

glutamate1alpha.rat  582  PVRYLEWSDIESIIAIAFSCLGILVTLFVTLIFVLYRDTPVVKSSSRELC
glutamate2.rat       557  PQEYIRWGDAWAVGPVTIACLGALATLFVLGVFVRHNATPVVKASGRELC
glutamate3.rat       566  PEDYIKWEDAWAIGPVTIACLGFLCTCIVITVFIKHNNTPLVKASGRELC
glutamate4.rat       577  PIVKLEWDSPWAVLPLFLAVVGIAATLFVVVTFVRYNDTPIVKASGRELS
glutamate5.rat       568  PVQYLRWGDPEPIAAVFACLGLLATLFVTVIFIIYRDTPVVKSSSRELC

glutamate1alpha.rat  632  YIILAGIFLGYVCPFTLIAKPTTTSCYLQRLLVGLSSAMCYSALVTKTNR
glutamate2.rat       607  YILLGGVFLCYCMTFVFIAKPSTAVCTLRRLGLGTAFSVCYSALLTKTNR
glutamate3.rat       616  YILLFGVSLSYCMTFFFIAKPSPVICALRRLGLGTSFAICYSALLTKTNC
glutamate4.rat       627  YVLLAGIFLCYATTFLMIAEPDLGTCSLRRIFLGLGMSISYAALLTKTNR
glutamate5.rat       618  YIILAGICLGYLCTFCLIAKPKQIYCYLQRIGIGLSPAMSYSALVTKTNR

glutamate1alpha.rat  682  IARILAGSKKKICTRKPRFMSAWAQVIIASILISVQLTLVVTLIIME-PP
glutamate2.rat       657  IARIFGGARE--GAQRPRFISPASQVAICLALISGQLLIVAAWLVVEAPG
glutamate3.rat       666  IARIFDGVKN--GAQRPKFISPSSQVFICLGLILVQIVMVSVWLILETPG
glutamate4.rat       677  IYRIFEQGKR--SVSAPRFISPASQLAITFILISLQLLGICVWFVVD-PS
glutamate5.rat       668  IARILAGSKKKICTKKPRFMSACAQLVIAFILICIQLGIIVALFIME-PP

glutamate1alpha.rat  731  MPILSYPSIKE------VYLICNTSNLGVVAPVGYNGLLIMSCTYYAFK
glutamate2.rat       705  TGKETAPERREV-----VTLRCNHRDASMLGSLAYNVLLIALCTLYAFK
glutamate3.rat       714  TRRYTLPEKRET-----VILKCNVKDSSMLISLTYDVVLVILCTVYAFK
glutamate4.rat       724  HSVVDFQDQRTLDPRFARGVLKCDISDLSLICLLGYSMLLMVTCTVYAIK
glutamate5.rat       717  DIMHDYPSIRE------VYLICNTTNLGVVTPLGYNGLLILSCTFYAFK

glutamate1alpha.rat  774  TRNVPANFNEAKYIAFTMYTTCIIWLAFVPIYF---GSNYKI----ITTC
glutamate2.rat       749  TRKCPENFNEAKFIGFTMYTTCIIWLAFLPIFYV-TSSDYRVQ--TTTMC
glutamate3.rat       758  TRKCPENFNEAKFIGFTMYTTCIIWLAFLPIFYV-TSSDYRVQ--TTTMC
glutamate4.rat       774  TRGVPETFNEAKPIGFTMYTTCIVWLAFIPIFFGTSQSADKLYIQTTTLT
glutamate5.rat       760  TRNVPANFNEAKYIAFTMYTTCIIWLAFVPIYF---GSNYKI----ITMC

glutamate1alpha.rat  817  FAVSLSVTVALGCMFTPKMYIIIAKPERNVRSAFTTSDVVRMHVGDGK--
glutamate2.rat       796  VSVSLSGSVVLGCLFAPKLHIILFQPQKNVVS--HRAPTSRFGSAAPR-A
glutamate3.rat       805  ISVSLSGFVVLGCLFAPKVHIVLFQPQKNVVT--HRLHLNRFSVSGTA-
glutamate4.rat       824  VSVSLSASVSLGMLYMPKVYIILFHPEQNVPK-RKRSLKAVVTAATM--
glutamate5.rat       803  FSVSLSATVALGCMFVPKVYIILAKPERNVRSAFTTSTVVRMHVGDGKSS

glutamate1alpha.rat  865  -LPCRSNTFLNIFRRKKPGAGNANSNGKSVSWSEPGGRQAPKGQHVWQRL
glutamate2.rat       843  SANLGQGSGSQFVPTVCNGREVVDSTTSSL----------------
glutamate3.rat       851  -TTYSQSSASTYVPTVCNGREVLDSTTSSL----------------
glutamate4.rat       870  -SNKFTQKGNFRPNGEAKSELCENLETPALATKQTYVTYTNHAI------
glutamate5.rat       853  SAASRSSSLVNLWKRRGGSSGETLSSNGKSVTWAQ--NEKSTRGQHLWQRL

glutamate1alpha.rat  914  SVHVKTNETACNQTAVIKPLTKSYQGSGKSLTFSDASTKTLYNVEEDNT
glutamate5.rat       901  SVHINKKENP-NQTAVIKPFPKSTENRGPGAAAGGGSGPGVAGAGNAGCT
```

FIGURE 1b.

```
glutamate1alpha.rat  964  P S A H F S P P S S P S M V V H R R G P P V A T T P - - P L P P H L T A E E T P L F L A D S - V I P
glutamate5.rat       950  A T G G P E P P D A G P K A L Y D V A E A E E S F P A A A R P R S P S P I S T L S H L A G S A G R T

glutamate1alpha.rat 1011  K G L P P P L P Q Q Q P Q Q P P P Q Q P P Q Q P K S L M D Q L Q G V V T N F G S G I P D F H A - V L
glutamate5.rat      1000  D D D A P S L H S E T A A R S S S S Q G - - - - - S L M E Q I S S V V T R F T A N I S E L N S M M L

glutamate1alpha.rat 1060  A G P G T P G N S L R S L Y P P P P P P Q H L Q M L P L H L S T F Q E E S I S P P G E D I D D D S E
glutamate5.rat      1045  S T A A T P G P P G T P I C S S Y L I P K E I Q - L P T T M T T F A E I Q - P L P A I E V T G G A Q

glutamate1alpha.rat 1110  R F K L L Q E F V Y E R E G N T E E D E L E E E E D L P T A S K L T P E D S P A L T P P S P F R D S
glutamate5.rat      1093  G A T G V S P A Q E T P T G A E - - - - - - - - - - - S A P G K P D L E E L V A L T P P S P F R D S

glutamate1alpha.rat 1160  V A S G S S V P S S P V S E S V L C T P P N V T Y A S V I L R D Y K Q S S S T L
glutamate5.rat      1132  V D S G S T T P N S P V S E S A L C I P S S P K Y D T L I I R D Y T Q S S S S L
```

FIGURE 1c.

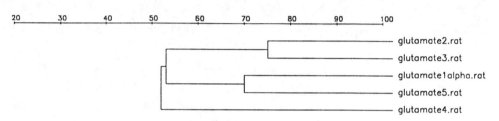

FIGURE 2. Phylogenetic tree of glutamate receptors. The phylogenetic tree was constructed according to the method of Feng and Doolittle (1990). The length of each "branch" of the tree correlates with the evolutionary distance between receptor sub-populations.

is greatly inhibited by guanine nucleotides or pertussis toxin pretreatment, suggesting that it interacts with G-protein mediated transduction events *in situ*.

Acknowledgment

The author appreciates the help and advice of Dr. J. Paul Burnett in the preparation of this chapter.

REFERENCES

Abe, T., Sugihara, H., Nawa, H., Shigemoto, R., Mizuno, N., and Nakanishi, S. (1992) Molecular characterization of a novel metabotropic glutamate receptor mGluR5 coupled to inositol phosphate/Ca^{2+} signal transduction. *J. Biol. Chem.* 267, 13361–13368.

Aramori, I. and Nakanishi, S. (1992) Signal transduction and pharmacological characteristics of a metabotropic glutamate receptor, mGluR1, in transfected CHO cells. *Neuron* 8, 757–765.

Condorelli, D. F., Dell'Albani, P., Amico, C., Casbona, G., Genazzani, A. A., Sortino, M. A., and Nicoletti, F. (1992) Development profile of metabotropic glutamate receptor mRNA in rat brain. *Mol. Pharmacol.* 41, 660–664.

Cha, J.-H. J., Makowiec, R. L., Penney, J. B., and Young, A. B. (1990) L-[^{3}H]Glutamate labels the metabotropic excitatory amino acid receptor in rodent brain. *Neurosci. Lett.* 113, 78–83.

Favaron, M., Rimland, J. M., and Manev, H. (1992) Depolarization- and agonist-regulated expression of neuronal metabotropic glutamate receptor 1 (mGluR1). *Life Sci.* 50, PL189-PL194.

Hayashi, Y., Tanabe, Y., Aramori, I., Masu, M., Shimamoto, K., Ohfune, Y., and Nakanishi, S. (1992) Agonist analysis of 2-(carboxycyclopropyl)glycine isomers for cloned metabotropic glutamate receptor subtypes expressed in Chinese hamster ovary cells. *Br. J. Pharmacol.* 107, 539–543.

Houamed, K. M., Kuijper, J. L., Gilbert, T. L., Haldeman, B. A., O'Hara, P. J., Mulvihill, E. R., Almers, W., and Hagen, F. S. (1991) Cloning, expression, and gene structure of a G protein-coupled glutamate receptor from rat brain. *Science* 252, 1318–1321.

Irving, A. J., Schofield, J. G., Watkins, J. C., Sunter, D. C., and Collingridge, G. L. (1990) 1S,3R-ACPD stimulates and L-AP3 blocks Ca^{2+} mobilization in rat cerebellar neurons. *Eur. J. Pharmacol.* 186, 363–365.

Martin, L. J., Blackstone, C. D., Huganir, R. L., and Price, D. L. (1992) Cellular localization of a metabotropic glutamate receptor in rat brain. *Neuron* 9, 259–270.

Masu, M., Tanabe, Y., Tsuchida, K., Shigemoto, R., and Nakanishi, S. (1991) Sequence and expression of a metabotropic glutamate receptor. *Nature* 349, 760–765.

Monaghan, D. T., Bridges, R. J., and Cotman, C. W. (1989) The excitatory amino acid receptors: their classes, pharmacology, and distinct properties in the function of the central nervous system. *Annu. Rev. Pharmacol. Toxicol.* 29, 365–402.

Nakanishi, S. (1992) Molecular diversity of glutamate receptors and implications for brain function. *Science* 258, 597–603.

Nicoletti, F., Meek, J. L., Iadorola, M. J., Chuang, D. M., Roth, B. L., and Costa, E. (1986a). Coupling of inositol phospholipid metabolism with excitatory amino acid recognition sites in rat hippocampus. *J. Neurochem.* 46, 40–46.

Nicoletti, F., Iadarola, M. J., Wroblewski, J. T., and Costa, E. (1986b). Excitatory amino acid recognition sites coupled with inositol phospholipid metabolism: Developmental changes and interaction with a_1-adrenoceptors. *Proc. Natl. Acad. Sci. U.S.A.* 83, 1931–1935.

Nicoletti, F., Magri, G., Ingrao, F., Bruno, V., Catania, M. V., Dell'Albani, P., Condorelli, D. F., and Avola, R. (1990) Excitatory amino acids stimulate inositol phospholipid hydrolysis and reduce proliferation in cultured astrocytes. *J. Neurochem.* 54, 771–777.

Pearce, B., Albrecht, J., Morrow, C., and Murphy, S. (1986) Astrocyte glutamate receptor activation promotes inositol phospholipid turnover and calcium flux. *Neurosci. Lett.* 72, 335–340.

Palmer, E., Monaghan, D. T., and Cotman, C. W. (1989) Trans-ACPD, a selective agonist of the phosphoinositide-coupled excitatory amino acid receptor. *Eur. J. Pharmacol.* 166, 585–587.

Pin, J.-P., Waeber, C., Prezeau, L., and Bockaert, J. (1992) Alternative splicing generates metabotropic glutamate receptors inducing diferent patterns of calcium release in *xenopus* oocytes. *Proc. Natl. Acad. Sci. U.S.A.* 89, 10331–10335.

Schoepp, D., Bockaert, J., and Sladeczek, F. (1990) Pharmacological and functional characteristics of metabotropic excitatory amino acid receptors. *Trends Pharmacol. Sci.* 11, 508–515.

Schoepp, D. D. and Hillman, C. C. (1990) Developmental and pharmacological characterization of quisqualate, ibotenate, and *trans*-1-amino-1,3-cyclopentanedicarboxylic acid stimulations of phosphoinositide hydrolysis in rat cortical brain slices. *Biogenic Amines* 7, 331–340.

Schoepp, D. D. and Johnson, B. G. (1993) Pharmacology of metabotropic glutamate receptor inhibition of cAMP formation in the adult rat hippocampus. *Neurochem. Int.*, 22, 277–283.

Schoepp, D. D., Johnson, B. G., True, R. A., and Monn, J. A. (1991a) Comparison of (1S,3R)-1-aminocyclopentane-1,3-dicarboxylic acid (1S,3R-ACPD)- and 1R,3S-ACPD-stimulated brain phosphoinositide hydrolsyis. *Eur. J. Pharmacol.— Mol. Pharmacol. Section* 207, 351–353.

Schoepp, D. D. and True, R. A. (1992) 1S,3R-ACPD-sensitive (metabotropic) [³H]glutamate receptor binding in membranes. *Neuorosci. Lett.* 145, 100–104.

Schoepp, D. D. and Conn, P. J. (1993) Metabotropic glutamate receptors in brain function and pathology. *Trends Pharmacol. Sci.*, 14, 13–20.

Shigemoto, R., Nakanishi, S., and Mizuno, N. (1992) Distribution of the mRNA for a metabotropic glutamate receptor (mGluR1) in the central nervous system: an *in situ* hybridization study in adult and developing rat. *J. Comp. Neurol.* 322, 121–135.

Sladeczek, F., Recasens, M., and Bockaert, J. (1988) A new mechanism for glutamate receptor action: phosphoinositide hydrolysis. *Trends Neurosci.* 11, 545–549.

Sladeczek, F., Pin, J.-P., Recasens, M., Bockaert, J., and Weiss, S. (1985) Glutamate stimulates inositol phosphate formation in striatal neurones. *Nature* 317, 717–719.

Sugiyama, H., Ito, I., and Hirono, C. (1987) A new type of glutamate receptor linked to inositol phospholipid metabolism. *Nature* 325, 531–533.

Tanabe, Y., Masu, M., Ishii, T., Shigemoto, R., and Nakanishi, S. (1992a) A family of metabotropic glutamate receptors. *Neuron* 8, 169–179.

Tanabe, Y., Nomura, A., Masu, M., Shigemoto, R., Mizuno, N., and Nakanishi, S. (1993) Signal transduction, pharmacological properties, and expression of patterns of two rat metabotropic glutamate receptors, mGluR3 and mGluR4. *J. Neurosci.*, 13, 1372–1378.

Thomsen, C., Mulvihill, E., Haldeman, B., and Suzdak, P. D. (1992a) A pharmacological characterization of the mGlu$_G$R-1a and the mGlu$_G$R-1b subtype of the metabotropic glutamate receptor expressed in a mammalian cell line. *Soc. Neuorosci. Abstr.* 18, 648.

Thomsen, C., Kristensen, P., Mulvihill, E., Haldeman, B., and Suzdek, P. D. (1992b) L-2-Amino-4-phosphonobutyrate (L-AP4) is an agonist at the type IV metabotropic glutamate receptor which is negatively coupled to adenylate cyclase. *Eur. J. Pharmacol.— Mol. Pharmacol. Section* 227, 361–362.

Willard, J. M. and Oswald, R. E. (1992) Interaction of the frog brain kainate receptor expression in Chinese Hamster Ovary cells with a GTP-binding protein. *J. Biol. Chem.* 267, 19112–19116.

14

Histamine Receptors

M. Ruat, E. Traiffort, R. Leurs, J. Tardivel-Lacombe, J. Diaz, J. M. Arrang, and J. C. Schwartz

1.14.0   Introduction

Three histamine receptor subtypes termed H_1, H_2, and H_3 have been identified so far and localized in brain and in peripheral tissues using classical biochemical, pharmacological, and radioligand binding approaches (Hill, 1990; Schwartz et al., 1991; Schwartz and Haas, 1992). It is widely accepted that these receptor subtypes interact with distinct G-proteins to elicit their functional responses. The introduction of the molecular genetic approach, which has been so successful in the field of receptors coupled to G-proteins, is just beginning to be applied to the isolation of histamine receptor genes. Very recently this strategy has led to the cloning of cDNAs encoding the H_1 and H_2 receptor proteins in several animal species (Table 1). The gene encoding the histamine H_3 receptor has not yet been isolated, however.

The H_1 receptor was the first member of this family to be pharmacologically characterized in the 1940s with the design of the first "antihistamines", now termed H_1-receptor antagonists, which was based upon the blockade of histamine-induced contraction of smooth muscles. Studies of the molecular properties, partial purification as well as precise localization of the H_1 receptor were recently made possible thanks to the development of specific reversible probes, i.e., [^{3}H]mepyramine or [^{125}I]iodobolpyramine, and of a photoaffinity ligand, [^{125}I]iodoazidophenpyramine (Ruat et al., 1988). These studies have revealed that the H_1 receptor subtype is constituted by a single protein (Ruat et al., 1992b). However, based upon biochemical data, the existence of an isoform of this protein in the heart was suggested (Ruat et al., 1990a).

After the suggestion of the existence of an additional histamine receptor subtype, pharmacologically distinct from the H_1 receptor (Ash and Schild, 1966), the utilization of the isolated guinea pig heart as the bioassay led to the definition of the H_2 receptor and the design of the first selective antagonists (Black et al., 1972). The pharmacological definition of H_1 and H_2 receptors has also led to drugs highly useful for treating life-threatening allergic and gastrointestinal disorders. Recently the design of a series of piperidinomethylphenoxycyanoguanidine derivatives (Hirschfeld et al., 1992) has led to a specific and highly potent [^{125}I]iodinated antagonist iodoaminopotentidine (^{125}I-APT) as a selective probe for the H_2 receptor. This radioligand allowed the first autoradiographic localization of the H_2 receptor in mammalian brain (Ruat et al., 1990b; Traiffort et al., 1992a). Biochemical studies of the H_2 receptor were also made possible using a derived photoaffinity probe (Ruat et al., 1990b).

Table 1. Molecular Properties of Cloned Histamine Receptors

Receptor	Species	Sequence	Second messenger	Ref.
H_1	Bovine	491 a.a.	—	Yamashita et al., 1991
	Guinea pig	488 a.a.	Pl hydrolysis; stimulation of arachidonic acid release	Leurs et al., 1993
H_2	Human	359 a.a.	Stimulation of adenylate cyclase	Gantz et al., 1991a
	Dog	359 a.a.	Stimulation of adenylate cyclase	Gantz et al., 1991b
	Rat	358 a.a.	Stimulation of adenylate cyclase; inhibition of arachidonic acid release	Ruat et al., 1991 Traiffort et al., 1992b

More recently, a third histamine receptor subtype, the H_3 receptor, was evidenced in rodent and human brain by the inhibition of histamine release and synthesis it mediates in various areas. Thus, the H_3 receptor was initially characterized as a presynaptic autoreceptor expressed at histaminergic nerve-endings (Arrang et al., 1983). With the design of (R)-methylhistamine and thioperamide, a highly potent and selective H_3-receptor agonist and antagonist respectively, the critical role of the H_3 receptor in the control of histaminergic neurons *in vivo* was established. The ligand ^{3}H-(R)-methylhistamine allowed the characterization of the receptor in membranes from central and peripheral tissues (Arrang et al., 1987). The autoradiographic localization of H_3 receptors performed in rat brain with this ligand, together with lesion data (Pollard et al., 1993) and pharmacological studies (Schwartz et al., 1990), revealed the presence of H_3 receptors on various cells, in addition to histaminergic neurons.

1.14.1.   Histamine H_1 Receptors

Selective labeling of the H_1 receptor protein was achieved in 1977 with the introduction of [^{3}H]mepyramine (Hill et al., 1977). This tritiated probe labels with a high affinity (Kd = 0.5 to 4 nM) a homogeneous population of sites identified unambiguously as typical H_1 receptors in various brain areas as well as in peripheral tissues (ileum, lungs). The affinity of this ligand derived from binding studies is in good agreement with that determined from a functional response, i.e., the contraction of the guinea pig ileum. [^{3}H]Mepyramine and several H_1 antagonists display a significantly lower affinity toward the rat or mouse H_1 receptor as compared to guinea pig or human H_1 receptors (Chang et al., 1979; Hill, 1980) (Table 2). These differences indicated structural heterogeneity among species. [^{125}I]Iodobolpyramine, a mepyramine derivative, labels guinea pig H_1 receptors with a high affinity (Kd = 0.15 nM) in central (Körner et al., 1986) and peripheral tissues (Ruat et al., 1990a). This compound has been useful for partial purification studies of the H_1 receptor protein solubilized with digitonin (Ruat et al., 1992a). Photoaffinity labeling studies of the H_1 receptor in brain and peripheral tissues, conducted with [^{125}I]iodoazidophenpyramine, a selective and highly potent related photoaffinity probe (Kd = 0.01 nM), have identified a 56 kDa peptide as being the ligand binding domain of the H_1 receptor. However, the identification of an apparently distinct 68 kDa protein in heart membranes suggests the occurrence of several isoforms of the H_1 receptor protein that cannot be distinguished pharmacologically (Ruat et al., 1990a).

Table 2.   Radioligands Used for Labeling Histamine Receptors

Radioligands	Species	Kd (nM)	Ref.
H₁ Receptors			
[³H]Mepyramine	Guinea pig	0.5	Hill et al., 1977
	Mouse, rat	4	Hill and Young, 1980
	Human	1	Chang et al., 1979
[¹²⁵I]Iodobolpyramine	Guinea pig	0.15	Körner et al., 1986
[¹²⁵I]Iodoazidophenpyramine*	Guinea pig	0.01	Ruat et al., 1988
	Rat	0.01	
H₂ Receptors			
[³H]Tiotidine	Guinea pig	3–30	Gajtkowski et al., 1983
[¹²⁵I]Iodoaminopotentidine	Guinea pig	0.3	Ruat et al., 1990b
	Human	0.3	Traiffort et al., 1992a
[¹²⁵I]Iodoazidopotentidine*	Guinea pig	—	Ruat et al., 1990b
H₃ Receptors			
[³H](R)α-Methylhistamine	Rat	0.5	Arrang et al., 1987
[¹²⁵I]Iodophenpropit	Rat	0.3	Jansen et al., 1992

*   Photoaffinity probes.

1.14.1.3   Bovine H₁ Receptors

Cloning of a bovine H₁ receptor was recently performed by Yamashita et al. by using an oocyte expression system (Yamashita et al., 1991). This technique has been successful for the cloning of several G-protein coupled receptors (Masu et al., 1987). Histamine induced Ca^{2+}-dependent inward Cl$^-$ currents in *Xenopus laevis* oocytes injected with poly(A)$^+$ mRNAs prepared from bovine adrenal medulla (Meyerhof et al., 1990; Sugama et al., 1991). This response was blocked by H₁ antagonists with a potency in agreement with that previously found at H₁ receptors characterized in bovine chromaffin cells in functional or binding studies. In order to clone the H₁ receptor, Yamashita et al. size-fractionated poly(A)$^+$ RNA from bovine adrenal medulla and tested various fractions in oocytes (Yamashita et al., 1991). Among the two peaks of poly(A)$^+$ mRNAs giving histamine-evoked currents in oocytes, one ranging from 2.5 to 3.5 kb was further selected for constructing a cDNA library in λZAPII and mRNA from several pools of independent clones were further tested for histamine-evoked currents in oocytes. The single positive pool was progressively subdivided into smaller pools until a single clone was obtained. Sequence analysis of the cDNA insert revealed an open reading frame encoding a protein of 491 amino acids. The encoded protein contains seven clusters of 20 to 25 hydrophobic residues representing putative transmembrane domains, clearly indicating its belonging to the superfamily of G-protein coupled receptors. In these transmembrane domains, the H₁ receptor shares homology with all other G-protein coupled receptors, particularly with the muscarinic receptors (40%). The histamine H₁ receptor possesses a very long third intracytoplasmic loop which presumably interacts with G-proteins, a structural feature previously reported for several receptors coupled to phospholipase C. It contains also three N-glycosylation sites located in the amino terminal region and in the second extracellular loop, and a high number of serine and threonine residues representing potential phosphorylation sites which may play an important role in regulating signal transduction.

In various receptors belonging to the same superfamily, the putative transmembrane helices have been shown to play an important role for ligand binding, and site-directed mutagenesis studies (Savarese and Fraser, 1992) suggested that an aspartate residue found in TM3 of all monoaminergic receptors is responsible for salt binding the positively charged amino group of amines. A corresponding residue is also found in the bovine H_1 as well as in all other cloned histamine receptors (Figure 1) and may play a similar role in histamine binding. A major difference between histamine and catecholamine receptors is the absence of serine residues in TM5 which, in the latter case, have been proposed to interact with the hydroxyl groups of the catechol moiety. Instead, a threonine and an asparagine residues are present in the H_1 receptor (Figure 2). To act as H_1 receptor agonists, molecules require the presence of a heteroaromatic nucleus including a nitrogen atom with a free pair of electrons on the position next to the ethylamine side chain. Therefore, the presence as a H_1-receptor feature of a proton-donating amino acid moiety might account for the previous empirical rule for H_1-receptor activation (Timmerman, 1992). Site-directed mutagenesis should clarify this issue.

Monkey kidney COS-7 cells transfected with a mammalian expression vector containing the full length bovine cDNA expressed [^{3}H]mepyramine binding sites with a Kd of 3.2 nM and a receptor density of 7 pmol/mg of protein (Yamashita et al., 1991). Ki values determined for mepyramine, doxepin, and (+) and (–) chlorpheniramine were in good agreement with those determined in a functional response. Second messenger coupling in these cells was not reported, although the induced chloride currents in the *Xenopus* oocytes, suggest a phospholipase C coupling.

1.14.1.8   Guinea Pig H_1 Receptors

In our laboratory (Traiffort et al., 1993), a guinea pig genomic library screened under high stringency hybridization conditions with a probe derived from the cDNA sequence of the bovine H_1 receptor led to the isolation of an intronless gene containing an open reading frame encoding a protein of 488 amino acid residues (Mr = 55,619) and showing 66% overall identity with the bovine H_1 receptor (Yamashita et al., 1991) and 90% homology within the TMs. Further comparison of the two proteins indicated that the amino terminal tail and the third intracellular loop of the guinea pig H_1 receptor were 8 amino acids longer and 11 amino acids shorter, respectively, than those of the bovine protein. Features that are conserved among these two proteins are the two consensus sequences for N-linked glycosylation present in the amino terminal tail. Several protein kinase C phosphorylation sites and a cAMP phosphorylation site are localized in the cytoplasmic loops of the guinea pig protein. Finally, the third cytoplasmic loop presents many serine or threonine residues that can be phosphorylated by various kinases.

Chinese hamster ovary (CHO) cells transfected with a plasmid containing the full length DNA sequence expressed [^{3}H]mepyramine binding sites ranging from 50 to 500 fmol/mg protein with a Kd of 0.4 nM, a value comparable to that previously characterized in brain membranes (Garbarg et al., 1992b). These sites were identified as representing the H_1 receptor since binding of [^{3}H]mepyramine on CHO(H_1) membranes was inhibited monophasically and in a stereoselective manner by various H_1 antagonists with the expected potencies.

As expected for a typical H_1 receptor (Schwartz et al., 1991; Hill and Donaldson, 1992), histamine potently stimulated the hydrolysis of phosphoinositides on CHO(H_1) cells with an EC_{50} of 1.4 μM, an effect competitively inhibited by mepyramine (Leurs et al., submitted). Moreover, histamine induced the release of [^{3}H]arachidonic acid with an EC_{50} of 1.3 μM via the activation of phospholipase A_2. In contrast to the hydrolysis of phosphoinositides, the stimulation of phospholipase A_2 was partially sensitive to pertussis toxin, a G_i- and G_o-protein inhibitor.

Finally, the H_1 receptor also interacted with the adenylate cyclase system. Whereas, histamine alone did not affect the cAMP levels in CHO(H_1) cells, it potentiated the cAMP production by forskoline with an EC_{50} of 0.4 μM. This effect was insensitive

FIGURE 1. Comparison of amino acid sequences of cloned receptors. Putative transmembranes domains corresponding to hydrophobic residues are indicated (TM1 to TM7). Conserved amino acids are indicated (*). (g-p: guinea pig; bov: bovine; hum: human).

to pertussis toxin and apparently independent of Ca^{2+} influx and protein kinase C activation. The exact mechanism of this potentiation awaits further clarification. These data indicate that in $CHO(H_1)$ cells, the H_1 receptor interacts with three major signal transduction pathways via the interaction with at least two different G-proteins (Leurs et al., 1993).

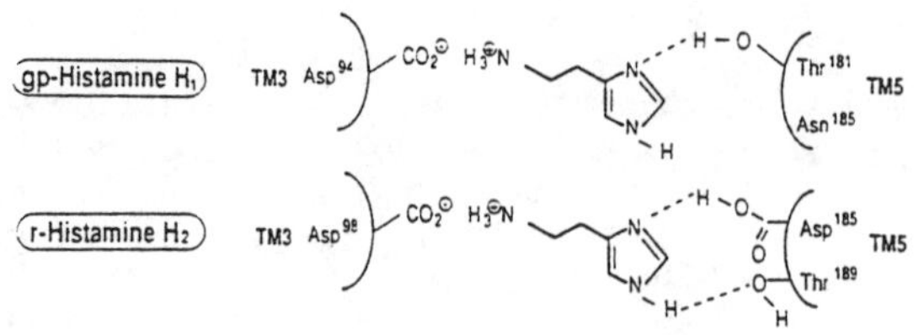

FIGURE 2. Models for the interaction of histamine with TM3 and TM5 of guinea pig (g-p) histamine H_1 and rat (r) histamine H_2 receptors.

Northern blot analysis of a variety of guinea pig tissues revealed two high-stringency hybridizing forms of mRNA for this H_1 receptor. One of ~3.3 kb was observed in all central and peripheral regions tested whereas the higher molecular weight species, of ~3.7 kb, was evidenced in lung and ileum, and was likely to represent another mRNA species of the H_1 receptor gene. Northern blot analysis of rat tissues performed with the same probe revealed a single band of ~3.3 kb. The signal was the strongest in hypothalamus and hippocampus, but was also easily detected in other central and peripheral regions. Regional distribution of H_1 receptor transcripts in guinea pig and rat brain tissues was consistent with the distribution of the H_1 receptor protein as previously determined in membrane-binding or autoradiographic studies using selective ligands (Pollard and Bouthenet, 1992). A high expression level was also observed in peripheral tissues where H_1-receptor mediated responses occur such as lung or heart.

In situ hybridization studies conducted in guinea pig brain are in good agreement with H_1 receptor localization already evidenced by autoradiography. For example, in cerebellum the presence of H_1 receptor mRNA in Purkinje cells may account for the high expression level of the protein in the molecular layer where extensive dendritic arborizations of these cells are found. In hippocampus as well, the high density of H_1 receptor mRNA found in Ammon's horn pyramidal cells, accounts for the high density of H_1 binding sites visualized by autoradiography in the radiatum and lacunosum molecular layers.

1.14.2   Histamine H_2 Receptors

The design of [^{125}I]iodoaminopotentidine as the first reliable probe for selectively labeling the H_2 receptor has led to the first autoradiographic localization of this receptor in rodent (Ruat et al., 1990) and human brain (Traiffort et al., 1992a). Tritiated probes such as [^{3}H]cimetidine, [^{3}H]ranitidine, or [^{3}H]impromidine had been unsuccessful in previous attempts at labeling this receptor (Garbarg et al., 1992b). [^{3}H]Tiotidine, which displays a Kd of 10 nM at the H_2 receptor (Gajtkowski et al., 1983), had been previously used to label this receptor in some preparations (Garbarg et al., 1992b) but its high nonspecific binding had restricted its use. [^{125}I]Iodoaminopotentidine labels a homogeneous population of binding sites in striatal, cortical, or hippocampal membranes of guinea pig brain as well as in human homogenates of post-mortem caudate nucleus. These sites were unambiguously identified as typical H_2 receptors with similar pharmacology in guinea pig and human brain. Photoaffinity labeling studies performed with [^{125}I]iodoazidopotentidine has led to the characterization of 32- and 58-kDa peptides as derived from the ligand recognition domain of the guinea pig H_2 receptor (Ruat et al., 1990b).

1.14.2.1   Human H_2 Receptors

The human H_2 receptor gene has been cloned in two steps (Gantz et al., 1991a). The polymerase chain reaction has been applied to the cDNA obtained from mRNA of human gastric fundic mucosa using degenerate primers based upon the nucleotide sequence of the cloned canine H_2 receptor gene. One resulting subcloned cDNA fragment had 83% homology to a region of the canine H_2 receptor gene. The amplified DNA was then used as a probe to screen a human genomic library and isolate a gene encoding a protein of 359 amino acid residues showing 87 and 82% homology with the canine and rat H_2 receptors, respectively. Northern blot analysis indicated the presence of a 4.8-kb transcript in human fundic mucosa. Histamine dose dependently increased cAMP accumulation (EC_{50} = 0.05 μM) in Colo-320 DM cells permanently transfected with the human H_2-receptor gene. This effect was blocked by 10 μM cimetidine, a H_2 antagonist. A detailed pharmacological study of the cloned receptor has not been reported, nor a precise distribution of H_2 transcripts in brain or in peripheral tissues.

1.14.2.5   Canine H_2 Receptors

The canine H_2 receptor gene was the first histamine receptor gene to be cloned (Gantz et al., 1991b). Cloning was performed using the polymerase chain reaction (PCR) with degenerate primers, according an approach based upon the high homology found between transmembrane domains of aminergic receptors. A DNA prepared from a partial-length clone obtained from canine gastric parietal cell cDNA, was used as a probe to isolate an intronless gene from a canine genomic library. When transfected with this gene, L-cells were shown to bind [³H]tiotidine and to express histamine-sensitive adenylate cyclase activity, thereby suggesting that the gene encoded a H_2 receptor. However, some doubt was left inasmuch as histamine receptor subtypes are poorly defined in the dog and a precise pharmacological characterization was not performed. A major 4.5-kb mRNA transcript was found in a fraction of fundic mucosal cells and also in brain indicating that expression of the encoded protein is not restricted to peripheral tissues.

The H_2 receptor is characterized by a very short third intracytoplasmic loop and a long C-terminal tail, two structural features usually encountered in receptors positively coupled to adenylate cyclase. The crucial aspartate residue found in TM3 (Asp^{98}) presumably interacts with the amino moiety of the histamine molecule as indicated by site-directed mutagenesis (Gantz et al., 1992). When this aspartate residue is replaced by an asparagine residue, increase in cellular cAMP content induced by histamine was completely abolished. Moreover, the mutated receptor failed to bind the antagonist [³H]tiotidine. A remarkable feature of the H_2 receptor is the presence of aspartate and threonine residues in TM5 (replaced in the H_1 receptor by threonine and asparagine, respectively), postulated to be important for hydrogen bonding with the nitrogen atoms of the imidazole ring of histamine (Gantz et al., 1991b; Birdsall, 1991). The mechanism for the interaction of histamine at this level is not entirely clear since the exact ionization state of the residues is not known, but the carboxyl and hydroxyl groups of these residues may function as a proton-accepting and proton-donating group, respectively, a model in agreement with the observation that tautomerism of the histamine imidazol ring is favorable for H_2 receptor activation (Ganellin, 1992; Leurs et al., 1991). It is interesting to note that this type of postulated interaction is similar to that taking place at the triad within the active site of serine proteases.

1.14.2.9   Rat H_2 Receptors

The rat H_2 receptor gene has been cloned by homology screening of a rat genomic library using oligonucleotides derived from the canine H_2 receptor sequence as probes

(Ruat et al., 1991). An intronless gene encoding a 358 amino acid protein with a deduced molecular weight of 40 kDa was isolated. This size is consistent with that (58 kDa) of the peptide evidenced by photoaffinity labeling of the H_2 receptor in brain (Ruat et al., 1990b). The rat receptor displays an 82% overall homology with the canine or human proteins. The rather high similarity of these three proteins, particularly at the level of TMs where over 90% homology is found, strongly suggests that they are species homologs. The sole gap is found in the second extracellular loop where an amino acid is lacking in the rat protein. An N-linked glycosylation site is found near the amino terminus and potential protein kinase C phosphorylation sites are observed in the third intracytoplasmic loop and in the C-terminal tail suggesting potential regulating sites. No potential protein kinase A phosphorylation site has been identified on any of the three proteins.

Convincing data concerning the histamine H_2 nature of the encoded protein came after stable expression of the rat gene in CHO cells (Traiffort et al., 1992b). The detailed pharmacological characterization of the expressed protein obtained with [^{125}I]iodoaminopotentidine corresponded to that of a typical H_2 receptor protein as indicated by the Kd value of the iodinated probe (Kd = 0.4 nM) as well as the Ki values of various antagonists which were similar to those derived from cerebral receptors (Ruat et al., 1990b; Traiffort et al., 1991, 1992a). Histamine potently stimulated cAMP accumulation in transfected cells, a response competitively blocked with the expected potency by ranitidine, a highly potent H_2 antagonist. Association of the H_2 receptor with stimulation of cAMP accumulation or adenylyl cyclase is well known and is thought to mediate most of the physiological responses triggered by this receptor. Unexpectedly, however, histamine potently (EC$_{50}$ = 0.05 μM) inhibited the arachidonic acid release from transfected CHO cells induced either by stimulation of constitutive purinergic receptors or by application of a Ca^{2+} ionophore (Traiffort et al., 1992b). This inhibitory effect, mimicked by H_2 agonists, and competitively blocked by ranitidine, was independent of changes in either cAMP or Ca^{2+} levels. Hence, transfected H_2 receptors may reduce arachidonic acid release through inhibition of phospholipase A_2, i.e., by a direct mechanism, similar to that responsible for receptor-linked inhibition of adenylyl cyclase. These results suggest that a single H_2 receptor subtype may be linked not only to adenylyl cyclase activation but also to reduction of phospholipase A_2 activity, a novel signaling pathway for H_2 receptors (Traiffort et al., 1992b).

Northern blot analysis of rat tissues revealed that a transcript of 6 kb is expressed in brain as well as in peripheral tissues (Ruat et al., 1991). The signal was easily detectable in rat brainstem but also in cerebral cortex, striatum, hippocampus, and also in hypothalamus where H_2 receptors have been found by binding studies (Ruat et al., 1990b). Among peripheral tissues, the stomach highly expresses the gene transcript, in agreement with the well-known control of gastric secretion mediated by the H_2 receptor. In guinea pig tissues, a transcript of 4.5 kb is also detected using the same approach (Traiffort et al., 1992b) in agreement with the previously known localization of the receptor protein in this species (Ruat et al., 1990b). Interestingly, a same mRNA species was found in both guinea pig stomach and heart, suggesting that the same molecular species mediates the actions of histamine in these tissues. This is consistent with the hypothesis that the marked differences in potency of some H_2 receptor antagonists in these tissues are related to bioavailability factors rather than to receptor heterogeneity. In guinea pig brain where labeling of the transcript was higher than in rat, the regional distribution of the transcript is in agreement with the H_2-receptor distribution determined by autoradiography or binding studies.

1.14.99   Other Histamine Receptors

1.14.99.1   Histamine H_3 Receptors

The selective binding of ^3H-(R)-methylhistamine at H_3-receptor sites is specifically modulated in the presence of guanylnucleotides, indicating that the H_3 receptor is coupled to its (so far unknown) effector system via a G protein (Arrang et al., 1990). Hence, it seems likely that the H_3 receptor belongs, like the two other histamine receptor subtypes, to the superfamily of proteins with 7 TMs. Structure-activity relationships among agonists show that small modifications of the histamine molecule have more dramatic effects on binding at H_3 than at either H_1 or H_2 receptors. This is consistent with the higher apparent potency of histamine at H_3 than at H_1 or H_2 receptors (Arrang et al., 1983) and may suggest a tighter binding of the amine, possibly with critical amino acid residues present in TM5 different from those involved in H_1 or H_2 receptors.

The pharmacological specificity of the H_3 receptor is high as shown by its stimulation or blockade by several agents displaying low, if any, activity at H_1 and H_2 receptors. Besides (R)-methylhistamine (Arrang et al., 1987), the (R)α,(S)βisomer of dimethylhistamine (Lipp et al., 1992b) and imetit or S-[2-(4-imidazolyl)ethyl]isothiourea (Garbarg et al., 1992a) behave as potent H_3-receptor agonists, whereas their activities at H_1 and H_2 receptors are very low. Regarding antagonists, thioperamide, i.e., N-cyclohexyl-4-(imidazol-4-yl)-1 piperidine carbothioamide behaves as a competitive antagonist with nanomolar affinity but displays negligible affinity at H_1 or H_2 receptors (Arrang et al., 1987).

Pharmacological data obtained at H_3 receptors with various H_1 or H_2 ligands do not allow any comparison between the putative homology shared by the H_3 receptor protein with either the H_1- or the H_2-receptor proteins (Lipp et al., 1992a). For example, both betahistine and impromidine, an H_1- and H_2-receptor agonist, respectively, behave as H_3-receptor antagonists (Arrang et al., 1985; 1983). More recently, [^{125}I]iodophenpropit, an imidazol derivative has been proposed as an H_3-antagonist ligand (Jansen et al., 1992).

REFERENCES

Arrang, J. M., Garbarg, M., and Schwartz, J. C. (1983) Autoinhibition of brain histamine release mediated by a novel class (H_3) of histamine receptor. *Nature* 302, 832–837.

Arrang, J. M., Garbarg, M., Quach, T. T., Dam Trung Tuong, M., Yeramian, E., and Schwartz, J. C. (1985) Actions of betahistine at histamine receptors in the brain. *Eur. J. Pharmacol.* 111, 73–84.

Arrang, J. M., Garbarg, M., Lancelot, J. C., Lecomte, J. M., Pollard, H., Robba, M., Schunack, W., and Schwartz, J. C. (1987) Highly potent and selective ligands for histamine H_3-receptors. *Nature* 327, 117–123.

Arrang, J. M., Roy, J., Morgat, J. L., Schunack, W., and Schwartz, J. C. (1990) Histamine H_3-receptor binding sites in rat brain membranes: modulation by guanine nucleotides and divalent cations. *Eur. J. Pharmacol.* 188, 219–227.

Ash, A. S. F. and Schild, H. O. (1966) Receptors mediating some actions of histamine. *Br. J. Pharmacol. Chemother.* 27, 427–439.

Birdsall, N. J. M. (1991) Cloning and structure-function of the H_2 histamine receptor. *Trends Pharmacol. Sci.* 12, 9–10.

Black, J. W., Duncan, W. A. M., Durant, C. J., Ganellin, C. R., and Parsons, M. E. (1972) Definition and antagonism of histamine H_2-receptors. *Nature* 236, 385–390.

Chang, R. S. L., Tran, V. T., and Snyder, S. H. (1979) Heterogeneity of histamine H_1-receptors: species variations in [^3H]mepyramine binding of brain membranes. *J. Neurochem.* 32, 1653–1663.

Gajtkowski, G. A., Norris, D. B., Rising, T. J., and Wood, T. P. (1983) Specific binding of [^{3}H]tiotidine to histamine H_2-receptors in guinea pig cerebral cortex. *Nature* 304, 65–67.

Ganellin, C. R. (1992) Pharmacochemistry of H_1 and H_2 receptors. In *The Histamine Receptor*, Schwartz, J. C. and Haas, H. L., Eds., Wiley-Liss, New York, pp. 1–56.

Gantz, I., Munzert, G., Tashiro, T., Schaffer, M., Wang, L., Delvalle, J., and Yamada, T. (1991a) Molecular cloning of the human histamine H_2 receptor. *Biochem. Biophys. Res. Commun.* 178, 1386–1392.

Gantz, I., Schaffer, M., Delvalle, J., Logsdon, C., Campbell, V., Uhler, M., and Yamada, T. (1991b) Molecular cloning of a gene encoding the histamine H_2 receptor. *Proc. Natl. Acad. Sci. U.S.A.* 88, 429–433.

Gantz, I., Delvalle, J., Wang, L. D., Tashiro, T., Munzert, G., Guo, Y. J., Konda, Y., and Yamada, T. (1992) Molecular basis for the interaction of histamine with the histamine H_2 receptor. *J. Biol. Chem.* 267, 20840–20843.

Garbarg, M., Arrang, J. M., Rouleau, A., Ligneau, X., Dam Trung Tuong, M., Schwartz, J. C., and Ganellin, C. R. (1992a) S-[2-(4-Imidazolyl)ethyl]isothiourea, a highly specific and potent histamine H_3-receptor agonist. *J. Pharmacol. Exp. Ther.* 263, 304–310.

Garbarg, M., Traiffort, E., Ruat, M., Arrang, J. M., and Sc2hwartz, J. C. (1992b) Reversible labeling of H_1, H_2 and H_3-receptors. In *The Histamine Receptor*, Schwartz, J. C. and Haas, H. L., Eds., Wiley-Liss, New York, pp. 73–95.

Hill, S. J., Young, J M., and Marrian, D. H. (1977) Specific binding of ^{3}H-mepyramine to histamine H_1-receptors in intestinal smooth muscle. *Nature* 270, 361–363.

Hill, S. J. and Young, J. M. (1980) Histamine H_1-receptors in the brain of the guinea pig and the rat: differences in ligand binding properties and regional distribution. *Br. J. Pharmacol.* 68, 687–696.

Hill, S. J. (1990) Distribution, properties and functional characteristics of three classes of histamine receptor. *Pharmacol. Rev.* 42, 45–83.

Hill, S. J. and Donaldson, J. (1992) The H_1 receptor and inositol phospholipid hydrolysis. In: *The Histamine Receptor*, Schwartz, J. C. and Haas, H. L., Eds., Wiley-Liss, New York, pp. 109–128.

Hirschfeld, J., Buschauer, A., Elz, S., Schunack, W., Ruat, M., Traiffort, E., and Schwartz, J. C. (1992) Iodoaminopotentidine and related compounds: A new class of ligands with high affinity and selectivity for the histamine H_2 receptor. *J. Med. Chem.* 35, 2231–2238.

Jansen, F. P., Rademaker, B., Bast, A., and Timmerman, H. (1992) The first radiolabeled histamine H_3 receptor antagonist, [^{125}I]iodophenpropit: saturable and reversible binding to rat cortex membranes. *Eur. J. Pharmacol.* 217, 203–205.

Körner, M., Bouthenet, M. L., Ganellin, C. R., Garbarg, M., Gros, C., Ife, R. J., Salès, N., and Schwartz, J. C. (1986) [^{125}I]Iodobolpyramine, a highly sensitive probe for histamine H_1-receptors in guinea pig brain. *Eur. J. Pharmacol.* 120, 151–160.

Leurs, R., Van der Goot, H., and Timmerman, H. (1991) Histaminergic agonists and antagonists. Recent developments. *Adv. Drug Res.* 20, 217–304.

Lipp, R., Stark, H., and Schunack, W. (1992a) Pharmacochemistry of H_3-receptors. In *The Histamine Receptor*, Schwartz, J. C. and Haas, H. L., Eds., Wiley-Liss, New York, pp. 57–72.

Lipp, R., Arrang, J. M., Garbarg, M., Luger, P., Schwartz, J. C., and Schunack, W. (1992b) Synthesis, absolute configuration, stereoselectivity and receptor-selectivity of (αR,βS)-,-dimethylhistamine, a novel highly potent histamine H_3-receptor agonist. *J. Med. Chem.*, 35, 4434–4441.

Masu, Y., Nakayama, K., Tamaki, H., Harada, Y., Kuno, M., and Nakanishi, S. (1987) cDNA cloning of bovine substance K receptor through oocyte expression system. *Nature* 329, 836–838.

Meyerhof, W., Schwarz, J. R., Hollt, V., and Richter, D. (1990) Expression of histamine H_1-receptors in *Xenopus* oocytes injected with messenger ribonucleic acid from bovine adrenal medulla: pertussis toxin insensitive activation of membrane chloride currents. *J. Neuroendoc.* 2, 547–553.

Pollard, H. and Bouthenet, M. L. (1992) Autoradiographic visualization of the three histamine receptor subtypes in the brain. In *The Histamine Receptor*, Schwartz, J. C. and Haas, H. L., Eds., Wiley-Liss, New York, pp. 179–192.

Pollard, H., Moreau, J., Arrang, J. M., and Schwartz, J. C. (1993) A detailed autoradiographic mapping of histamine H_3-receptors in rat brain areas. *Neuroscience*, 52, 169–189.

Ruat, M., Bouthenet, M. L., Schwartz, J. C., and Ganellin, C. R. (1990a) Unique histamine H_1-receptor isoform in heart. *J. Neurochem.* 55, 379–385.

Ruat, M., Körner, M., Garbarg, M., Gros, C., Schwartz, J C., Tertiuk, W., and Ganellin, C. R. (1988) Characterization of histamine H_1-receptor binding peptides in guinea pig brain using [^{125}I]iodoazidophenpyramine, an irreversible specific photoaffinity probe. *Proc. Natl. Acad. Sci. U.S.A.* 85, 2743–2747.

Ruat, M., Traiffort, E., Bouthenet, M. L., Schwartz, J. C., Hirschfeld, J., Buschauer, A., and Schunack, W. (1990b) Reversible and irreversible labeling and autoradiographic localization of the cerebral H_2-receptor using [^{125}I]iodinated probes. *Proc. Natl. Acad. Sci. U.S.A.* 87, 1658–1662.

Ruat, M., Traiffort, E., Arrang, J. M., Leurs, R., and Schwartz, J. C. (1991) Cloning and tissue expression of a rat histamine H_2-receptor gene. *Biochem. Biophys. Res. Commun.* 179, 1470–1478.

Ruat, M., Traiffort, E., Garbarg, M., Schwartz, J. C., Demonchaux, P., Ife, R. J., Tertiuk, W., and Ganellin, C. R. (1992a) Design of an affinity matrix for purification of the histamine H_1 receptor from guinea pig cerebellum. *J. Neurochem.* 58, 350–356.

Ruat, M., Traiffort, E., and Schwartz, J. C. (1992b) Biochemical properties of histamine receptors. In *The Histamine Receptor*, Schwartz, J. C. and Haas, H. L., Eds., Wiley-Liss, New York, pp. 97–107.

Savarese, T. M. and Fraser, C. M. (1992) *In vitro* mutagenesis and the search for structure-function relationships among G protein-coupled receptors. *Biochem. J.* 283, 1–19.

Schwartz, J. C., Arrang, J. M., Garbarg, M., and Pollard, H. (1990) A third histamine receptor subtype: characterization, localization and functions of the H_3 receptor. *Agents Actions* 30, 13–23.

Schwartz, J. C., Arrang, J. M., Garbarg, M., Pollard, H., and Ruat, M. (1991) Histaminergic transmission in the mammalian brain. *Physiol. Rev.* 71, 1–52.

Schwartz, J. C. and Haas, H. L., Eds. (1992) *The Histamine Receptor*. Receptor Biochemistry and Methodology, Vol. 16. Wiley-Liss, New York.

Sugama, K., Yamashita, M., Fukui, H., Seiji, I., and Wada, H. (1991) Functional expression of H_1-histaminergic receptors in *Xenopus laevis* oocytes injected with bovine adrenal medullary mRNA. *Jpn. J. Pharmacol.* 55, 287–290.

Timmerman, H. (1992) Cloning of the H_1-histamine receptor. *Trends Pharmacol. Sci.* 13, 6–7.

Traiffort, E., Ruat, M., and Schwartz, J. C. (1991) Interaction of mianserin, amitriptyline and haloperidol with guinea pig cerebral histamine H_2 receptors studied with [^{125}I]iodoaminopotentidine. *Eur. J. Pharmacol. — Mol. Pharmacol. Sect.* 207, 143–148.

Traiffort, E., Pollard, H., Moreau, J., Ruat, M., Schwartz, J. C., Martinez-Mir, M. I., and Palacios, J. M. (1992a) Pharmacological characterization and autoradiographic localization of histamine H_2 receptors in human brain identified with [^{125}I]iodoaminopotentidine. *J. Neurochem.* 59, 290–299.

Traiffort, E., Ruat, M., Arrang, J. M., Leurs, R., Piomelli, D., and Schwartz, J. C. (1992b) Expression of a cloned rat histamine H_2 receptor mediating inhibition of arachidonate release and activation of cAMP accumulation. *Proc. Natl. Acad. Sci. U.S.A.* 89, 2649–2653.

Yamashita, M., Fukui, H., Sugama, K., Horio, Y., Ito, S., Mizuguchi, H., and Wada, H. (1991) Expression cloning of a cDNA encoding the bovine histamine H_1 receptor. *Proc. Natl. Acad. Sci. U.S.A.* 88, 11515–11519.

Traiffort, E., Leurs, R., Arrang, J. N., Tardivel-Lacombe, J., Diaz, J., Schwartz, J. C., and Ruat, M. (1993) Guinea pig histamine H_1 receptor. I. Gene cloning, characterization and tissue expression revealed by in situ hybridization. *J. Neurochem.*, in press.

Leurs, R., Traiffort, E., Arrang, J. M., Tardivel-Lacombe, J., Ruat, M., and Schwartz, J. C. (1993) Guinea pig histamine H_1 receptor. II. Stable expression in chinese hamster ovary cells reveals the interaction with three major signal transduction pathways. *J. Neurochem.*, in press.

15

Hormone Receptors

Rolf Sprengel and Carola Eva

1.15.0   Introduction

All peptide hormones interact with specific receptors at the cell surface of their target cells. Peptide hormones, in this chapter, will be defined as products of the endocrine glands with major effects in the endocrine system. In general, two types of receptors can be distinguished in function and structure. The first type are receptors composed of an extracellular, short membrane spanning a cytoplasmatic domain, which sometimes carries tyrosine kinase activity (see Volume 4). The second type are single polypeptides with seven putative membrane spanning segments which couple to G proteins. The G protein-coupled receptors mediate their physiological effect by activating second messenger pathways like cAMP and phosphoinositol pathways via coupling to G proteins. According to their primary amino acid sequence, these receptors can be divided into three separate subfamilies, each of which will be discussed separately:

A. Glycoprotein hormone receptors with minimal amino acid similarities to other G protein-coupled receptors.

B. Hormone receptors with minimal amino acid similarity to other G protein-coupled receptors.

C. Hormone receptors with moderate amino acid similarities to other G protein-coupled receptors.

A.   Glycoprotein Hormone Receptors with Minimal Amino Acid Similarities to Other G Protein-Coupled Receptors

The glycoprotein hormone receptors can be considered as a unique group among members of the G protein-coupled receptor family. Based on a phylogenetic sequence analysis, they appear to be the oldest members of this receptor family. They interact with large complex molecules (28 to 38 kDa in size) called the glycoprotein hormones. This term is usually applied to the pituitary hormones lutropin (luteinizing hormone, LH), follitropin (follicle stimulating hormone, FSH), thyrotropin (thyroid stimulating hormone, TSH) and the placental hormone choriogonadotropin (hCG). Other glycosylated polypeptides with hormone activity, like erythropoietin, are not included in this hormone family (Pierce and Parsons, 1981). The term gonadotropin is used for LH, hCG, and FSH which play major roles in the reproductive activities. Each hormone interacts specifically with its receptors in nanomolar concentrations, with the exception of LH and hCG. These hormones elicit identical biological responses through the activation of the same receptor.

The immediate response of the target cell to the binding of glycoprotein hormones is an increase in adenylyl cyclase activity mediated by intracellular membrane-associated G_s-protein. The resulting increase of cAMP ultimately leads to an increase in steroid synthesis and secretion. The glycoprotein hormones have the most complex chemical structures of the classical protein hormones. They are heteromeric molecules composed of a common α-subunit and a noncovalently associated β-subunit. The β-subunit is specific for each hormone, although they show a high degree of sequence homology especially in the regions which are supposed to interact with the α-subunit. LH and hCG share common biological properties and their β-subunits possess a high degree of sequence homology, with one obvious difference being a proline- and serine-rich carboxy-terminal extension of the hCG β-subunit. Each subunit shows disulfide bridges with 10 half cysteines in the a and 12 half cysteines in the β-subunit (Pierce and Parsons, 1981; Ryan et al., 1988) A further complexity of the hormones is achieved by the asparagine- and serine-linked carbohydrate side groups. Two asparagine-linked groups are found in the α-subunit. In the β-subunit the number of asparagine-linked oligosaccharides varies in position and number (Sairam, 1989) The serine-linked oligosaccharides, which are present in the carboxy-terminal extension of the hCG b-subunit, are not required for biological activity of the hormone because they are absent from the LH molecule. However, the asparagine-linked side chains seems to be important for the functional activity of the hormone. Enzymatically deglycosylated hormones and recombinantly expressed deglycosylated hormones are still active in receptor binding assays but fail to activate the hormone-mediated second messenger pathways (Matzuk et al., 1989; Sairam, 1990). Deglycosylated FSH with antagonistic activity was detected in sera from women treated with gonadotropin releasing hormone (GnRH) analogs (Dahl et al., 1988). Therefore, it is likely that native deglycosylated hormones (antihormones) might be involved in the regulation of receptor activation.

Molecular cloning identified the glycoprotein hormone receptors as a unique group within the superfamily of G protein-coupled receptors (McFarland et al., 1989; Loosfelt et al., 1989). Like all others, they contain seven hydrophobic transmembrane spanning segments which are important in signal transduction. However, in contrast to other G protein-coupled receptors, the glycoprotein hormone receptors are synthesized as precursor molecules with an amino-terminal signal peptide, and they contain a large extracellular domain, 300 to 350 amino acids in length. This sequence displays 14 sequence repeats built on a motif similar to a motif found in the family of leucine-rich glycoproteins (Patthy, 1987). This motif appears to form amphiphilic peptide surfaces involved in the protein-protein interaction. As experimentally indicated, this large extracellular domain mediates the hormone binding (Braun et al., 1991; Tsai-Morris et al., 1990; Xie et al., 1990). For the LH-CG receptor, amino-terminal leucine-rich repeats 1 to 8 seem to be responsible for the hormone-specific binding. In the case of the FSH receptor, amino acids of repeat 1 to 11 are involved (Braun et al., 1991). The TSH receptor needs the entire extracellular domain, with multiple discontinuous contact sites especially localized in repeat 7 and 8 for TSH-specific binding (Nagayama et al., 1991a; Nagayama et al., 1991b). Repeats 12 and 13 of the extracellular domain are highly variable in length and sequence and they might function as a hinge region which separates the hormone

```
HorA.FSH.human              1 - - - - M A L L L V S L L A F - L S L G S G C H H R I C H C S N R V F L C Q E S K V T E I P S D L P
HorA.FSH.rat                1 - - - - M A L L L V S L L A L - L G T G S G C H H W L C H C S N R V F L C Q D S K V T E I P T D L P
HorA.FSH.sheep              1 - - - - M A L F L V A L L A F - L S L G S G C H H R L C H C S N G V F L C Q D S K V T E M P S D L P
HorA.LH-CG.human            1 - - - - M K Q R F S P L Q - - - - L L K L L L L L Q A P L - P R A L R - R L C P E P C N C V P D G A
HorA.LH-CG.mouse            1 - - - - M G R R V P A L R Q L - L V L A M L V L K Q S Q L H S P E L S G S R C P E P C D C A P D G A
HorA.LH-CG.porcine          1 - - - - M R R R S L A L R - - - L L L A L L L L - P P P L - P Q T L L G A P C P E P C S C R P D G A
HorA.LH-CG.rat              1 - - - - M G R R V P A L R Q L - L V L A V L L L K P S Q L Q S R E L S G S R C P E P C D C A P D G A
HorA.TSH.canine             1 - - - - - - M R P P P L L H L A L L L A L P R S L G G K G C P S P P C E C H Q E D D F R V T C K D I
HorA.TSH.human              1 - - - - - - M R P A D L L Q L V L L L D L P R D L G G M G C S S P P C E C H Q E E D F R V T C K D I
HorA.TSH.rat                1 - - - - - - M R P G S L L Q L T L L L A L P R S L W G R G C T S P P C E C H Q E D D F R V T C K E L
HorB.Calcitonin.human       1 - - - - - - M R F T F T S R C L A L F L - - - L L N H P T P - - - - I L P A F S N Q T Y - P T I -
HorB.Calcitonin.porcine     1 - - - - - - M R F T L T R W C L T L F I - - - F L N R P L P - - - - V L P D S A D G A H T P T L -
HorB.GLP-1.rat              1 - - - - - - M L L T Q L H C P Y L L L L - - - L V V L S C - - - - - L P K A - P S A Q V M D F L -
HorB.PTH.human              1 - - - - - - M G T A R I A P G L A L L L C C P V L S S A Y A - - - - L V D A D D V M T K E E Q I F
HorB.PTH.opposum           1 - - - - - - M G A P R I S H S L A L L L C C S V L S S V Y A - - - - L V D A D D V I T K E E Q I I
HorB.PTH.rat                1 - - - - - - M G A A R I A P S L A L L L C C P V L S S A Y A - - - - L V D A D D V F T K E E Q I F
HorB.Secretin.rat           1 - - - - - - M L S T M R P R L S L L - L L - - - R L L L - - - - - - L T K A A H T V G V P P R L -
HorB.VIP.rat                1 - - - - - - M R P P S P P H V R W L C V L - - - A G A L A C A - - - - L R P A G S Q A A S P Q H E -
HorC.ACTH.human             1 - - - - - - - - - M K H I I - - - - - - - - - - - - - N S Y E N I N N T A R N -
HorC.ADH.human              1 - - - - - - - - M L M A S T T S A V - - - - - - - - - - - - - - P G H P S L P S L P S N -
HorC.ATII-1.bovine          1 - - - - - - - - - M I L N - - - - - - - - - - - - - - - - - - - S S T E D G I K R I Q D -
HorC.ATII-1.human           1 - - - - - - - - - M I L N - - - - - - - - - - - - - - - - - - - S S T E D G I K R I Q D -
HorC.ATII-1.rat             1 - - - - - - - - - M A L N - - - - - - - - - - - - - - - - - - - S S A E D G I K R I Q D -
HorC.ATII-1B.rat            1 - - - - - - - - - M T L N - - - - - - - - - - - - - - - - - - - S S T E D G I K R I Q D -
HorC.AV1a.rat               1 M S F P R G S Q D R S V G N S S P W W P L T T - - - - - - - - - - - - - E G S N G S Q E A A R L -
HorC.AV2.rat                1 - - - - - - - - - - M L L V S T V S A V - - - - - - - - - - - - - P G L F S P P S S P S N -
HorC.Bombesin.guineapig     1 - - - - - - - - - M S Q K Q P Q S P N Q T L - - - - - - - - I S I T N D T E S S S S V V S N -
HorC.Bombesin.human         1 - - - - - - - - - M A L N - - D C F L L N L - - - - - - - E V D H F M H C N I S - S H S A D -
HorC.Bombesin.mouse         1 - - - - - - - - - M A P N - - N C S H L N L - - - - - - - D V D P F L S C N D T F N Q S L S -
HorC.CCKa.rat               1 - - - - - - - - - M S H S P A R Q H L V E S S - - - R M D V V D S L L M N G S N I T P P C E -
HorC.CCKb.rat               1 - - - - - - - - - M E L L K L - N R S V Q G P - - - G P G S G S S L C R P G V S L L N S S S -
HorC.GPRN.human             1 - - - - - - - - - M D L H L F D Y A E P G N F - - - - - S D I S W P C N S S D C I V V D T V -
HorC.Gastrin.canine         1 - - - - - - - - - M E L L K L - N R S A Q G S - - - G A G P G A S L C R A G G A L L N S S G -
HorC.Gastrin.rat            1 - - - - - - - - - M E L L K L - N S S V Q G P - - - G P G S G S S L C H P G V S L L N S S S -
HorC.GnRH.human             1 - - - - - - - - - - M A N S A S P E Q - - - - - - - - - - - - - N Q N H C S A I N N S I -
HorC.GnRH.mouse             1 - - - - - - - - - - M A N N A S L E Q - - - - - - - - - - - - - D P N H C S A I N N S I -
HorC.GnRH.rat               1 - - - - - - - - - - M A N N A S L E Q - - - - - - - - - - - - - D Q N H C S A I N N S I -
HorC.MSH.human              1 - - - - - - - - - M A V Q G S Q R R L L - - - - - - - - - - - - - G S L N S T P T A I P Q -
HorC.MSH.mouse              1 - - - - - - - - - M S T Q E P Q K S L V - - - - - - - - - - - - - G S L N S N A T - - S H -
HorC.NMB.human              1 - - - - - - - - - M P S K - - S L S N L S V - - - - - T T G A N E S G S V P E G W E R D F L -
HorC.NMB.rat                1 - - - - - - - - - M P P R - - S L P N L S L - - - - - P T E A S E S E L E P E V W E N D F L -
HorC.Oxy.human              1 M E - - - - - - - G A L A A N W S A E A - - - - - - - - - - - A N A S A A P P G A E G -
HorC.RDC1.canine            1 - - - - - - - - - M D L H L F D Y A E P G N F - - - - - S D I S W P C N S S D C I V V D T V -
HorC.Somat1.human           1 - - - - - - - M F P N G T A S S P S S S P S P S P G S C G E G G G S R G P G A G A A D - G M E -
HorC.Somat1.mouse           1 - - - - - - - M F P N G T A S S P S S S P S P S P G S C G E G A C S R G P G S G A A D - G M E -
HorC.Somat1.rat             1 - - - - - - - M F P N G T A P S P T S S P S S S P G G C G E G V C S R G P G S G A A D - G M E -
HorC.Somat2.bovine          1 - - - - - - - - - M D L V S E - L N E T Q P - - - - - - - - - - W L T T P F D L N - G S V -
HorC.Somat2.human           1 - - - - - - - - - M D M A D E P L N G S H T - - - - - - - - - - W L S I P F D L N - G S V -
HorC.Somat2.mouse           1 - - - - - - - - - M E M S S E Q L N G S Q V - - - - - - - - - - W V S S P F D L N - G S L -
HorC.Somat2.rat             1 - - - - - - - - - M E L T S E Q F N G S Q V - - - - - - - - - - W I P S P F D L N - G S L -
HorC.Somat3.mouse           1 - - - - - - - - - M A T V T Y P S S E P M T - - - - - - - - - - - L D P G N T S S T W P L D -
HorC.Somat3.rat             1 - - - - - - - - - M A A V T Y P S S V P T T - - - - - - - - - - - L D P G N A S S A W P L D -
HorC.TA.rat                 1 - - - - - - - - - - - - - M A G - - - - - - - - - - - - N C S W E A H S T N Q N -
HorC.TRH.mouse              1 - - - - - - - - - M E - - - - - - - - - - - - - - - - N D T V S E M N Q T E L -
HorC.TRH.rat                1 - - - - - - - - - M E - - - - - - - - - - - - - - - - N E T V S E L N Q T E L -
HorC.mas.human              1 - - - - - - - - - - - M D G - - - - - - - - - - - - - S N V T S F V V E E P T -
HorC.mas.rat                1 - - - - - - - - - - - M D Q - - - - - - - - - - - - - S N M T S F - A E E K A -
HorC.mrg.human              1 - - - - - - - - - - M V W G K I C W F S Q - - - - - - - - - - - - - R A G W T V F A E S Q I -
```

FIGURE 1. Amino acid sequences of a selected number of hormone receptors, the glycoprotein hormone receptors, and the secretin-receptor family.

```
HorA.FSH.human           46  R N A I E L R F V L T K L R V I Q K G A F S G F G D L E K I E I S Q N D V L E V I E A D V F S N L P
HorA.FSH.rat             46  R N A I E L R F V L T K L - - - - - - R V I P K G S F A G F G D L E K I E I S Q N D V L E V I E A
HorA.FSH.sheep           46  R D A V E L R F V L T K L - - - - - - R V I P E G A F S G F G D L E K I E I S Q N D V L E V I E A
HorA.LH-CG.human         41  L R A P A P R P S - T R L S L A Y L P V K V I P S Q S F R G L N E V I K I E I S Q I D S L E R I E A
HorA.LH-CG.mouse         46  L R C P G P R A G L A R L S L T Y L P V K V I P S Q A F R G L N E V V K I E I S Q S D S L E R I E A
HorA.LH-CG.porcine       42  L R C P G P R A G L S R L S L T Y L P I K V I P S Q A F R G L N E V V K I E I S Q S D S L E K I E A
HorA.LH-CG.rat           46  L R C P G P R A G L A R L S L T Y L P V K V I P S Q A F R G L N E V V K I E I S Q S D S L E R I E A
HorA.TSH.canine          45  H R I P T L P P S T Q T L K F I E T Q L K T I P S R A F S N L P N I S R - - - - - - - - - - - - -
HorA.TSH.human           45  Q R I P S L P P S T Q T L K L I E T H L R T I P S H A F S N L P N I S R I Y V S I D V T L Q Q L E S
HorA.TSH.rat             45  H Q I P S L P P S T Q T L K L I E T H L K T I P S L A F S S L P N I S R I Y L S I D A T L Q R L E P
HorB.Calcitonin.human    35  - - - - - - - E P K P F L - - - - - - - - - - - - - - Y V V G R K K M M D A Q Y K C Y D R M - -
HorB.Calcitonin.porcine  36  - - - - - - - E P E P F L - - - - - - - - - - - - - - Y I L G K Q R M L E A Q H R C Y D R M - -
HorB.GLP-1.rat           34  - - - - - - - - F E K W - - - - - - - - - - - - - - K L Y - - - S D - - - - - Q C H H N L - -
HorB.PTH.human           40  L L H R A Q A Q C E K R L - - - - - - - K E V L Q R P A S I M E S D K G W T S A S T S G K P R K D
HorB.PTH.opposum         40  L L R N A Q A Q C E Q R L - - - - - - - K E V L R V P - E L A E S A K D W - - M S R S A K T K K E
HorB.PTH.rat             40  L L H R A Q A Q C D K L L - - - - - - - K E V L H T A A N I M E S D K G W T P A S T S G K P R K E
HorB.Secretin.rat        33  - - - - - - - - C D V R - - - - - - - - - - - - - - R V L L E E R A - - - - - H C L Q Q L S K
HorB.VIP.rat             37  - - - - - - - - C E Y L - - - - - - - - - - - - - - Q L I E I Q R Q - - - - - Q C L E E - - -
HorC.ACTH.human          18  - - - - - - - - - - - - - - - - - - - - - - - - - - N - - S D - - - - - - C P R V - - -
HorC.ADH.human           23  - - - - - - - - - - - - - - - - - - - - - - - - - - S S - - Q E R P L - D T R D P L L A R - -
HorC.ATII-1.bovine       17  - - - - - - - - - - - - - - - - - - - - - - - - - - D C - - P K A G - R H N Y I F - - - - -
HorC.ATII-1.human        17  - - - - - - - - - - - - - - - - - - - - - - - - - - D C - - P K A G - R H N Y I F - - - - -
HorC.ATII-1.rat          17  - - - - - - - - - - - - - - - - - - - - - - - - - - D C - - P K A G - R H S Y I F - - - - -
HorC.ATII-1B.rat         17  - - - - - - - - - - - - - - - - - - - - - - - - - - D C - - P K A G - R H N Y I F - - - - -
HorC.AV1a.rat            36  - - - - - - - - - - - - - - - - - - - - - - - - - - G E - - G D S P L G D V R N E E L A K - -
HorC.AV2.rat             23  - - - - - - - - - - - - - - - - - - - - - - - - - - S S - - Q E E L L - D D R D P L L V R - -
HorC.Bombesin.guineapig  30  - - - - - - - - - - - - - - - - - - - - - - - - - - D T - - T N K G - W T G D N S P G I - E
HorC.Bombesin.human      28  - - - - - - - - - - - - - - - - - - - - - - - - - - L P - - V N D D - - W S H P G I L - - - -
HorC.Bombesin.mouse      29  - - - - - - - - - - - - - - - - - - - - - - - - - - P P - - K M D N - W F H P G F I - - - -
HorC.CCKa.rat            35  - - - - - - - - - - - - - - - - - - - - - - - - - - L G - L E N E T L F C L D Q P Q P S K E W
HorC.CCKb.rat            34  - - - - - - - - - - - - - - - - - - - - - - - - - - A G N L S C D P - P R I R G T G T R E L
HorC.GPRN.human          33  - - - - - - - - - - - - - - - - - - - - - - - - - - M C - - P N M P - N K S V L L - - - - -
HorC.Gastrin.canine      34  - - - - - - - - - - - - - - - - - - - - - - - - - - A G N L S C E P - - P R L R G A G T R E L
HorC.Gastrin.rat         34  - - - - - - - - - - - - - - - - - - - - - - - - - - A G N L S C E P - - P R I R G T G T R E L
HorC.GnRH.human          22  - - - - - - - - - - - - - - - - - - - - - - - - - - P L - - M Q G N L - P T L T L S G K - -
HorC.GnRH.mouse          22  - - - - - - - - - - - - - - - - - - - - - - - - - - P L - - I Q G K L - P T L T V S G K - -
HorC.GnRH.rat            22  - - - - - - - - - - - - - - - - - - - - - - - - - - P L - - T Q G K L - P T L T L S G K - -
HorC.MSH.human           24  - - - - - - - - - - - - - - - - - - - - - - - - - - L - - - G L A A N Q T G A R C L E V - - -
HorC.MSH.mouse           22  - - - - - - - - - - - - - - - - - - - - - - - - - - L - - - G L A T N Q S E P W C L Y V - - -
HorC.NMB.human           31  - - - - - - - - - - - - - - - - - - - - - - - - - - P A - - S D G T - - T T E L V I R - - - -
HorC.NMB.rat             31  - - - - - - - - - - - - - - - - - - - - - - - - - - P D - - S D G T - - T A E L V I R - - - -
HorC.Oxy.human           26  - - - - - - - - - - - - - - - - - - - - - - - - - - N R - - T A G P - - P R R N E A L A R - -
HorC.RDC1.canine         33  - - - - - - - - - - - - - - - - - - - - - - - - - - L C - - P N M P - N K S V L L - - - - -
HorC.Somat1.human        40  - - - - - - - - - - - - - - - - - - - - - - - - - - E P - - G R N A - S Q N G T L S E G Q G
HorC.Somat1.mouse        40  - - - - - - - - - - - - - - - - - - - - - - - - - - E P - - G R N A - S Q N G T L S E G Q G
HorC.Somat1.rat          40  - - - - - - - - - - - - - - - - - - - - - - - - - - E P - - G R N S - S Q N G T L S E G Q G
HorC.Somat2.bovine       25  - - - - - - - - - - - - - - - - - - - - - - - - - - G A - - A N I S - - N Q T E P Y Y D L A S
HorC.Somat2.human        26  - - - - - - - - - - - - - - - - - - - - - - - - - - V S - - T N T S - - N Q T E P Y Y D L T S
HorC.Somat2.mouse        26  - - - - - - - - - - - - - - - - - - - - - - - - - - G P - - S N G S - - N Q T E P Y Y D M T S
HorC.Somat2.rat          26  - - - - - - - - - - - - - - - - - - - - - - - - - - G P - - S N G S - - N Q T E P Y Y D M T S
HorC.Somat3.mouse        27  - - - - - - - - - - - - - - - - - - - - - - - - - - T T - - L G N T - S A G A S L T G L A V
HorC.Somat3.rat          27  - - - - - - - - - - - - - - - - - - - - - - - - - - T S - - L G N A - S A G T S L A G L A V
HorC.TA.rat              16  - - - - - - - - - - - - - - - - - - - - - - - - - - K - - - M C - - - - - - - - P G - - -
HorC.TRH.mouse           15  - - - - - - - - - - - - - - - - - - - - - - - - - - Q - - - P Q A A - - - V A L E Y Q - - - -
HorC.TRH.rat             15  - - - - - - - - - - - - - - - - - - - - - - - - - - P - - - P Q V A - - - V A L E Y Q - - - -
HorC.mas.human           16  - - - - - - - - - - - - - - - - - - - - - - - - - - N - - I S - - - - - - - - - - T G - - -
HorC.mas.rat             15  - - - - - - - - - - - - - - - - - - - - - - - - - - M - - N T - - - - - - - - - - S S - - -
HorC.mrg.human           24  - - - - - - - - - - - - - - - - - - - - - - - - - - S - - L S C S L C L H S G D Q E A - - -
```

FIGURE 1b.

```
HorA.FSH.human           96   K L H E I R I E K A N N L L Y I T P E A F Q N L P N L Q Y L L I S N T G I K H L P D V H K I H S L Q
HorA.FSH.rat             89   D V F S N L P K L H E I R I E K A N N V L Y I N P E A F Q N L P S L R Y L L I S N T G I K H L P A V
HorA.FSH.sheep           89   N V F S N L P K L H E I R I E K A N N L L Y I D P D A F Q N L P N L R Y L L I S N T G I K H L P A V
HorA.LH-CG.human         90   N A F D N L L N L S E I L I Q N T K N L R Y I E P G A F I N L P R L K Y L S I C N T G I R K F P D V
HorA.LH-CG.mouse         96   N A F D N L L N L S E I L I Q N T K N L L Y I E P G A F T N L P R L K Y L S I C N T G I R T L P D V
HorA.LH-CG.porcine       92   N A F D N L L N L S E I L I Q N T K N L V Y I E P G A F T N L P R L K Y L S I C N T G I R K L P D V
HorA.LH-CG.rat           96   N A F D N L L N L S E L L I Q N T K N L L Y I E P G A F T N L P R L K Y L S I C N T G I R T L P D V
HorA.TSH.canine          81   - - - - - - - - - - - I E I R N T R S L T S I D P D A L K E L P L L K F L G I F N T G L G V F P D V
HorA.TSH.human           95   H S F Y N L S K V T H I E I R N T R N L T Y I D P D A L K E L P L L K F L G I F N T G L K M F P D L
HorA.TSH.rat             95   H S F Y N L S K M T H I E I R N T R S L T Y I D P D A L T E L P L L K F L G I F N T G L R I F P D L
HorB.Calcitonin.human    60   - - - Q Q L P A Y Q G E G - - - - - - - - - - - P Y C N R T W D G W L C W D D T P A G - - - - - V L
HorB.Calcitonin.porcine  61   - - - Q K L P P Y Q G E G - - - - - - - - - - - L Y C N R T W D G W S C W D D T P A G - - - - - V L
HorB.GLP-1.rat           49   - - - S L L P P P T E L V - - - - - - - - - - - C N R T F D K Y S C W P D T P P N - - - - - T T
HorB.PTH.human           82   K A S G K L Y P E S E E D K E A P T G S R Y R G R P C L P E W D H I L C W P L G A P G - - - - - E V
HorB.PTH.opposum         79   K P A E K L Y P Q A E E S R E V S D R S R L Q D G E C L P E W D N I V C W P A G V P G - - - - - K V
HorB.PTH.rat             82   K A S G K F Y P E S K E N K D V P T G S R R R G R P C L P E W D N I V C W P L G A P G - - - - - E V
HorB.Secretin.rat        53   E K K G A L G P E T A S G - - - - - - - - - - - - C E G L W D N M S C W P S S A P A - - - - - R T
HorB.VIP.rat             54   - - - A Q L E N E T - T G - - - - - - - - - - - - C S K M W D N L T C W P T T P R G - - - - - Q A
HorC.ACTH.human          25   - - V L P E E I F F T I S - - - - - - - - - - - - I V G V L E N L I V L L A V F K N K N L Q A P M
HorC.ADH.human           39   - - A E L A L L S I V F V - - - - - - - - - - - - A V A L S N G L V L A A L A R R G R R - - - - G
HorC.ATII-1.bovine       29   - - I M I P T L Y S I I F - - - - - - - - - - - - V V G I F G N S L V V I V I Y F Y M K - - - - -
HorC.ATII-1.human        29   - - V M I P T L Y S I I F - - - - - - - - - - - - V V G I F G N S L V V I V I Y F Y M K - - - - -
HorC.ATII-1.rat          29   - - V M I P T L Y S I I F - - - - - - - - - - - - V V G I F G N S L V V I V I Y F Y M K - - - - -
HorC.ATII-1B.rat         29   - - V M I P T L Y S I I F - - - - - - - - - - - - V V G I F G N S L V V I V I Y F Y M K - - - - -
HorC.AV1a.rat            53   - - L E I A V L A V I F V - - - - - - - - - - - - V A V L G N S S V L L A L - H R T P R - - - - K
HorC.AV2.rat             39   - - A E L A L L S T I F V - - - - - - - - - - - - A V A L S N G L V L G A L I R R G R R - - - - G
HorC.Bombesin.guineapig  46   A L C A I Y I T Y A V I I - - - - - - - - - - - - S V G I L G N A I L I K V F F K T K S - - - - -
HorC.Bombesin.human      41   - - Y V I P A V Y G V I I - - - - - - - - - - - - L I G L I G N I T L I K I F C T V K S - - - - -
HorC.Bombesin.mouse      42   - - Y V I P A V Y G L I I - - - - - - - - - - - - V I G L I G N I T L I K I F C T V K S - - - - -
HorC.CCKa.rat            55   Q S A L Q I L L Y S I I F - - - - - - - - - - - - L L S V L G N T L V I T V L I R N K R - - - - -
HorC.CCKb.rat            53   E M A I R I T L Y A V I F - - - - - - - - - - - - L M S V G G N V L I I V V L G L S R R - - - - -
HorC.GPRN.human          45   - - Y T L S F I Y I F I F - - - - - - - - - - - - V I G M I A N S V V V W V N I Q A K T - - - - -
HorC.Gastrin.canine      53   E L A I R V T L Y A V I F - - - - - - - - - - - - L M S V G G N V L I I V V L G L S R R - - - - -
HorC.Gastrin.rat         53   E L A I R I T L Y A V I F - - - - - - - - - - - - L M S I G G N M L I I V V L G L S R R - - - - -
HorC.GnRH.human          37   - - I R V T V T F F L F L - - - - - - - - - - - - L S A T F N A S F L L K L Q K W T Q K K E K G K
HorC.GnRH.mouse          37   - - I R V T V T F F L F L - - - - - - - - - - - - L S T A F N A S F L L K L Q K W T Q K R K K G K
HorC.GnRH.rat            37   - - I R V T V T F F L F L - - - - - - - - - - - - L S T A F N A S F L V K L Q R W T Q K R K K G K
HorC.MSH.human           39   - - S I S D G L F L S L G - - - - - - - - - - - - L V S L V E N A L V V A T I A K N R N L H S P M
HorC.MSH.mouse           37   - - S I P D G L F L S L G - - - - - - - - - - - - L V S L V E N V L V V I A I T K N R N L H S P M
HorC.NMB.human           44   - - C V I P S L Y L L I I - - - - - - - - - - - - T V G L L G N I M L V K I F I T N S A - - - - -
HorC.NMB.rat             44   - - C V I P S L Y L I - - - - - - - - - - - - - - S V G L L G N I M L V K I F L T N S T - - - - -
HorC.Oxy.human           41   - - V E V A V L C L I L L - - - - - - - - - - - - L A L S G N A C V L L A L - R T T R Q - - - - K
HorC.RDC1.canine         45   - - Y T L S F I Y I F I F - - - - - - - - - - - - V I G M I A N S V V V W V N I Q A K T - - - - -
HorC.Somat1.human        57   S A I L I S F I Y S V V C - - - - - - - - - - - - L V G L C G N S M V I Y V I L R Y A K - - - - -
HorC.Somat1.mouse        57   S A I L I S F I Y S V V C - - - - - - - - - - - - L V G L C G N S M V I Y V I L R Y A K - - - - -
HorC.Somat1.rat          57   S A I L I S F I Y S V V C - - - - - - - - - - - - L V G L C G N S M V I Y V I L R Y A K - - - - -
HorC.Somat2.bovine       42   N V V L - T F I Y F V V C - - - - - - - - - - - - I I G L C G N T L V I Y V I L R Y A K - - - - -
HorC.Somat2.human        43   N A V L - T F I Y F V V C - - - - - - - - - - - - I I G L C G N T L V I Y V I L R Y A K - - - - -
HorC.Somat2.mouse        43   N A V L - T F I Y F V V C - - - - - - - - - - - - V V G L C G N T L V I Y V I L R Y A K - - - - -
HorC.Somat2.rat          43   N A V L - T F I Y F V V C - - - - - - - - - - - - V V G L C G N T L V I Y V I L R Y A K - - - - -
HorC.Somat3.mouse        44   S G I L I S L V Y L V V C - - - - - - - - - - - - V V G L L G N S L V I Y V V L R H T S - - - - -
HorC.Somat3.rat          44   S G I L I S L V Y L V V C - - - - - - - - - - - - V V G L L G N S L V I Y V V L R H T S - - - - -
HorC.TA.rat              21   - - M S E A L - E L Y S R - - - - - - - - - - - - G F L T I E Q I A T L P - - - - - - - - - -
HorC.TRH.mouse           26   - - - V V T I L L V V I I C - - - - - - - - - - - G L G I V G N I M V V L V V M R T K H - - - - -
HorC.TRH.rat             26   - - - V V T I L L V V V I C - - - - - - - - - - - G L G I V G N I M V V L V V M R T K H - - - - -
HorC.mas.human           21   - - R N A S V - G - - - - - - - - - - - - - - - - - - N A H R - - - - - - - - - - - - -
HorC.mas.rat             20   - - R N A S L - G - - - - - - - - - - - - - - - - - - T S H P - - - - - - - - - - - - -
HorC.mrg.human           39   - - Q N P N L V S Q L C G - - - - - - - - - - - - - V F L Q N E T N E T I H M Q M S M A V G Q Q A L
```

FIGURE 1c.

```
HorA.FSH.human           146  KVLLDIQDNINIHTIERNSFVGLSFESVILWLNKNGIQEIHNCAFNGTQL
HorA.FSH.rat             139  HKIQSLQ-KVLLDIQDNINIHIVARNSFMGLSFESVILWLSKNGIEEIHN
HorA.FSH.sheep           139  HKIQSLQ-KVLLDIQDNINIHTVERNSFMGLSFESMIVWLSKNGIQEIHN
HorA.LH-CG.human         140  TKVFSSESNFILEICDNLHITTIPGNAFQGMNNESVTLKLYGNGFEEVQS
HorA.LH-CG.mouse         146  SKISSSEFNFILEICDNLYITTIPGNAFQGMNNESITLKLYGNGFEEVQS
HorA.LH-CG.porcine       142  TKIFSSEFNFILEICDNLHITTVPANAFQGMNNESITLKLYGNGFEEIQS
HorA.LH-CG.rat           146  TKISSSEFNFILEICDNLHITTIPGNAFQGMNNESVTLKLYGNGFEEVQS
HorA.TSH.canine          120  TKVYSTDVFFILEITDNPYMASIPANAFQGLCNETLTLKLYNNGFTSIQG
HorA.TSH.human           145  TKVYSTDIFFILEITDNPYMTSIPVNAFQGLCNETLTLKLYNNGFTSVQG
HorA.TSH.rat             145  TKIYSTDVFFILEITDNPYMTSVPENAFQGLCNETLTLKLYNNGFTSIQG
HorB.Calcitonin.human     91  SYQFCPDYFPDFDPSEK--VTKYCDEK-----GVWFKHPENNRTWSNYTM
HorB.Calcitonin.porcine   92  AEQYCPDYFPDFDAAEK--VTKYCGED-----GDWYRHPESNISWSNYTM
HorB.GLP-1.rat            78  ANISCPWYLPWYHKVQHRLVFKRCGPD-----GQWVRGP-RGQSWRDASQ
HorB.PTH.human           127  VAVPCPDYIYDFNHKGH--AYRRCDRN-----GSWELVPGHNRTWANYSE
HorB.PTH.opposum         124  VAVPCPDYIYDFNHKGR--AYRRCDSN-----GSWELVPGHNRTWANYSE
HorB.PTH.rat             127  VAVPCPDYIYDFNHKGH--AYRRCDRN-----GSWEVVPGHNRTWANYSE
HorB.Secretin.rat         85  VEVQCPKFLLMLSNKNG-SLFRNCTQD-----G-W---S-ETFPRP-DLA
HorB.VIP.rat              82  VVLDCPLIFQLFAPIHGYNISRSCTEE-----G-W---S-QLEPGPYHIA
HorC.ACTH.human           60  YFFICSLAISDMLGSLYKILENILI---------ILRNMGYLKPRGSF--
HorC.ADH.human            70  HWAPIHVFIGHLCLADLAVALFQVL---------PQLAWKATDRF-RG--
HorC.ATII-1.bovine        59  LKTVASVFLLNLALADLCFLLTLPL---------WAVYTAMEYRW-PF--
HorC.ATII-1.human         59  LKTVASVFLLNLALADLCFLLTLPL---------WAVYTAMEYRW-PF--
HorC.ATII-1.rat           59  LKTVASVFLLNLALADLCFLLTCPL---------WAVYTAMEYRW-PF--
HorC.ATII-1B.rat          59  LKTVASVFLLNLALADLCFLLTLPL---------WAVYTAMEYRW-PF--
HorC.AV1a.rat             83  T-SRMHLFIRHLSLADLAVAFFQVL---------PQLCWTSP-SF-RG--
HorC.AV2.rat              70  RWAPMHVFISHLCLADLAVALFQVL---------PQLAWDATDRF-HG--
HorC.Bombesin.guineapig   78  MQTVPNIFITSLALGDLLLLLTCVP---------VDATHYLAEGW-LF--
HorC.Bombesin.human       71  MRNVPNLFISSLALGDLLLLITCAP---------VDASRYLADRW-LF--
HorC.Bombesin.mouse       72  MRNVPNLFISSLALGDLLLLVTCAP---------VDASKYLADRW-LF--
HorC.CCKa.rat             87  MRTVTNIFLLSLAVSDLMLCLFCMP---------FNLIPNLLKDF-IF--
HorC.CCKb.rat             85  LRTVTNAFLLSLAVSDLLLAVACMP---------FTLLPNLMGTF-IF--
HorC.GPRN.human           75  TGYDTHCYILNLAIADLWVVLTIPV---------WVVSLVQHNQW-PM--
HorC.Gastrin.canine       85  LRTVTNAFLLSLAVSDLLLAVACMP---------FTLLPNLMGTF-IF--
HorC.Gastrin.rat          85  LRTVTNAFLLSLAVSDLLLAVACMP---------FTLLPNLMGTF-IF--
HorC.GnRH.human           72  KLSRMKLLLKHLTLANLLETLIVMP---------LDGMWNITVQW-YA--
HorC.GnRH.mouse           72  KLSRMKVLLKHLTLANLLETLIVMP---------LDGMWNITVQW-YA--
HorC.GnRH.rat             72  KLSRMKVLLKHLTLANLLETLIVMP---------LDGMWNITVQW-YA--
HorC.MSH.human            74  YCFICCLALSDLLVSGTNVLETAVI---------LLLEAGALVARAAV--
HorC.MSH.mouse            72  YYFICCLALSDLMVSVSIVLETTII---------LLLEVGILVARVAL--
HorC.NMB.human            74  MRSVPNIFISNLAAGDLLLLLTCVP---------VDASRYFFDEW-MF--
HorC.NMB.rat             74  MRSVPNIFISNLAAGDLLLLLTCVP---------VDASRYFFDEW-VF--
HorC.Oxy.human            71  H-SRLFFFMKHLSIADLVVAVFQVL---------PQLLWDITFRF-YG--
HorC.RDC1.canine          75  TGYDTHCYILNLAIADLWVVVTIPV---------WVVSLVQHNQW-PM--
HorC.Somat1.human         89  MKTATNIYILNLAIADELLMLSVPF---------LVTSTLLRH--W-PF--
HorC.Somat1.mouse         89  MKTATNIYILNLAIADELLMLSVPF---------LVTSTLLRH--W-PF--
HorC.Somat1.rat           89  MKTATNIYILNLAIADELLMLSVPF---------LVTSTLLRH--W-PF--
HorC.Somat2.bovine        73  MKTITNIYILNLAIADELFMLGLPF---------LAMQVALVH--W-PF--
HorC.Somat2.human         74  MKTITNIYILNLAIADELFMLGLPF---------LAMQVALVH--W-PF--
HorC.Somat2.mouse         74  MKTITNIYILNLAIADELFMLGLPF---------LAMQVALVH--W-PF--
HorC.Somat2.rat           74  MKTITNIYILNLAIADELFMLGLPF---------LAMQVALVH--W-PF--
HorC.Somat3.mouse         76  SPSVTSVYILNLALADELFMLGLPF---------LAAQNALSY--W-PF--
HorC.Somat3.rat           76  SPSVTSVYILNLALADELFMLGLPF---------LAAQNALSY--W-PF--
HorC.TA.rat               43  PPAVTNYIFLLLCLCGLVGNGLVLW---------FFGFSIKRTPFSIY--
HorC.TRH.mouse            56  MRTPTNCYLVSLAVADLMVLVAAGL---------PNITDSIYGSW-VY--
HorC.TRH.rat              56  MRTATNCYLVSLAVADLMVLVAAGL---------PNITDSIYGSW-VY--
HorC.mas.human            31  QIPIVHWVIMSISPVGFVENGILLW---------FLCFRMRRNPFTVY--
HorC.mas.rat              30  PIPIVHWVIMSISPLGFVENGILLW---------FLCFRMRRNPFTVY--
HorC.mrg.human            74  PLNIIAPKAVLVSLCGVLLNGTVFW---------LLCCGA-TNPYMVY--
```

FIGURE 1d.

```
HorA.FSH.human          196   D A V N L S D N N N L E E L P N D V F H G A S G P V I L D I - S R T R I H S L P S Y G L E N L K K L
HorA.FSH.rat            188   C A F N G T Q L D E L N L S D N N N L E E L P N D V F Q G A - S G P V I L D I S R T K V H S L P N H
HorA.FSH.sheep          188   C A F N G T Q L D E L N L S D N S N L E E L P N D V F Q G A - S G P V I L D I S R T R I R S L P S Y
HorA.LH-CG.human        190   H A F N G T T L T S L E L K E N V H L E K M H N G A F R G A - T G P K T L D I S S T K L Q A L P S Y
HorA.LH-CG.mouse        196   H A F N G T T L I S L E L K E N I Y L E K M H S G T F Q G A - T G P S I L D V S S T K L Q A L P S H
HorA.LH-CG.porcine      192   H A F N G T T L I S L E L K E N A H L K K M H N D A F R G A - R G P S I L D I S S T K L Q A L P S Y
HorA.LH-CG.rat          196   H A F N G T T L I S L E L K E N I Y L E K M H S G A F Q G A - T G P S I L D I S S T K L Q A L P S H
HorA.TSH.canine         170   H A F N G T K L D A V Y L N K N K Y L S A I D K D A F G G V Y S G P T L L D V S Y T S V T A L P S K
HorA.TSH.human          195   Y A F N G T K L D A V Y L N K N K Y L T V I D K D A F G G V Y S G P S L L D V S Q T S V T A L P S K
HorA.TSH.rat            195   H A F N G T K L D A V Y L N K N K Y L T A I D K D A F G G V Y S G P T L L D V S S T S V T A L P S K
HorB.Calcitonin.human   134   C - - - - N A F T P E K L K N A Y V L Y Y L A I - - - V G H - - - S L S I F T L V I S L G I F V F F
HorB.Calcitonin.porcine 135   C - - - - N A F T P D K L Q N A Y I L Y Y L A I - - - V G H - - - S L S I L T L L I S L G I F M F L
HorB.GLP-1.rat          122   C Q M D D D E I E V - Q K G V A K M Y S S Y Q V M Y T V G Y - - - S L S L G A L L L A L V I L L G L
HorB.PTH.human          170   C - - - - V K F L T N E T R E R E V F D R L G M I Y T V G Y - - - S V S L A S L T V A V L I L A Y F
HorB.PTH.opposum        167   C - - - - V K F L T N E T R E R E V F D R L G M I Y T V G Y - - - S I S L G S L T V A V L I L G Y F
HorB.PTH.rat            170   C - - - - L K F M T N E T R E R E V F D R L G M I Y T V G Y - - - S M S L A S L T V A V L I L A Y F
HorB.Secretin.rat       123   C G V - N I N N S F N E R R H A - Y L L K L K V M Y T V G Y - - - S S S L A M L L V A L S I L C S F
HorB.VIP.rat            122   C G L N D R A S S L D E Q Q Q T K F Y N T V K T G Y T I G Y - - - S L S L A S L L V A M A I L S L F
HorC.ACTH.human         99    - - - - - E T T A D D I I D S L F V L S L L G S I F S L S - - - - V I A A D R Y I T I F H A L R Y -
HorC.ADH.human          108   - - - - - P D A L C R A V K Y L Q M V G M Y A S S Y M I L - - - - A M T L D R H R A I C R P M L A -
HorC.ATII-1.bovine      97    - - - - - G N Y L C K I A S A S V S F N L Y A S V F L L T - - - - C L S I D R Y L A I V H P M K S -
HorC.ATII-1.human       97    - - - - - G N Y L C K I A S A S V S F N L Y A S V F L L T - - - - C L S I D R Y L A I V H P M K S -
HorC.ATII-1.rat         97    - - - - - G N H L C K I A S A S V T F N L Y A S V F L L T - - - - C L S I D R Y L A I V H P M K S -
HorC.ATII-1B.rat        97    - - - - - G N H L C K I A S A S V S F N L Y A S V F L L T - - - - C L S I D R Y L A I V H P M K S -
HorC.AV1a.rat           119   - - - - - P D W L C R V V K H L Q V F A M F A S A Y M L V - - - - V M T A D R Y I A V C H P L K T -
HorC.AV2.rat            108   - - - - - P D A L C R A V K Y L Q M V G M Y A S S Y M I L - - - - A M T L D R H R A I C R P M L A -
HorC.Bombesin.guineapig 116   - - - - - G R I G C K V L S F I R L T S V G V S V F T L T - - - - I L S A D R Y K A V V K P L E R -
HorC.Bombesin.human     109   - - - - - G R I G C K L I P F I Q L T S V G V S V F T L T - - - - A L S A D R Y K A I V R P M D I -
HorC.Bombesin.mouse     110   - - - - - G R I G C K L I P F I Q L T S V G V S V F T L T - - - - A L S A D R Y K A I V R P M D I -
HorC.CCKa.rat           125   - - - - - G S A V C K T T T Y F M G T S V S V S T F N L V - - - - A I S L E R Y G A I C R P L Q S -
HorC.CCKb.rat           123   - - - - - G T V I C K A I S Y L M G V S V S V S T L N L V - - - - A I A L E R Y S A I C R P L Q A -
HorC.GPRN.human         113   - - - - - G E L T C K V T H L I F S I N L F S G I F F L T - - - - C M S V D R Y L S I T Y F T N T -
HorC.Gastrin.canine     123   - - - - - G T V V C K A V S Y L M G V S V S V S T L S L V - - - - A I A L E R Y S A I C R P L Q A -
HorC.Gastrin.rat        123   - - - - - G T V I C K A V S Y L M G V S V S V S T L N L V - - - - A I A L E R Y S A I C R P L Q A -
HorC.GnRH.human         110   - - - - - G E L L C K V L S Y L K L F S M Y A P A F M M V - - - - V I S L D R S L A I T R P L A L -
HorC.GnRH.mouse         110   - - - - - G E F L C K V L S Y L K L F S M Y A P A F M M V - - - - V I S L D R S L A I T Q P L A V -
HorC.GnRH.rat           110   - - - - - G E F L C K V L S Y L K L F S M Y A P A F M M V - - - - V I S L D R S L A V T Q P L A V -
HorC.MSH.human          113   - - - - - L Q Q L D N V I D V I T C S S M L S S L C F L G - - - - A I A V D R Y I S I F Y A L R Y -
HorC.MSH.mouse          111   - - - - - V Q Q L D N L I D V L I C G S M V S S L C F L G - - - - I I A I D R Y I S I F Y A L R Y -
HorC.NMB.human          112   - - - - - G K V G C K L I P V I Q L T S V G V S V F T L T - - - - A L S A D R Y R A I V N P M D M -
HorC.NMB.rat            112   - - - - - G K L G C K L I P A I Q L T S V G V S V F T L T - - - - A L S A D R Y R A I V N P M D M -
HorC.Oxy.human          108   - - - - - P D L L C R L V K Y L Q V V G M F A S T Y L L L - - - - L M S L D R C L A I C Q P L R S -
HorC.RDCl.canine        113   - - - - - G E L T C K I T H L I F S I N L F S I F F L T - - - - C M S V D R Y L S I T Y F A S T -
HorC.Somat1.human       126   - - - - - G A L L C R L V L S V D A V N M F T S I Y C L T - - - - V L S V D R Y V A V H P I K A -
HorC.Somat1.mouse       126   - - - - - G A L L C R L V L S V D A V N M F T S I Y C L T - - - - V L S V D R Y V A V H P I K A -
HorC.Somat1.rat         126   - - - - - G A L L C R L V L S V D A V N M F T S I Y C L T - - - - V L S V D R Y V A V H P I K A -
HorC.Somat2.bovine      110   - - - - - G K A I C R V V M T V D G I N Q F T S I F C L T - - - - V M S I D R Y L A V H P I K S -
HorC.Somat2.human       111   - - - - - G K A I C R V V M T V D G I N Q F T S I F C L T - - - - V M S I D R Y L A V H P I K S -
HorC.Somat2.mouse       111   - - - - - G K A I C R V V M T V D G I N Q F T S I F C L T - - - - V M S I D R Y L A V H P I K S -
HorC.Somat2.rat         111   - - - - - G K A I C R V V M T V D G I N Q F T S I F C L T - - - - V M S I D R Y L A V H P I K S -
HorC.Somat3.mouse       113   - - - - - G S L M C R L V M A V D G I N Q F T S I F C L T - - - - V M S V D R Y L A V V H P T R S -
HorC.Somat3.rat         113   - - - - - G S L M C R L V M A V D G I N Q F T S I F C L T - - - - V M S V D R Y L A V V H P T R S -
HorC.TA.rat             82    - - - - - F L H L A S A D G I Y L F S K A V I A L - L N M - - - - G T F L G S F P D Y V R R V S R -
HorC.TRH.mouse          94    - - - - - G Y V G C L C I T Y L Q Y L G I N A S S C S I T - - - - A F T I E R Y I A I C H P I K A -
HorC.TRH.rat            94    - - - - - G Y V G C L C I T Y L Q Y L G I N A S S C S I T - - - - A F T I E R Y I A I C H P I K A -
HorC.mas.human          70    - - - - - I T H L S I A D I S L L F C I F I L S I D Y A L - - - - D Y E L S S G H Y Y T I V T L S -
HorC.mas.rat            69    - - - - - I T H L S I A D I S L L F C I F I L S I D Y A L - - - - D Y E L S S G H Y Y T I V T L S -
HorC.mrg.human          112   - - - - - I L H L V A A D V I Y L C C S A V G F L Q V T L - - - - - L T Y H G V V F F I P D F L A -
```

FIGURE 1e.

```
HorA.FSH.human          245  R A R S T Y N L K K L P T L E K L V A L M E A S L T Y P S H C C A F A N W R R Q I S - - - - - - - -
HorA.FSH.rat            237  G L E N L K K L R A R S T Y R L K K L P N L D K F V T L M E A S L T Y P S H C C A F A N L K R - - -
HorA.FSH.sheep          237  G L E N L K K L R A K S T Y H L K K L P S L E K F V T L V E A S L T Y P S H C C A F A N W R R - - -
HorA.LH-CG.human        239  G L E S I Q R L I A T S S Y S L K K L P S K Q T F V N L L R A T L H Y P S H C C A F - - - - - - - -
HorA.LH-CG.mouse        245  G L E S I Q T L I A T S S Y S L K T L P S R E K F T S L L V A T L T Y P S H C C A F - - - - - - - -
HorA.LH-CG.porcine      241  G L E S I Q T L I A T S S Y S L K K L P S R E K F T N L L D A T L T Y P S H C C A F - - - - - - - -
HorA.LH-CG.rat          245  G L E S I Q T L I A L S S Y S L K T L P S K E K F T S L L V A T L T Y P S H C C A F - - - - - - - -
HorA.TSH.canine         220  G L E H L K E L I A R N T W T L K K L P L S L S F L H L T R A D L S Y P S H C C A F K N Q K K I R G
HorA.TSH.human          245  G L E H L K E L I A R N T W T L K K L P L S L S F L H L T R A D L S Y P S H C C A F K N Q K K I R G
HorA.TSH.rat            245  G L E H L K E L I A K N T W T L K K L P L S L S F L H L T R A D L S Y P S H C C A F K N Q K K I R G
HorB.Calcitonin.human   174  R K L T T I F P L N W K Y R K A L S L G C Q R V T L H - K N M F L T Y - - - - - - - - - - - -
HorB.Calcitonin.porcine 175  R - - - - - - - - - - - - - - S I S C Q R V T L H - K N M F L T Y - - - - - - - - - - - -
HorB.GLP-1.rat          168  R - - - - - - - - - - - - K L H C T R N Y I H - G N L F A S F - - - - - - - - - - - -
HorB.PTH.human          213  R - - - - - - - - - - - - R L H C T R N Y I H - M H L F L S F - - - - - - - - - - - -
HorB.PTH.opposum        210  R - - - - - - - - - - - - R L H C T R N Y I H - M H L F V S F - - - - - - - - - - - -
HorB.PTH.rat            213  R - - - - - - - - - - - - R L H C T R N Y I H - M H M F L S F - - - - - - - - - - - -
HorB.Secretin.rat       168  R - - - - - - - - - - - - R L H C T R N Y I H - M H L F V S F - - - - - - - - - - - -
HorB.VIP.rat            169  R - - - - - - - - - - - - K L H C T R N Y I H - M H L F M S F - - - - - - - - - - - -
HorC.ACTH.human         139  - - - - - - - - - - - - H S I - V T M R R T V V V L T V I W
HorC.ADH.human          148  - - - - - - - - - - - - Y R H G S G A H W N R P V L - V A W
HorC.ATII-1.bovine      137  - - - - - - - - - - - - R L R - R T M L V A K V T C I I I W
HorC.ATII-1.human       137  - - - - - - - - - - - - R L R - R T M L V A K V T C I I I W
HorC.ATII-1.rat         137  - - - - - - - - - - - - R L R - R T M L V A K V T C I I I W
HorC.ATII-1B.rat        137  - - - - - - - - - - - - R L R - R T M L V A K V T C I I I W
HorC.AV1a.rat           159  - - - - - - - - - - - - L Q Q - - P A R R S R L M I A T S W
HorC.AV2.rat            148  - - - - - - - - - - - - Y R H G G G A R W N R P V L - V A W
HorC.Bombesin.guineapig 156  - - - - - - - - - - - - Q P S - N A I L K T C A K A G C I W
HorC.Bombesin.human     149  - - - - - - - - - - - - Q A S - H A L M K I C L K A A F I W
HorC.Bombesin.mouse     150  - - - - - - - - - - - - Q A S - H A L M K I C L K A A L I W
HorC.CCKa.rat           165  - - - - - - - - - - - - R V W - Q T K S H A L K V I A A T W
HorC.CCKb.rat           163  - - - - - - - - - - - - R V W - Q T R S H A A R V I L A T W
HorC.GPRN.human         153  - - - - - - - - - - - - P S S - R K K M V R R V V C I L V W
HorC.Gastrin.canine     163  - - - - - - - - - - - - R V W - Q T R S H A A R V I I A T W
HorC.Gastrin.rat        163  - - - - - - - - - - - - R V W - Q T R S H A A R V I L A T W
HorC.GnRH.human         150  - - - - - - - - - - - - K S N - - - S K V G Q S M V G L A W
HorC.GnRH.mouse         150  - - - - - - - - - - - - Q S N - - - S K L E Q S M I S L A W
HorC.GnRH.rat           150  - - - - - - - - - - - - Q S K - - - S K L E R S M T S L A W
HorC.MSH.human          153  - - - - - - - - - - - - H S I - V T L P R A P R A V A A I W
HorC.MSH.mouse          151  - - - - - - - - - - - - H S I - V T L P R A R R A V V G I W
HorC.NMB.human          152  - - - - - - - - - - - - Q T S - G A L L R T C V K A M G I W
HorC.NMB.rat            152  - - - - - - - - - - - - Q T S - G V V L W T S L K A V G I W
HorC.Oxy.human          148  - - - - - - - - - - - - L R R - - - - R T D R L A V L A T W
HorC.RDC1.canine        153  - - - - - - - - - - - - S S R - R K K V V R R A V C V L V W
HorC.Somat1.human       166  - - - - - - - - - - - - A R Y - R R P T V A K V V N L G V W
HorC.Somat1.mouse       166  - - - - - - - - - - - - A R Y - R R P T V A K V V N L G V W
HorC.Somat1.rat         166  - - - - - - - - - - - - A R Y - R R P T V A K V V N L G V W
HorC.Somat2.bovine      150  - - - - - - - - - - - - A K W - R R P R T A K M I N V A V W
HorC.Somat2.human       151  - - - - - - - - - - - - A K W - R R P R T A K M I T M A V W
HorC.Somat2.mouse       151  - - - - - - - - - - - - A K W - R R P R T A K M I N V A V W
HorC.Somat2.rat         151  - - - - - - - - - - - - A K W - R R P R T A K M I N V A V W
HorC.Somat3.mouse       153  - - - - - - - - - - - - A R W - R T A P V A R T V S R A V W
HorC.Somat3.rat         153  - - - - - - - - - - - - A R W - R T A P V A R M V S A A V W
HorC.TA.rat             121  - - - - - - - - - - - - I V G - L C T F F A G V S L L P A I
HorC.TRH.mouse          134  - - - - - - - - - - - - Q F L - C T F S R A K K I I I F V W
HorC.TRH.rat            134  - - - - - - - - - - - - Q F L - C T F S R A K K I I I F V W
HorC.mas.human          110  - - - - - - - - - - - - V T F - L F G Y N T G L Y L L T A I - - - - - - - - - - - -
HorC.mas.rat            109  - - - - - - - - - - - - V T F - L F G Y N T G L Y L L T A I - - - - - - - - - - - -
HorC.mrg.human          151  - - - - - - - - - - - - I L S - P F S F Q V C L C L L V A I - - - - - - - - - - - -
```

FIGURE 1f.

```
HorA.FSH.human        287  - - - - - - - - - E L H P I C N K S I L R Q E V D Y M T Q A R G - - - - - - - - - - - - Q R S S
HorA.FSH.rat          284  - - - - - - Q I S E L H P I C N K S I L R Q D I D D M T Q I G D - - - - - - - - - - Q R V S
HorA.FSH.sheep        284  - - - - - - Q T S D L H P I C N K S I L R Q E V D D M T Q A R G - - - - - - - - - - Q R I S
HorA.LH-CG.human      281  - - - - - - - - - R N L P T K E L N F S H S I S E N F S K Q C E - - - - - - - - - - - - S T V R
HorA.LH-CG.mouse      287  - - - - - - - - - R N L P K K E Q N F S F S I F E N F S K Q C E - - - - - - - - - - - - S T V R
HorA.LH-CG.porcine    283  - - - - - - - - - R N L P T K E Q N F S F S I F K N F S K Q C E - - - - - - - - - - - - S T A R
HorA.LH-CG.rat        287  - - - - - - - - - R N L P K K E Q N F S F S I F E N F S K Q C E - - - - - - - - - - - - S T V R
HorA.TSH.canine       270  I L E S L M C N E S S I R S L R Q R K S V N T L N G P F D Q E Y E E Y L G D S H A G Y K D N S Q F Q
HorA.TSH.human        295  I L E S L M C N E S S M Q S L R Q R K S V N A L N S P L H Q E Y E E N L G D S I V G Y K E K S K F Q
HorA.TSH.rat          295  I L E S L M C N E S S I R N L R E R K S V N V M R G P V Y Q E Y E E G L G D N H V G Y K Q N S K F Q
HorB.Calcitonin.human 208  - - - - - - - - I L N S M I I I H L V E V V P N G E L V - - - - - - - - - - - - - - - -
HorB.Calcitonin.porcine 193  - - - - - - - - V L N S I I I I V H L V V I V P N G E L V - - - - - - - - - - - - - - - -
HorB.GLP-1.rat        186  - - - - - - - - V L K A G S V L V I D W L L K T R Y S Q K - - - - - - - - - - - - - I G D D
HorB.PTH.human        231  - - - - - - - - M L R A V S I F V K D A V L Y S G A T L D - - - - - - - - - - - - - - E A E R
HorB.PTH.opposum      228  - - - - - - - - M L R A V S I F I K D A V L Y S G V S T D - - - - - - - - - - - - - - E I E R
HorB.PTH.rat          231  - - - - - - - - M L R A A S I F V K D A V L Y S G F T L D - - - - - - - - - - - - - - E A E R
HorB.Secretin.rat     186  - - - - - - - - I L R A L S N F I K D A V L - - - F S S D - - - - - - - - - - - - - - - -
HorB.VIP.rat          187  - - - - - - - - I L R A T A V F I K D M A L - - - F N S G - - - - - - - - - - - - - - - -
HorC.ACTH.human       156  - - - - - - - - - - T F C T G T G I - - - - T M V - - - - - - - - - - - - - - - - - - -
HorC.ADH.human        165  - - - - - - - - A F S L L L S L P Q L F I F A - - Q R N - - - - - - - - - - - - - - - - -
HorC.ATII-1.bovine    154  - - - - - - - - L L A G L A S L P T I - I H R N V F F I - - - - - - - - - - - - - - - -
HorC.ATII-1.human     154  - - - - - - - - L L A G L A S L P A I - I H R N V F F I - - - - - - - - - - - - - - - -
HorC.ATII-1.rat       154  - - - - - - - - L M A G L A S L P A V - I H R N V Y F I - - - - - - - - - - - - - - - -
HorC.ATII-1B.rat      154  - - - - - - - - L M A G L A S L P A V - I Y R N V Y F I - - - - - - - - - - - - - - - -
HorC.AV1a.rat         175  - - - - - - - - V L S F I L S T P Q Y F I F S V I E I E - - - - - - - - - - - - - - - -
HorC.AV2.rat          165  - - - - - - - - A F S L L L S L P Q L F I F A - - Q R D - - - - - - - - - - - - - - - - -
HorC.Bombesin.guineapig 173  - - - - - - - - I M S M I F A L P E A - I F S N V H T L - - - - - - - - - - - - - - - -
HorC.Bombesin.human   166  - - - - - - - - I I S M L L A I P E A - V F S D L H P F - - - - - - - - - - - - - - - -
HorC.Bombesin.mouse   167  - - - - - - - - I V S M L L A I P E A - V F S D L H P F - - - - - - - - - - - - - - - -
HorC.CCKa.rat         182  - - - - - - - - C L S F T I M T P Y P - I Y S N L V P F - - - - - - - - - - - - - - - -
HorC.CCKb.rat         180  - - - - - - - - L L S G L L M V P Y P - V Y T M V Q P - - - - - - - - - - - - - - - - -
HorC.GPRN.human       170  - - - - - - - - L L A F C V S L P D T - Y Y L K T V T S - - - - - - - - - - - - - - - -
HorC.Gastrin.canine   180  - - - - - - - - M L S G L L M V P Y P - V Y T A V Q P - - - - - - - - - - - - - - - - -
HorC.Gastrin.rat      180  - - - - - - - - L L S G L L M V P Y P - V Y T V V Q P - - - - - - - - - - - - - - - - -
HorC.GnRH.human       165  - - - - - - - - I L S S V F A G P Q L Y I F R M I H L A - - - - - - - - - - - - - - - -
HorC.GnRH.mouse       165  - - - - - - - - I L S I V F A G P Q L Y I F R M I Y L A - - - - - - - - - - - - - - - -
HorC.GnRH.rat         165  - - - - - - - - I L S I V F A G P Q L Y I F R M I Y L V - - - - - - - - - - - - - - - -
HorC.MSH.human        170  - - - - - - - - - V A S V V F S T - - - - L F I - - - - - - - - - - - - - - - - - - - -
HorC.MSH.mouse        168  - - - - - - - - - M V S I V S S T - - - - L F I - - - - - - - - - - - - - - - - - - - -
HorC.NMB.human        169  - - - - - - - - V V S V L L A V P E A - V F S E V A R I - - - - - - - - - - - - - - - -
HorC.NMB.rat          169  - - - - - - - - V V S V L L A V P E A - V F S E V A R I - - - - - - - - - - - - - - - -
HorC.Oxy.human        162  - - - - - - - - L G C L V A S A P Q V H I F S L R E V - - - - - - - - - - - - - - - - -
HorC.RDC1.canine      170  - - - - - - - - L L A F C V S L P D T - Y Y L K T V T S - - - - - - - - - - - - - - - -
HorC.Somat1.human     183  - - - - - - - - V L S L L V I L P I V - V F S R T - A A - - - - - - - - - - - - - - - -
HorC.Somat1.mouse     183  - - - - - - - - V L S L L V I L P I V - V F S R T - A A - - - - - - - - - - - - - - - -
HorC.Somat1.rat       183  - - - - - - - - V L S L L V I L P I V - V F S R T - A A - - - - - - - - - - - - - - - -
HorC.Somat2.bovine    167  - - - - - - - - G V S L L V I L P I M - I Y A G L - R S - - - - - - - - - - - - - - - -
HorC.Somat2.human     168  - - - - - - - - G V S L L V I L P I M - I Y A G L - R S - - - - - - - - - - - - - - - -
HorC.Somat2.mouse     168  - - - - - - - - C V S L L V I L P I M - I Y A G L - R S - - - - - - - - - - - - - - - -
HorC.Somat2.rat       168  - - - - - - - - G V S L L V I L P I M - I Y A G L - R S - - - - - - - - - - - - - - - -
HorC.Somat3.mouse     170  - - - - - - - - V A S A V V V L P V V - V F S G V - - - - - - - - - - - - - - - - - - -
HorC.Somat3.rat       170  - - - - - - - - V A S A V V V L P V V - V F S G V - - - - - - - - - - - - - - - - - - -
HorC.TA.rat           138  - - - - - - - - - S I E R C V S V - - - - I F P - - - - - - - - - - - - - - - - - - - -
HorC.TRH.mouse        151  - - - - - - - - - A F T S I Y C M L W F - F L L D L N I S - - - - - - - - - - - - - - -
HorC.TRH.rat          151  - - - - - - - - - A F T S I Y C M L W F - F L L D L N I S - - - - - - - - - - - - - - -
HorC.mas.human        127  - - - - - - - - - S V E R C L S V - - - - L Y P - - - - - - - - - - - - - - - - - - - -
HorC.mas.rat          126  - - - - - - - - - S V E R C L S V - - - - L Y P - - - - - - - - - - - - - - - - - - - -
HorC.mrg.human        168  - - - - - - - - - S T E R C V C V - - - - L F P - - - - - - - - - - - - - - - - - - - -
```

FIGURE 1g.

HorA.FSH.human	314	L A E D N E S S Y S R G F D M T - - - - - - - - - - - - - - - - - Y T E F D - - Y D L C N E - V V D
HorA.FSH.rat	314	L I - D D E P S Y G K G S D M M - - - - - - - - - - - - - - - - - Y N E F D - - Y D L C - N - E V V
HorA.FSH.sheep	314	L A E D D E P S Y A K G F D M M - - - - - - - - - - - - - - - - - Y S E F D - - Y D L C - S - E V V
HorA.LH-CG.human	308	K - - - - - - - - - - - - - S E - - - - - - - - - - - - - - - - - L S G W D - - Y E Y G F C - L P K
HorA.LH-CG.mouse	314	E A N N E T L Y S A I F E E N E - - - - - - - - - - - - - - - - - L S G W D - - Y D Y D F C - S P K
HorA.LH-CG.porcine	310	R P N N E T L Y S A I F A E S E - - - - - - - - - - - - - - - - - L S D W D - - Y D Y G F C - S P K
HorA.LH-CG.rat	314	K A D N E T L Y S A I F E E N E - - - - - - - - - - - - - - - - - L S G W D - - Y D Y G F C - S P K
HorA.TSH.canine	320	D T D S N S H Y Y V F F E E Q E D E I L G F G Q E L K N P Q E E T L Q A F D S H Y D Y T V C G G N E
HorA.TSH.human	345	D T H N N A H Y Y V F F E E Q E D E I I G F G Q E L K N P Q E E T L Q A F D S H Y D Y T I C G D S E
HorA.TSH.rat	345	E G P S N S H Y Y V F F E E Q E D E I I G F G Q E L K N P Q E E T L Q A F D S H Y D Y T V C G D N E
HorB.Calcitonin.human	229	- R R D
HorB.Calcitonin.porcine	214	- K R D
HorB.GLP-1.rat	211	L S V S - V C L S D G A
HorB.PTH.human	256	L T E E E L R A I A Q A P P P - P A T A A A G
HorB.PTH.opposum	253	I T E E E L R A F T E - - - P - P P A D K A G
HorB.PTH.rat	256	L T E E E L H I I A Q V P P P - P A A A A V G
HorB.Secretin.rat	204	- D V T - Y C - - D A H
HorB.VIP.rat	205	- E I D - H C - - S E A
HorC.ADH.human	183	- V E G G S G - V
HorC.ATII-1.bovine	173	- E N T N
HorC.ATII-1.human	173	- E N T N
HorC.ATII-1.rat	173	- E N T N
HorC.ATII-1B.rat	173	- E N T N
HorC.AV1a.rat	195	- V N N G T K - T
HorC.AV2.rat	183	- V G N G S G - V
HorC.Bombesin.guineapig	192	- R D P N K N M T
HorC.Bombesin.human	185	- H E E S T N Q T
HorC.Bombesin.mouse	186	- H V K D T N Q T
HorC.CCKa.rat	201	- T K N N N Q T
HorC.CCKb.rat	198	- V - G P R
HorC.GPRN.human	189	- A S N N
HorC.Gastrin.canine	198	- A G G A R
HorC.Gastrin.rat	198	- V - G P R
HorC.GnRH.human	185	- D S S G Q T K V
HorC.GnRH.mouse	185	- D G S G P T - V
HorC.GnRH.rat	185	- D G S G P A - V
HorC.NMB.human	188	- S S L D N S S
HorC.NMB.rat	188	- G S S D N S S
HorC.Oxy.human	181	- A D G - V
HorC.RDC1.canine	189	- A S N N
HorC.Somat1.human	201	- N S D G
HorC.Somat1.mouse	201	- N S D G
HorC.Somat1.rat	201	- N S D G
HorC.Somat2.bovine	185	- N Q W G
HorC.Somat2.human	186	- N Q W G
HorC.Somat2.mouse	186	- N Q W G
HorC.Somat2.rat	186	- N Q W G
HorC.Somat3.mouse	186	- P R G
HorC.Somat3.rat	186	- P R G
HorC.TRH.mouse	170	- T Y K N A V
HorC.TRH.rat	170	- T Y K D A I

FIGURE 1h.

HorA.FSH.human   344 V T C S P K P D A F N P C E D I M G Y N I L R V L I W F I S I L A I T G N I I V L V I L T T S Q Y K
HorA.FSH.rat   342 D V T C S P K P D A F N P C E D I M G Y N I L R V L I W F I S I L A I T G N T T V L V V L T T S Q Y
HorA.FSH.sheep   343 D V T C S P E P D A F N P C E D I M G Y D I L R V L I W F I S I L A I T G N I L V L V I L I T S Q Y
HorA.LH-CG.human   325 T P R C A P E P D A F N P C E D I M G Y D F L R V L I W L I N I L A I M G N M T V L F V L L T S R Y
HorA.LH-CG.mouse   344 T L Q C T P E P D A F N P C E D I M G Y A F L R V L I W L I N I L A I F G N L T V L F V L L T S R Y
HorA.LH-CG.porcine 340 T L Q C A P E P D A F N P C E D I M G Y D F L R V L I W L I N I L A I M G N V T V L F V L L T S H Y
HorA.LH-CG.rat   344 T L Q C A P E P D A F N P C E D I M G Y A F L R V L I W L I N I L A I F G N L T V L F V L L T S R Y
HorA.TSH.canine   370 D M V C T P K S D E F N P C E D I M G Y K F L R I V V W F V S L L A L L G N V F V L I V L L T S H Y
HorA.TSH.human   395 D M V C T P K S D E F N P C E D I M G Y K F L R I V V W F V S L L A L L G N V F V L L I L L T S H Y
HorA.TSH.rat   395 D M V C T P K S D E F N P C E D I M G Y K F L R I V V W F V S P M A L L G N V F V L F V L L T S H Y
HorB.Calcitonin.human   232 P V S C K I L H F F H Q - - - - - - Y M M A C N Y F W M L C E G I Y L H T L I V V A V F T E K Q R
HorB.Calcitonin.porcine 217 P P I C K V L H F F H Q - - - - - - Y M M S C N Y F W M L C E G V Y L H T L I V V S V F A E G Q R
HorB.GLP-1.rat   222 V A G C R V A T V I M Q - - - - - - Y G I I A N Y C W L L V E G V Y L Y S L L S I T T F S E K S F
HorB.PTH.human   278 Y A G C R V A V T F F L - - - - - - Y F L A T N Y Y W I L V E G L Y L H S L I F M A F F S E K K Y
HorB.PTH.opposum   272 F V G C R V A V T V F L - - - - - - Y F L T T N Y Y W I L V E G L Y L H S L I F M A F F S E K K Y
HorB.PTH.rat   278 Y A G C R V A V T F F L - - - - - - Y F L A T N Y Y W I L V E G L Y L H S L I F M A F F S E K K Y
HorB.Secretin.rat   212 K V G C K L V M I F F Q - - - - - - Y C I M A N Y A W L L V E G L Y L H T L L A I S F F S E R K Y
HorB.VIP.rat   213 S V G C K A A V V F F Q - - - - - - Y C V M A N F F W L L V E G L Y L Y T L L A V S F F S E R K Y
HorC.ACTH.human   167 - - - - - - I F S H H - - - - - - - V P T V I T F T S L F P L M - - - - - - - - - L V - - - -
HorC.ADH.human   190 T - D C - - W A C F A E - - - - - - - P W G R R T Y V T W I A L M - V F V A P T L G I A - - - -
HorC.ATII-1.bovine 177 I T V C - - A F H Y E - - - - - - - - S Q N - S T L P V G L G L T - K N I L - - - G F L - - - -
HorC.ATII-1.human   177 I T V C - - A F H Y E - - - - - - - - S Q N - S T L P I G L G L T - K N I L - - - G F L - - - -
HorC.ATII-1.rat   177 I T V C - - A F H Y E - - - - - - - - S R N - S T L P I G L G L T - K N I L - - - G F L - - - -
HorC.ATII-1B.rat   177 I T V C - - A F H Y E - - - - - - - - S Q N - S T L P I G L G L T - K N I L - - - G F V - - - -
HorC.AV1a.rat   202 Q - D C - - W A T F I Q - - - - - - - P W G T R A Y V T W M T S G - V F V A P V V V L G - - - -
HorC.AV2.rat   190 F - D C - - W A R F A E - - - - - - - P W G L R A Y V T W I A L M - V F V A P A L G I A - - - -
HorC.Bombesin.guineapig 200 S E W C - - A F - Y P - - - - - - - - V S E - K L L Q E I H A L L - S F L V - - - F Y I - - - -
HorC.Bombesin.human   193 F I S C - - - A P Y P - - - - - - - - H S N - E L H P K I H S M A - S F L V - - - F Y V - - - -
HorC.Bombesin.mouse   194 F I S C - - - A P Y P - - - - - - - - H S N - E L H P K I H S M A - S F L V - - - F Y V - - - -
HorC.CCKa.rat   208 A N M C - - R F L L P S - - - - - - - D A M Q Q S W Q T F L L L I - L F L L P G I V M V - - - -
HorC.CCKb.rat   202 V L Q C - - M H R W P S - - - - - - - A R V Q Q T W S V L L L L L - L F F I P G V V I A - - - -
HorC.GPRN.human   193 E T Y C - - R S F Y P - - - - - - - - E H S I K E W L I G M E L V - S V V L - - - G F A - - - -
HorC.Gastrin.canine 203 A L Q C - - V H R W P S - - - - - - - A R V R Q T W S V L L L L L - L F F V P G V V M A - - - -
HorC.Gastrin.rat   202 V L Q C - - M H R W P S - - - - - - - A R V R Q T W S V L L L M L - L F F I P G V V M A - - - -
HorC.GnRH.human   193 F S Q C V T H C S F S Q - - - - - - - W W - H Q A F Y N F F T F S C L F I I P L F I M L - - - -
HorC.GnRH.mouse   192 F S Q C V T H C S F P Q - - - - - - - W W - H Q A F Y N F F T F G C L F I I P L L I M L - - - -
HorC.GnRH.rat   192 F S Q C V T H C S F P Q - - - - - - - W W - H E A F Y N F F T F S C L F I I P L L I M L - - - -
HorC.MSH.human   181 - - - - - - - A Y Y D H - - - - - - - - V A V L I C L V V F F L A M - - - - - - - - - L V - -
HorC.MSH.mouse   179 - - - - - - - T Y Y K H - - - - - - - - T A V L L C L V T F F L A M - - - - - - - - - L A - -
HorC.NMB.human   195 F T A C - - - I P Y P - - - - - - - - Q T D - E L H P K I H S V L - I F L V - - - Y F L - - - -
HorC.NMB.rat   195 F T A C - - - I P Y P - - - - - - - - Q T D - E L H P K I H S V L - I F L V - - - Y F L - - - -
HorC.Oxy.human   185 F - D C - - W A V F I Q - - - - - - - P W G P K A Y I T W I T L A - V Y I V P V I V L A - - - -
HorC.RDC1.canine   193 E T Y C - - R S F Y P - - - - - - - - E H S V K E W L I S M E L V - S V V L - - - G F A - - - -
HorC.Somat1.human   205 T V A C - - N M L M P - - - - - - - - E P A - Q R W L V G F V L Y - T F L M - - - G F L - - - -
HorC.Somat1.mouse   205 T V A C - - N M L M P - - - - - - - - E P A - Q R W L V G F V L Y - T F L M - - - G F L - - - -
HorC.Somat1.rat   205 T V A C - - N M L M P - - - - - - - - E P A - Q R W L V G F V L Y - T F L M - - - G F L - - - -
HorC.Somat2.bovine 189 R S S C - - T I N W P - - - - - - - - G E S - G A W Y T G F I I Y - A F I L - - - G F L - - - -
HorC.Somat2.human   190 R S S C - - T I N W P - - - - - - - - G E S - G A W Y T G F I I Y - T F I L - - - G F L - - - -
HorC.Somat2.mouse   190 R S S C - - T I N W P - - - - - - - - G E S - G A W Y T G F I I Y - A F I L - - - G F L - - - -
HorC.Somat2.rat   190 R S S C - - T I N W P - - - - - - - - G E S - G A W Y T G F I I Y - A F I L - - - G F L - - - -
HorC.Somat3.mouse   189 M S T C - - H M Q W P - - - - - - - - E P A - A A W R T A F I I Y - M A A L - - - G F F - - - -
HorC.Somat3.rat   189 M S T C - - H M Q W P - - - - - - - - E P A - A A W R T A F I I Y - T A A L - - - G F F - - - -
HorC.TA.rat   149 - - - - - - M W Y W R - - - - - - - - R R P K R L S A G V C A L L - - - - - - - - - W L - - - -
HorC.TRH.mouse   176 V V S C - - G Y K I S R - - - - - - - N Y Y S P I Y L M D F G V F - - Y V V P M I L A T - - - -
HorC.TRH.rat   176 V I S C - - G Y K I S R - - - - - - - N Y Y S P I Y L M D F G V F - - Y V M P M I L A T - - - -
HorC.mas.human   138 - - - - - - - I W Y R C - - - - - - - - H R P K Y Q S A L V C A L L - - - - - - - - - W A - - - -
HorC.mas.rat   137 - - - - - - - I W Y R C - - - - - - - - H R P K H Q S A F V C A L L - - - - - - - - - W A - - - -
HorC.mrg.human   179 - - - - - - - I W Y R C - - - - - - - - H R P K Y T S N V V C T L I - - - - - - - - - W G - - - -

FIGURE 1i.

```
HorA.FSH.human          394  L T V P R F L M C N L A F A D L C I G I Y L L L I A S V D I - - - - - - - - - - - H T K S Q Y H N Y
HorA.FSH.rat            392  K L T V P R F L M C N L A F A D L C I G I Y L L L I A S V D - - - - - - - - - - - I H T K S Q Y H N
HorA.FSH.sheep         393  K L T V P R F L M C N L A F A D L C I G I Y L L L I A S V D - - - - - - - - - - - V H T K S Q Y H N
HorA.LH-CG.human        375  K L T V P R F L M C N L S F A D F C M G L Y L L L I A S V D - - - - - - - - - - - S Q T K G Q Y Y N
HorA.LH-CG.mouse        394  K L T V P R F L M C N L S F A D F C M G L Y L L L I A S V D - - - - - - - - - - - S Q T K G Q Y Y N
HorA.LH-CG.porcine      390  K L T V P R F L M C N L S F A D F C M G L Y L L L I A S V D - - - - - - - - - - - A Q T K G Q Y Y N
HorA.LH-CG.rat          394  K L T V P R F L M C N L S F A D F C M G L Y L L L I A S V D - - - - - - - - - - - S Q T K G Q Y Y N
HorA.TSH.canine         420  K L T V P R F L M C N L A F A D F C M G M Y L L L I A S V D - - - - - - - - - - - L Y T H S E Y Y N
HorA.TSH.human          445  K L N V P R F L M C N L A F A D F C M G M Y L L L I A S V D - - - - - - - - - - - L Y T H S E Y Y N
HorA.TSH.rat            445  K L T V P R F L M C N L A F A D F C M G V Y L L L I A S V D - - - - - - - - - - - L Y T H T E Y Y N
HorB.Calcitonin.human   275  - - - L R W Y Y L L G W G F P - - - - - - - - L V P T T I H - - - - - - - A I T R A V Y F N
HorB.Calcitonin.porcine 260  - - - L W W Y H V L G W G F P - - - - - - - - L I P T T A H - - - - - - - A I T R A V L F N
HorB.GLP-1.rat          265  - - - F S L Y L C I G W G S P - - - - - - - - L L F V I P W - - - - - - - V V V K C L F E N
HorB.PTH.human          321  - - - L W G F T V F G W G L P - - - - - - - - A V F V A V W - - - - - - - V S V R A T L A N
HorB.PTH.opposum        315  - - - L W G F T L F G W G L P - - - - - - - - A V F V A V W - - - - - - - V T V R A T L A N
HorB.PTH.rat            321  - - - L W G F T I F G W G L P - - - - - - - - A V F V A V W - - - - - - - V G V R A T L A N
HorB.Secretin.rat       255  - - - L Q A F V L L G W G S P - - - - - - - - A I F V A L W - - - - - - - A I T R H F L E N
HorB.VIP.rat            256  - - - F W G Y I L I G W G V P - - - - - - - - S V F I T I W - - - - - - - T V V R I Y F E D
HorC.ACTH.human         188  - - - F I L C L Y V H - - - - - - - - - - - - - - - - - - - - - - - - -
HorC.ADH.human          223  - - - A C Q V L - - - - - - - - - - - - - - - - - - - - - - - - - - - - -
HorC.ATII-1.bovine      206  - - - F P F L I I L T - - - - - - - - - - - - - - - - - - - - - - - - -
HorC.ATII-1.human       206  - - - F P F L I I L T - - - - - - - - - - - - - - - - - - - - - - - - -
HorC.ATII-1.rat         206  - - - F P F L I I L T - - - - - - - - - - - - - - - - - - - - - - - - -
HorC.ATII-1B.rat        206  - - - F P F L I I L T - - - - - - - - - - - - - - - - - - - - - - - - -
HorC.AV1a.rat           235  - - - T C Y G F I C Y - - - - - - - - - - - - - - - - - - - - - - - - -
HorC.AV2.rat            223  - - - A C Q V L - - - - - - - - - - - - - - - - - - - - - - - - - - - - -
HorC.Bombesin.guineapig 228  - - - I P L S I I S V - - - - - - - - - - - - - - - - - - - - - - - - -
HorC.Bombesin.human     221  - - - I P L S I I S V - - - - - - - - - - - - - - - - - - - - - - - - -
HorC.Bombesin.mouse     222  - - - I P L A I I S V - - - - - - - - - - - - - - - - - - - - - - - - -
HorC.CCKa.rat           242  - - - V A Y G L I S L E L Y Q G I K F D A S Q K K S A K E K K P S T G - - - - - - - - - - - S
HorC.CCKb.rat           236  - - - V A Y G L I S R E L Y L G L H F D G E N D S E T Q S R A R N Q G G L P G G A A P G P V H Q N G
HorC.GPRN.human         223  - - - V P F S I I A V - - - - - - - - - - - - - - - - - - - - - - - - -
HorC.Gastrin.canine     237  - - - V A Y G L I S R E L Y L G L R F D - - E D S D S E S R V R S Q G G L R G G A G P G P A P P N G
HorC.Gastrin.rat        236  - - - V A Y G L I S R E L Y L G L R F D G D N D S D T Q S R V R N Q G G L P G G T A P G P V H Q N G
HorC.GnRH.human         229  - - - I C N A K I I F - - - - - - - - - - - - - - - - - - - - - - - - -
HorC.GnRH.mouse         228  - - - I C N A K I I F - - - - - - - - - - - - - - - - - - - - - - - - -
HorC.GnRH.rat           228  - - - I C N A K I I F - - - - - - - - - - - - - - - - - - - - - - - - -
HorC.MSH.human          202  - - - L M A V L Y V H - - - - - - - - - - - - - - - - - - - - - - - - -
HorC.MSH.mouse          200  - - - L M A I L Y A H - - - - - - - - - - - - - - - - - - - - - - - - -
HorC.NMB.human          223  - - - I P L A I I S I - - - - - - - - - - - - - - - - - - - - - - - - -
HorC.NMB.rat            223  - - - I P L V I I S I - - - - - - - - - - - - - - - - - - - - - - - - -
HorC.Oxy.human          218  - - - T C Y G L I S F - - - - - - - - - - - - - - - - - - - - - - - - -
HorC.RDC1.canine        223  - - - I P F C V I A V - - - - - - - - - - - - - - - - - - - - - - - - -
HorC.Somat1.human       234  - - - L P V G A I C L - - - - - - - - - - - - - - - - - - - - - - - - -
HorC.Somat1.mouse       234  - - - L P V G A I C L - - - - - - - - - - - - - - - - - - - - - - - - -
HorC.Somat1.rat         234  - - - L P V G A I C L - - - - - - - - - - - - - - - - - - - - - - - - -
HorC.Somat2.bovine      218  - - - V P L T I I C L - - - - - - - - - - - - - - - - - - - - - - - - -
HorC.Somat2.human       219  - - - V P L T I I C L - - - - - - - - - - - - - - - - - - - - - - - - -
HorC.Somat2.mouse       219  - - - V P L T I I C L - - - - - - - - - - - - - - - - - - - - - - - - -
HorC.Somat2.rat         219  - - - V P L T I I C L - - - - - - - - - - - - - - - - - - - - - - - - -
HorC.Somat3.mouse       218  - - - G P L L V I C L - - - - - - - - - - - - - - - - - - - - - - - - -
HorC.Somat3.rat         218  - - - G P L L V I C L - - - - - - - - - - - - - - - - - - - - - - - - -
HorC.TA.rat             170  - - - L S F L V T S I - - - - - - - - - - - - - - - - - - - - - - - - -
HorC.TRH.mouse          209  - - - V L Y G F I A R I L F - - - - - - - - - - - - - - - - - - - - - - - - -
HorC.TRH.rat            209  - - - V L Y G F I A R I L F - - - - - - - - - - - - - - - - - - - - - - - - -
HorC.mas.human          159  - - - L S C L V T T M - - - - - - - - - - - - - - - - - - - - - - - - -
HorC.mas.rat            158  - - - L S C L V T T M - - - - - - - - - - - - - - - - - - - - - - - - -
HorC.mrg.human          200  - - - L P F C I N I V - - - - - - - - - - - - - - - - - - - - - - - - -
```

FIGURE 1j.

```
HorA.FSH.human          433  A I D W Q T G A G C D A A G F F T V F A S E - - - - - - - - - L S V Y T L T A I T L E R W H T I T
HorA.FSH.rat            431  Y A I D W Q T G A G C D A A G F F T V F A S - - - - - - - - - E L S V Y T L T A I T L E R W H T I
HorA.FSH.sheep          432  Y A I D W Q T G A G C D A A G F F T V F A S - - - - - - - - - E L S V Y T L T A I T L E R W H T I
HorA.LH-CG.human        414  H A I D W Q T G S G C S T A G F F T V L A S - - - - - - - - - E L S V Y T L T V I T L E R W H T I
HorA.LH-CG.mouse        433  H A I D W Q T G S G C S A A G F F T V F A S - - - - - - - - - E L S V Y T L T V I T L E R W H T I
HorA.LH-CG.porcine      429  H A I D W Q T G N G C S V A G F F T V F A S - - - - - - - - - E L S V Y T L T V I T L E R W H T I
HorA.LH-CG.rat          433  H A I D W Q T G S G C A A G F F T V F A S - - - - - - - - - E L S V Y T L T V I T L E R W H T I
HorA.TSH.canine         459  H A I D W Q T G P G C N T A G F F T V F A S - - - - - - - - - E L S V Y T L T V I T L E R W Y A I
HorA.TSH.human          484  H A I D W Q T G P G C N T A G F F T V F A S - - - - - - - - - E L S V Y T L T V I T L E R W Y A I
HorA.TSH.rat            484  H A I D W Q T G P G C N T A G F F T V F A S - - - - - - - - - E L S V Y T L T V I T L E R W Y A I
HorB.Calcitonin.human   303  D N C - W L - S V - - E T H L L Y I I H G P - - - - - - - - - V M A A L V V N F F F L L - - N I V
HorB.Calcitonin.porcine 288  D N C - W L - S V - - D T N L L Y I I H G P - - - - - - - - - V M A A L V V N F F F L L - - N I L
HorB.GLP-1.rat          293  V Q C - W T S N D - N M G F W W I L R I P - - - - - - - - - V L L A I L I N F F I F A - - R I I
HorB.PTH.human          349  T G C - W D - L S - - S G N K K W I I Q V P - - - - - - - - - I L A S I V L N F I L F I - - N I V
HorB.PTH.opposum        343  T E C - W D - L S - - S G N K K W I I Q V P - - - - - - - - - I L A A I V V N F I L F I - - N I I
HorB.PTH.rat            349  T G C - W D - L S - - S G H K K W I I Q V P - - - - - - - - - I L A S V V L N F I L F I - - N I I
HorB.Secretin.rat       283  T G C - W D I N A - N A S V W W I R G P - - - - - - - - - V I L S I L I N F I F F I - - N I L
HorB.VIP.rat            284  F G C - W D T I I - N S S L W W I I K A P - - - - - - - - - I L L S I L V N F V L F I - - C I I
HorC.ACTH.human         196  - - - - - - - - - - - - - - M F L L A R S - - - - - - - - - H T R K I S T L P R - - - - - - - -
HorC.ADH.human          228  - - - - - - - - - - - - I F R E I H A S L V P G P - - - S E R P G G R R R G R - - - - - - R T
HorC.ATII-1.bovine      214  - - - - - - - - - - - - - S Y T L I W K - - - - - - - - - T L - K K A Y - - - - - - - - - - -
HorC.ATII-1.human       214  - - - - - - - - - - - - - S Y T L I W K - - - - - - - - - A L - K K A Y - - - - - - - - - - -
HorC.ATII-1.rat         214  - - - - - - - - - - - - - S Y T L I W K - - - - - - - - - A L - K K A Y - - - - - - - - - - -
HorC.ATII-1B.rat        214  - - - - - - - - - - - - - S Y T L I W K - - - - - - - - - A L - K K A Y - - - - - - - - - - -
HorC.AV1a.rat           243  - - - - - - - - - - H I W R N I R G K T A S S R H S K G D K G S G E A V G P F H K G L L V T
HorC.AV2.rat            228  - - - - - - - - - - - - I F R E I H A S L V P G P - - - S E R A G T P Q R A P - - - - - - D R
HorC.Bombesin.guineapig 236  - - - - - - - - - - - - - Y Y S L I A R - - - - - - - - - T L - Y K S T L N I P T - - - - - E
HorC.Bombesin.human     229  - - - - - - - - - - - - - Y Y Y F I A K - - - - - - - - - N L - I Q S A Y N L P V - - - - - E
HorC.Bombesin.mouse     230  - - - - - - - - - - - - - Y Y Y F I A R - - - - - - - - - N L - I Q S A Y N L P V - - - - - E
HorC.CCKa.rat           275  S T R Y - - - - - - E D S D G C Y L Q K S R P P R K - - - - - L E L - Q Q L S S G S G G S R L N R I
HorC.CCKb.rat           283  G C R P V T S V A G E D S D G C C V Q L P R S R - - - - - - L E M - T T L T T P T P G P V P G - -
HorC.GPRN.human         231  - - - - - - - - - - - - F Y F L L A R - - - - - - - - - A I - S A S S - - - - - - - - - - -
HorC.Gastrin.canine     282  S C R P E G G L A G E D G D G C Y V Q L P R S R Q T - - - - - L E L - S A L T A P T P G P G G G - -
HorC.Gastrin.rat        283  G C R H V T - V A G E D N D G C Y V Q L P R S R - - - - - - L E M - T T L T T P T P G P G L A - -
HorC.GnRH.human         237  - - - - - - - - - - - T L T R V L H - - - - - - - - - Q D P H E L Q L N - - - - - - - - - -
HorC.GnRH.mouse         236  - - - - - - - - - - - A L T R V L H - - - - - - - - - Q D P R K L Q M N - - - - - - - - -
HorC.GnRH.rat           236  - - - - - - - - - - - A L T R V L H - - - - - - - - - Q D P R K L Q L N - - - - - - - - -
HorC.MSH.human          210  - - - - - - - - - - - - M L A R A C Q - - - - - - - - - H A Q G I A R L H K - - - - - - - -
HorC.MSH.mouse          208  - - - - - - - - - - - - M F T R A C Q - - - - - - - - - H V Q G I A Q L H K - - - - - - - -
HorC.NMB.human          231  - - - - - - - - - - - - - Y Y Y H I A K - - - - - - - - - T L - I K S A H N L P G - - - - - E
HorC.NMB.rat            231  - - - - - - - - - - - - - Y Y Y H I A K - - - - - - - - - T L - I R S A H N L P G - - - - - E
HorC.Oxy.human          226  - - - - - - - - - - K I W Q N L R L K T A A - - - A G A E A P E G A A A G - - D G G R V A L
HorC.RDC1.canine        231  - - - - - - - - - - - - F Y C L L A R - - - - - - - - - A I - S A S S - - - - - - - - - - -
HorC.Somat1.human       242  - - - - - - - - - - - - - C Y V L I I A - - - - - - - - - K M - R M V A L K A - - - - - - - -
HorC.Somat1.mouse       242  - - - - - - - - - - - - - C Y V L I I A - - - - - - - - - K M - R M V A L K A - - - - - - - -
HorC.Somat1.rat         242  - - - - - - - - - - - - - C Y V L I I A - - - - - - - - - K M - R M V A L K A - - - - - - - -
HorC.Somat2.bovine      226  - - - - - - - - - - - - - C Y L F I I I - - - - - - - - - K V - K S S G I R V - - - - - - - -
HorC.Somat2.human       227  - - - - - - - - - - - - - C Y L F I I I - - - - - - - - - K V - K S S G I R V - - - - - - - -
HorC.Somat2.mouse       227  - - - - - - - - - - - - - C Y L F I I I - - - - - - - - - K V - K S S G I R V - - - - - - - -
HorC.Somat2.rat         227  - - - - - - - - - - - - - C Y L F I I I - - - - - - - - - K V - K S S G I R V - - - - - - - -
HorC.Somat3.mouse       226  - - - - - - - - - - - - - C Y L L I V V - - - - - - - - - K V - R S T T R R V R A P S C Q W V
HorC.Somat3.rat         226  - - - - - - - - - - - - - C Y L L I V V - - - - - - - - - K V - R S T T R R V R A P S C Q W V
HorC.TA.rat             178  - - - - - - - - - - - - H N Y F C M F - - - - - - - - - L G H E A - S G T A - - - - - - - -
HorC.TRH.mouse          220  - - - - - - - - - - - - L N P I P S D P K E N - - - - - S K M W K N D S I H Q N K N L N L N A
HorC.TRH.rat            220  - - - - - - - - - - - - L N P I P S D P K E N - - - - - S K T W K N D S T H Q N K N M N L N T
HorC.mas.human          167  - - - - - - - - - - - - E Y V M C I D - - - - - - - - - R E E E S H S R N D - - - - - - - -
HorC.mas.rat            166  - - - - - - - - - - - - E Y V M C I D - - - - - - - - - S G E E S H S Q S D - - - - - - - -
HorC.mrg.human          208  - - - - - - - - - - - - K S L F L T Y - - - - - - - - - T K H - - - - V K A - - - - - - - -
```

FIGURE 1k.

```
HorA.FSH.human          473   H A M Q L D C K V Q L R H A A S V M V M G W I F A F A A A L F P I F G I S S Y M K V S I C L P M D I
HorA.FSH.rat            471   T H A M Q L E C K V Q L R H A A S V M V L G W T F A F A A A L F P I F G I S S Y M K V S I C L P M D
HorA.FSH.sheep          472   T H A M Q L E C K V H V R H A A S I M L V G W V F A F A V A L F P I F G I S S Y M K V S I C L P M D
HorA.LH-CG.human        454   T Y A I H L D Q K L R L R H A I L I M L G G W L F S S L I A M L P L V G V S N Y M K V S I C F P M D
HorA.LH-CG.mouse        473   T Y A V Q L D Q K L R L R H A I P I M L G G W I F S T L M A T L P L V G V S S Y M K V S I C L P M D
HorA.LH-CG.porcine      469   T Y A I Q L D Q K L R L R H A I P I M L G G W L F S T L I A M L P L V G V S S Y M K V S I C L P M D
HorA.LH-CG.rat          473   T Y A V Q L D Q K L R L R H A I P I M L G G W L F S T L I A T M P L V G I S N Y M K V S I C L P M D
HorA.TSH.canine         499   T F A M R L D R K I R L R H A Y A I M V G G W V C C F L L A L L P L V G I S S Y A K V S I C L P M D
HorA.TSH.human          524   T F A M R L D R K I R L R H A C A I M V G G W V C C F L L A L L P L V G I S S Y A K V S I C L P M D
HorA.TSH.rat            524   T F A M R L D R K I R L R H A Y T I M A G G W V S C F L L A L L P M V G I S S Y A K V S I C L P M D
HorB.Calcitonin.human   337   R V L V T K M R E T H E A E S H M - - - Y L K A V K A T M I L V P L L G I - Q F V V F P W R P S N K
HorB.Calcitonin.porcine 322   R V L V K K L K E S Q E A E S H M - - - Y L K A V R A T L I L V P L L G V - Q F V V L P W R P S T P
HorB.GLP-1.rat          328   H L L V A K L R A H Q M H Y A D Y K - - F - R L A R S T L T L I P L L G V - H E V V F A F V T D E H
HorB.PTH.human          383   R V L A T K L R E T N A G R C D T R Q Q Y R K L L K S T L V L M P L F G V - H Y I V F M A T P Y T E
HorB.PTH.opposum        377   R V L A T K L R E T N A G R C D T R Q Q Y R K L L K S T L V L M P L F G V - H Y I V F M A T P Y T E
HorB.PTH.rat            383   R V L A T K L R E T N A G R C D T R Q Q Y R K L L R S T L V L M P L F G V - H Y T V F M A L P Y T E
HorB.Secretin.rat       318   R I L M R K L R T Q E T R G S E T N - H Y K R L A K S T L L L I P L F G I - H Y I V F A F S P E D -
HorB.VIP.rat            319   R I L V Q K L R P P D I G K N D S S - P Y S R L A K S T L L L I P L F G I - H Y V M F A F F P D N F
HorC.ACTH.human         213   - - - A N - - - - - - - - M K G A I T L T I L L G V F I F C W A P F V L H V L L M T F C P S N - - -
HorC.ADH.human          254   G S P G E G A H V S A A V A K T V R M T L V I V V V Y V L C W A P F F L V Q L W A A W D P E - - - -
HorC.ATII-1.bovine      227   - - - E I Q K N K P R K D - D I F K I I L A I V L F F F F S W V P H Q I F T F M D V L I Q L G L I R
HorC.ATII-1.human       227   - - - E I Q K N K P R N D - D I F K I I M A I V L F F F F S W I P H Q I F T F L D V L I Q L G I I R
HorC.ATII-1.rat         227   - - - E I Q K N K P R N D - D I F R I I M A I V L F F F F S W V P H Q I F T F L D V L I Q L G V I H
HorC.ATII-1B.rat        227   - - - K I Q K N T P R N D - D I F R I I M A I V L F F F F S W V P H Q I F T F L D V L I Q L G I I R
HorC.AV1a.rat           279   P C V S S V K S I S R A K I R T V K M T F V I V S A Y I L C W A P F F I V Q M W S V W D E N F - - -
HorC.AV2.rat            254   - S P S E G A H V S A A M A K T V R M T L V I V I V L A I P F F L V Q L W A A W D P E - - - -
HorC.Bombesin.guineapig 255   E Q S H A R K Q V E S R K - R I A K T V L V L V A L F A L C W L P N H L L N L Y H S F T H K A Y E -
HorC.Bombesin.human     248   G N I H V K K Q I E S R K - R L A K T V L V F V G L F A F C W L P N H V I Y L Y R S Y H S E - V -
HorC.Bombesin.mouse     249   G N I H V K K Q I E S R K - R L A K T V L V F V G L F A F C W L P N H V I Y L Y R S Y H S E - V -
HorC.CCKa.rat           313   R S S S S A A N L I A K K - R V I R M L I V I V V L F F L C W M P I F S A N A W R A Y D T V S - - -
HorC.CCKb.rat           323   - P R P N Q A K L L A K K - R V V R M L L V I V L L F F L C W L P V Y S V N T W R A F D G P G - - -
HorC.GPRN.human         244   - - - D Q E K H S S R - - - - - - K I I F S Y V V V F L V C W L P Y H V A V L L D I F S I L H Y I P
HorC.Gastrin.canine     324   - P R P Y Q A K L L A K K - R V V R M L L V I V V L F F L C W L P L Y S A N T W R A F D S S G - - -
HorC.Gastrin.rat        322   - S - A N Q A K L L A K K - R V V R M L L V I V L L F F L C W L P I Y S A N T W C A F D G P G - - -
HorC.GnRH.human         253   - - - Q S K N N I P R A R L K T L K M T V A F A T S F T V C W T P Y Y V L G I W Y W F D P E M - - -
HorC.GnRH.mouse         252   - - - Q S K N N I P R A R L R T L K M T V A F A T S F V V C W T P Y Y V L G I W Y W F D P E M - - -
HorC.GnRH.rat           252   - - - Q S K N N I P R A R L R T L K M T V A F G T S F V I C W T P Y Y V L G I W Y W F D P E M - - -
HorC.MSH.human          227   - - - R Q R P V H Q G F G L K G A V T L T I L L G I F F L C W G P F F L H L T L I V L C P E H - - -
HorC.MSH.mouse          225   - - - R R R S I R Q G F C L K G A A T L T I L L G I F F L C W G P F F L H L L L I V L C P Q H - - -
HorC.NMB.human          250   Y N E H T K K Q M E T R K - R L A K I V L V F V G C F I F C W F P N H I L Y M Y R S F N Y N E - I -
HorC.NMB.rat            250   Y N E H T K K Q M E T R K - R L A K I V L V F V G C F V F C W F P N H I L Y L Y R S F N Y K E - I -
HorC.Oxy.human          257   A R V S S V K L I S K A K I R T V K M T F I I V L A F I V C W T P F F F V Q M W S V W D A N - - - -
HorC.RDC1.canine        244   - - - D Q E K Q S S R - - - - - - K I I F S Y V V V F L V C W L P Y H V V V L L D I F S I L H Y I P
HorC.Somat1.human       258   - - - G W Q Q R K R S E R - K I T L M V M M V V M V F V I C W M P F Y V V Q L - - - - V N V F - - -
HorC.Somat1.mouse       258   - - - G W Q Q R K R S E R - K I T L M V M M V V M V F V I C W M P F Y V V Q L - - - - V N V F - - -
HorC.Somat1.rat         258   - - - G W Q Q R K R S E R - K I T L M V M M V V M V F V I C W M P F Y V V Q L - - - - V N V F - - -
HorC.Somat2.bovine      242   - - - G S S K R K K S E K - K V T R M V S I V V A V F I F C W L P F Y I F N V S S V S V A I S - - -
HorC.Somat2.human       243   - - - G S S K R K K S E K - K V T R M V S I V V A V F I F C W L P F Y I F N V S S V S M A I S - - -
HorC.Somat2.mouse       243   - - - G S S K R K K S E K - K V T R M V S I V V A V F I F C W L P F Y I F N V S S V S V A I S - - -
HorC.Somat2.rat         243   - - - G S S K R K K S E K - K V T R M V S I V V A V F I F C W L P F Y I F N V S S V S V A I S - - -
HorC.Somat3.mouse       250   Q A P A C Q R R R R S E R - R V T R M V V A V V A L F V L C W M P F Y L L N I V N V V C P L P - - -
HorC.Somat3.rat         250   Q A P A C Q R R R R S E R - R V T R M V V A V V A L F V L C W M P F Y L L N I V N V V C P L P - - -
HorC.TA.rat             194   - - - C L N M D I S L G I L L F F L F C P L M V L P C L A L I L H V E C R A R R R Q R S A K L - - -
HorC.TRH.mouse          250   T N R C F N S T V S S R K - Q V T K M L A V V V I L F A L L W M P Y R T L V V V N S F L S - - - - -
HorC.TRH.rat            250   T N R C F N S T V S S R K - Q V T K M L A V V V I L F A L L W M P Y R T L V V V N S F L S - - - - -
HorC.mas.human          184   - - - C R A V I I F I A I L S F L V F T P L M L V S S T I L V V K I R - K N T W A S H S S K L - - -
HorC.mas.rat            183   - - - C R A V I I F I A I L S F L V F T P L M L V S S T I L V V K I R - K N T W A S H S S K L - - -
HorC.mrg.human          221   - - - C - V I F L K L S G L F H A I L S L V M C V S S L T L L I R F L C C S Q - Q Q K A T R V - - -
```

FIGURE 1l.

HorA.FSH.human	523	D S P L S Q L Y V M S L L V L N V L A F V V I C G C Y I H I Y L T V R N P N I V S S S S D T R I A K
HorA.FSH.rat	521	I D S P L S Q L Y V M A L L V L N V L A F V V I C G C Y T H I Y L T V R N P T I V S S S S D T K I A
HorA.FSH.sheep	522	I D S P L S Q L Y V M S L L V L N V L A F V V I C G C Y T H I Y L T V R N P N I T S S S S D T K I A
HorA.LH-CG.human	504	V E T T L S Q V Y I L T I L I L N V V A F L I I C A C Y I K I Y F A V R N P E L M A T N K D T K I A
HorA.LH-CG.mouse	523	V E S T L S Q V Y I L S I L L L N A V A F V V I C A C Y V R I Y F A V Q N P E L T A P N K D T K I A
HorA.LH-CG.porcine	519	V E T T L S Q V Y I L T I L I L N V V A F I I I C A C Y I K I Y F A V Q N P E L M A T N K D T K I A
HorA.LH-CG.rat	523	V E S T L S Q V Y I L S I L I L N V V A F V V I C A C Y I R I Y F A V Q N P E L T A P N K D T K I A
HorA.TSH.canine	549	T E T P L A L A Y I I L V L L L N I V A F I I V C S C Y V K I Y I T V R N P Q Y N P G D K D T K I A
HorA.TSH.human	574	T E T P L A L A Y I F V L T L N I V A F V V C C H V K I Y I T V R N P Q Y N P G D K D T K I A
HorA.TSH.rat	574	T D T P L A L A Y I A L V L L L N V V A F V I V C S C Y V K I Y I T V R N P Q Y N P R D K D T K I A
HorB.Calcitonin.human	383	M L G K I Y D Y V M H - - - S L - - I H F Q G F - - - F V A T I Y C F C N N E V Q T T V K R Q W A Q
HorB.Calcitonin.porcine	368	L L G K I Y D Y V V H - - - S L - - I H F Q G F - - - F V A I I Y C F C N H E V Q G A L K R Q W N Q
HorB.GLP-1.rat	374	A Q G T L R S T K L F F D L F F - - S S F Q G L - - - L V A V L Y C F L N K E V Q A E L L R R W R R
HorB.PTH.human	432	V S G T L W Q V Q M H Y E M L F - - N S F Q G F - - - F V A I I Y C F C N G E V Q A E I K K S W S R
HorB.PTH.opposum	426	V S G I L W Q V Q M H Y E M L F - - N S F Q G F - - - F V A I I Y C F C N G E V Q A E I K K S W S R
HorB.PTH.rat	432	V S G T L W Q I Q M H Y E M L F - - N S F Q G F - - - F V A I I Y C F C N G E V Q A E I R K S W S R
HorB.Secretin.rat	365	A - - - - M E V Q L F F E L A L - - G S F Q G L - - - V V A V L Y C F L N G E V Q L E V Q K K W R Q
HorB.VIP.rat	367	K - - - - A Q V K M V F E L V V - - G S F Q G F - - - V V A I L Y C F L N G E V Q A E L R R K W R R
HorC.ACTH.human	249	- - - P Y C A C Y M S L F Q V N G M L I M C N A - - V I D P F I Y A F R S P E L R D A F K K - - - -
HorC.ADH.human	300	- - - - - A P L E G A P F V L L M L L A S L N S - - C T N P W I Y A S F S S S V S S E - L R S - - -
HorC.ATII-1.bovine	273	- D C K I E D I V D T A M P I T C L A Y F N N - - C L N P L F Y G F L G K K F K K Y F L Q L - - -
HorC.ATII-1.human	273	- D C R I A D I V D T A M P I T C I A Y F N N - - C L N P L F Y G F L G K K F K K Y F L Q L - - -
HorC.ATII-1.rat	273	- D C K I S D I V D T A M P I T C I A Y F N N - - C L N P L F Y G F L G K K F K K Y F L Q L - - -
HorC.ATII-1B.rat	273	- D C E I A D I V D T A M P I T C I A Y F N N - - C L N P L F Y G F L G K K F K K Y F L Q L - - -
HorC.AV1a.rat	326	- - - I W T D S E N P S I T I T A L L A S L N S - - C C N P W I Y M F S G H L L Q D C V Q S - - -
HorC.AV2.rat	299	- - - - - A P L E R P P F V L L M L L A S L N S - - C T N P W I Y A S F S S S V S S E - L R S - - -
HorC.Bombesin.guineapig	303	- - - D S S A I H F I V T I F S R V L A F S N S - - C V N P F A L Y W L S K T F Q K Q F K A Q - - -
HorC.Bombesin.human	295	- - - D T S M L H F V T S I C A R L L A F T N S - - C V N P F A L Y L L S K S F R K Q F N T Q - - -
HorC.Bombesin.mouse	296	- - - D T S M L H F V T S I C A R L L A F T N S - - C V N P F A L Y L L S K S F R K Q F N T Q - - -
HorC.CCKa.rat	359	- - - A E K H L S G T P I S F I L L L S Y T S S - - C V N P I I Y C F M N K R F R L G F M A T - -
HorC.CCKb.rat	368	- - - A Q R A L S G A P I S F I H L L S Y V S A - - C V N P L V Y C F M H R R F R Q A C L D T - - -
HorC.GPRN.human	285	F T C R L E H A L F T A L H V T Q C L S L V H C - - C V N P V L Y S F I N R N Y - R Y - - E L - -
HorC.Gastrin.canine	369	- - - A H R A L S G A P I S F I H L L S Y A S A - - C V N P L V Y C F M H R R F R Q A C L E T - - -
HorC.Gastrin.rat	366	- - - A H R A L S G A P I S F I H L L S Y A S A - - C V N P L V Y C F M H R R F R Q A C L D T - - -
HorC.GnRH.human	297	- - - L - N R L S D P V N H F F F L F A F L N P - - C F D P L I Y G Y F S - - - - - - - - - - - - -
HorC.GnRH.mouse	296	- - - L - N R V S E P V N H F F F L F A F L N P - - C F D P L I Y G Y F S - - - - - - - - - - - - -
HorC.GnRH.rat	296	- - - L - N R V S E P V N H F F F L F G F L N P - - C F D P L I Y G Y F S - - - - - - - - - - - - -
HorC.MSH.human	271	- - - P T C G C I F K N F N L F L A L L I I C N A - - I I D P L I Y A F H S Q E L R R T L K E - - - -
HorC.MSH.mouse	269	- - - P T C S C I F K N F N L F L L L L I V L S S - - T V D P L I Y A F R S Q E L R M T L K E - - -
HorC.NMB.human	297	- - - D P S L G H M I V T L V A R V L S F G N S - - C V N P F A L Y L L S E S F R R H F N S Q - - -
HorC.NMB.rat	297	- - - D P S L G H M I V T L V A R V L S F S N S - - C V N P F A L Y L L S E S F R K H F N S Q - - -
HorC.Oxy.human	303	- - - - - A P K E A S A F I I V M L L A S L N S - - C C N P W I Y M L F T G H L F H E L V Q R - - -
HorC.RDC1.canine	285	F T C Q L E N F L F T A L H V T Q C L S L V H C - - C V N P V L Y S F I N R N Y - R Y - - E L - - -
HorC.Somat1.human	297	- - - - A E Q D D A T V S Q L S V I L G Y A N S - - C A N P I L Y G F L S D N F K R S F Q R I L - -
HorC.Somat1.mouse	297	- - - - A E Q D D A T V S Q L S V I L G Y A N S - - C A N P I L Y G F L S D N F K R S F Q R I L - -
HorC.Somat1.rat	297	- - - - A E Q D D A T V S Q L S V I L G Y A N S - - C A N P I L Y G F L S D N F K R S F Q R I L - -
HorC.Somat2.bovine	285	- - - - P T P A L K G M F D F V V V L T Y A N S - - C A N P I L Y A F L S D N F K K S F Q N V L - -
HorC.Somat2.human	286	- - - - P T P A L K G M F D F V V V L T Y A N S - - C A N P I L Y A F L S D N F K K S F Q N V L - -
HorC.Somat2.mouse	286	- - - - P T P A L K G M F D F V V I L T Y A N S - - C A N P I L Y A F L S D N F K K S F Q N V L - -
HorC.Somat2.rat	286	- - - - P T P A L K G M F D F V V I L T Y A N S - - C A N P I L Y A F L S D N F K K S F Q N V L - -
HorC.Somat3.mouse	296	- - - - E E P A F F G L Y F L V V A L P Y A N S - - C A N P I L Y G F L S Y R F K Q G F R R I L L R
HorC.Somat3.rat	296	- - - - E E P A F F G L Y F L V V A L P Y A N S - - C A N P I L Y G F L S Y R F K Q G F R R I L L R
HorC.TA.rat	238	- - - N H V V L A I V S V F L V S S I Y L G I D - - W F L F W V F Q I P A P F P E Y V T D L - - - -
HorC.TRH.mouse	294	- - - - S P F Q E N W F L L F C R I C I Y L N S - - A I N P V I Y N L M S Q K F R A A F R K L - - -
HorC.TRH.rat	294	- - - - S P F Q E N W F L L F C R I C I Y L N S - - A I N P V I Y N L M S Q K F R A A F R K L - - -
HorC.mas.human	227	- - - Y I V I M V T I I I F L I F A M P M R L L - - Y L L Y Y E Y W S T F G N L H H I S L L - - - -
HorC.mas.rat	226	- - - Y I V I M V T I I I F L I F A M P M R V L - - Y L L Y Y E Y W S T F G N L H H I S L L - - - -
HorC.mrg.human	263	- - - Y A V V Q I S A P M F L L W A L P L S V A - - P L I T - D F - K M F V T T S Y L I S L - - - -

FIGURE 1m.

```
HorA.FSH.human          573 RMAMLIFTDFLCMAPISFFAISASLKVPLITVSKAKILLVLFHPINSCAN
HorA.FSH.rat            571 KRMATLIFTDFLCMAPISFFAISASLKVPLITVSKAKILLVLFYPINSCA
HorA.FSH.sheep          572 KRMAMLIFTDFLCMAPISFFAISASLKVPLITVSKSKILLVLFYPINSCA
HorA.LH-CG.human        554 KKMAILIFTDFTCMAPISFFAISAAFKVPLITVTNSKVLLVLFYPINSCA
HorA.LH-CG.mouse        573 KKMAILIFTDFTCMAPISFFAISAAFKVPLITVTNSKVLLVLFYPVNSCA
HorA.LH-CG.porcine      569 KKMAVLIFTDFTCMAPISFFAISAALKVPLITVTNSKVLLVLFYPVNSCA
HorA.LH-CG.rat          573 KKMAILIFTDFTCMAPISFFAISAAFKVPLITVTNSKILLVLFYPVNSCA
HorA.TSH.canine         599 KRMAVLIFTDFMCMAPISFYALSALMNKPLITVTNSKILLVLFYPLNSCA
HorA.TSH.human          624 KRMAVLIFTDFICMAPISFYALSAILNKPLITVSNSKILLVLFYPLNSCA
HorA.TSH.rat            624 KRMAVLIFTDFMCMAPISFYALSALMNKPLITVTNSGVLLVLFYPLNSCA
HorB.Calcitonin.human   425 FKIQ-WN--QRW----------GRRPSNRSARAAAA-------------
HorB.Calcitonin.porcine 410 YQAQRWA--GRR----------STRAANAAAATAAA-------------
HorB.GLP-1.rat          419 WQEGKAL--Q-----------EERMASSHGSHMAPA-------------
HorB.PTH.human          477 WTLALDF--KRKARSGSSSYSYGPMVSHTSVTNVGPRVGLGLPL-----S
HorB.PTH.opposum        471 WTLALDF--KRKARSGSSTYSYGPMVSHTSVTNVGPRGGLALSL-----S
HorB.PTH.rat            477 WTLALDF--KRKARSGSSSYSYGPMVSHTSVTNVGPRAGLSLPL-----S
HorB.Secretin.rat       406 WH---L-Q-----------EFPLRPVAFNNSF-----------------
HorB.VIP.rat            408 WH---L-Q-----------GVLGWSSKSQHPWGG---------------
HorC.ACTH.human         290 ----MIF------------CSRYW------------------------
HorC.ADH.human          339 ----LLC------------CAR-GRTPPSLGPQ---------------
HorC.ATII-1.bovine      317 ----LKY------------IPPKAKSHSNLSTK---------------
HorC.ATII-1.human       317 ----LKY------------IPPKAKSHSNLSTK---------------
HorC.ATII-1.rat         317 ----LKY------------IPPKAKSHSSLSTK---------------
HorC.ATII-1B.rat        317 ----LKY------------IPPTAKSHAGLSTK---------------
HorC.AV1a.rat           368 ----FPC------------CHS---MAQKFAKD---------------
HorC.AV2.rat            338 ----LLC------------CAQ-RHTTHSLGPQ---------------
HorC.Bombesin.guineapig 345 ----LFC------------CKGELPEPPLAATP---------------
HorC.Bombesin.human     337 ----LLC------------CQPGLIIRSHS--T---------------
HorC.Bombesin.mouse     338 ----LLC------------CQPGLMNRSHS--T---------------
HorC.CCKa.rat           401 ---FP-C------------CPNPGPPGVRGEVG---------------
HorC.CCKb.rat           410 ---CARC------------CPRP-PRARPQPLP---------------
HorC.GPRN.human         327 ----MKA------------FIFKYSAKTGL-TK---------------
HorC.Gastrin.canine     411 ---CARC------------CPRP-PRARPRPLP---------------
HorC.Gastrin.rat        408 ---CARC------------CPRP-PRARPRPLP---------------
HorC.GnRH.human         328 ----------------------------L-------------------
HorC.GnRH.mouse         327 ----------------------------L-------------------
HorC.GnRH.rat           327 ----------------------------L-------------------
HorC.MSH.human          312 ----VLT------------CS--W------------------------
HorC.MSH.mouse          310 ----VLL------------CS--W------------------------
HorC.NMB.human          339 ----LCC------------GRKSYQERGTSYLL---------------
HorC.NMB.rat            339 ----LCC------------GQKSYPERSTSYLL---------------
HorC.Oxy.human          343 ----FLC------------CSASYLKGRRLGET---------------
HorC.RDC1.canine        327 ----MKA------------FIFKYSAKTGL-TK---------------
HorC.Somat1.human       339 ---CLSW------------MDNAAEEPVDYYAT---------------
HorC.Somat1.mouse       339 ---CLSW------------MDNAAEEPVDYYAT---------------
HorC.Somat1.rat         339 ---CLSW------------MDNAAEEPVDYYAT---------------
HorC.Somat2.bovine      327 ---CLVK------------V-SGTDDGERSDSK---------------
HorC.Somat2.human       328 ---CLVK------------V-SGTDDGERSDSK---------------
HorC.Somat2.mouse       328 ---CLVK------------V-SGTEDGERSDSK---------------
HorC.Somat2.rat         328 ---CLVK------------V-SGAEDGERSDSK---------------
HorC.Somat3.mouse       340 PSRRIRS------------QEPGSGPPEKTEEE---------------
HorC.Somat3.rat         340 PSRRVRS------------QEPGSGPPEKTEEE---------------
HorC.TA.rat             279 ----CIC------------INSSA-----------------------
HorC.TRH.mouse          335 ----CNC------------KQKPTEKAANYSVA---------------
HorC.TRH.rat            335 ----CNC------------KQKPTEKAANYSVA---------------
HorC.mas.human          268 ----FST------------INSSA-----------------------
HorC.mas.rat            267 ----FST------------INSSA-----------------------
HorC.mrg.human          302 ----FLI------------INSSA-----------------------
```

FIGURE 1n.

HorA.FSH.human	623	P F L Y A I F T K N F R R D F F I L L S K C G C Y E M Q A Q I Y R T E T S S T V H N T H P R N G H -
HorA.FSH.rat	621	N P F L Y A I F T K N F R R D F F I L L S K F G C Y E M Q A Q I Y R T E T S S A T H N F H A R K S H
HorA.FSH.sheep	622	N P F L Y A I F T R N F R R D F F I L L S K F G C Y E V Q A Q T Y R S E T S F T A H N F H P R N G H
HorA.LH-CG.human	604	N P F L Y A I F T K T F Q R D F F L L L S K F G C C K R R A D P L Y R R K D F S A Y T S N C K N G -
HorA.LH-CG.mouse	623	N P F L Y A V F T K A F Q R D F F L L L S R F G C C K H R A E - L Y R R K E F S A C T F N S K N G -
HorA.LH-CG.porcine	619	N P F L Y A I F T K A F R R D F F L L L S K S G C C K H Q A E - L Y R R K D F S A Y - - - C K N G -
HorA.LH-CG.rat	623	N P F L Y A I F T K A F Q R D F L L L L S R F G C C K R R A E - L Y R R K E F S A Y T S N C K N G -
HorA.TSH.canine	649	N P F L Y A I F T K A F Q R D V F I L L S K F G I C K R Q A Q A Y R G G Q R V S P K N S A G I Q I Q K
HorA.TSH.human	674	N P F L Y A I F T K A F Q R D V F I L L S K F G I C K R Q A Q A Y R G G Q R V P P K N S T D I Q V Q K
HorA.TSH.rat	674	N P F L Y A I F T K A F Q R D V F I L L S K F G L C K H Q A Q A Y Q A Q R V C P N N N T G I Q I Q K
HorB.Calcitonin.human	448	- - - - - - - A - - - A E A G D I P I Y I C H Q E P R N E P A N N Q - - - G E E - - - - - - - - - - -
HorB.Calcitonin.porcine	434	- - - - - - - A A A L A E T V E I P V Y I C H Q E P R E E P A G E E P V V E V E - - - - - - - - - - -
HorB.GLP-1.rat	442	- - - - - - - G T C H G D P C E K L Q L M S A G S S S G T G C E P S A K T - - - - - - - - - - - - - -
HorB.PTH.human	520	P R L L P - - T A T T N G H P Q L P G H A K P G T P A L E T L E T T P P A M A A P K D D G F L N G S
HorB.PTH.opposum	514	P R L A P G A G A S A N G H H Q L P G Y V K H G S I S E N S L P S S G P E - P G T K D D G Y L N G -
HorB.PTH.rat	520	P R - L P - - P A T T N G H S Q L P G H A K P G A P A T E T - E T L P V T M A V P K D D G F L N G S
HorB.Secretin.rat	423	- - - - - - - S N A T N G P T H S - - - T K A S T E Q S R S I - - - P R A - - - - - - - - - - - -
HorB.VIP.rat	427	- - - - - - - S N G A T C S T Q V S M L T R V S P S A R R S S S F Q A E V - - - - - - - - - - - - -
HorC.ADH.human	355	- - - - - - - D E S C T T A S S S L A K D T S S -
HorC.ATII-1.bovine	334	- - - - - - - M S T L S Y R P S E N G N S S T -
HorC.ATII-1.human	334	- - - - - - - M S T L S Y R P S D N V S S S T -
HorC.ATII-1.rat	334	- - - - - - - M S T L S Y R P S D N M S S S A -
HorC.ATII-1B.rat	334	- - - - - - - M S T L S Y R P S D N M S S S A -
HorC.AV1a.rat	382	- - - - - - - D S D S M S R K T D F L F -
HorC.AV2.rat	354	- - - - - - - D E S C A T A S S S L M K D T P S -
HorC.Bombesin.guineapig	362	- - - - - - - L N S L A V M G R V S G T E N T H I S E I G V A S F - - - - - - - - - - - - -
HorC.Bombesin.human	352	- - - - - - - G R S T T C M T S L K S T N P - - - S V A T F S - L - - - - - - - - - - - - -
HorC.Bombesin.mouse	353	- - - - - - - G R S T T C M T S F K S T N P - - - S - A T F S - L - - - - - - - - - - - - -
HorC.CCKa.rat	418	- - - - - - - E E E D G R T I R A L L S R Y S -
HorC.CCKb.rat	427	- - - - - - - D E D P P T P S I A S L S R L S -
HorC.GPRN.human	343	- - - - - - - L I D A S - R V S E T E Y S A L -
HorC.Gastrin.canine	428	- - - - - - - D E D P P T P S I A S L S R L S -
HorC.Gastrin.rat	425	- - - - - - - D E D P P T P S I A S L S R L S -
HorC.NMB.human	356	- - - - - - - S S S A V R M T S L K S N A K - - - N M V T N S V L - - - - - - - - - - - -
HorC.NMB.rat	356	- - - - - - - S S S A V R M T S L K S N A K - - - N V V T N S V L - - - - - - - - - - - -
HorC.Oxy.human	360	- - - - - - - S A S K K S N S S S F V L S H R S S S Q R S C S Q P S T A - - - - - - - - - -
HorC.RDC1.canine	343	- - - - - - - L I D A S - R V S E T E Y S A L -
HorC.Somat1.human	357	- - - - - - - A L K S R A Y S V E D F Q P E N - - - L E S G G V F - - - - - - - - - - - -
HorC.Somat1.mouse	357	- - - - - - - A L K S R A Y S V E D F Q P E N - - - L E S G G V F - - - - - - - - - - - -
HorC.Somat1.rat	357	- - - - - - - A L K S R A Y S V E D F Q P E N - - - L E S G G V F - - - - - - - - - - - -
HorC.Somat2.bovine	344	- - - - - - - Q D K S R L N E T T E T Q R T L - - - L N - - G D L - - - - - - - - - - - -
HorC.Somat2.human	345	- - - - - - - Q D K S R L N E T T E T Q R T L - - - L N - - G D L - - - - - - - - - - - -
HorC.Somat2.mouse	345	- - - - - - - Q D K S R L N E T T E T Q R T L - - - L N - - G D L - - - - - - - - - - - -
HorC.Somat2.rat	345	- - - - - - - Q D K S R L N E T T E T Q R T L - - - L N - - G D L - - - - - - - - - - - -
HorC.Somat3.mouse	361	- - - - - - - E D E E E E E R R E E E E R R M Q R G Q E M N G R L S Q I A Q A G T S G Q Q P R P C T
HorC.Somat3.rat	361	- - - - - - - E D E E E E E R R E E E E R R M Q R G Q E M N G R L S Q I A Q P G P S G Q Q Q R P C T
HorC.TRH.mouse	352	- - - - - - - L N Y S V I K E S D R F S T E L E D I T V T D T Y V - - - - - - - - - - - - -
HorC.TRH.rat	352	- - - - - - - L N Y S V I K E S D R F S T E L D D I T V T D T Y V - - - - - - - - - - - - -

FIGURE 1o.

binding domain and the transmembrane spanning domain of the receptor. The extracellular domain contains several putative N-linked glycosylation sites. Two of the sites are conserved among the three receptors (Figure 2). For the TSH receptor, it was shown by site-directed mutagenesis that these sites are important in the expression of functional TSH receptor (Russo et al., 1991), although they seem to be not important for hormone binding as suggested by TSH binding to *in vitro* translated nonglycosylated TSH-receptor (Akamizu et al., 1990b). The alignment further reveals, in leucine rich repeat 1 and 14, 8 conserved cysteine residues. Therefore, the formation of disulfide bonds might be crucial for the conformational integrity of the large extracellular domain.

HorA.FSH.human	672	C S S A P R - - - V T S G S T Y I L V P L S H L A Q N - - - - - - - - - - - - - - - - -
HorA.FSH.rat	671	C S S A P R - - - V T - - N S Y V L V P L N H S S Q N - - - - - - - - - - - - - - - - -
HorA.FSH.sheep	672	C P P A P R - - - V T N G S N Y T L I P L R H L A K N - - - - - - - - - - - - - - - - -
HorA.LH-CG.human	653	F T G S N K - - - P S Q S T L K L S T L H C Q G T A L - - - - - - - - - - - - - - - - -
HorA.LH-CG.mouse	671	F P R S S K - - - P S Q A A L K L S I V H C Q Q P T P - - - - - - - - - - - - - - - - -
HorA.LH-CG.porcine	664	F T G S N K - - - P S R S T L K L T T L Q C Q Y S T V - - - - - - - - - - - - - - - - -
HorA.LH-CG.rat	671	F P G A S K - - - P S Q A T L K L S T V H C Q Q P I P - - - - - - - - - - - - - - - - -
HorA.TSH.canine	699	V T R D M R Q S L P N M Q D E Y E L L E N S H L T P N K Q G G Q I S K E Y N Q T V L - - - -
HorA.TSH.human	724	V T H D M R Q G L H N M E D V Y E L I E N S H L T P K K Q G G Q I S E E Y M Q T V L - - - -
HorA.TSH.rat	724	I P Q D T R Q S L P N V Q D T Y E P L G S S H L T P K L Q G R I S E E Y T Q T A L - - - -
HorB.Calcitonin.human	475	- S A E I I P - - L N I I E Q E S S A -
HorB.Calcitonin.porcine	467	- G V E V I A - - M E V L E Q E T S A -
HorB.GLP-1.rat	472	- - S L A S S - - L P R L A D S P T -
HorB.PTH.human	568	C S G L D E E - - A S G P E R P P A L L Q E E W E T V M - - - - - - - - - - - - - - - - -
HorB.PTH.opposum	562	- S G L Y E P - - M V G - E Q P P P L L E E E R E T V M - - - - - - - - - - - - - - - - -
HorB.PTH.rat	566	C S G L D E E - - A S G S A R P P P L L Q E E W E T V M - - - - - - - - - - - - - - - - -
HorB.Secretin.rat	447	- - S I I -
HorB.VIP.rat	457	- - S L V -
HorC.ATII-1.bovine	350	- - - - - - - - - - K K P A P C I E V E -
HorC.ATII-1.human	350	- - - - - - - - - - K K P A P C F E V E -
HorC.ATII-1.rat	350	- - - - - - - - - - K K P A S C F E V E -
HorC.ATII-1B.rat	350	- - - - - - - - - - K K S A S F F E V E -
HorC.Bombesin.guineapig	388	- - - - - - - - - - I G R P M K K E E N R V -
HorC.Bombesin.human	374	- - - - - - - - - - I N G N I C H E R Y V -
HorC.Bombesin.mouse	374	- - - - - - - - - - I N R N I C H E G Y V -
HorC.CCKa.rat	434	- - - - - - - - - - Y S H M S T S A P P P -
HorC.CCKb.rat	443	- - - - - - - - - - Y T T I S T L G P G -
HorC.GPRN.human	358	- - - - - - - - - - E Q N A K -
HorC.Gastrin.canine	444	- - - - - - - - - - Y T T I S T L G P G -
HorC.Gastrin.rat	441	- - - - - - - - - - Y T T I S T L G P G -
HorC.NMB.human	379	- - - - - - - - - - L N G H S M K Q E M A M -
HorC.NMB.rat	379	- - - - - - - - - - L N G H S T K Q E I A L -
HorC.RDC1.canine	358	- - - - - - - - - - E Q N A K -
HorC.Somat1.human	380	- - - - - - - - - - R N G T C T S R I T T L -
HorC.Somat1.mouse	380	- - - - - - - - - - R N G T C A S R I S T L -
HorC.Somat1.rat	380	- - - - - - - - - - R N G T C A S R I S T L -
HorC.Somat2.bovine	365	- - - - - - - - - - Q T S I -
HorC.Somat2.human	366	- - - - - - - - - - Q T S I -
HorC.Somat2.mouse	366	- - - - - - - - - - Q T S I -
HorC.Somat2.rat	366	- - - - - - - - - - Q T S I -
HorC.Somat3.mouse	404	G T A K E Q Q L L P Q E A T A G D K A S T L S H L -
HorC.Somat3.rat	404	G T A K E Q Q L L P Q E A T A G D K A S T L S H L -
HorC.TRH.mouse	378	- - - - - - - - - - S T T K V S F D D T C L A S E N - - - - - - - - - - - - - - - - - - -
HorC.TRH.rat	378	- - - - - - - - - - S T T K V S F D D T C L A S E K N G P S S C T Y G Y S L T A K Q E K I

FIGURE 1p.

Ligand binding to the extracellular domain of the receptor is exceptional since in most other G protein-coupled receptors the ligand-binding pocket appears to lie within the plane of the lipid bilayer and hence involves amino acid residues in the transmembrane regions (Strader et al., 1989). For the glycoprotein hormones and their receptors, the ligand-receptor interaction seems to be different. Two models can be envisaged for the signal transduction mechanism. In one model, hormone binding leads to a conformational change of the entire receptor. In the second model, hormone binding triggers receptor stimulation by the correct positioning of the hormone into the receptor activation site, which is defined by the arrangement of the membrane spanning segments. The findings of Ji and Ji (1991a; 1991b) indicate that the transmembrane spanning domain of the LH-CG receptor displays low affinity binding and stimulates cAMP production after hormone activation.

In this context, it should be mentioned that autoantibodies against the TSH receptor can be detected in patients with a history of Graves' hyperthyroidism and

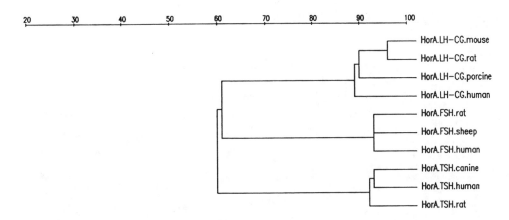

FIGURE 2. Phylogenetic trees of selected members of each subfamily, (A) a selected number of hormone receptors, (B) the glycoprotein hormone receptors, and (C) the secretin-receptor family. The tree was obtained by comparing the whole polypeptide sequences of the cloned receptors. The software used is called TREE. It is based on the progressive alignment derived by Feng and Doolittle (Feng and Doolittle, 1987), The names and the appropriate reference for the receptors are given in Table 1. The total length of vertical lines between different receptors give their putative phylogenetic distance in arbitrary units. The average amino acid identity between individual receptors or receptor subfamilies is indicated in percent at several branch points.

frank hypothyroidism where the autoantibodies act as TSH agonists (stimulatory antibody) and as TSH antagonists (blocking antibodies), respectively (Rees et al., 1988). These two types of antibodies can be envisaged as antibodies which distinguish functionally different receptor domains. However, the binding sites for most autoantibodies appear to be not identical to the TSH binding site of the receptor (Nagayama and Rapoport, 1992).

The carboxy-terminal half of the glycoprotein receptor polypeptides contains seven hydrophobic segments of membrane-spanning length and displays sequence homology to most members of the G protein-coupled receptor family. Although the overall sequence similarity is low, certain conserved amino acid residues are also present in the glycoprotein hormone receptors. For example (1) the aspartic acid residue within transmembrane spanning segment 2; (2) the asparagine in segment 7; (3) the conserved proline residues in transmembrane segment 4, 6, and 7, which were suggested to induce bends in transmembrane helices to facilitate the interaction of adjacent helices (O'Dowd et al., 1989); and (4) two cysteine residues, thought to form a disulfide bridge between the second and third extracellular loops (Dixon et al., 1987). A third conserved cysteine residue, which is also found in FSH-, TSH-, and LH-CG-receptors, immediately follows transmembrane segment 7. This cysteine has been implicated as the residue that is palmitoylated to anchor the receptor to the plasma membrane (O'Dowd et al., 1989).

The genes for the rat LH-CG, human TSH and rat FSH receptor were identified as single copy genes which span more than 60 kb. The rat LH-CG receptor consists of 11 exons and 10 introns, whereas the human TSH and the rat FSH receptor are split into 10 exons and 9 introns (Gross et al., 1991; Heckert et al., 1992; Koo et al., 1991). The extracellular hormone binding domain is encoded by the first nine (ten for the LH receptor) exons and part of the last exon. The leucine-rich repeats of the extracellular domain are determined as monomers or multimeric units by separate exons. The transmembrane and intracellular domains are encoded by the last exon.

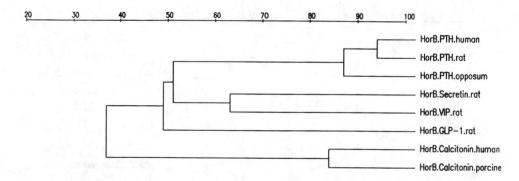

FIGURE 2b.

In none of the receptor genes was an intron in the 5'-untranslated part of the genes. Promoter regions and transcription start sites were mapped immediately upstream of the coding regions; in positions −80 and −98 relative to the translational start site for the rat FSH-receptor (Heckert et al., 1992) and in position −157 for the human TSH-receptor (Gross et al., 1991). Huhtanieni et al. (1992) determined the transcriptional start sites for the mouse FSH and LH-CG receptors are in pos −534 and −310. Wang et al. (1992) showed that a 1370 bp rat LH-CG receptor promoter fragment (pos. −1 to −1370) confers specific expression and negative regulation of gene transcription by cAMP in Leydig cell. In the rat TSH receptor gene, a "minimal" region, exhibiting promoter activity, tissue specificity, and negative regulation by TSH, was mapped between position −195 and −39 which is a highly conserved region in the rat and human TSH-receptor genes (Ikuyama et al., 1992).

Molecular cloning of the different receptors has been a useful tool to investigate the hormone binding and second messenger pathways activated by the individual receptors. For this analysis, the cDNAs of the LH-CG, FSH, and TSH receptors were expressed in heterologous systems like COS-7, Chinese hamster ovary (CHO), and human embryonic kidney (HEK) 293 cells. It was demonstrated that the recombinantly expressed receptors interact specifically with the corresponding hormone and display hormone binding affinities between $K_d = 0.18$ and 0.60 nM. The hormone concentrations which were necessary to induce half-maximal stimulation of cyclic AMP production were in the range of 0.01 to 0.30 nM depending on the receptor and the cell lines used (Loosfelt et al., 1989; McFarland et al., 1989; Parmentier et al., 1989; Sprengel et al., 1990).

1.15.11   Luteinizing Hormone-Choriogonadotropin (LH-CG) Receptors

LH is released from the pituitary under the influence of GnRH and progesterone. HCG is synthesized in the placenta during pregnancy. Both hormones act on the same target cells but the carboxy-terminal extension of the hCG β-subunit increases the half life of hCG (Fares et al., 1992). In the female the target cell for LH and hCG are the Theca cells, the large follicle granulosa cells, and the luteal cells in the ovary. LH stimulates ovulation and is the major hormone involved in the regulation of progesterone synthesis and secretion by the corpus lutem. In the male LH acts on the Leydig cells in testis, where the hormone stimulates steroidogenesis.

The key experiment which elucidated the molecular design of the glycoprotein hormone receptors was the successful purification of the LH-CG receptor. In

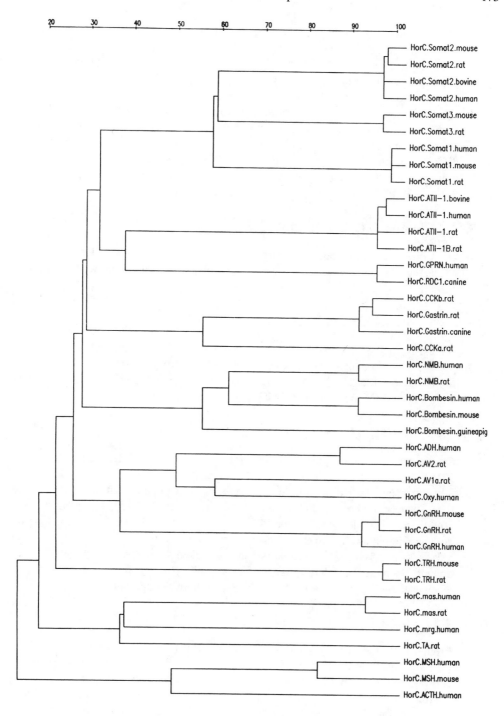

FIGURE 2c.

1989, McFarland and co-workers (McFarland et al., 1989) succeeded in determining peptide sequences of LH-CG receptor which was purified from ovaries of pseudopregnant rats. Using these peptide sequences the rat cDNA clone for the receptor was isolated. At the same time, Loosfelt et al.(1989) used monoclonal antibodies against LH-CG receptor of porcine testis to isolate the LH-CG receptor cDNA from a lambda gt10 expression library. Using the nucleotide sequences of the rat and porcine receptor, the mouse (Gudermann et al., 1992a) and human LH-CG receptors were isolated (Frazier et al., 1990; Jia et al., 1991; Minegish et al., 1990). The sequence analysis of the cloned receptor cDNAs revealed that the LH-CG receptor is expressed as a precursor molecule containing an amino-terminal signal peptide. The mature receptor is composed of an extracellular domain and a carboxy-terminal seven membrane spanning domain. The 7 membrane spanning domain displays sequence similarities to other G protein-coupled receptors. The different receptors show between 84 and 92% overall sequence identity and the most highly conserved regions are the putative transmembrane segments (91 to 94% similarity).

1.15.11.1 Human LH-CG Receptors

DNA probes corresponding to human and porcine LH-CG receptor cDNA helped to assign the LH-CG receptor gene to human chromosome 2p21 (Rousseau-Merck et al., 1990b).

1.15.11.7 Porcine LH-CG Receptors

As discussed above, Loosfelt used monoclonal antibodies against LH-CG receptor of porcine testis to isolate the LH-CG receptor cDNA from a lambda gt10 expression library (Loosfelt et al., 1989). As discussed below, several alternative spliced forms were detected for the porcine LH-CG receptor mRNAs.

1.15.11.10 Rat LH-CG Receptors

Ovaries from pseudopregnant rat and ovaries and testes from adult cycling rats contain LH/CG-receptor transcripts of 6.7, 4.3, 2.6, and 1.2 kb (McFarland et al., 1989; Segaloff et al., 1990). For the LH-CG receptor, mRNA different alternative splice forms were reported (Loosfelt et al., 1989; Segaloff et al., 1990; Tsai-Morris et al., 1990). These encode the whole (or different parts of) the extracellular LH-CG binding domain without any transmembrane spanning segment. In addition to these truncated receptors, putative LH-CG receptors were described where part of the extracellular domain seems to be deleted (Segaloff et al., 1990). It remains to be worked out, whether these "receptor variants" are of any biological importance.

1.15.11.11 Mouse LH-CG Receptors

From functional studies using the murine LH-CG receptor, Gudermann et al. (1992a) concluded that the LH-CG receptor has the capability to activate, in addition to the adenylyl cyclase, a second intracellular signaling pathway leading to stimulation of phospholipase C and resulting in formation of inositol phosphates and elevations in intracellular Ca^{2+}. In these studies, the murine LH receptor was expressed in L cells and after hormone stimulation an intracellular increase of phosphoinositide hydrolysis was observed ($EC_{50} = 2.400$ pM hCG). This was accompanied by an increase in intracellular Ca^{2+}. When they injected the same receptor cDNA into defolliculated *Xenopus* oocytes, they observed a similar hCG and LH dependent increase of intracellular Ca^{2+}, as measured by activated Cl⁻ current, using the two-microelectrode voltage-clamp method (Gudermann et al., 1992b).

1.15.12   Follicle Stimulating Hormone (FSH) Receptors

FSH is released from the anterior pituitary under the influence of GnRH and estrogens. It acts on granulosa cells from small- and medium-sized ovarian follicles promoting the development of follicles and it regulates the estrogen production. In testis, it acts on Sertoli cells supporting the gametogenesis.

1.15.12.1   Human FSH Receptors

The cDNA of the rat LH-CG receptor was used to identify, by homology screening, the cDNA of the human FSH receptor (Minegish et al., 1991). As predicted from the cDNAs, the FSH receptor is a single 75-kDa polypeptide with about 350 amino acid residues in the extracellular domain which contains several N-linked glycosylation sites. This domain is connected to a structure containing seven putative transmembrane segments which displays sequence similarity to G protein-coupled receptors. Thus, the FSH receptor is identical in its structural design to the LH-CG receptor. These receptors share 50% sequence similarity in their large extracellular domains and 80% identity across the seven transmembrane segments (Figure 2). Expression of the cloned cDNA in mammalian cells conferred FSH-dependent cAMP accumulation. The selectivity for FSH was attested by the fact that the related human glycoprotein hormones hCG and human TSH do not stimulate adenylyl cyclase in FSH receptor-expressing HEK 293 cells even when these hormones are present at high concentrations.

1.15.12.5   Sheep FSH Receptors

The sheep FSH receptor has been cloned by Yarney et al. (unpublished observation).

1.15.12.10   Rat FSH Receptors

The cDNA of the rat LH-CG receptor was used to identify, by homology screening, the cDNA of the rat FSH receptor (Sprengel et al., 1990). Northern blot analysis with total RNA from rat ovaries, testes, and cultured Sertoli and granulosa cells with the FSH receptor probe revealed two predominant hybridization signals in the range of 6.0 to 5.0 and 2.5 to 2.6 kb, as well as minor signals around 4. 0 and 1.5 kb (Heckert and Griswold, 1991; LaPolt et al., 1992; Tilly et al., 1992). Alternatively spliced forms, as reported or the LH-CG and the TSH receptor mRNAs were not reported for the FSH receptor. The chromosomal localization of the FSH receptor gene is still unknown.

1.15.13   Thyroid Stimulating Hormone (TSH) Receptors

TSH is secreted from thyrotroph cells of the anterior pituitary under the control of the thyroid releasing hormone (TRH). Its target is the thyroid follicle cell of the thyroid gland. There it stimulates the synthesis of thyroid hormones, hypertrophy, and hyperplasia. The cDNA sequences for the human (Frazier et al., 1990; Libert et al., 1989a; Misrahi et al., 1990; Nagayama et al., 1989) and rat TSH receptor (Akamizu et al., 1990a) were isolated by homology screen with the TSH receptor or LH-CG receptor cDNA probes. Like the LH-CG and the FSH receptor, the TSH receptor is synthesized as a precursor molecule. The mature receptor consists of approximately

400 amino acid residues amino-terminal hormone binding domain connected to the carboxyl-terminal domain of about 350 amino acids that contains seven putative membrane segments. The hormone binding domain and the transmembrane spanning part of the receptor show about 35 and about 75% amino acid identity, respectively, when compared to gonadotropin receptors.

Thyroid cells contain one or two major TSH-receptor transcripts, 4.9 kb in the dog (Parmentier et al., 1989), 3.9 to 4.0 kb in human (Frazier et al., 1990; Libert et al., 1989a; Misrahi et al., 1990), and 4.4 to 3.0 kb and 5.6 to 3.3 kb in FRTL-5 cells (Akamizu et al., 1990a, Frazier et al., 1990). Libert et al. described the isolation of a cDNA encoding a TSH receptor variant with partially deleted extracellular domain (Libert et al., 1990b). The biological importance of this "receptor form" is still unknown.

1.15.13.1 Human TSH Receptors

The TSH receptor gene was localized on human chromosome 14q31 using probes corresponding to the 5′ and 3′ parts of human TSH-receptor cDNA (Rousseau-Merck et al., 1990a; Libert et al., 1990c). Van Sande et al. (1990) found that CHO cells stably transfected with human TSH receptor cDNA respond between 0.1 and 1 mU/ml to TSH for cyclic AMP accumulation and between 1 and 10 mU/ml TSH for inositol monophosphate increase. These data indicate that the TSH receptor is coupled to cyclic AMP and the inositol phosphate cascades.

1.15.13.6 Canine TSH Receptors

Libert and Vassart used an independent and new approach — the screening for related members of a protein family by the polymerase chain reaction — to identify new members of the G protein-coupled receptor family (Libert et al., 1989b). Using the polymerase chain reaction, this group revealed the primary sequence of the dog TSH receptor (Parmentier et al., 1989).

1.15.13.10 Rat TSH Receptors

The cDNA sequence for the rat TSH receptor was isolated by homology screen by Akamizu et al. (1990a).

1.15.13.11 Mouse TSH Receptors

The TSH receptor gene was localized to mouse chromosome 12 using rat TSH-receptor cDNA (Akamizu et al., 1990a).

B. Hormone Receptors with Moderate Amino Acid Similarity to Other G Protein-Coupled Receptors

The recent cloning of G protein-coupled receptors using expression libraries revealed the primary structures of the secretin receptor (Ishihara et al., 1991), the calcitonin receptor (Lin et al., 1991), and the parathyroid hormone-parathyroid hormone-related peptide (PTH-PTHrP) receptor (Jüppner et al., 1991). The analysis of the polypeptide sequence of the different receptors showed that these receptors contain seven putative membrane spanning segments, but share only moderate sequence homology with other members of the G protein-coupled receptors. Thus, it was suggested that these receptors represent a new subfamily of the G protein-coupled receptors. This subfamily is now growing as indicated by the recent cloning of novel receptors for vasoactive intestinal polypeptide (VIP) and for the glucagon-like peptide 1 (GLP-1). Within this subfamily, the receptor sequences are related with amino acid sequence similarities in the range of 33 to 40% between the secretin-, VIP-, GLP1-, and the PTH-PTHr-P-receptor. The calcitonin-receptor shares about 27% sequence identity with the other members of this subfamily. The highest sequence identity for these proteins resides in the putative transmembrane spanning segments.

Like the glycoprotein hormone receptors, these receptors are expressed as precursor molecules with an amino-terminal signal peptide. The amino-terminal extracellular domain is unique to each receptor and contains several putative glycosylation sites and four cysteines which are perfectly conserved between members of this subfamily (Thorens, 1992). Calcitonin, VIP, secretin, and PTH receptors are positively coupled to the adenylyl cyclase activity. Interestingly, sequences with similar characteristics were found in the short third cytoplasmic loop of these hormone receptors, suggesting that amino acid residues of the third cytoplasmatic loop may play an important role in receptor-G_s protein interaction (Lin et al., 1991).

Vasoactive intestinal peptide (VIP), secretin, glucagon, peptide histidine isoleucine (PHI), peptide histidine methionine (PHM), pituitary adenylyl cyclase activating polypeptide (PACAP), and gastric inhibitory peptide (GIP) constitute a family of structurally related hormones (Gozes and Brenneman, 1989; Rosselin, 1986). They have various pharmacological effects on a variety of tissues including pancreas, liver, intestine, lung, heart, and kidney (Robberecht et al., 1990). In addition, these hormones can be found in the central and the peripheral nervous system acting as neurotransmitters and neuroregulators.

Diverse functions of these peptide hormones are mediated by the interaction with their respective receptors (Rosselin, 1986). Since the binding of VIP to its receptor is displaced by secretin, PACAP, and PHI, and vice versa, it has been postulated that receptors for these structurally homologous peptides may be also related (Couvineau et al., 1990; Robberecht et al., 1990). In addition, VIP, secretin, calcitonin, and glucagon increase intracellular cAMP levels and GTP regulates the interaction of hormones and their receptors, suggesting that receptors for these hormones are coupled to the cyclase-simulating G_s protein (Couvineau et al., 1990; Roth et al., 1984).

1.15.21 Parathyroid Hormone (PTH) Receptors

Parathyroid hormone (PTH) is an 84 amino-acid containing protein that is secreted by the chief cells of the parathyroid gland. PTH is a single polypeptide with a molecular weight of 9300 Da with no cysteine residues and hence no disulfide bridges. Together with calcitonin and vitamin D, it belongs to the calcium/phosphorus regulating hormones. The major target organs for PTH actions are kidney and bone where it regulates calcium and phosphorus metabolism. Parathyroid hormone related peptide (PTHrP) shares 8 to 13 amino-terminal residues with PTH. Both peptides appear to bind to the same receptor which was described as a protein molecule of 80 kDa (Jüppner et al., 1988; Orloff et al., 1989). PTHrP seems to cause hypercalcemia in some cancer patients (Mangin et al., 1988; Strewler et al., 1987; Suva et al., 1987) and seems to stimulate multiple signal pathways — cAMP, inositol phosphates, and calcium (Abou-Samra et al., 1989; Bringhurst et al., 1989; Hruska et al., 1987; Miyauchi et al., 1990; Rappaport and Stern, 1986; Yamaguchi et al., 1987).

1.15.21.1 Human PTH Receptors

The sequence of the human PTH receptor has been deposited in Genbank (Shipani et al., unpublished).

1.15.21.13 Opossum PTH Receptors

Jüppner et al. (1991) described a cDNA encoding a 585-amino acid parathyroid hormone-parathyroid hormone-related peptide (PTH-PTHrP) receptor. This cDNA was isolated by COS-7 cells transfected with an opossum kidney cell cDNA library and screened by ^{125}I-PTHrP(1-36) binding followed by photoemulsion autoradiography. The isolated PTH-PTHrP-receptor is predicted to contain ten hydrophobic regions with seven putative membrane spanning regions and an amino-terminal hydrophobic signal peptide. The primary structure of the receptor has a striking homology with the calcitonin-, the secretin-, and the VIP-receptor (Figure 1). As for all members of this family, the designation of the membrane spanning segments is speculative and the unique sequence of this receptor subfamily mandates a detailed analysis of its functional and topological domains.

1.15.21.10 Rat PTH Receptors

The rat homolog of the opossum receptor was isolated by Abou-Samra and co-workers from a rat osteosarcoma (ROS 17/2.8) cells cDNA library by expression cloning (Abou-Samra et al., 1992). The rat bone PTH receptor is 78% identical to the opossum kidney receptor which indicates conservation of this receptor across distant mammalian species (Abou-Samra et al., 1992). Southern blot analysis of rat genomic DNA suggested that a single copy gene encodes the PTH receptor and supported their finding that bone and kidney share one PTH receptor. A single mRNA species migrating between the 28s and 18s ribosomal RNA is expressed from this gene in rat kidney and rat osteosarcoma cells.

The functional integrity of the kidney and the bone receptors were demonstrated in heterologous cell expression studies. In transiently transfected COS-7 cells which express more than 10^6 opossum PTH-PTHrP-receptors per cell, the expressed opossum receptor bound PTH(1-34) and PTHrP(1-36) with equal affinity ($K_D = 4$ nM). PTH (3-34) and PTH (7-34) bound with lower affinities and the K_D values determined were in the range of 17 and 230 nM, respectively. When the opossum PTH receptor was stably expressed in porcine renal cells (LLC-PK$_1$) at densities of 10,000 to 100,000 receptors per cell, the K_D for PTH (1-34) was about 0.3 nM which is comparable to the K_d of the receptor in native opossum kidney cells. The lower affinity of the PTH-PTHrP receptor in COS-7 cells might reflect the inefficient G protein coupling in those cells. For the rat bone PTH receptor expressed in COS-7 cells, a similar pharmacological profile was determined with K_d values of about 10 nM for PTH (1-34) and PTHrP (1-36).

Both peptides PTH (1-34) and PTHrP (1-36) stimulated cAMP accumulation (4-fold) in COS-7 cells transfected with the kidney PTH receptor cDNA. They had similar ED$_{50}$ values (ED$_{50}$ = 0.4 nM) for both ligands. These values are comparable to ED$_{50}$s of native receptors in opossum kidney cells. When expressed in COS cells, Abou-Samra and co-workers observed that the rat bone or opossum kidney PTH receptor mediate PTH and PTHrP stimulation of both adenylyl cyclase and phospholipase C. In COS-7 cells expressing the rat PTH-PTHrP-receptor, they observed a GTP-dependent stimulation of the adenylyl cyclase (8- to 14-fold) indicating a coupling of the PTH receptor to the adenylyl cyclase through a G protein, presumably G$_s$. In addition, they found an increase of intracellular Ca^{2+} in COS-7 cells expressing the rat PTH receptor after treatment of the cells with PTH (1-34) and PTHrP (1-36) in doses as low as 5 nM. Hence, the PTH receptor is another example that a single receptor can activate dual messenger pathways. These properties could explain the diversity of hormone action without the need to postulate other receptor subtypes (Abou-Samra et al., 1992).

1.15.22   Vasoactive Intestinal Peptide (VIP) Receptors

VIP is a 28 amino acid polypeptide that regulates several functions in the central and peripheral nervous system (Said, 1986). VIP is believed to act through the stimulation of high-affinity receptors that are positively coupled to adenylyl cyclase activity (Gozes and Brenneman, 1989). VIP receptors are expressed in brain and in various peripheral tissues, including lung, liver, and intestine (Gozes and Brenneman, 1989). Moreover, VIP is thought to affect several immunological functions by acting directly on specific receptors located on B and T lymphocytes (Finch et al., 1989).

The pharmacological profile of VIP receptors still remains to be elucidated. Since binding studies to VIP receptors has been so far determined in tissues that often contain receptors for other VIP-related molecules, the ligand specificity of the VIP receptor has not been accurately determined (Robberecht et al., 1991). Moreover, no selective peptidic or nonpeptidic antagonists of VIP receptors are currently available.

Recently, two independent groups have purified two VIP receptor polypeptides from porcine liver and guinea pig lung. Both proteins differ dramatically in molecular weight, 53, and 18 kDa, respectively (Brugger et al., 1991; Couvineau et al., 1990). The molecular basis of the difference between these two VIP-binding proteins is not known.

On the other hand, two distinct cDNA clones encoding hormone receptors have been recently isolated from a human leukemic pre-B cells cDNA library (Sreedharan et al., 1991) and a rat lung cDNA library (Ishihara et al., 1992). Both of the encoded proteins display the pharmacological profile of VIP receptors and may be co-expressed in the same tissues (see below). However, these two receptors have no significant sequence identity.

The proposed human VIP receptor has a canine homolog (RDC1) which is inactive in VIP-specific binding (Cook et al., 1992) and might not be considered as a VIP receptor (Nagata et al., 1992). Therefore, we prefer to designate the proposed human VIP receptor "GPRN1" according to Sreedharan et al. (1991). Whether the GPRN1 and rat VIP receptors constitute two distinct VIP receptor subtypes with different functional and pharmacological properties remains to be determined.

1.15.22.10   Rat VIP Receptors

A cDNA clone encoding a VIP receptor has been recently isolated from a rat lung cDNA library by low stringency hybridization using the rat secretin receptor cDNA as probe (Ishihara et al., 1992). The isolated clone encodes a 459 amino acid protein with several structural motifs common to G protein-coupled receptors. A 30 amino acid signal peptide is present in the amino-terminal region of the encoded protein, and the mature receptor probably consists of 429 amino acids with a molecular weight of approximately 49 kDa. The rat VIP receptor (48%) (Ishihara et al., 1992) displays a high sequence homology to the rat secretin receptor and the parathyroid hormone-parathyroid hormone-related peptide receptor (Figure 1) suggesting that these receptors interact with related polypeptides. Furthermore, it shows 33 and 39% identity with the porcine calcitonin receptor (Lin et al., 1991) and the opossum parathyroid receptor and parathyroid hormone-related peptide, respectively (Jüppner et al., 1991). On the other hand, the amino acid sequence of the rat VIP receptor shows no similarity with other members of the G protein-coupled receptors, including the GPRN1 cDNA proposed to encode the human VIP receptor.

Membranes of COP cells expressing the recombinant VIP receptor bind ^{125}I-VIP in a saturable manner. Two classes of VIP binding sites were found with an apparent K_d of 0.173 nM and 21 nM. Competition studies indicate that unlabeled peptides structurally related to VIP displace the ^{125}I VIP binding with a rank of order consistent with the endogenous VIP receptor. PACAP-38 and PACAP-27 were slightly more potent than VIP in displacing the binding of the iodinated ligand (IC$_{50}$ were 1, 2.5, and 3 nM, respectively). Helodermin and PHI display an IC$_{50}$ of 6 nM while secretin was 100 times less potent that VIP, and glucagon was ineffective. VIP significantly stimulated intracellular cAMP formation in COSGs1 cells transiently expressing the VIP receptor. The order of potency of the structurally related peptides was well correlated with their potency in inhibiting the VIP binding, with PACAO-38 > PACAP-27 > VIP > Helodermin = PHM » secretin. In accordance to the inability of glucagon to bind the cloned VIP receptor it barely stimulates the accumulation of cAMP.

The tissue distribution of VIP receptor transcript is in good agreement with the distribution of VIP receptor obtained by binding studies (Gozes and Brenneman, 1989). Northern hybridization revealed the presence of one species of VIP receptor mRNA, approximately 5.5 kb long highly expressed in lung, liver, and intestine. VIP mRNA was also detected in all subregions of rat brain, where two additional bands of 2.4 and 1.3 kb were found. *In situ* hybridization also indicated that the prominent expression of VIP receptor mRNA in brain was observed in the cerebral cortex, hippocampus, and mitral cell layer of olfactory bulb, while weak hybridization signals were widely distributed in the subcortical regions and cerebellar cortex.

1.15.23   Secretin Receptors

Secretin, a 27 amino acid peptide, was the first hormone to be discovered in 1902 by W. Bayliss and E. Starling. It stimulates the secretion of bicarbonate, enzyme, and potassium ions from the pancreas (Bayliss and Starling, 1902). Several tissues, including pancreas, heart, kidney, intestine, and brain, bear secretin-specific receptors that are positively coupled to the stimulation of adenylyl cyclase (Christophe et al., 1984; Fremeau et al., 1986; Gespach et al., 1986; Roth et al., 1984). High and low affinity secretin binding sites have been identified in rat pancreatic acini (Bissonnette et al., 1984). Numerous attempts to purify and biochemically characterize the secretin receptor have been unsuccessful owing to its low expression levels. However, chemical cross-linking analysis of the receptor indicated that the secretin receptor in rat and gastric gland and pancreatic acini has a molecular weight of 62 and 51 kDa, respectively (Bawab et al., 1988; Gossen et al., 1989).

1.15.23.10   Rat Secretin Receptors

The secretin receptor was the first receptor for the hormones of the secretin/glucagon family to be cloned in 1991 (Ishihara et al., 1991). The cDNA clone was isolated by expression cloning from a library of NG108-15 cell line. The encoded 449 amino acid protein contains a 23 amino acid signal sequence at the amino-terminal and the mature peptide consists of 427 amino acids with a calculated molecular weight of approximately 49 kDa. A two-site model was found to be most consistent to analyze saturation data of ^{125}I-secretin binding to membranes of COS-7 cells expressing the recombinant receptor.

The high-affinity component showed an apparent K_d of 0.57 nM, while the low-affinity component showed a K_d of 14.5 nM. This binding profile is very similar to the binding of ^{125}I-secretin to membranes of NG 108-15 cells expressing the endogenous receptor (Ishihara et al., 1991). Displacement studies using unlabeled peptides revealed an order of potency typical of the binding to secretin receptor, with secretin most potent (IC_{50} = 4.5 nM), followed by the carboxy-terminal peptide of secretin (secretin $_{5-27}$) and VIP, that were 1000-fold less potent than secretin in inhibiting the binding of ^{125}I-secretin to the membranes (IC_{50} = 9.5 and 4.5 mM, respectively).

The effector coupling mechanism of the expressed secretin receptor was evaluated in a COSG$_s$ cell line which overexpresses the a subunit of the G$_s$-protein. As predicted by biochemical studies in rat gastric gland (Gespach et al., 1986) secretin potently stimulates the accumulation of cAMP, with an EC_{50} of 1 nM. VIP was approximately 60 times less active, while glucagon was ineffective. One 2.5-kb mRNA for the secretin receptor was identified in heart, stomach, and pancreas but not in lung, liver, kidney, and brain.

1.15.24 Glucagon-Like Peptide11 (GLP-1) Receptors

The two gut hormones, glucagon-like peptide-1 (GLP-1) and gastric inhibitory peptide (GIP), potentiate the effect of glucose on insulin secretion and are thus called glucoincretins. GLP-1 is part of the preproglucagon molecule (Kreymann et al., 1987) that is proteolytically processed in L cells to GLP-1 (1-37) and GLP-1 (7-36) amide and GLP-1-(7-37) (Bell et al., 1983; Mojsov et al., 1986). Only the truncated forms are biologically active in concentrations as low as 1 to 10 pM. For maximum activity at their target cells, the glucoincretins require the presence of glucose at or above the physiological concentration of about 5 mM. The physiological effect of GLP-1 is mediated by the stimulation of adenylyl cyclase (Drucker et al., 1987; Göke et al., 1989). Therefore, it was speculated that the receptors for these hormones belong to the G protein-coupled receptor family.

1.15.24.10 Rat GLP-1 Receptors

A cDNA encoding the rat GLP-1 receptor was successfully isolated from COS-7 cells transiently expressing a pancreatic islet cDNA plasmid library. Individual cells expressing the GLP-1 receptor were labeled by binding to ^{125}I-GLP-1 and identified by photographic emulsion autoradiography (Thorens, 1992). The sequence analysis of the GLP-1 receptor cDNA revealed one major open reading frame coding for a 463-amino acid polypeptide. The search for sequence identities showed that the GLP-1-receptor is related to the secretin-, calcitonin-, VIP-, and PTH-PTHrP-receptors. Like these, the GLP1-receptor contains an amino-terminal signal peptide and seven hydrophobic putative transmembrane spanning segments. In Northern blot analysis two mRNAs of 2.7 and 3.6 kb were detected in pancreatic islets, lung, and stomach. The receptor seems not to be expressed in brain, liver, thymus, muscle, intestine, and colon.

In COS7 cells, the cloned GLP-1 receptor created high-affinity binding sites for ^{125}I-GLP-1 (K_d = 0.6 nM) which is comparable to the binding observed for the native receptor in rat insulinoma cells (K_d = 0.12 nM). Binding of ^{125}I-GLP-1 was displaced by unlabeled GLP with 50% displacement achieved at 0.5 to 1.0 nM. Other peptide hormones of related structure such as secretin, calcitonin, VIP, and GIP were not able to displace binding. GLP-1 binding was partially inhibited by high concentrations of glucagon (1 m*M*). Stimulation of adenylyl cyclase was demonstrated by GLP-1 mediated cAMP accumulation in COS-7 cells using 0.1 and 1 n*M* GLP-1.

1.15.25   Calcitonin Receptors

Calcitonin is a 32-amino acid hormone that regulates the serum calcium concentration by increasing renal calcium excretion and inhibiting osteoclast-mediated bone reabsorption (Coop, 1972). In mammals, it is secreted by specialized C cells which are found primarily in the thyroid gland, but also at a small number of extrathyroidal sites. Biochemical studies demonstrated that calcitonin receptor is a G protein-coupled receptor which is functionally coupled to the stimulation of adenylyl cyclase (Gilman, 1984). The receptor was also shown to be coupled to a second signal transduction pathway (i.e., intracellular calcium mobilization) via a pertussis toxin-sensitive G protein (Chakraborty et al., 1991; Zaidi et al., 1990).

1.15.25.1   Human Calcitonin Receptors

The human calcitonin receptor was cloned by Gorn et al. (1992).

1.15.25.7   Porcine Calcitonin Receptors

Lin et al. (1991) isolated a cDNA encoding the porcine calcitonin receptor by expressing a cDNA library from porcine kidney epithelial cell line. They revealed the primary structure of the porcine calcitonin receptor as 482 amino acid polypeptide with seven hydrophobic domains and a putative signal peptide at the amino-terminus. The mature receptor contains a long extracellular amino-terminus with three potential N-glycosylation sites and a short intracytoplasmatic loop between transmembrane spanning regions 5 and 6. As mentioned above, it shows a high sequence homology with the receptors of the secretin/glucagon peptide family.

High-affinity binding with ^{125}I-salmon calcitonin was demonstrated using membranes of COS-7 cells expressing the porcine calcitonin receptor. The apparent K_d was approximately 6 n*M*. Bovine parathyroid hormone did not compete for binding of the iodinated salmon calcitonin. The cloned receptor is functionally coupled to an increase of intracellular cAMP. In response to calcitonin, COS-7 cells transfected with calcitonin receptor exhibit a fourfold increase of cAMP concentration. In addition to the stimulation of the adenylyl cyclase, the cloned calcitonin receptor serves another second messenger pathway. The recombinantly expressed receptor can stimulate both the cAMP and inositol phosphate pathways, with EC_{50} values differing between the two pathways by one order of magnitude when expressed in HEK 293 cells (Chabre et al., 1992).

C.   Hormone Receptors with Significant Amino Acid Similarities to Other G Protein-Coupled Receptors

The G protein-coupled peptide hormone receptors which are similar in amino acid sequence and primary structure to most members of the G protein-coupled receptor superfamily constitute the majority of the hormone receptors. Modern molecular biological methods, in particular the use of recombinant DNA techniques, have resulted in the cloning of multiple members of the peptide hormone receptors. Some of them have been isolated by homology screening and some were cloned by expression screening of cDNA libraries. The receptor molecules which were identified are usually polypeptides 350 to 500 amino acids in length and contain 7 highly hydrophobic regions of 20 to 25 amino acids each. At the extracellular amino-terminal part of the receptors several potential N-linked glycosylation sites can be detected. At the intracellular carboxy-terminus tyrosine, serine, and threonine residues serve as potential phosphorylation sites for cellular regulation of receptor activity. A number of short sequences are highly conserved and occur within the putative transmembrane-spanning helices and intracellular loop regions. The third intracellular loop between transmembrane helix 5 and 6 is highly variable in length and sequence. This part of the receptor has been implicated in G protein-coupling (Kobilka et al., 1988; O'Dowd et al., 1988). Two cysteine residues that are thought to form a disulfide bridge between the second and third extracellular loops are present in most receptors. In the carboxy-terminal tail of the receptors, a third conserved cysteine residue has been implicated as a site that is palmitoylated to anchor the receptor to the plasma membrane (O'Dowd et al., 1989). The peptide hormone receptors described in this chapter follow this structural dogma for G protein-coupled receptors. In multiple alignments and phylogenetic comparisons, they show sequence identities in the range of 16 to 45%. Endothelin receptors (Arai et al., 1990; Sakurai et al., 1990) are discussed in Chapter 9.

1.15.31   Somatostatin-1 Receptors

Somatostatin is a peptide hormone that is widely distributed in the body. It is a tetradecapeptide that acts on multiple organs including brain, pituitary, gut, exocrine and endocrine pancreas, adrenals, thyroid, and kidneys to suppress the release of many hormones, including growth hormone, insulin, glucagon, gastrin, and secretin (Reichlin, 1983a, b). It is also described as a neurotransmitter modulating neuronal activity (Miyoshi et al., 1989; Wang et al., 1989; White et al., 1991; Lamberts, 1988; Reubi et al., 1988).

Somatostatin exerts its biological effects by binding to specific high-affinity receptors (He et al., 1990; Reubi, 1984; Reubi, 1985; Tran et al., 1985). The results of earlier studies have suggested the existence of at least two somatostatin receptor subtypes (Reubi, 1984; 1985; Tran et al., 1985). They show similar affinities for the native peptides somatostain-14 and somatostatin-28, but different affinities for various synthetic somatostatin analogs. Somatostatin-14 is a member of the somatostatin-like peptides that also include the amino-terminal extended form, somatostatin-28. These two bioactive forms are derived from tissue-specific proteolytic processing of prosomatostatin, the 92 amino acid precursor of somatostatin-14 and -18.

1.15.31.1   Human Somatostatin-1 Receptors

For the high-affinity somatostatin receptor it was suggested that this receptor is coupled to GTP-binding proteins (Law et al., 1991; Luini et al., 1985; Mahy et al., 1988a, b; Marin et al., 1991). Therefore, one approach to identify the gene for the somatostatin receptor was to screen for receptor cDNAs by PCR-mediated homology screen. From human pancreatic islet RNA, Yamada et al. amplified successfully two different cDNA fragments encoding somatostatin receptor isoform-1 and somatostatin receptor isoform-2 (Yamada et al., 1992). The corresponding intronless human genes were subsequently isolated, and the identity of the receptors was analyzed by transient expression of the full length cDNA in COS-7 cells. Transfected cells showed specific high-affinity binding for ^{125}I-Tyr^{11}somatostatin-14 which was not present in untransfected cells. Further functional analysis of stably transfected Chinese hamster ovary cells expressing isoform-1 and isoform-2 revealed that both receptors have higher affinity for somatostatin-14 than somatostatin-28, with inhibitory concentrations of 1.5 and 4.7 nM, respectively; these values are similar to those previously reported for the high affinity somatostatin receptor (Tran et al., 1985). RNA blotting studies show that the human isoform-1 mRNA is 4.8 kb in size. It expressed at high levels in jejunum, stomach, pancreatic islets, and at low levels in brain and kidney.

1.15.31.10   Rat Somatostatin-1 Receptors

The rat homolog of the somatostatin-1 receptor was isolated in 1991 by PCR (Meyerhof et al., 1991). At that time, this receptor was still described as an orphan receptor expressed mainly in rat hypothalamus and cortex with a single mRNA species of about 3.8 kb.

1.15.31.11   Mouse Somatostatin-1 Receptors

The intronless genes for the mouse somatostatin-1 receptor were characterized together with the human receptor gene (Yamada et al., 1992). The pharmacology of the mouse receptors was not determined. The somatostatin receptors from rat and mouse show between 97 and 99% amino acid identity to the corresponding human homolog.

1.15.32   Somatostatin-2 Receptors

1.15.32.1   Human Somatostatin-2 Receptors

As noted above, Yamada et al. amplified two different cDNA fragments encoding somatostatin-1 and somatostatin-2 receptors (Yamada et al., 1992). The tissue distribution of the somatostatin-2 receptor is quite different than the somatostatin-1 receptor, and the two 8.5- and 2.5-kb transcripts are highly expressed in brain and kidney. In pancreatic islets isoform-2 was also present. The DNA analysis of the two human receptor isoforms revealed that the two different somatostatin receptors contain 391 and 369 amino acids, respectively. They are members of the superfamily of G protein-coupled receptors. There is 46% amino acid identity between sequences of receptor isoform-1 and isoform-2. The presence of multiple somatostatin receptors suggests that the biological effects of somatostatin are mediated by a family of receptors that are expressed in a tissue-specific manner.

1.15.32.4   Bovine Somatostatin-2 Receptors

The cloning and sequencing of the somatostatin-2 receptor has been reported by Xin et al., (unpublished; GenBank L06613).

1.15.32.10   Rat Somatostatin-2 Receptors

The rat somatostatin-2 receptor was isolated by an expression-cloning strategy (Kluxen et al., 1992). The pharmacological profile determined for the rat somatostatin-2 receptor is similar to that of the human homolog. The two major rat mRNA species (which are 2.6 and 2.7 kb in size) are strongly expressed in hippocampus.

1.15.32.11   Mouse Somatostatin-2 Receptors

The intronless genes for the mouse somatostatin-2 was characterized together with the human receptor genes (Yamada et al., 1992). The pharmacology of the mouse receptor was not determined. The somatostatin receptors from rat and mouse show between 97 and 99% amino acid identity to the corresponding human homolog.

1.15.33   Somatostatin-3 Receptors

1.15.33.10   Rat Somatostatin-3 Receptors

The rat somatostatin-3 receptor was cloned by Meyerhof et al. (1992).

1.15.33.11   Mouse Somatostatin-3 Receptors

The mouse somatostatin-3 receptor was cloned by Yasuda et al. (1992).

1.15.41   Angiotensin (AT) II-I Receptors

Angiotensin II, the effector peptide of the renin-angiotensin system, elicits a wide variety of responses which affect blood pressure and electrolyte balance. Angiotensin II activates specific receptors on various target organs, including arteriolar smooth muscle, adrenal cortex, adrenal medulla, kidney, liver, brain, and pituitary. Pharmacological inhibition of angiotensin II receptors was first achieved in 1971 with the discovery of saralasin, the first specific peptide antagonist of angiotensin II. The subsequent development of nonpeptide angiotensin II antagonists have allowed the identification of at least two distinct angiotensin II receptors, named AT_1 and AT_2 receptors (Catt and Abbott, 1991; Timmermans et al., 1991). AT_1 receptor has high affinity for the selective nonpeptide antagonists DuP753 (Chiu et al., 1989; Wong et al., 1990) or EXP6803 (Dudley et al., 1990) while the AT_2 receptor binds with high affinity the peptide antagonists CGP42112A, WL19 (Whitebread et al., 1989) or EXP655 (Wong et al., 1990) and the nonpeptide antagonist PD123177 (Whitebread et al., 1989). By contrast, the peptide agonists angiotensin II and angiotensin I, as well as the nonselective peptide antagonist saralasin, exhibit no selectivity toward either subtype. Angiotensin III has a slightly higher affinity for the AT_2 sites. Binding of ligands to the AT_1 receptors, but not to AT_2 receptors is sensitive to sulfhydryl-reducing agents and is inhibited by millimolar concentrations of dithiothreitol (Whitebread et al., 1989).

The AT_1 receptor is coupled through G proteins to effectors including phospholipase C and adenylyl cyclase (Douglas, 1987). Conversely, the AT_2 receptor does not appear to couple to G proteins and may be structurally distinct from the G protein-coupled receptor superfamily (Speth and Kim, 1990).

AT_1 and AT_2 receptors also differ markedly in their distribution between tissues (Timmermans et al., 1991). AT_1 receptors are distributed mainly in vascular smooth muscle and zona glomerulosa cells. By contrast AT_2 sites are abundant in adrenal medulla and in the uterus of various species.

Numerous attempts to purify the AT receptors have been unsuccessful owing to its instability and low expression levels (Catt and Abbott, 1991). Molecular cloning of bovine and rat AT_1 receptors was first reported in 1991 by two independent groups, both using expression cloning strategies (Murphy et al., 1991; Sasaki et al., 1991). This was followed by the isolation of the human receptor cDNA and gene by several laboratories (Bergsma et al., 1992; Furuta et al., 1992; Takayanagi et al., 1992). More recent reports indicate that a distinct subtype of AT_1 receptor, named AT_{1B}, may also exist (Kakar et al., 1992b; Iwai and Inagami, 1992).

1.15.41.1 Human AT II-1 Receptors

The molecular cloning and sequencing of the human gene encoding the AT_1 receptor was reported in 1992 (Furuta et al., 1992). The entire coding region is contained in one exon. Beyond the stop codon two putative polyadenylation signals are located in the 1.4-kb 3' flanking region, at positions +1947 and +2154, respectively. This is in agreement with Murphy's observation for the rat gene (Murphy et al., 1991). Six ATTTA nucleotide sequences were found in the 3' flanking region, which may cause instability of the human AT_1 receptor mRNA. An exon-intron junction was identified at nucleotide position -47-48, suggesting that the gene may contain an intron in the 5' untranslated region.

Two independent groups also reported the cloning of the cDNA encoding the human AT_1 receptor (Bergsma et al., 1992; Takayanagi et al., 1992). The amino acid sequence is identical in all publications. Human clones encode an open reading frame of 359 amino acids that share 94.4 and 95.3% identity to the rat and the bovine sequences, respectively. All the structural motifs typical of G protein-coupled receptors and previously indicated for the rat and bovine receptors are conserved in the human receptor. Human AT_1 receptor contains one potential N-glycosylation site in the amino-terminal extracellular domain and two in the second extracellular loop. Each of the four extracellular domains contains the cysteine residue that might form disulfide bonds essential for ligand binding. In the rat and bovine receptor these amino acid residues are well conserved. A high degree of nucleotide sequence conservation was also observed within the 5'- and the 3'-untranslated regions, suggesting that those untranslated regions may be required for some regulatory functions.

Binding studies performed in COS-7 cells transiently expressing the human AT_1 receptor revealed a pharmacological profile that was consistent with the AT_1 receptor. The rank of potency of ligands in displacing the binding of ^{125}I-angiotensin II was (Sar^1, Ile^8) Ang II > (Sar^1, Ala^8) Ang II > Ang II > Ang III > Ang I. Likewise, SKF108566 and Dup 753, but not the AT_2 selective antagonists WL19 and CGP42112A, displaced ^{125}I-(Sar^1, Ile^8) Ang II binding at nanomolar concentration.

Xenopus oocytes were used to examine the functional coupling of the human AT_1 receptor to a second messenger system (Bergsma et al., 1992). In oocytes injected with human AT_1 receptor mRNA, angiotensin II elicits a rapid change in Ca^{2+}-dependent chloride current. The electrophysiological response to angiotensin II treatment is significantly inhibited by two nonpeptide AT_1 receptor antagonists SKF 108566 and Dup 753, but not by the AT_2 receptor antagonist WL18.

The distribution of AT_1 receptor in human tissues was performed by Northern blot analysis. Bergsma and co-workers identified a single mRNA transcript, approximately 2.4 kb long, with a pattern of expression similar to the tissue distribution reported for the rat AT_1 receptor, with the exception of the heart where the human tissue displays a much higher level of expression (Bergsma et al., 1992). The 2.4-kb mRNA species was also found in adrenal cortex-derived tumors but not in adrenal medulla-derived tumors (Takayanagi et al., 1992). This is consistent with the observation that the AT_1 receptor in adrenal is mainly distributed in the cortex and the AT_2 receptor in the medulla (Timmermans et al., 1991).

1.15.41.4 Bovine AT II -1 Receptors

The bovine AT_1 receptor was isolated from cultured bovine adrenal zona glomerulosa cells by mammalian cell expression cloning (Sasaki et al., 1991). The isolated clone encodes for an open reading frame of 359 amino acid residues with a relative molecular mass of 41,093 Da. The encoded protein contains 7 transmembrane domains with 20 to 30% sequence identity with those of other G protein-coupled receptors and the other common features described for G protein-coupled receptors. Cysteine residues are present in each of the four extracellular receptor domains. These residues may be important to form the 2 disulfide bridges required for ligand-binding conformation as well as to confer sensitivity to AT_1 receptor to reducing agents.

Binding studies on membranes of COS-7 cells transfected with the recombinant receptor demonstrated a specific binding to ^{125}I angiotensin II with a K_d of 0.28 nM. Displacement studies using peptidic or nonpeptidic agonists and antagonists revealed an order of potency typical of the binding to the AT_1 receptor, with (Sar^1, Ile^8)-angiotensin II most potent, followed by (Sar^1, Ala^8)-angiotensin II, angiotensin II, angiotensin III, Dup753, and angiotensin I. Furthermore, the AT_2 selective antagonist EXP655 was virtually inactive in displacing the ^{125}I-angiotensin II binding. Dithiothreitol (1 mM) reduced the binding of ^{125}I-angiotensin II by about 50%. COS-7 cells transiently expressing bovine AT_1 receptor responded to angiotensin II or III (100 nM) with a transient increase in the concentration of intracellular calcium and inositol phosphate production. Changes in intracellular cAMP content were not evaluated.

Tissue distribution of the receptor transcript was investigated by Northern blot analysis. A single 3.3-kb mRNA species was found to be highly expressed in bovine adrenal cortex. Weak hybridization signals were also detected in adrenal medulla and renal medulla but not in the renal cortex. A large 3.3-kb transcript was identified in cultured bovine adrenocortical cells that is negatively regulated by the treatment with the adrenal-targeting hormones ACTH and insulin (Takayanagi et al., 1992).

1.15.41.7 Porcine AT II-1 Receptors

The porcine AT II-1 has been cloned but not yet published (Itazaki et al., in press).

1.15.41.10 Rat AT II-1 Receptors

The rat AT_1 receptor was isolated by expression cloning from rat aortic vascular smooth muscle, a tissue that is rich in AT_1 receptor (Murphy et al., 1991). The rat AT_1 receptor displays significant homologies to the bovine receptor in terms of its structure, pharmacology, biochemistry, and tissue distribution. The rat cDNA encodes a 359 amino acid protein which is 92% homologous with the bovine receptor, indicating that the gene is conserved among mammalian species. Like the bovine AT_1 receptor, the rat protein contains three potential N-glycosylation sites and four cysteine residues in the extracellular regions.

Radioligand binding studies indicated that the expressed rat AT_1 receptor has a pharmacological profile that is similar to the bovine AT_1 receptor. Membranes from COS-7 cells transiently expressing the recombinant receptor bound with high affinity the nonselective compound (Sar^1, Ile^8)-angiotensin II (K_i: 0.46 nM) and the AT_1 selective antagonist Dup753 (K_i: 6.3 nM), while the AT_2 selective antagonist PD123177 was ineffective (K_i > 10,000 nM). The order of potency of agents binding was similar to the bovine receptor: angiotensin II > angiotensin III > angiotensin I, but both angiotensin II and angiotensin III bind to the rat vascular receptor with an affinity that is one order of magnitude higher than its K_i at the bovine receptor. This probably reflects differences in assay conditions rather than intrinsic pharmacological differences. In CHO-K1 cells constitutively expressing the rat AT_1 receptor, angiotensin II (100 nM) was found to stimulate inositol phosphate production and intracellular calcium mobilization.

Interestingly, two distinct RNA species (2.3 and 3.5 kb) were detected for the rat AT_1 receptor. Both transcripts are highly expressed in vascular smooth muscle. The smaller transcript is the most abundant and is expressed in several rat tissues with the exception of testes and brain. The 3.5-kb transcript is expressed in liver, kidney, adrenal, and lung. The reason of the apparent discrepancy between the size and distribution of bovine and rat mRNAs still remains to be elucidated. These data have been largely confirmed by another laboratory (Iwai et al., 1991). Iwai and co-workers isolated from a rat kidney cDNA library prepared from 16-week old spontaneous hypertensive rats an AT_1 receptor cDNA identical to the cDNA reported by Murphy et al., 1991. Northern blot analysis reveals a 2.3-kb hybridization transcript in adrenal, liver, and kidney. By using the polymerase chain reaction, Iwai et al. (1991) also demonstrated that the expression level of the rat AT_1 receptor mRNA is positively modulated by low sodium diet feeding, but not by the administration of captopril or continuous infusion of angiotensin II.

1.15.42 Angiotensin (AT) II-1B Receptors

1.15.42.10 Rat AT II-1B Receptors

A variety of data suggest that further distinctions within the AT_1 receptors may exist (Catt and Abbott, 1991). Kakar et al. (1992b) and Iwai and Inagami (1992) recently reported the molecular cloning of a second form of the rat AT_1 receptor, which exhibits high similarity with the rat AT_1 receptor previously isolated, relative to amino acid sequence, binding to angiotensin II analogs, and functional coupling to intracellular calcium mobilization. The novel AT_1 receptor was called subtype AT_{1B}, to distinguish it from the previously isolated cDNA, named AT_{1A} by these authors. The AT_{1B} receptor cDNA was isolated from a rat anterior pituitary expression library and a rat adrenal cDNA library. The rat AT_{1B} cDNA encodes for a 359 amino acid protein which exhibits a very high level of identity with the amino acid sequence of AT_{1A} receptor (95%). It is likely that AT_{1A} and AT_{1B} derive from different genes, since divergence between the nucleotide and amino acid sequences do not appear to derive from alternatively spliced RNA. In addition, Kakar et al. (1992b) reported that at least two fragments of rat genomic DNA digested with multiple restriction enzymes hybridize with the AT_{1B} receptor cDNA probe.

The AT_{1B} binding shares a similar pharmacological profile to AT_{1A} sites, with the exception of angiotensin I which is approximately tenfold more potent on AT_{1A} receptors (Murphy et al., 1992; Sasaki et al., 1991). Rank of potency of ligands was (Sar^1, Ile^8) angiotensin II > angiotensin II > angiotensin III. The receptor clearly is an AT_1 receptor since it binds with high affinity Dup 753 ($IC_{50} = 3$ nM), whereas PD 123319 is inactive at 1 mM. Like the AT_{1A} receptor, the AT_{1B} receptor is functionally coupled to increases in intracellular calcium concentration.

The tissue distribution of the rat AT_{1B} receptor mRNA was compared to the expression of the AT_{1A} receptor by using reverse transcriptase/polymerase chain reaction. Those data indicate that anterior pituitary, adrenal, and uterus express primarily AT_{1B} mRNA, whereas aortic vascular muscle cells, lung, and ovary express primarily AT_{1A} mRNA. Both transcripts are similarly expressed in spleen, kidney and liver. Estrogen treatment suppress AT_{1B} but not AT_{1A} mRNA levels in the pituitary gland. The expression of the two AT_1 subtypes was also studied in the brain using reverse transcriptase/polymerase chain reactions. AT_{1B} mRNA was predominant in subfornical organ (SFO) and organum vasculosum of the lamina terminalis (OVLT), the two regions that mediate angiotensin II-induced drinking behavior, and also in cerebellum. AT_{1A} mRNA was predominant in the hypothalamus. Thus, the two AT_1 receptor subtypes established as residing in peripheral tissues also are found in the central nervous system where the AT_{1B} subtype may modulate drinking behavior (Kakar et al., 1992a).

1.15.43   GPRN/RDC1 Receptors

1.15.43.1   Human GPRN Receptors

In 1991, Shreedharan et al. (1991) reported the cloning of the putative human VIP receptor from Nalm 6 human leukemic pre-B cells and from HT-29 human colon carcinoma cells, two cell lines that possess specific receptors for VIP. The cDNA clone, named GPRN1, has a 1281 bp nucleotide sequence encoding a protein of 362 amino acids. The amino acid sequence of the GPRN1 product contains seven hydrophobic segments and several features common to G protein-coupled receptor (Figure 2), including three potential N-glycosylation sites (Asn-13, Asn-22, and Asn-39) in the amino-terminal region and several serine and threonine residues as possible phosphorylation sites in the carboxyl-terminal tail. In addition, this receptor contains two cysteine residues in the second and third extracellular domains that may form a disulfide bond important for ligand-binding conformation. Human GPRN1 displays a high amino acid sequence homology (94%) with the dog RDC1 receptor (Libert et al., 1980a) and lower degree of homology with the human Met-Leu-Phe receptor (19%; Boulay et al., 1990), substance K receptor (19%; Masu et al., 1987), and rat mas receptor (13%; Young et al., 1988). No significant sequence identity was found with the secretin receptor cDNA.

Binding studies to membranes prepared from GPRN1-transfected COS-6 cells revealed that ^{125}I-VIP binds a single class of high-affinity binding sites with a K_d of 2.5 nM. The displacement of ^{125}I-VIP specific binding by secretin and PHI required a 50-fold higher concentration for half maximum inhibition than displacement by VIP, while glucagon was even less effective.

CHO K1 cells constitutively expressing the GPRN1-encoded protein responded to VIP with a threefold increase in intracellular cAMP concentration. The calculated EC_{50} for VIP-induced stimulation of adenylyl cyclase was in the low nanomolar range. Northern blot analysis revealed the presence of one species of GPRN1 mRNA in Nalm6, HT-29 cells and in several tissues including rat brain, colon, heart, lung, kidney, spleen, and small intestine.

1.15.43.6   Canine RDC1 Receptors

The RDC1 receptor is a 361 amino acid orphan receptor (Libert et al., 1990a) belonging to the G protein-coupled receptor superfamily originally cloned from a dog thyroid cDNA library (Libert et al., 1989a). The gene encoding the dog RDC1 receptor contains an intron at least 600 bp long in the 5′ untranslated region and an AT rich palindromic sequence in the 3′ untranslated region which may cause instability of the receptor mRNA (Cook et al., 1992). The RDC1 receptor gene has been localized to human chromosome region 2q37 (Libert et al., 1991). The high degree of both nucleotide and amino acid sequence identity (91 and 94%, respectively) with the human GPRN receptor suggested that RDC1 is the canine homolog of the human GPRN receptor.

1.15.51   Gonadotropin-Releasing Hormone (GnRH) Receptors

The gonadotropin-releasing hormone (GnRH) is a key mediator in the integration of the neural and endocrine systems. Normal reproduction depends on the pulsatile release of physiological concentrations of GnRH. GnRH binds to specific high-affinity pituitary receptors and triggers the secretion of the gonadotropins and high levels of agonist lead to a suppression of gonadotropin secretion. The capacity of GnRH analogs both to activate and to inhibit the hypothalamic-pituitary-gonadal axis has led to their wide clinical utility in the treatment of a variety of disorders ranging from infertility to prostatic carcinoma. The GnRH activated receptor stimulates phospholipase C via a G protein and generates inositol-1.4.5-triphosphate and diacylglycerol. These second messengers, in turn, release calcium from intracellular stores and activate protein kinase C (Clayton, 1989; Huckle and Conn, 1988; Huckle et al., 1988).

1.15.51.1   Human GnRH Receptors

The human GnRH receptor has been cloned Kakar et al. (1992).

1.15.51.10   Rat GnRH Receptors

The rat GnRH receptor has been cloned Eidne et al. (1992).

1.15.51.11   Mouse GnRH Receptors

To identify parts of the GnRH receptor, Tsutsumi et al. (1992) used a PCR-based strategy. For this, degenerate oligonucleotides corresponding to conserved motifs of G protein-coupled receptors and RNA from the GnRH receptor-expressing mouse gonadotrope cell line, aT3-1 (Windle et al., 1990) was used. The functional activity and ligand specificity of the isolated mouse GnRH receptor was tested by injecting *Xenopus* oocytes with the sense RNA derived from a full length mouse GnRH receptor cDNA. When exposed to GnRH, these oocytes demonstrated a large depolarizing response. This response was due to the activation of the oocyte's calcium-dependent chloride current by calcium released from intracellular stores after GnRH stimulation. The depolarizing response was characteristic for GnRH (Sealfon et al., 1990) and identical to the response obtained when polyA+ RNA from the mouse gonadotrope cell line, aT3-1, was injected. The pharmacological response of the cloned GnRH receptor was in agreement with the native receptor (Sealfon et al., 1990). The GnRH agonist [D-Ser(t-Bu)^6,Pro^9-NHEt]GnRH elicited a depolarizing current in mouse GnRH-injected oocytes. In the presence of equimolar weak GnRH antagonist [D-Phe^{2,6},Pro3]GnRH, there was a 60% reduction in the response to GnRH, in comparison to GnRH alone. Two potent GnRH antagonists completely abolished the GnRH mediated current. Displacement of ^{125}I [D-Ala^6,NaMe-Leu^7,Pro^9-NHet]GnRH (GnRH-A) by GnRH-A on oocyte membranes containing the cloned GnRH receptor, revealed a dissociation constant of 4.5 nM which was similar to a K_d of 2.9 nM obtained using membranes of aT3-1 cells (Horn et al., 1991).

The nucleotide sequence of the isolated GnRH-receptor cDNA predicted that the GnRH receptor is a single polypeptide, 327 amino acids in length. Among the G protein-coupled receptors the GnRH receptor is one of the smallest members and unlike any others it lacks the polar cytoplasmatic tail. The receptor protein is most likely glycosylated at one of its three putative N-linked glycosylation sites. The solubilized binding subunit of the GnRH receptor was reported to be in the range of 50–60 kDa (Iwashita et al., 1988). The mRNA of the mouse GnRH-receptor is about 4.8 kb in length and is expressed in clustered cells in the anterior pituitary. In GnRH-neuron derived cells (GT-1), corticotroph (AtT-20), or somatolactotroph (GH3) cell lines, no GnRH receptor mRNA was detected.

1.15.52   Cholecystokin a (CCKa) Receptors

1.15.52.10   Rat CCKa Receptors

CCK receptors have been divided into two major subtypes on the basis of pharmacological studies. CCKa receptors are localized in the gastrointestinal tract but are also
present in both the peripheral and central nervous systems. The CCKa receptor was
cloned from rat pancreas by a PCR approach (Wank et al., 1992a). The receptor is a
444 amino acid protein. When expressed in *Xenopus* oocytes, the receptor displays the
appropriate pharmacological properties.

1.15.53   Cholecystokin b (CCKb) Receptors

1.15.53.10   Rat CCKb Receptors

The CCKb receptor is a 452 amino acid protein that has been identified in the rat gastrointestinal
system and brain. The receptor is 48% identical to the CCKa receptor (Wank et al., 1992b).

1.15.54   Gastrin Receptors

Gastrin stimulates acid secretion from gastric parietal cells. The molecule is structurally related to CCK.

1.15.54.6   Canine Gastrin Receptors

The canine gastrin receptor is a 453 amino acid protein with significant homology to CCKa
and CCKb receptors (Kopin et al., 1992). The expressed receptor displays the pharmacological characteristics of the native gastrin receptor and was found to stimulate PI turnover.

1.15.54.10   Rat Gastrin Receptors

1.15.61   Thyrotropin-Releasing Hormone (TRH)
Receptors

Thyrotropin-releasing hormone (TRH) was initially isolated from the hypothalamus.
It was described as an important releasing factor for the anterior pituitary gland. TRH
is distributed throughout the central and peripheral nervous systems as well as in
extraneural tissues (Hokfelt et al., 1989; Lechan and Segerson, 1989). The receptor for
TRH was detected in the pituitary (Hinkle, 1989) and in brain (Sharif, 1989). There
was some indirect evidence that a G protein couples the TRH-receptor to phospholipase C (Straub and Gershengorn, 1986) and transmembrane signaling studies showed
that the TRH mediates its effect via the inositol phospholipid-calcium-protein kinase
C transduction pathway (Gershengorn, 1989). Therefore, there was the hypothesis that
the TRH-receptor is a member of the large family of G protein-coupled receptors.

1.15.61.10   Rat TRH Receptors

The rat homolog of the mouse TRH receptor was characterized by Zhao et al. (1992). Its
cDNA was isolated from GH4C1 cells and DNA sequence analysis revealed a TRH

receptor molecule of 412 amino acid residues which shows 96% similarity to the mouse TRH receptor and which contains additional 19 amino acid residues at its carboxy-terminus. A single TRH-receptor mRNA of about 4 kb was identified in rat GH3 cells.

1.15.61.11   Mouse TRH Receptors

As for many other membrane-bound receptor proteins, conventional biochemical techniques failed to purify the TRH-receptor protein (Johnson et al., 1984; Phillips and Hinkle, 1989). Therefore, Straub and co-workers decided to use *Xenopus laevis* oocytes in an expression-cloning strategy to identify a cDNA encoding the mouse pituitary TRH-receptor (Straub et al., 1990). For this, pools from a directional, size-selected lambda ZAP expression library were transcribed *in vitro*, injected into oocytes and assayed for TRH responsiveness by voltage clamp recording. After several serial divisions of the pools, they identified a single clone which conferred a TRH-dependent response in oocytes. Those oocytes were able to bind TRH and the competitive antagonist chlordiazepoxide with appropriate affinities of $K_i = 30$ nM and $K_i = 3$ mM, respectively. These values were determined by the inhibition of [^{3}H]methyl-TRH binding and were close to those obtained from TRH binding studies in GH$_3$ pituitary cells (Gershengorn and Paul, 1986). The electrophysiological responses to TRH were dose dependent and 10 nM TRH evoked the half-maximal response. COS-1 cells transiently expressing the mouse TRH receptor bind TRH and chlordiazepoxide with similar affinities to those described above. In the transfected COS-1 cells, the stimulation of inositol phosphate formation exhibited an EC$_{50}$ of approximately 10 nM. The mouse TRH-receptor cDNA encodes a protein of 393 amino acids that shows similarities to G protein-coupled receptors with conserved cysteine residues in the extracellular loops, proline residues in some putative membrane spanning segments, and N-linked glycosylation residues in the amino-terminal part of the receptor.

1.15.62   Antidiuretic Hormone (ADH) Receptors

1.15.62.1   Human ADH Receptors

Hypothalamic release of ADH modulates renal excretion of water. The human ADH receptor is a 371 amino acid protein that was identified by a genomic expression approach (Birnbaumer et al., 1992). The sequence is most homologous to the AV2 receptor and belongs to a subfamily of receptors that also includes the AV1a and oxytocin receptors.

1.15.63   Arginine-Vasopressin 1a (AV1a) Receptors

1.15.63.10   Rat AV1a Receptors

Arginine vasopressin has a variety of actions including inhibition of diuresis, smooth muscle contraction, stimulation of liver glycogenolysis, and modulation of pituitary ACTH release. The AV1a receptor for arginine-vasopressin is a 394 amino acid protein identified in hepatic tissues (Morel et al., 1992).

1.15.64   Arginine-Vasopressin 2 (AV2) Receptors

1.15.64.10   Rat AV2 Receptors

A rat kidney arginine vasopressin receptor (designated AV2) was cloned by Lolait et al. (1992). The receptor is a 370 amino acid protein with highest sequence homology to the human ADH receptor. Specific radioligand binding was most consistent with the charateristics of the V2 AVP receptor. The receptor message was only identified in the kidney and the gene was located on the long arm of the X chromosome. This locus is close to the gene for nephrogenic diabetes insipidus, suggesting a possible link between this disease and AV2 receptor expression.

1.15.65   Oxytocin Receptors

1.15.65.1   Human Oxytocin Receptors

Oxytoxin is responsible for the induction of uterine contractions at the onset of labor. The human oxytocin receptor is a 388 amino acid protein (Kimura et al., 1992). When expressed in *Xenopus* oocytes, inactivation of the receptor induces an inward membrane current. The receptor density appears to increase in the myometrium at the time of labor.

1.15.71   Neuromedin B (NMB) Bombesin Receptors

Neuromedin B is a bombesin-like peptide that appears to exert its physiological effects via a specific G protein-coupled receptor.

1.15.71.1   Human NMB Receptors

The human NMB receptor is a 390 amino acid protein (Corjay et al., 1991).

1.15.71.10   Rat NMB Receptors

The rat NMB receptor is a 390 amino acid protein that shows higher affinity for NMB than for gastrin releasing peptide (GRP) (Wada et al., 1991).

1.15.72   Gastrin Releasing Peptide (GRP) Bombesin Receptors

1.15.72.1   Human GRP Receptors

The human GRP receptor is a 384 amino acid protein (Corjay et al., 1991).

1.15.72.9   Guinea Pig GRP Receptors

The guinea pig GRP receptor is a 399 amino acid protein (Corjay et al., 1991).

1.15.72.11   Mouse GRP Receptors

The mouse GRP receptor is a 384 amino acid protein (Spindel et al., 1990).

1.15.81   Melanocyte-Stimulating Hormone (MSH) Receptors

The melanocortins, melanocyte-stimulating hormone (MSH) and adrenocorticotropic hormone (ACTH), are expressed primarily in the pituitary. They are synthesized as huge precursor polypeptide, proopiomelanocortin (POMC) which is the precursor for a large and complex family of peptides with a variety of hormonal activities. Three major activities of these peptides include regulation of adrenal glucocortical and aldosterone production (through ACTH), control of melanocyte growth and pigment production (through α-MSH, β-MSH, β-lipotropic hormone, and ACTH), and analgesia (through β-endorphin). In addition, these peptides have a variety of biological activities in other areas including brain, the pituitary, and the immune system. Binding sites for ACTH and MSH are found throughout the brain (Tatro, 1990) and α-MSH regulates temperature control in the septal region of the brain and stimulates prolactin release from the pituitary (Ellerkmann et al., 1992). Receptors for the melanocortins in the nervous system and the immune system also exhibit distinct pharmacological properties when compared with those on melanocytes and adrenocortical cells, which suggests the existence of multiple melanocortin receptors (Tatro et al., 1990a; Tatro et al., 1990b). MSH and ACTH mediate an activation of the adenylyl cyclase via G protein (Mertz and Catt, 1991). Therefore, the melanocortin receptors were expected to belong to the G protein-coupled receptors coupling to G_s protein.

The primary structure of mouse and human MSH receptors was deduced from the nucleotide sequences (see below). The human MSH-receptor amino acid sequence was 76% identical and colinear with the murine MSH receptor sequence. Looking at the primary protein sequences, it became obvious that the MSH receptors belong to a new superfamily of the G protein-coupled receptors. The MSH receptors are small (315 to 317 amino acids) containing a very short amino-terminal extracellular and a short carboxy-terminal intracellular domain. Unusually short is the fourth and fifth transmembrane spanning domain and the second extracellular loop. In addition, one or both of the conserved cysteine residues, thought to form a disulfide bond between the first and second extracellular loops, are missing in this receptor subfamily. The murine MSH-receptor is encoded predominantly by a single 4-kb mRNA species and the human MSH-receptor by 3-kb messenger RNA. These transcripts are present in melanoma samples and primary melanocytes.

1.15.81.1   Human MSH Receptors

Parts of the human MSH receptor gene were first isolated (Mountjoy et al., 1992) from a human melanoma cell containing large numbers of MSH binding sites (Tatro et al., 1990a) using PCR based homology screen (Libert et al., 1989b). The DNA from the human PCR fragment of the MSH receptor was used to isolate the mouse MSH-receptor cDNA and the intronless human ACTH- and MSH-receptor gene (Mountjoy et al., 1992). The genomic human MSH receptor sequence was not analyzed for functionality.

1.15.81.11   Mouse MSH Receptors

For the cloned mouse MSH-receptor, functionality was demonstrated in HEK 293 cells transfected with the cloned mouse MSH-receptor cDNA. In HEK 293 cells expressing the cloned MSH-receptor, a specific hormone-dependent accumulation of cyclic AMP was observed. A two- to threefold increase of the intracellular cAMP was measured after stimulation of the cells using different MSH agonists. The median effective concentration values determined for α- and βMSH were 2 nM, for ACTH 8 nM and for NDP-MSH 0.028 nM whereas γMSH had little or no activity.

1.15.82   Adrenocorticotropic Hormone (ACTH) Receptors

1.15.82.1   Human ACTH Receptors

The human intronless ACTH receptor gene was isolated in parallel to the gene for the MSH receptor (Mountjoy et al., 1992). For the analysis of the human ACTH-receptor, the cloned receptor was recombinantly expressed in Cloudman S91 cells. and the stimulation of the receptor with 1 nM ACTH produced a 10-fold increase in intracellular cAMP. In HEK 293 cells the mouse ACTH-receptor was silent, and in Cloudman S91 cells a pharmacological characterization of the ACTH-receptor turned out to be impossible due to the endogenous MSH-receptors in those cells which respond to ACTH concentrations above 2 nM. A detailed pharmacological analysis of the cloned ACTH receptor still has to be performed. The nucleotide sequence of the human ACTH receptor showed that the human ACTH-receptor is a small G protein-coupled receptor (297 amino acids in length) with 39% conserved amino acid residues compared to the human MSH-receptor sequence. Like the MSH receptors, the ACTH receptor has a short amino- and carboxy-terminal receptor domain and a short extracellular loop between the putative transmembrane segments four and five. The transcript of the ACTH receptor is about 4 kb and can be detected in the adrenal gland from human and rhesus macaque. *In situ* hybridization localized the expression of the human ACTH-receptor to adrenal cortex, primarily across the zona fasciculata and the cortical half of the zona glomerulosa.

1.15.91   mas Oncogene Receptors

A major concern in the angiotensin field has been the relationship between the mas oncogene and the angiotensin receptor. The mas oncogene was first isolated in human (Young et al., 1986) and later in rats (Young et al., 1988). It was found to encode a protein sharing conserved structural motifs with G protein-coupled receptor superfamily.

1.15.91.1   Human mas Oncogene Receptors

The human mas oncogene predicted to encode a peptide receptor with mitogenic activity functionally coupled to the inositol signaling pathways (Hanley and Jackson, 1987). In 1988, Jackson demonstrated that, when expressed in neuronal cell line NG115-401L or in *Xenopus* oocytes, the mas oncogene determined an increased functional response to angiotensin II, with intracellular calcium mobilization or chloride channel opening (Jackson et al., 1988). These structural and functional features of the mas oncogene product suggested that this gene might be the gene for the angiotensin receptor. However, several key findings are not reconcilable to this hypothesis. First, the tissue distribution of the angiotensin II binding and the mas oncogene expression is very different (Young et al., 1988). Second, the binding of angiotensin peptides to mas oncogene protein was not demonstrated. Third, although the mas oncogene product responds selective and dose-dependently to exogenous angiotensins, the pharmacological profile of the physiological response is atypical (angiotensin III is more potent than angiotensin II) and classical antagonists for angiotensin II and angiotensin III receptors did not inhibit the angiotensin-induced responses in mas-expressing oocytes.

Table 1.

Receptor	Second messenger	Species cloned	Chromosomal location	a.a. Sequence	Accession number	Primary reference
A. Glycoprotein Hormone Receptors with Minimal Amino Acid Similarities to Other G Protein-Coupled Receptors						
LH-CG		Human		685	M63108	Minegish et al., 1990
	+AC	Porcine		696	M29526	Loosfelt et al., 1989
	+AC	Rat		700	M26199	McFarland et al., 1989
	+AC+PI	Mouse		700	M81310	Gudermann et al., 1992
FSH	+AC	Human		695	M95489	Minegish et al., 1991
		Sheep		695	L07302	Yarney et al., 1992
	+AC	Rat		692	L02842	Sprengel et al., 1990
TSH	+AC+PI	Human	14q31	764	M32215	Misrahi et al., 1990
	+AC	Canine		764	M29957	Parmentier et al., 1989
	+AC	Rat		764	M34842	Akamizu et al., 1990a
	+AC	Mouse	12	—	—	Akamizu et al., 1990a
B. Hormone Receptors with Moderate Amino Acid Similarity to Other G Protein-Coupled Receptors						
PTH	+AC	Human		593	L04308	Schipani et al. (unpublished)
	+AC+PI	Rat		591	M77184	Abou-Samra et al., 1992
	+AC+PI	Opposum		585	M74445	Jueppner et al., 1991
VIP	+AC	Rat		459	M86835	Ishihara et al., 1992
Secretin	+AC	Rat		449	X59132	Ishihara et al., 1991
GLP-1	+AC	Rat		485	M97797	Thorens et al., 1992
Calcitonin		Human		490	L00587	Gorn et al., 1992
	+AC+PI	Porcine		482	M74420	Lin et al., 1991
C. Hormone Receptors with Structural Features Common to Most G Protein-Coupled Receptors						
Somatostatin-1		Human		391	M81829	Yamada et al., 1992
		Rat		391	X62314	Meyerhof et al., 1991
		Mouse		391	M81831	Yamada et al., 1992
Somatostatin-2		Human		369	M81830	Yamada et al., 1992
		Bovine		368	L06613	Xin et al. (unpublished)
		Rat		369	M96817	Kluxen et al., 1992
		Mouse		369	M81832	Yamada et al., 1992
Somatostatin-3		Rat		428	X63574	Meyerhof et al., 1992
		Mouse		428	M91000	Yasuda et al., 1992
AT II-1	+PI	Human		359	M87290	Bergsma et al., 1992
	+PI	Bovine		359	X62294	Sasaki et al., 1991
		Porcine		359	D11340	Itazaki et al. (in press)
	+PI	Rat		359	X62295	Murphy et al., 1991
AT II-1B	+PI	Rat		359	X64052	Iwai et al., 1992
GPRN/RDC1	+AC	Human	2q37	362	M64749	Sreedharan et al., 1991
		Canine		362	X14048	Libert et al., 1990a
GnRH	+PI	Human		328	L03380	Kakar et al., 1992
		Rat		327	X68980	Eidne et al., 1992
		Mouse		327	M93108	Tsutsumi et al., 1992
CCKa		Rat		444	M88096	Wank et al., 1992
CCKb		Rat		452	M99418	Wank et al., 1992
Gastrin	+PI	Canine		453	M87834	Kopin et al., 1992
		Rat		450	D12817	Nagata et al., 1992
TRH	+PI	Rat		412	M90308	Zhao et al., 1992
		Mouse		393	M59811	Straub et al., 1990
ADH		Human		371	Z11687	Birnbaumer et al., 1992
AV-1a		Rat		394	Z11690	Morel et al., 1992
AV-2		Rat	X (long arm)	370	Z11932	Lolait et al., 1992
Oxytocin		Human		388	X64878	Kimura et al., 1992

Table 1. **(Continued)**

Receptor	Second messenger	Species cloned	Chromosomal location	a.a. Sequence	Accession number	Primary reference
C. Hormone Receptors with Structural Features Common to Most G Protein-Coupled Receptors						
GRP/Bombesin 1991		Human		384	M73481	Corjay et al.,
		Guinea pig		399	X67126	Gorbulev et al.,
1992						
		Mouse		384	M57922	Battey et al., 1991
NMB 1991		Human		390	P28336	Corjay et al.,
		Rat		390	P24053	Wada et al., 1991
MSH 1992	+AC	Human		317	X65634	Mountjoy et al.,
		Mouse		315	X65635	Mountjoy et al.,
1992						
ACTH 1992	+AC	Human		297	X65633	Mountjoy et al.,
mas oncogene 1986		Human		325	M13150	Young et al.,
		Rat		324	J03823	Young et al.,
1988						
TA (Thoracic aorta)		Rat		343	M35297	Ross et al., 1990
mrg 1991		Human		377	A39485	Monnot et al.,

1.15.91.10   Rat mas Oncogene

As noted above the rat mas oncogene receptor was first cloned by Young et al. (1988).

1.15.92   Thoracic Aorta (TA) Receptors

1.15.92.10   Rat TA Receptors

A novel putative G protein-coupled receptor closely related to the mas gene was recently cloned. The rat thoracic aorta (TA) gene (Ross et al., 1990) product displays a 34% sequence identity with the mas gene. As previously found for the mas gene, the expression of this clone in COS-7 cells fails to increase the number of angiotensin II binding sites. The RTA product does not possess any functional property of the angiotensin receptor.

1.15.93   mrg Receptors

1.15.93.1   Human mrg Receptors

The mas related gene *mrg* encodes a 378 amino acid protein which is 35 and 29% identical to the mas product and the RTA product, respectively (Monnot et al., 1991). Injection of the *mrg* mRNA in *Xenopus* oocytes was found to potentiate the response of endogenous angiotensin receptors to angiotensin II. These data strongly suggest that the mas-related genes (i.e., RTA and mrg) encode for 7 transmembrane segment receptors that are not angiotensin receptors but may interfere with the endogenous angiotensin receptor signal pathway. The low sequence homology of the cloned AT_1 receptors with the mas gene family products (<10% overall homology) has definitively proven that these receptors are different.

Acknowledgments

We thank Ilpo Huhtaniemi, Simon David Sprengel, David Laurie, and acknowledge Stephen Perowtha for figures and tables. Support from P. H. Seeburg and the "Convenzione Università di Torino and Fidia S.p.A." is gratefully acknowledged. We apologize to any fellow colleagues whose work we failed to cite.

REFERENCES

Abou-Samra, A.-B., Jüppner, H., Force, T., Freeman, M. W., Kong, X. F., Schipani, E., Urena, P., Richards, J., Bonventre, J. V., Potts, J. J., Kronenberg, H. M., and Segre, G. V. (1992) Expression cloning of a common receptor for parathyroid hormone and parathyroid hormone-related peptide from rat osteoblast-like cells: a single receptor stimulates intracellular accumulation of both cAMP and inositol trisphosphates and increases intracellular free calcium. *Proc. Natl. Acad. Sci. U.S.A.* 89, 2732–2736.

Abou-Samra, A.-B., Jüppner, H., Westerberg, D., Potts, J. J., and Segre, G. V. (1989) Parathyroid hormone causes translocation of protein kinase-C from cytosol to membranes in rat osteosarcoma cells. *Endocrinology* 124, 1107–1113.

Akamizu, T., Ikuyama, S., Saji, M., Kosugi, S., Kozak, C., McBride, O. W., and Kohn, L. D. (1990a) Cloning, chromosomal assignment, and regulation of the rat thyrotropin receptor: expression of the gene is regulated by thyrotropin, agents that increase cAMP levels, and thyroid autoantibodies. *Proc. Natl. Acad. Sci. U.S.A.* 87, 5677–5681.

Akamizu, T., Kosugi, S., and Kohn, L. D. (1990b) Thyrotropin receptor processing and interaction with thyrotropin. *Biochem. Biophys. Res. Commun.* 169, 947–952.

Arai, H., Hori, S., Aramori, I., Ohkubo, H., and Nakanishi, S. (1990) Cloning and expression of a cDNA encoding an endothelin receptor. *Nature* 348, 730–732.

Battey, J. F., Way, J. M., Corjay, M. H., Shapira, H., Kusano, K., Harkins, R., Wu, J. M., Slattery, T., Mann, E., and Feldman, R. I. (1991) Molecular cloning of the bombesin/gastrin-releasing peptide receptor from Swiss 3T3 cells. *Proc. Natl. Acad. Sci. U.S.A.* 88, 395–399.

Bawab, W., Gespach, C., Marie, J. C., Chastre, E., and Rosselin, G. (1988) Pharmacology and molecular identification of secretin receptors in rat gastric glands. *Life Sci.* 42, 791–798.

Bayliss, W. M. and Starling, E. H. (1902) Mechanism of pancreatic secretion. *J. Physiol.* 28, 325–334.

Bell, G. I., Sanchez, P. R., Laybourn, P. J., and Najarian, R. C. (1983) Exon duplication and divergence in the human preproglucagon gene. *Nature* 304, 368–371.

Bergsma, D. J., Ellis, C., Kumar, C., Nuthulaganti, P., Kersten, H., Elshourbagy, N., Griffin, E., Stadel, J. M., and Aiyar, N. (1992) Cloning and characterization of a human angiotensin II type 1 receptor. *Biochem. Biophys. Res. Commun.* 183, 989–995.

Birnbaumer, M., Seibold, A., Gilbert, S., Ishido, M., Barberis, C., Antaramian, A., Brabet, P., and Rosenthal, W. (1992) Molecular cloning of the receptor for human antidiuretic hormone. *Nature* 357, 333–335.

Bissonnette, B. M., Collen, M. J., Adachi, H., Jensen, R. T., and Gardner, J. D. (1984) Receptors for vasoactive intestinal peptide and secretin on rat pancreatic acini. *Am. J. Physiol.* 246, G710-G717.

Boulay, F., Tardif, M., Brouchon, L., and Vignais, P. (1990) The human N-formylpeptide receptor. Characterization of two cDNA isolates and evidence for a new subfamily of G protein-coupled receptors. *Biochemistry* 29, 11123–11133.

Braun, T., Schofield, P. R., and Sprengel, R. (1991) Amino-terminal leucine-rich repeats in gonadotropin receptors determine hormone selectivity. *EMBO J.* 10, 1885–1890.

Bringhurst, F. R., Zajac, J. D., Daggett, A. S., Skurat, R. N., and Kronenberg, H. M. (1989) Inhibition of parathyroid hormone responsiveness in clonal osteoblastic cells expressing a mutant form of 3′,5′-cyclic adenosine monophosphate-dependent protein kinase. *Mol. Endocrinol.* 3, 60–67.

Brugger, C. H., Stallwood, D., and Paul, S. (1991) Isolation of a low molecular mass vasoactive intestinal peptide binding protein. *J. Biol. Chem.* 266, 18358–18362.

Catt, K. and Abbott, A. (1991) Molecular cloning of angiotensin II receptors may presage further receptor subtypes. *Trends Pharmacol. Sci.* 12, 279–281.

Chabre, O., Conklin, B. R., Lin, H. Y., Lodish, H. F., Wilson, E., Ives, H. E., Catanzariti, L., Hemmings, B. A., and Bourne, H. R. (1992) A recombinant calcitonin receptor independently stimulates 3',5'-cyclic adenosine monophosphate and Ca^{2+}/inositol phosphate signaling pathways. *Mol. Endocrinol.* 6, 551–656.

Chakraborty, M., Chatterjee, D., Kellokumpu, S., Rasmussen, H., and Baron, R. (1991) Cell cycle-dependent coupling of the calcitonin receptor to different G proteins. *Science* 251, 1078–1082.

Chiu, A. T., Herblin, W. F., McCall, D. E., Ardecky, R. J., Carini, D. J., Duncia, J. V., Pease, L. J., Wong, P. C., Wexler, R. R., Johnson, A. L., et al. (1989) Identification of angiotensin II receptor subtypes. *Biochem. Biophys. Res. Commun.* 165, 196–203.

Christophe, J., Waelbroeck, M., Chatelain, P., and Robberecht, P. (1984) Heart receptors for VIP, PHI and secretin are able to activate adenylate cyclase and to mediate inotropic and chronotropic effects. Species variations and physiopathology. *Peptides* 5, 341–353.

Clayton, R. N. (1989) Gonadotrophin-releasing hormone: its actions and receptors. *J. Endocrinol.* 120, 11–19.

Cook, J. S., Wolsing, D. H., Lameh, J., Olson, C. A., Correa, P. E., Sadee, W., Blumenthal, E. M., and Rosenbaum, J. S. (1992) Characterization of the RDC1 gene which encodes the canine homolog of a proposed human VIP receptor. Expression does not correlate with an increase in VIP binding sites. *FEBS Lett.* 300, 149–152.

Coop, D. H. (1972) Evolution of calcium regulation in vertebrates. *Clin. Endocrinol. Metab.* 1, 21–32.

Corjay, M. H., Dobrzanski, D. J., Way, J. M., Viallet, J., Shapira, H., Worland, P., Sausville, E. A., and Battey, J. F. (1991) Two distinct bombesin receptor subtypes are expressed and functional in human lung carcinoma cells. *J. Biol. Chem.* 266, 18771–18779.

Couvineau, A., Voisin, T., Guijarro, L., and Laburthe, M. (1990) Purification of vasoactive intestinal peptide receptor from porcine liver by a newly designed one-step affinity chromatography. *J. Biol. Chem.* 265, 13386–13390.

Dahl, K. D., Bicsak, T. A., and Hsueh, A. J. (1988) Naturally occurring antihormones: secretion of FSH antagonists by women treated with a GnRH analog. *Science* 239, 72–74.

Dixon, R. A., Sigal, I. S., Rands, E., Register, R. B., Candelore, M. R., Blake, A. D., and Strader, C. D. (1987) Ligand binding to the beta-adrenergic receptor involves its rhodopsin-like core. *Nature* 326, 73–77.

Douglas, J. G. (1987) Angiotensin receptor subtypes of the kidney cortex. *Am. J. Physiol.* 253, F1–F7.

Drucker, D. J., Philippe, J., Mojsov, S., Chick, W. L., and Habener, J. F. (1987) Glucagon-like peptide I stimulates insulin gene expression and increases cyclic AMP levels in a rat islet cell line. *Proc. Natl. Acad. Sci. U.S.A.* 84, 3434–3438.

Dudley, D. T., Panek, R. L., Major, T. C., Lu, G. H., Bruns, R. F., Klinkefus, B. A., Hodges, J. C., and Weishaar, R. E. (1990) Subclasses of angiotensin II binding sites and their functional significance. *Mol. Pharmacol.* 38, 370–377.

Eidne, K. A., Sellar, R. E., Couper, G., Anderson, L., and Taylor, P. L. (1992) Molecular cloning and characterization of the rat pituitary gonadotropin-releasing hormone (GnRH) receptor. *Mol. Cell. Endocrinol.* 90, 5–9.

Ellerkmann, E., Nagy, G. M., and Frawley, L. S. (1992) Alpha-melanocyte-stimulating hormone is a mammotrophic factor released by neurointermediate lobe cells after estrogen treatment. *Endocrinology* 130, 133–138.

Fares, F. A., Suganuma, N., Nishimori, K., LaPolt, P. S., Hsueh, A. J., and Boime, I. (1992) Design of a long-acting follitropin agonist by fusing the C-terminal sequence of the chorionic gonadotropin beta subunit to the follitropin beta subunit. *Proc. Natl. Acad. Sci. U.S.A.* 89, 4304–4308.

Feng, D. F. and Doolittle, R. F. (1987) Progressive sequence alignment as a prerequisite to correct phylogenetic trees. *J. Mol. Evol.* 25, 351–353.

Finch, R. J., Sreedharan, S. P., and Goetzl, E. J. (1989) High-affinity receptors for vasoactive intestinal peptide on human myeloma cells. *J. Immunol.* 142, 1977–1981.

Frazier, A. L., Robbins, L. S., Stork, P. J., Sprengel, R., Segaloff, D. L., and Cone, R. D. (1990) Isolation of TSH and LH/CG receptor cDNAs from human thyroid: regulation by tissue specific splicing. *Mol. Endocrinol.* 4, 1264–1276.

Fremeau, R. J., Korman, L. Y., and Moody, T. W. (1986) Secretin stimulates cyclic AMP formation in the rat brain. *J. Neurochem.* 46, 1947–1955.

Furuta, H., Guo, D. F., and Inagami, T. (1992) Molecular cloning and sequencing of the gene encoding human angiotensin II type 1 receptor. *Biochem. Biophys. Res. Commun.* 183, 8–13.

Gershengorn, M. C. (1989) Mechanism of signal transduction by TRH. *Ann. N.Y. Acad. Sci.* 553, 191–196.

Gershengorn, M. C. and Paul, M. E. (1986) Evidence for tight coupling of receptor occupancy by thyrotropin-releasing hormone to phospholipase C-mediated phosphoinositide hydrolysis in rat pituitary cells: use of chlordiazepoxide as a competitive antagonist. *Endocrinology* 119, 833–839.

Gespach, C., Bataille, D., Vauclin, N., Moroder, L., Wunsch, E., and Rosselin, G. (1986) Secretin receptor activity in rat gastric glands. Binding studies, cAMP generation and pharmacology. *Peptides* 1, 155–163.

Gilman, A. G. (1984) G proteins and dual control of adenylate cyclase. *Cell* 36, 577–579.

Göke, R., Trautmann, M. E., Haus, E., Richter, G., Fehmann, H. C., Arnold, R., and Göke, B. (1989) Signal transmission after GLP-1(7–36)amide binding in RINm5F cells. *Am. J. Physiol.* 257, G397–G401.

Gorn, A. H., Lin, H. Y., Auron, P. E., Flannery, M. R., Tapp, D. R., Manning, C. A., Lodish, H. F., and Goldring, S. R. (1992) Cloning, characterization, and expression pf a human calcitonin receptor from an ovarian carcinoma cell line. *J. Clin. Invest.* 90, 1726–1735.

Gossen, D., Poloczek, P., Svoboda, M., and Christophe, J. (1989) Molecular architecture of secretin receptors: the specific covalent labelling of a 51 kDa peptide after cross-linking of [^{125}I]iodosecretin to intact rat pancreatic acini. *FEBS Lett.* 243, 205–208.

Gozes, I. and Brenneman, D. E. (1989) VIP: molecular biology and neurobiological function. *Mol. Neurobiol.* 3, 201–236.

Gross, B., Misrahi, M., Sar, S., and Milgrom, E. (1991) Composite structure of the human thyrotropin receptor gene. *Biochem. Biophys. Res. Commun.* 177, 679–687.

Gudermann, T., Birnbaumer, M., and Birnbaumer, L. (1992a) Evidence for dual coupling of the murine luteinizing hormone receptor to adenylyl cyclase and phosphoinositide breakdown and Ca2+ mobilization. Studies with the cloned murine luteinizing hormone receptor expressed in L cells. *J. Biol. Chem.* 267, 4479–4488.

Gudermann, T., Nichols, C., Levy, F. O., Birnbaumer, M., and Birnbaumer, L. (1992b) Ca2+ mobilization by the LH receptor expressed in Xenopus oocytes independent of 3′,5′-cyclic adenosine monophosphate formation: evidence for parallel activation of two signaling pathways. *Mol. Endocrinol.* 6, 272–278.

Hanley, M. R. and Jackson, T. (1987) Substance K receptor: return of the magnificent seven. *Nature* 329, 766–767.

He, H. T., Rens-Domiano, S., Martin, J. M., Law, S. F., Borislow, S., Woolkalis, M., Manning, D., and Reisine, T. (1990) Solubilization of active somatostatin receptors from rat brain. *Mol. Pharmacol.* 37, 614–621.

Heckert, L. L., Daley, I. J., and Griswold, M. D. (1992) Structural organization of the follicle-stimulating hormone receptor gene. *Mol. Endocrinol.* 6, 70–80.

Heckert, L. L. and Griswold, M. D. (1991) Expression of follicle-stimulating hormone receptor mRNA in rat testes and Sertoli cells. *Mol. Endocrinol.* 5, 670–677.

Higgins, D. G. and Sharp, P. M. (1989) Fast and sensitive multiple sequence alignments on a microcomputer. *Comput. Appl. Biosci.* 5, 151–153.

Hinkle, P. M. (1989) Pituitary TRH receptors. *Ann. N.Y. Acad. Sci.* 553, 176–187.

Hokfelt, T., Tsuruo, Y., Ulfhake, B., Cullheim, S., Arvidsson, U., Foster, G. A., Schultzberg, M., Schalling, M., Arborelius, L., Freedman, J., Post, C., and Visser, T. (1989) Distribution of TRH-like immunoreactivity with special reference to coexistence with other neuroactive compounds. *Ann. N.Y. Acad. Sci.* 553, 76–105.

Horn, F., Bilezikjian, L. M., Perrin, M. H., Bosma, M. M., Windle, J. J., Huber, K. S., Blount, A. L., Hille, B., Vale, W., and Mellon, P. L. (1991) Intracellular responses to gonadotropin-releasing hormone in a clonal cell line of the gonadotrope lineage. *Mol. Endocrinol.* 5, 347–355.

Hruska, K. A., Moskowitz, D., Esbrit, P., Civitelli, R., Westbrook, S., and Huskey, M. (1987) Stimulation of inositol trisphosphate and diacylglycerol production in renal tubular cells by parathyroid hormone. *J. Clin. Invest.* 79, 230–239.

Huckle, W. R. and Conn, P. M. (1988) Molecular mechanism of gonadotropin releasing hormone action. II. The effector system. *Endocr. Rev.* 9, 387–395.

Huckle, W. R., McArdle, C. A., and Conn, P. M. (1988) Differential sensitivity of agonist- and antagonist-occupied gonadotropin-releasing hormone receptors to protein kinase C activators. A marker for receptor activation. *J. Biol. Chem.* 263, 3296–3302.

Huhtanieni, I. T., Eskola, V., Pakarinen, P., Matikainen, T., and Sprengel, R. (1992) The murine luteinizing hormone and follicle-stimulating hormone receptor genes: transcription initiation sites, putative promoter sequences and promoter activity. *Mol. Cell. Endocrinol.* 88, 55–66.

Ikuyama, S., Niller, H. H., Shimura, H., Akamizu, T., and Kohn, L. D. (1992) Characterization of the 5′-flanking region of the rat thyrotropin receptor gene. *Mol. Endocrinol.* 6, 793–804.

Ishihara, T., Nakamura, S., Kaziro, Y., Takahashi, T., Takahashi, K., and Nagata, S. (1991) Molecular cloning and expression of a cDNA encoding the secretin receptor. *EMBO J.* 10, 1635–1641.

Ishihara, T., Shigemoto, R., Mori, K., Takahashi, K., and Nagata, S. (1992) Functional expression and tissue distribution of a novel receptor for vasoactive intestinal polypeptide. *Neuron* 8, 811–819.

Iwai, N. and Inagami, T. (1992) Identification of two subtypes in the rat type I angiotensin II receptor. *FEBS Lett.* 298, 257–260.

Iwai, N., Yamano, Y., Chaki, S., Konishi, F., Bardhan, S., Tibbetts, C., Sasaki, K., Hasegawa, M., Matsuda, Y., and Inagami, T. (1991) Rat angiotensin II receptor: cDNA sequence and regulation of the gene expression. *Biochem. Biophys. Res. Commun.* 177, 299–304.

Iwashita, M., Hirota, K., Izumi, S. I., Chen, H. C., and Catt, K. J. (1988) Solubilization and characterization of the rat pituitary gonadotropin-releasing hormone receptor. *J. Mol. Endocrinol.* 1, 187–196.

Jackson, T. R., Blair, L. A., Marshall, J., Goedert, M., and Hanley, M. R. (1988) The mas oncogene encodes an angiotensin receptor. *Nature* 335, 437–440.

Ji, I. and Ji, T. H. (1991a) Exons 1–10 of the rat LH receptor encode a high affinity hormone binding site and exon 11 encodes G-protein modulation and a potential second hormone binding site. *Endocrinology* 128, 2648–2650.

Ji, I. H. and Ji, T. H. (1991b) Human choriogonadotropin binds to a lutropin receptor with essentially no N-terminal extension and stimulates cAMP synthesis. *J. Biol. Chem.* 266, 13076–13079.

Jia, X. C., Oikawa, M., Bo, M., Tanaka, T., Ny, T., Boime, I., and Hsueh, A. J. (1991) Expression of human luteinizing hormone (LH) receptor: interaction with LH and chorionic gonadotropin from human but not equine, rat, and ovine species. *Mol. Endocrinol.* 5, 759–768.

Johnson, W. A., Nathanson, N. M., and Horita, A. (1984) Solubilization and characterization of thyrotropin-releasing hormone receptors from rat brain. *Proc. Natl. Acad. Sci. U.S.A.* 81, 4227–4231.

Jüppner, H., Abou-Samra, A.-B., Freeman, M., Kong, X. F., Schipani, E., Richards, J., Kolakowski, L. J., Hock, J., Potts, J. J., Kronenberg, H. M., and Serge, G. V. (1991) A G protein-linked receptor for parathyroid hormone and parathyroid hormone-related peptide. *Science* 254, 1024–1026.

Jüppner, H., Abou-Samra, A.-B., Uneno, S., Gu, W. X., Potts, J. J., and Segre, G. V. (1988) The parathyroid hormone-like peptide associated with humoral hypercalcemia of malignancy and parathyroid hormone bind to the same receptor on the plasma membrane of ROS 17/2.8 cells. *J. Biol. Chem.* 263, 8557–8560.

Kakar, S. S., Musgrove, L. C., Devor, D. C., Sellers, J. C., and Neill, J. D. (1992) Cloning of human gonadotropin releasing hormone (GnRH) receptor. *Biochem. Biophys. Res. Commun.* 189, 289–295.

Kakar, S. S., Riel, K. K., and Neill, J. D. (1992a) Differential expression of angiotensin II receptor subtype mRNAs (AT-1A and AT-1B) in the brain. *Biochem. Biophys. Res. Commun.* 185, 688–692.

Kakar, S. S., Sellers, J. C., Devor, D. C., Musgrove, L. C., and Neill, J. D. (1992b) Angiotensin II type-1 receptor subtype cDNAs: differential tissue expression and hormonal regulation. *Biochem. Biophys. Res. Commun.* 183, 1090–1096.

Kimura, T., Tanizawa, O., Mori, K., Brownstein, M. J., and Okayama, H. (1992) Structure and expression of a human oxytocin receptor. *Nature* 356, 526–529.

Kluxen, F. W., Bruns, C., and Lubbert, H. (1992) Expression cloning of a rat brain somatostatin receptor cDNA. *Proc. Natl. Acad. Sci. U.S.A.* 89, 4618–4622.

Kobilka, B. K., Kobilka, T. S., Daniel, K., Regan, J. W., Caron, M. G., and Lefkowitz, R. J. (1988) Chimeric alpha 2-beta 2-adrenergic receptors: delineation of domains involved in effector coupling and ligand binding specificity. *Science* 240, 1310–1316.

Koo, Y. B., Ji, I., Slaughter, R. G., and Ji, T. H. (1991) Structure of the luteinizing hormone receptor gene and multiple exons of the coding sequence. *Endocrinology* 128, 2297–3208.

Kopin, A. S., Lee, Y. M., McBride, E. W., Miller, L. J., Lu, M., Lin, H. Y., Kolakowski, L. J., and Beinborn, M. (1992) Expression cloning and characterization of the canine parietal cell gastrin receptor. *Proc. Natl. Acad. Sci. U.S.A.* 89, 3605–3609.

Kreymann, B., Williams, G., Ghatei, M. A., and Bloom, S. R. (1987) Glucagon-like peptide-1 7–36, a physiological incretin in man. *Lancet* 2, 1300–1304.

Lamberts, S. W. (1988) The role of somatostatin in the regulation of anterior pituitary hormone secretion and the use of its analogs in the treatment of human pituitary tumors. *Endocr. Rev.* 9, 417–436.

LaPolt, P. S., Tilly, J. L., Aihara, T., Nishimori, K., and Hsueh, A. J. (1992) Gonadotropin-induced up- and down-regulation of ovarian follicle-stimulating hormone (FSH) receptor gene expression in immature rats: effects of pregnant mare's serum gonadotropin, human chorionic gonadotropin, and recombinant FSH. *Endocrinology* 130, 1289–1295.

Law, S. F., Manning, D., and Reisine, T. (1991) Identification of the subunits of GTP-binding proteins coupled to somatostatin receptors. *J. Biol. Chem.* 266, 17885–17897.

Lechan, R. M. and Segerson, T. P. (1989) Pro-TRH gene expression and precursor peptides in rat brain. Observations by hybridization analysis and immunocytochemistry. *Ann. N.Y. Acad. Sci.* 553, 29–59.

Libert, F., Lefort, A., Gerard, C., Parmentier, M., Perret, J., Ludgate, M., Dumont, J. E., and Vassart, G. (1989a) Cloning, sequencing and expression of the human thyrotropin (TSH) receptor: evidence for binding of autoantibodies. *Biochem. Biophys. Res. Commun.* 165, 1250–1255.

Libert, F., Parmentier, M., Lefort, A., Dinsart, C., Van-Sande, J., Maenhaut, C., Simons, M. J., Dumont, J. E., and Vassart, G. (1989b) Selective amplification and cloning of four new members of the G protein-coupled receptor family. *Science* 244, 569–572.

Libert, F., Parmantier, M., Lefort, A., Dumont, J. E., and Vassart, G. (1990a) Complete nucleotide sequence of a putative G protein-coupled receptor: RDC1. *Nucleic Acids Res.* 18, 7.

Libert, F., Parmentier, M., Maenhaut, C., Lefort, A., Gerard, C., Perret, J., Van, S. J., Dumont, J. E., and Vassart, G. (1990b) Molecular cloning of a dog thyrotropin (TSH) receptor variant. *Mol. Cell Endocrinol.* 68, R15–R17.

Libert, F., Passage, E., Lefort, A., Vassart, G., and Mattei, M. G. (1990c) Localization of human thyrotropin receptor gene to chromosome region 14q3 by in situ hybridization. *Cytogenet. Cell Genet.* 54, 82–83.

Libert, F., Passage, E., Parmentier, M., Simons, M. J., Vassart, G., and Mattei, M. G. (1991) Chromosomal mapping of A1 and A2 adenosine receptors, VIP receptor, and a new subtype of serotonin receptor. *Genomics* 11, 225–227.

Lin, H. Y., Harris, T. L., Flannery, M. S., Aruffo, A., Kaji, E. H., Gorn, A., Kolakowski, L. J., Lodish, H. F., and Goldring, S. R. (1991) Expression cloning of an adenylate cyclase-coupled calcitonin receptor. *Science* 254, 1022–1024.

Lolait, S. J., O'Carroll, A. M., McBride, O. W., Konig, M., Morel, A., and Brownstein, M. J. (1992) Cloning and characterization of a vasopressin V2 receptor and possible link to nephrogenic diabetes insipidus. *Nature* 357, 336–339.

Loosfelt, H., Misrahi, M., Atger, M., Salesse, R., Thi, M. T. V. H., Jolivet, A., Guichon-Mantel, A., Sar, S., Jallal, B., Garnier, J., and Milgrom, E. (1989) Cloning and sequencing of porcine LH-hCG receptor cDNA: variants lacking transmembrane domain. *Science* 245, 525–528.

Luini, A., Lewis, D., Guild, S., Corda, D., and Axelrod, J. (1985) Hormone secretagogues increase cytosolic calcium by increasing cAMP in corticotropin-secreting cells. *Proc. Natl. Acad. Sci. U.S.A.* 82, 8034–8038.

Mahy, N., Woolkalis, M., Manning, D., and Reisine, T. (1988a) Characteristics of somatostatin desensitization in the pituitary tumor cell line AtT-20. *J. Pharmacol. Exp. Ther.* 247, 390–396.

Mahy, N., Woolkalis, M., Thermos, K., Carlson, K., Manning, D., and Reisine, T. (1988b) Pertussis toxin modifies the characteristics of both the inhibitory GTP binding proteins and the somatostatin receptor in anterior pituitary tumor cells. *J. Pharmacol. Exp. Ther.* 246, 779–785.

Mangin, M., Webb, A. C., Dreyer, B. E., Posillico, J. T., Ikeda, K., Weir, E. C., Stewart, A. F., Bander, N. H., Milstone, L., Barton, D. E., et al. (1988) Identification of a cDNA encoding a parathyroid hormone-like peptide from a human tumor associated with humoral hypercalcemia of malignancy. *Proc. Natl. Acad. Sci. U.S.A.* 85, 597–601.

Marin, P., Delumeau, J. C., Tence, M., Cordier, J., Glowinski, J., and Premont, J. (1991) Somatostatin potentiates the alpha 1-adrenergic activation of phospholipase C in striatal astrocytes through a mechanism involving arachidonic acid and glutamate. *Proc. Natl. Acad. Sci. U.S.A.* 88, 9016–9020.

Matzuk, M. M., Keene, J. L., and Boime, I. (1989) Site specificity of the chorionic gonadotropin N-linked oligosaccharides in signal transduction. *J. Biol. Chem.* 264, 2409–2414.

McFarland, K. C., Sprengel, R., Phillips, H. S., Köhler, M., Rosemblit, N., Nikolics, K., Segaloff, D. L., and Seeburg, P. H. (1989) Lutropin-choriogonadotropin receptor: an unusual member of the G protein-coupled receptor family. *Science* 245, 494–499.

Mertz, L. M. and Catt, K. J. (1991) Adrenocorticotropin receptors: functional expression from rat adrenal mRNA in Xenopus laevis oocytes. *Proc. Natl. Acad. Sci. U.S.A.* 88, 8525–8529.

Meyerhof, W., Paust, H.-J., Schönrock, C., and Richter, D. (1991) Cloning of a cDNA encoding a novel putative G protein-coupled receptor expressed in specific brain regions. *DNA Cell Biol.* 10, 689–894.

Meyerhof, W., Wulfsen, I., Schoenrock, C., Fehr, S., and Richter, D. (1992) Molecular cloning of a somatostatin-28 receptor and comparison of its expression pattern with that of a somatostatin-14 receptor in rat brain. *Proc. Natl. Acad. Sci. U.S.A.* 89, 10267–10271.

Minegish, T., Nakamura, K., Takakura, Y., Ibuki, Y., and Igarashi, M. (1991) Cloning and sequencing of human FSH receptor cDNA. *Biochem. Biophys. Res. Commun.* 175, 1125–1130.

Minegish, T., Nakamura, K., Takakura, Y., Miyamoto, K., Hasegawa, Y., Ibuki, Y., and Igarashi, M. (1990) Cloning and sequencing of human LH/hCG receptor cDNA. *Biochem. Biophys. Res. Commun.* 172, 1049–1054.

Misrahi, M., Loosfelt, H., Atger, M., Sar, S., Guichon-Mantel, A., and Milgrom, E. (1990) Cloning, sequencing and expression of human TSH receptor. *Biochem. Biophys. Res. Commun.* 166, 394–403.

Miyauchi, A., Dobre, V., Rickmeyer, M., Cole, J., Forte, L., and Hruska, K. A. (1990) Stimulation of transient elevations in cytosolic Ca^{2+} is related to inhibition of Pi transport in OK cells. *Am. J. Physiol.* 259, F485–F493.

Miyoshi, R., Kito, S., Katayama, S., and Kim, S. U. (1989) Somatostatin increases intracellular Ca^{2+} concentration in cultured rat hippocampal neurons. *Brain Res.* 489, 361–364.

Mojsov, S., Heinrich, G., Wilson, I. B., Ravazzola, M., Orci, L., and Habener, J. F. (1986) Preproglucagon gene expression in pancreas and intestine diversifies at the level of post-translational processing. *J. Biol. Chem.* 261, 11880–11889.

Monnot, C., Weber, V., Stinnakre, J., Bihoreau, C., Teutsch, B., Corvol, P., and Clauser, E. (1991) Cloning and functional characterization of a novel mas-related gene, modulating intracellular angiotensin II actions. *Mol. Endocrinol.* 5, 1477–1487.

Morel, A., O'Carroll, A. M., Brownstein, M. J., and Lolait, S. J. (1992) Molecular cloning and expression of a rat V1a arginine vasopressin receptor. *Nature* 356, 523–526.

Mountjoy, K. G., Robbins, L. S., Mortrud, M. T., and Cone, R. D. (1992) The Cloning of a Family of Genes that Encode the Melanocortin Receptors. *Science* 257, 1248–1251.

Murphy, P. M., Ozcelik, T., Kenney, R. T., Tiffany, H. L., McDermott, D., and Francke, U. (1992) A structural homologue of the N-formyl peptide receptor. Characterization and chromosome mapping of a peptide chemoattractant receptor family. *J. Biol. Chem.* 267, 7637–7643.

Murphy, T. J., Alexander, R. W., Griendling, K. K., Runge, M. S., and Bernstein, K. E. (1991) Isolation of a cDNA encoding the vascular type-1 angiotensin II receptor. *Nature* 351, 233–236.

Nagata, S., Ishihara, T., Robberecht, P., Libert, F., Parmentier, M., Christophe, J., and Vassart, G. (1992) RDC1 may not be VIP receptor. *Trends Pharmacol. Sci.* 13, 102–103.

Nagayama, Y., Kaufman, K. D., Seto, P., and Rapoport, B. (1989) Molecular cloning, sequence and functional expression of the cDNA for the human thyrotropin receptor. *Biochem. Biophys. Res. Commun.* 165, 1184–1190.

Nagayama, Y. and Rapoport, B. (1992) The thyrotropin receptor 25 years after its discovery: new insight after its molecular cloning. *Mol. Endocrinol.* 6, 145–156.

Nagayama, Y., Russo, D., Wadsworth, H. L., Chazenbalk, G. D., and Rapoport, B. (1991a) Eleven amino acids (Lys-201 to Lys-211) and 9 amino acids (Gly-222 to Leu-230) in the human thyrotropin receptor are involved in ligand binding. *J. Biol. Chem.* 266, 14926–14930.

Nagayama, Y., Wadsworth, H. L., Chazenbalk, G. D., Russo, D., Seto, P., and Rapoport, B. (1991b) Thyrotropin-luteinizing hormone/chorionic gonadotropin receptor extracellular domain chimeras as probes for thyrotropin receptor function. *Proc. Natl. Acad. Sci. U.S.A.* 88, 902–905.

O'Dowd, B. F., Hnatowich, M., Caron, M. G., Lefkowitz, R. J., and Bouvier, M. (1989) Palmitoylation of the human beta 2-adrenergic receptor. Mutation of Cys341 in the carboxyl tail leads to an uncoupled nonpalmitoylated form of the receptor. *J. Biol. Chem.* 264, 7564–7569.

O'Dowd, B. F., Hnatowich, M., Regan, J. W., Leader, W. M., Caron, M. G., and Lefkowitz, R. J. (1988) Site-directed mutagenesis of the cytoplasmic domains of the human beta 2-adrenergic receptor. Localization of regions involved in G protein-receptor coupling. *J. Biol. Chem.* 263, 15985–15992.

O'Dowd, B. F., Lefkowitz, R. J., and Caron, M. G. (1989) Structure of the adrenergic and related receptors. *Annu. Rev. Neurosci.* 12, 67–83.

Orloff, J. J., Wu, T. L., and Stewart, A. F. (1989) Parathyroid hormone-like proteins: biochemical responses and receptor interactions. *Endocr. Rev.* 10, 476–495.

Parmentier, M., Libert, F., Maenhaut, C., Lefort, A., Gerard, C., Perret, J., Van Sande, J., Dumont, J. E., and Vassart, G. (1989) Molecular cloning of the thyrotropin receptor. *Science* 246, 1620–1622.

Patthy, L. (1987) Detecting homology of distantly related proteins with consensus sequences. *J. Mol. Biol.* 198, 567–577.

Phillips, W. J. and Hinkle, P. M. (1989) Solubilization and characterization of pituitary thyrotropin-releasing hormone receptors. *Mol. Pharmacol.* 35, 533–540.

Pierce, J. G. and Parsons, T. F. (1981) Glycoprotein hormones:structure and function. *Annu. Rev. Biochem.* 50, 465–495.

Rappaport, M. S. and Stern, P. H. (1986) Parathyroid hormone and calcitonin modify inositol phospholipid metabolism in fetal rat limb bones. *J. Bone Miner. Res.* 1, 173–179.

Rees, S. B., McLachlan, S. M., and Furmaniak, J. (1988) Autoantibodies to the thyrotropin receptor. *Endocr. Rev.* 9, 106–121.

Reichlin, S. (1983a) Somatostatin. *N. Engl. J. Med.* 309, 1495–1501.

Reichlin, S. (1983b) Somatostatin (second of two parts). *N. Engl. J. Med.* 309, 1556–1563.

Reubi, J. C. (1984) Evidence for two somatostatin-14 receptor types in rat brain cortex. *Neurosci. Lett.* 49, 259–263.

Reubi, J. C. (1985) New specific radioligand for one subpopulation of brain somatostatin receptors. *Life Sci.* 36, 1829–1836.

Reubi, J. C., Lamberts, S. W., and Maurer, R. (1988) Somatostatin receptors in normal and tumoral tissue. *Horm. Res.* 29, 65–69.

Robberecht, P., Cauvin, A., Gourlet, P., and Christophe, J. (1990) Heterogeneity of VIP receptors. *Arch. Int. Pharmacodyn. Ther.* 303, 51–66.

Robberecht, P., Gourlet, P., Cauvin, A., Buscail, L. P. D. N., Arimura, A., and Christophe, J. (1991) PACAP and VIP receptors in rat liver membranes. *Am. J. Physiol.* 260, G97–G102.

Ross, P. C., Figler, R. A., Corjay, M. H., Barber, C. M., Adam, N., Harcus, D. R., and Lynch, K. R. (1990) RTA, a candidate G protein-coupled receptor: cloning, sequencing, and tissue distribution. *Proc. Natl. Acad. Sci. U.S.A.* 87, 3052–3056.

Rosselin, G. (1986) The receptors of the VIP family peptides (VIP, secretin, GRF, PHI, PHM, GIP, glucagon and oxyntomodulin). Specificities and identity. *Peptides* 1, 89–100.

Roth, B. L., Beinfeld, M. C., and Howlett, A. C. (1984) Secretin receptors on neuroblastoma cell membranes: characterization of ^{125}I-labeled secretin binding and association with adenylate cyclase. *J. Neurochem.* 42, 1145–1152.

Rousseau-Merck, M., Misrahi, M., Loosfelt, H., Atger, M., Milgrom, E., and Berger, R. (1990a) Assignment of the human thyroid stimulating hormone receptor (TSHR) gene to chromosome 14q31. *Genomics* 8, 233–236.

Rousseau-Merck, M. F., Misrahi, M., Atger, M., Loosfelt, H., Milgrom, E., and Berger, R. (1990b) Localization of the human luteinizing hormone/choriogonadotropin receptor gene (LHCGR) to chromosome 2p21. *Cytogenet. Cell Genet.* 54, 77–79.

Russo, D., Chazenbalk, G. D., Nagayama, Y., Wadsworth, H. L., and Rapoport, B. (1991) Site-directed mutagenesis of the human thyrotropin receptor: role of asparagine-linked oligosaccharides in the expression of a functional receptor. *Mol. Endocrinol.* 5, 29–33.

Ryan, R. J., Charlesworth, M. C., McCormick, D. J., Milius, R. P., and Keutmann, H. T. (1988) The glycoprotein hormones: recent studies of structure-function relationships. *FASEB J.* 2, 2661–2669.

Said, S. I. (1986) Vasoactive intestinal peptide. *J. Endocrinol. Invest.* 9, 119–200.

Sairam, M. R. (1989) Role of carbohydrates in glycoprotein hormone signal transduction. *FASEB J.* 3, 1915–1926.

Sairam, M. R. (1990) Complete dissociation of gonadotropin receptor binding and signal transduction in mouse Leydig tumour cells. Obligatory role of glycosylation in hormone action. *Biochem. J.* 265, 667–674.

Sakurai, T., Yanagisawa, M., Takuwa, Y., Miyazaki, H., Kimura, S., Goto, K., and Masaki, T. (1990) Cloning of a cDNA encoding a non-isopeptide-selective subtype of the endothelin receptor. *Nature* 348, 732–735.

Sasaki, K., Yamano, Y., Bardhan, S., Iwai, N., Murray, J. J., Hasegawa, M., Matsuda, Y., and Inagami, T. (1991) Cloning and expression of a complementary DNA encoding a bovine adrenal angiotensin II type-1 receptor. *Nature* 351, 230–233.

Sealfon, S. C., Gillo, B., Mundamattom, S., Mellon, P. L., Windle, J. J., Landau, E., and Roberts, J. L. (1990) Gonadotropin-releasing hormone receptor expression in Xenopus oocytes. *Mol. Endocrinol.* 4, 119–124.

Segaloff, D. L., Sprengel, R., Nikolics, K., and Ascoli, M. (1990) Structure of the lutropin/choriogonadotropin receptor. *Recent Prog. Horm. Res.* 46, 261–301.

Sharif, N. A. (1989) Quantitative autoradiography of TRH receptors in discrete brain regions of different mammalian species. *Ann. N.Y. Acad. Sci.* 553, 147–175.

Speth, R. C. and Kim, K. H. (1990) Discrimination of two angiotensin II receptor subtypes with a selective agonist analogue of angiotensin II, p-aminophenylalanine 6 angiotensin II. *Biochem. Biophys. Res. Commun.* 169, 997–1006.

Spindel, E. R., Giladi, E., Brehm, P., Goodman, R. H., and Segerson, T. P. (1990) Cloning and functional characterization of a complementary DNA encoding the murine fibroblast bombesin/gastrin-releasing peptide receptor. *Mol. Endocrinol.* 4, 1956–1963.

Sprengel, R., Braun, T., Nikolics, K., Segaloff, D. L., and Seeburg, P. H. (1990) The testicular receptor for follicle stimulating hormone: structure and functional expression of cloned cDNA. *Mol. Endocrinol.* 4, 525–530.

Sreedharan, S. P., Robichon, A., Peterson, K. E., and Goetzl, E. J. (1991) Cloning and expression of the human vasoactive intestinal peptide receptor. *Proc. Natl. Acad. Sci. U.S.A.* 88, 4986–4990.

Strader, C. D., Sigal, I. S., and Dixon, R. A. (1989) Genetic approaches to the determination of structure-function relationships of G protein-coupled receptors. *Trends Pharmacol. Sci.* 84, 4384–4388.

Straub, R. E., Frech, G. C., Joho, R. H., and Gershengorn, M. C. (1990) Expression cloning of a cDNA encoding the mouse pituitary thyrotropin-releasing hormone receptor. *Proc. Natl. Acad. Sci. U.S.A.* 87, 9514–9518.

Straub, R. E. and Gershengorn, M. C. (1986) Thyrotropin-releasing hormone and GTP activate inositol trisphosphate formation in membranes isolated from rat pituitary cells. *J. Biol. Chem.* 261, 2712–2717.

Strewler, G. J., Stern, P. H., Jacobs, J. W., Eveloff, J., Klein, R. F., Leung, S. C., Rosenblatt, M., and Nissenson, R. A. (1987) Parathyroid hormonelike protein from human renal carcinoma cells. Structural and functional homology with parathyroid hormone. *J. Clin. Invest.* 80, 1803–1807.

Suva, L. J., Winslow, G. A., Wettenhall, R. E., Hammonds, R. G., Moseley, J. M., Diefenbach, J. H., Rodda, C. P., Kemp, B. E., Rodriguez, H., Chen, E. Y., et al. (1987) A parathyroid hormone-related protein implicated in malignant hypercalcemia: cloning and expression. *Science* 237, 893–896.

Takayanagi, R., Ohnaka, K., Sakai, Y., Nakao, R., Yanase, T., Haji, M., Inagami, T., Furuta, H., Gou, D. F., Nakamuta, M., et al. (1992) Molecular cloning, sequence analysis and expression of a cDNA encoding human type-1 angiotensin II receptor. *Biochem. Biophys. Res. Commun.* 183, 910–916.

Tatro, J. B. (1990) Melanotropin receptors in the brain are differentially distributed and recognize both corticotropin and alpha-melanocyte stimulating hormone. *Brain Res.* 536, 124–132.

Tatro, J. B., Atkins, M., Mier, J. W., Hardarson, S., Wolfe, H., Smith, T., Entwistle, M. L., and Reichlin, S. (1990a) Melanotropin receptors demonstrated in situ in human melanoma. *J. Clin. Invest.* 85, 1825–1832.

Tatro, J. B., Entwistle, M. L., Lester, B. R., and Reichlin, S. (1990b) Melanotropin receptors of murine melanoma characterized in cultured cells and demonstrated in experimental tumors in situ. *Cancer Res.* 50, 1237–1242.

Thorens, B. (1992) Expression Cloning of the Pancreatic beta-Cell Receptor for the Gluco-Incretin Hormone Glucagon-Like Peptide-1. *Proc. Natl. Acad. Sci. U.S.A.* 89, 8641–8645.

Tilly, J. L., LaPolt, P. S., and Hsueh, A. J. (1992) Hormonal regulation of follicle-stimulating hormone receptor messenger ribonucleic acid levels in cultured rat granulosa cells. *Endocrinology* 130, 1296–1302.

Timmermans, P. B., Wong, P. C., Chiu, A. T., and Herblin, W. F. (1991) Nonpeptide angiotensin II receptor antagonists. *Trends Pharmacol. Sci.* 12, 55–62.

Tran, V. T., Beal, M. F., and Martin, J. B. (1985) Two types of somatostatin receptors differentiated by cyclic somatostatin analogs. *Science* 228, 492–495.

Tsai-Morris, C. H., Buczko, E., Wang, W., and Dufau, M. L. (1990) Intronic nature of the rat luteinizing hormone receptor gene defines a soluble receptor subspecies with hormone binding activity. *J. Biol. Chem.* 265, 19385–19388.

Tsutsumi, M., Zhou, W., Millar, R. P., Mellon, P. L., Roberts, J. L., Flanagan, C. A., Dong, K., Gillo, B., and Sealfon, S. C. (1992) Cloning and Functional Expression of a Mouse Gonadotropin-Releasing Hormone Receptor. *Mol. Endocrinol.* 6, 1163–1169.

Van Sande, J., Raspe, E., Perret, J., Lejeune, C., Maenhaut, C., Vassart, G., and Dumont, J. E. (1990) Thyrotropin activates both the cyclic AMP and the PIP2 cascades in CHO cells expressing the human cDNA of TSH receptor. *Mol. Cell Endocrinol.* 74, R1–R6.

Wada, E., Way, J., Shapira, H., Kusano, K., Lebacq, V. A., Coy, D., Jensen, R., and Battery, J. (1991) cDNA cloning, characterization, and brain region-specific expression of a neuromedin-B-preferring bombesin receptor. *Neuron* 6, 421–430.

Wang, H., Nelson, S., Ascoli, M., and Segaloff, D. L. (1992) The 5′-flanking region of the rat luteinizing hormone/chorionic gonadotropin receptor gene confers Leydig cell expression and negative regulation of gene transcription by 3′,5′-cyclic adenosine monophosphate. *Mol. Endocrinol.* 6, 320–326.

Wang, H. L., Bogen, C., Reisine, T., and Dichter, M. (1989) Somatostatin-14 and somatostatin-28 induce opposite effects on potassium currents in rat neocortical neurons. *Proc. Natl. Acad. Sci. U.S.A.* 86, 9616–9620.

Wank, S. A., Harkins, R., Jensen, R. T., Shapira, H. A. D. W., and Slattery, T. (1992a) Purification, molecular cloning, and functional expression of the cholecystokinin receptor from rat pancreas. *Proc. Natl. Acad. Sci. U.S.A.* 89, 3125–3129.

Wank, S. A., Pisegna, J. R., and de Weerth, A. (1992b) Brain and Gastrointestinal Cholecystokinin Receptor Family — Structure and Functional Expression. *Proc. Natl. Acad. Sci. U.S.A.* 89, 8691–8695.

White, R. E., Schonbrunn, A., and Armstrong, D. L. (1991) Somatostatin stimulates Ca(2+)-activated K+ channels through protein dephosphorylation. *Nature* 351, 570–573.

Whitebread, S., Mele, M., Kamber, B., and De Gasparo, M. (1989) Preliminary biochemical characterization of two angiotensin II receptor subtypes. *Biochem. Biophys. Res. Commun.* 163, 284–291.

Windle, J. J., Weiner, R. I., and Mellon, P. L. (1990) Cell lines of the pituitary gonadotrope lineage derived by targeted oncogenesis in transgenic mice. *Mol. Endocrinol.* 4, 597–603.

Wong, P. C., Hart, S. D., Zaspel, A. M., Chiu, A. T., Ardecky, R. J., Smith, R. D., and Timmermans, P. B. (1990) Functional studies of nonpeptide angiotensin II receptor subtype-specific ligands: DuP 753 (AII-1) and PD123177 (AII-2). *J. Pharmacol. Exp. Ther.* 255, 584–592.

Xie, Y. B., Wang, H., and Segaloff, D. L. (1990) Extracellular domain of lutropin/choriogonadotropin receptor expressed in transfected cells binds choriogonadotropin with high affinity. *J. Biol. Chem.* 265, 21411–21414.

Yamada, Y., Post, S. R., Wang, K., Tager, H. S., Bell, G. I., and Seino, S. (1992) Cloning and functional characterization of a family of human and mouse somatostatin receptors expressed in brain, gastrointestinal tract, and kidney. *Proc. Natl. Acad. Sci. U.S.A.* 89, 251–255.

Yamaguchi, D. T., Hahn, T. J., Iida, K. A., Kleeman, C. R., and Muallem, S. (1987) Parathyroid hormone-activated calcium channels in an osteoblast-like clonal osteosarcoma cell line. cAMP-dependent and cAMP-independent calcium channels. *J. Biol. Chem.* 262, 7711–7718.

Yasuda, K., Rens-Domaino, S., Breder, C. D., Law, S. F., Saper, C. B., Reiseine, T., and Bell, G. I. (1992) Cloning of a novel somatostatin receptor, SSTR3, coupled to adenyl cyclase. *J. Biol. Chem.* 28, 20422–20428.

Young, D., O'Neill, K., Jessell, T., and Wigler, M. (1988) Characterization of the rat mas oncogene and its high-level expression in the hippocampus and cerebral cortex of rat brain. *Proc. Natl. Acad. Sci. U.S.A.* 85, 5339–5342.

Young, D., Waitches, G., Birchmeier, C., Fasano, O., and Wigler, M. (1986) Isolation and characterization of a new cellular oncogene encoding a protein with multiple potential transmembrane domains. *Cell* 45, 711–719.

Zaidi, M., Datta, H. K., Moonga, B. S., and MacIntyre, I. (1990) Evidence that the action of calcitonin on rat osteoclasts is mediated by two G proteins acting via separate post-receptor pathways. *J. Endocrinol.* 126, 473–481.

Zhao, D., Yang, J., Jones, K. E., Gerald, C., Suzuki, Y., Hogan, P. G., Chin, W. W., and Tashjian, A. J. (1992) Molecular cloning of a complementary deoxyribonucleic acid encoding the thyrotropin-releasing hormone receptor and regulation of its messenger ribonucleic acid in rat GH cells. *Endocrinology* 130, 3529–3536.

16

5-Hydroxytryptamine Receptors

Stephen J. Peroutka, M.D., Ph.D.

1.16.0 Introduction

5-Hydroxytryptamine (5-HT) receptors consist of at least 3 distinct types of molecular structures (Table 1): G protein-coupled receptors, ligand-gated ion channels (Volume 2, Chapter 16), and transporters (Volume 3, Chapter 16). Prior to the introduction of molecular biological techniques, the classification of 5-HT receptors was based predominantly on the pharmacological properties of the receptors. For example, 5-HT$_1$ receptors were defined as membrane binding sites which displayed nanomolar affinity for ^{3}H-5-HT (Peroutka and Snyder, 1979). Subsequently, Bradley et al. (1986) defined "5-HT$_1$-like" receptors by their susceptibility to antagonism by methiothepin and/or methysergide, resistance to antagonism by 5-HT$_2$ antagonists, and potent agonism by 5-carboxamidotryptamine (5-CT). Thus, these classification systems are dependent upon the availability of selective pharmacological agents.

More recently, molecular biological data have unequivocally confirmed the existence of multiple 5-HT receptors. Indeed, the multiplicity of 5-HT receptor subtypes, both within and between species, has exceeded most of the predictions that might have been made on the basis of pharmacological data. Within the group of G protein-coupled 5-HT receptors (Table 2), the evolutionary relationships between the known 5-HT receptor subtypes were determined by a phylogenetic tree analysis (Figure 1) (Peroutka, 1992). The aligned sequences of all identified G protein-coupled 5-HT receptors (Figure 2) were compared and a phylogenetic tree was constructed using the method of Feng and Doolittle (1990). The length of each branch (Figure 1) correlates with the evolutionary distance between receptor subpopulations. Thus, G protein-coupled 5-HT receptors have differentiated into 2 clearly discernible major branches. The low level of homology (approximately 25%) between the 2 major branches suggests that 5-HT$_1$ and 5-HT$_2$ receptors diverged from a common ancestor gene early in evolution, prior to the differentiation of vertebrates and invertebrates.

The 5-HT$_1$ receptor family or branch includes 5-HT$_{1A}$, 5-HT$_{1B}$, 5-HT$_{1D}$, 5-HT$_{1E}$, and 5-HT$_{1F}$ (or 5-HT$_{1E\beta}$) receptors, the 5-HT receptors identified in *drosophila* (Witz et al., 1990; Saudou et al., 1992) and mouse 5-HT$_5$ receptors (Plassat et al., 1992). The mouse 5-HT$_5$ receptor appears to have differentiated early in evolution. The mammalian 5-HT$_1$ receptors then differentiated from the *drosophila* 5-HT receptor branch of the phylogenetic tree. The next evolutionary differentiation occurs when 5-HT$_{1A}$

Table 1. Overview of 5-Hydroxytryptamine Receptors

G Protein-coupled receptors

5-HT$_1$ Family — 5-HT$_{1A}$, 5-HT$_{1B}$, 5-HT$_{1D}$, 5-HT$_{1E}$, 5-HT$_{1F}$, 5-HT$_{dro1}$, 5-HT$_{dro2A}$, 5-HT$_{dro2B}$, 5-HT$_5$

5-HT$_2$ Family — 5-HT$_2$, 5-HT$_{1C}$, 5-HT$_{2F}$

Ligand-gated ion channels

5-HT$_3$

Transporters

5-HT Uptake site

receptors branch from a receptor group which eventually evolved into 5-HT$_{1B}$, 5-HT$_{1D}$, 5-HT$_{1E}$, and 5-HT$_{1F}$ receptors. Finally, for all 5-HT receptors identified, the interspecies variation is minimal (i.e., >90% identity between species homologs) as indicated by the very short branches linking these subtypes in the phylogenetic tree.

The 5-HT$_2$ branch of G protein-coupled 5-HT receptors includes 5-HT$_2$, 5-HT$_{1C}$, and 5-HT$_{2F}$ receptors. These receptors share a significant number of molecular biological, pharmacological, and biochemical characteristics (Peroutka, 1990), as might have been predicted by the evolutionary similarity. For all identified members of the 5-HT$_2$ receptor family, the interspecies variation is minimal as indicated by the short branches linking these receptor subtypes in the phylogenetic tree (Figure 1).

Table 2. 5-Hydroxytryptamine

Receptor	Second messenger	Species cloned	Chromosomal location	a.a. Sequence	Accession number	Primary reference
5-HT$_{1A}$	–AC	Human	5q11.2-q13	421	P08908	Kobilka et al., 1987
		Rat 1		422	P19327	Albert et al., 1990
		Rat 2		422	—	Fujiwara et al., 1991
		Mouse	Distal 13	—	—	Oakey et al., 1991
5-HT$_{1B}$	–AC	Human	6q13	390	M89478	Jin et al., 1992
		Rat		386	M89954	Voigt et al., 1991
		Mouse		386	M85151	Maroteaux et al., 1992
5-HT$_{1D}$	–AC	Human	1p34.3-36.3	377	R15137	Hamblin et al., 1991
		Canine		377	P11614	Maenhaut et al., 1991
		Guinea pig		—	—	Weydert et al., 1992
		Rat		374	M89953	Hamblin et al., 1991
		Mouse		—	—	Weydert et al., 1992
5-HT$_{1E}$	–AC	Human		365	Z11166	Levy et al., 1992
5-HT$_{1F}$	–AC	Human		366	L05597	Lovenberg et al., 1992
		Rat		366	L05596	Lovenberg et al., 1992
		Mouse		366	S46903	Amlaiky et al., 1992
5-HT$_1$		Snail		509	???GEN8947	Sugamori et al., 1993
5-HT$_{dro1}$	+AC	Drosophila 3(100 A)		564	P20905	Witz et al., 1990
5-HT$_{dro2A}$	–AC	Drosophila 2(56 A-B)		834	S18153	Saudou et al., 1992
5-HT$_{dro2B}$	–AC	Drosophila 2(56 A-B)		645	S18154	Saudou et al., 1992
5-HT$_5$	—	Mouse	—	357	Z18278	Plassat et al., 1992
5-HT$_2$	+PI	Human	13q14-q21	471	JS0615	Yang et al., 1991
		Rat	—	471	P14842	Pritchett et al., 1988
		Hamster	—	471	P18599	Chambard et al., 1990
		Mouse	14	471	—	Yang et al., 1991
5-HT$_{1C}$	+PI	Human	—'	458	JS0616	Saltzman et al., 1991
		Rat	—	460	P08909	Julius et al., 1988
		Mouse	X	459	—	Yu et al., 1991
5-HT$_{2F}$	+PI	Rat	—	479	X66842	Kursar et al., 1992

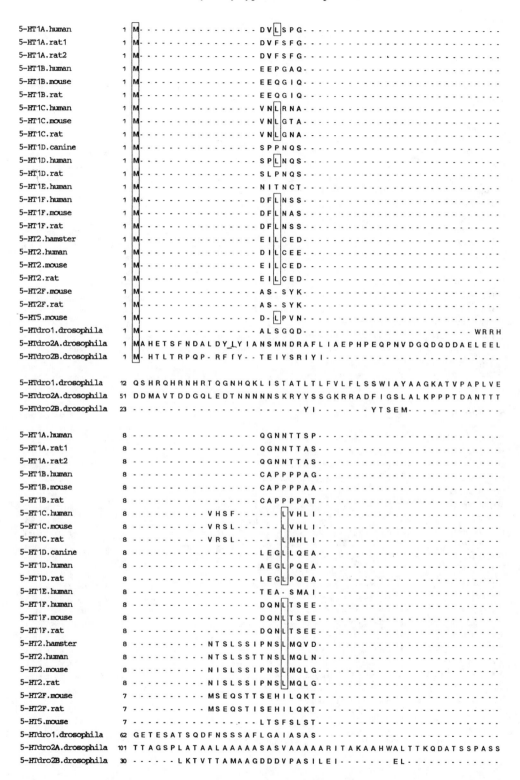

FIGURE 1. Alignment of the amino acid sequences of 5-hydroxytryptamine receptors. The sequences were obtained from commercial protein databases such as GenBank and EMBL. The references for each sequence are cited in the text.

```
5-HT1A.human         16 - - - - - - - P A P F E T G G N T - - - - - - - - - - - - - - - - - - - - - - - T G I S D V T V S
5-HT1A.rat1          16 - - - - - - - Q E P F G T G G N V - - - - - - - - - - - - - - - - - - - - - - - T S I S D V T F S
5-HT1A.rat2          16 - - - - - - - Q E P F G T G G N V - - - - - - - - - - - - - - - - - - - - - - - T S I S D V T F S
5-HT1B.human         16 - - - - - - - S E T W V P Q A N L S S A P - - - - - - - - - S Q N C S A K D Y I Y Q D S I S L P
5-HT1B.mouse         16 - - - - - - - S Q T G V P L T N L - - - - - - - - - - S H N C S A D G Y I Y Q D S I A L P
5-HT1B.rat           16 - - - - - - - S Q T G V P L A N L - - - - - - - - - - S H N C S A D D Y I Y Q D S I A L P
5-HT1C.human         17 - - - - - - - G - - - L L V W Q C D I S V - - - - - - - - - S P V A A I V T D I F N T S D G G R
5-HT1C.mouse         17 - - - - - - - G - - - L L V W Q F D I S I - - - - - - - - - S P V A A I V T D T F N S S D G G R
5-HT1C.rat           17 - - - - - - - G - - - L L V W Q F D I S I - - - - - - - - - S P V A A I V T D T F N S S D G G R
5-HT1D.canine        16 - - - - - - - S N R S L N A T E - - - - - - - - - - - - - - T P E A W G P E T L Q A L -
5-HT1D.human         16 - - - - - - - S N R S L N A T E - - - - - - - - - - - - - - T S E A W D P R T L Q A L -
5-HT1D.rat           16 - - - - - - - S N R S L N A T - - - - - - - - - - - - - - - - G A W D P E V L Q A L -
5-HT1E.human         15 - - - - - - - R P K T I T E - - - - - - - - - - - - - - - - - - - - - - - - - - - - -
5-HT1F.human         16 - - - - - - - L L N R M P S - - - - - - - - - - - - - - - - - - - - - - - - - - - - -
5-HT1F.mouse         16 - - - - - - - L L N R M P S - - - - - - - - - - - - - - - - - - - - - - - - - - - - -
5-HT1F.rat           16 - - - - - - - L L N R M P S - - - - - - - - - - - - - - - - - - - - - - - - - - - - -
5-HT2.hamster        23 - - - - - - G D S G L Y R N D F N S R D - - - - - - - - - - A N S S D A S N W T I D G E N R T N
5-HT2.human          23 - - - - - - D D T R L Y S N D F N S G E - - - - - - - - - - A N T S D A F N W T V D S E N R T N
5-HT2.mouse          23 - - - - - - D D S R L Y P N D F N S R D - - - - - - - - - - A N T S E A S N W T I D A E N R T N
5-HT2.rat            23 - - - - - - D G P R L Y H N D F N S R D - - - - - - - - - - A N T S E A S N W T I D A E N R T N
5-HT2F.mouse         22 - - - - - - C D H L I L T N R S G L E T - - - - - - - - - D S V A E E M K Q T V E G Q G H T -
5-HT2F.rat           22 - - - - - - C D H L I L T D R S G L K A - - - - - - - - - E S A A E E M K Q T A E N Q G N T -
5-HT5.mouse          15 - - - - - - P S S L E P N R S L - - - - - - - - - - - - - - - D T E V L R P S R P F L S A
5-HTdro1.drosophila  87 - - - - - - S T G S G S G S G S G S G S - - - - - - - - - - G S G S G S Y G L A S M N S S P I A
5-HTdro2A.drosophila 151 P A L Q L I D M D N N Y T N V A V G L G A M L L N D T L L L E G N D S S L F G E M L A N R S G H L D
5-HTdro2B.drosophila 54 P A I L L - - N E S L F I E L N G N L T Q L V - - - - - - - - D T T S N L S Q I V W N R S V N G N
```

```
5-HT1A.human         35 Y - - - - - - - - - - - - - - - - - - - - - - - - - - - - - - - - - - - - - Q V I T S L L
5-HT1A.rat1          35 Y - - - - - - - - - - - - - - - - - - - - - - - - - - - - - - - - - - - - - Q V I T S L L
5-HT1A.rat2          35 Y - - - - - - - - - - - - - - - - - - - - - - - - - - - - - - - - - - - - - Q V I T S L L
5-HT1B.human         48 W - - - - - - - - - - - - - - - - - - - - - - - - - - - - - - - - - - - - - K V L L V M L
5-HT1B.mouse         44 W - - - - - - - - - - - - - - - - - - - - - - - - - - - - - - - - - - - - - K V L L V A L
5-HT1B.rat           44 W - - - - - - - - - - - - - - - - - - - - - - - - - - - - - - - - - - - - - K V L L V A L
5-HT1C.human         46 - - - - - F K F P D G V - - - - - - - - - - - - - - - - - - - - - - - - - - Q N W P A L S
5-HT1C.mouse         46 L - - - - F Q F P D G V - - - - - - - - - - - - - - - - - - - - - - - - - - Q N W P A L S
5-HT1C.rat           46 L - - - - F Q F P D G V - - - - - - - - - - - - - - - - - - - - - - - - - - Q N W P A L S
5-HT1D.canine        38 - - - - - - - - - - - - - - - - - - - - - - - - - - - - - - - - - - - - - - K I S L A L L
5-HT1D.human         38 - - - - - - - - - - - - - - - - - - - - - - - - - - - - - - - - - - - - - - K I S L A V V
5-HT1D.rat           35 - - - - - - - - - - - - - - - - - - - - - - - - - - - - - - - - - - - - - - R I S L V V V
5-HT1E.human         22 - - - - - - - - - - - - - - - - - - - - - - - - - - - - - - - - - - - - - - K M L I C M T
5-HT1F.human         23 - - - - - - - - - - - - - - - - - - - - - - - - - - - - - - - - - - - - - - K I L V S L T
5-HT1F.mouse         23 - - - - - - - - - - - - - - - - - - - - - - - - - - - - - - - - - - - - - - K I L V S L T
5-HT1F.rat           23 - - - - - - - - - - - - - - - - - - - - - - - - - - - - - - - - - - - - - - K I L V S L T
5-HT2.hamster        55 L S F E G Y L P P T C L - - - - - - - - - - - - - - - - S I L H L Q E K N W S A L L
5-HT2.human          55 L S C E G C L S P S C L - - - - - - - - - - - - - - - - S L L H L Q E K N W S A L L
5-HT2.mouse          55 L S C E G Y L P P T C L - - - - - - - - - - - - - - - - S I L H L Q E K N W S A L L
5-HT2.rat            55 L S C E G Y L P P T C L - - - - - - - - - - - - - - - - S I L H L Q E K N W S A L L
5-HT2F.mouse         53 - - - - - - - - - - - - - - - - - - - - - - - - - - - - - - - - - - - - V H W A A L L
5-HT2F.rat           53 - - - - - - - - - - - - - - - - - - - - - - - - - - - - - - - - - - - - V H W A A L L
5-HT5.mouse          39 F - - - - - - - - - - - - - - - - - - - - - - - - - - - - - - - - - - - - - R V L V L T L
5-HTdro1.drosophila  119 I V S Y Q G I T S S N L G D S N T T L V P L S D T P L L L E E F A A G E F V L P P L T S I F V S I V
5-HTdro2A.drosophila 201 L I N G T G G L N V T T - - - - - - - - - - - - - - - - - S K V A E D D F T Q L L R M A V T S V L
5-HTdro2B.drosophila 93 G N S N T F D L V D D E - - - - - - - - - - - - - - - - - Q E R A A V E F W L L V K M I A M A V V
```

FIGURE 1b.

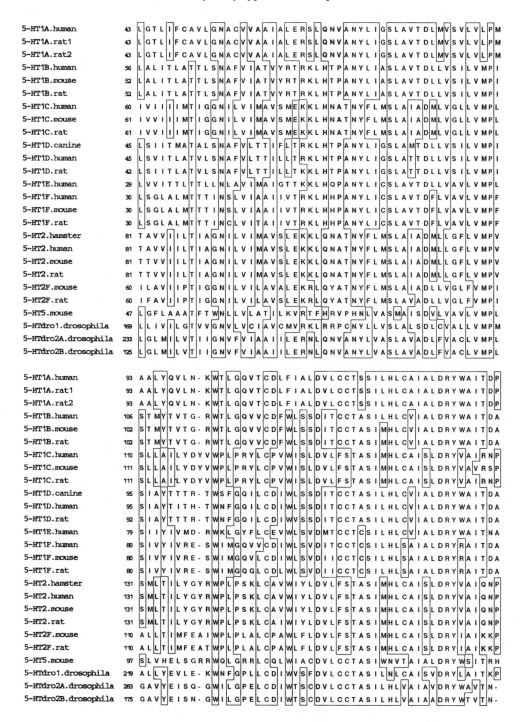

FIGURE 1c.

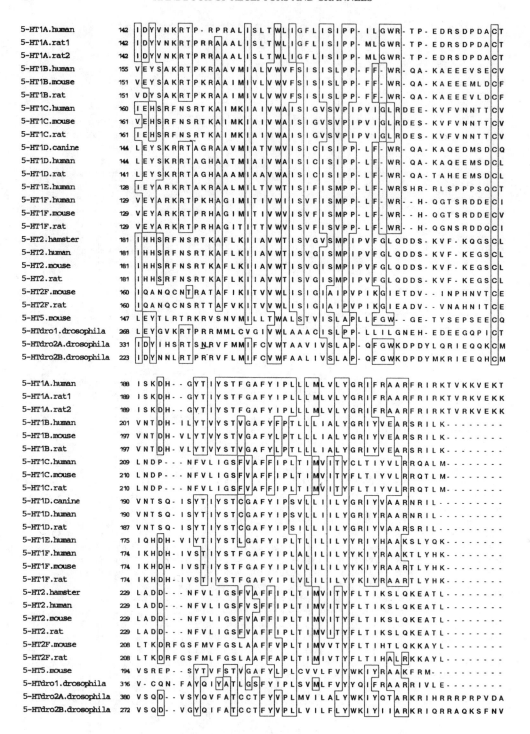

FIGURE 1d.

```
5-HT1A.human           236  G A D T R - - - - - - - - - - - - - - - - - - - - - - - - - - - - - - - - - - - - - - - - - - - - -
5-HT1A.rat1            237  G A G T S - - - - - - - - - - - - - - - - - - - - - - - - - - - - - - - - - - - - - - - - - - - - -
5-HT1A.rat2            237  G A G T S - - - - - - - - - - - - - - - - - - - - - - - - - - - - - - - - - - - - - - - - - - - - -
5-HTdro2A.drosophila   428  A V N N N Q P D G G A A T D T K L H R L R L R L G R F S T A K S K T G S A V G V S G P A S G G R A L
5-HTdro2B.drosophila   320  T L T E T D C D - S A V R E L K K E R S K R R A E R - - - K R L E A G E R T P V D G D G M G G Q - -


5-HTdro2A.drosophila   478  G L V D G N S T N T V N T V E D T E F S S S N V D S K S R A G V E A P S T S G N Q I A T V S H L V A
5-HTdro2B.drosophila   364  - - L Q R R T R K R M R I C - - - - F G R N T N T A N V V A G S E G A V A R S M A A I A V - D F A S


5-HT1A.human           241  - - - - - - - - - - - - - - - - - - - - - - - - - - - - - - - - - - - - - - - - - - H G A S P A P Q
5-HT1A.rat1            242  - - - - - - - - - - - - - - - - - - - - - - - - - - - - - - - - - - - - - - - - - - L G T S S A P P
5-HT1A.rat2            242  - - - - - - - - - - - - - - - - - - - - - - - - - - - - - - - - - - - - - - - - - - L G T S S A P P
5-HT1B.human           242  - - - - - - - - - - - - - - - - - - - - - - - - - - - - - - - - - - - - - - - - - - Q T P N R T
5-HT1B.mouse           238  - - - - - - - - - - - - - - - - - - - - - - - - - - - - - - - - - - - - - - - - - - Q T P N K T
5-HT1B.rat             238  - - - - - - - - - - - - - - - - - - - - - - - - - - - - - - - - - - - - - - - - - - Q T P N K T
5-HT1C.human           248  - - - - - - - - - - - - - - - - - - - - - - - - - - - - - - - - - - - - - - - - - - L L H G H T
5-HT1C.mouse           249  - - - - - - - - - - - - - - - - - - - - - - - - - - - - - - - - - - - - - - - - - - L L R G H T
5-HT1C.rat             249  - - - - - - - - - - - - - - - - - - - - - - - - - - - - - - - - - - - - - - - - - - L L R G H T
5-HT1D.canine          230  - - - - - - - - - - - - - - - - - - - - - - - - - - - - - - - - - - - - - - - - - - N P P S L Y
5-HT1D.human           230  - - - - - - - - - - - - - - - - - - - - - - - - - - - - - - - - - - - - - - - - - - N P P S L Y
5-HT1D.rat             227  - - - - - - - - - - - - - - - - - - - - - - - - - - - - - - - - - - - - - - - - - - N P P S L Y
5-HT1E.human           216  - - - - - - - - - - - - - - - - - - - - - - - - - - - - - - - - - - - - - - - - - - R G S S R H
5-HT1F.human           215  - - - - - - - - - - - - - - - - - - - - - - - - - - - - - - - - - - - - - - - - - - R Q A S R I
5-HT1F.mouse           215  - - - - - - - - - - - - - - - - - - - - - - - - - - - - - - - - - - - - - - - - - - R Q A S R M
5-HT1F.rat             215  - - - - - - - - - - - - - - - - - - - - - - - - - - - - - - - - - - - - - - - - - - R Q A S R M
5-HT2.hamster          268  - - - - - - - - - - - - - - - - - - - - - - - - - - - - - - - - - - - - - - - - - - C V S D L S
5-HT2.human            268  - - - - - - - - - - - - - - - - - - - - - - - - - - - - - - - - - - - - - - - - - - C V S D L G
5-HT2.mouse            268  - - - - - - - - - - - - - - - - - - - - - - - - - - - - - - - - - - - - - - - - - - C V S D L S
5-HT2.rat              268  - - - - - - - - - - - - - - - - - - - - - - - - - - - - - - - - - - - - - - - - - - C V S D L S
5-HT2F.mouse           250  - - - - - - - - - - - - - - - - - - - - - - - - - - - - - - - - - - - - - - - - - - V K N K P P
5-HT2F.rat             250  - - - - - - - - - - - - - - - - - - - - - - - - - - - - - - - - - - - - - - - - - - V R N R P P
5-HTdro1.drosophila    356  - - - - - - - - - - - - - - - - - - - - - - - - - - - - - - - - - - - - - - - - - - E K R A Q T H
5-HTdro2A.drosophila   528  L A K Q Q G K S T A K S S A A V N G M A P S G R Q E D D G Q R P E H G E Q E D R E E L E D Q D E Q V
5-HTdro2B.drosophila   407  L A - - - - - - - - - - - - - - - - I T R E E T E F S T S N Y - - - - - - - - - D N K S H A


5-HT1A.human           249  P K K S V N G E - - - - - - - - - - - - - - - - - - - - - - - - - - - S G S R N W R L G V E S K A
5-HT1A.rat1            250  P K K S L N G Q - - - - - - - - - - - - - - - - - - - - - - - - - - - P G S G D W R R C A E N R A
5-HT1A.rat2            250  P K K S L N G Q - - - - - - - - - - - - - - - - - - - - - - - - - - - P G S G D W R R C A E N R A
5-HT1B.human           248  G K R L T R A Q - - - - - - - - - - - - - - - - - - - - - - - L - - - - - - - - - - - - - - - -
5-HT1B.mouse           244  G K R L T R A Q - - - - - - - - - - - - - - - - - - - - - - - L - - - - - - - - - - - - - - - -
5-HT1B.rat             244  G K R L T R A Q - - - - - - - - - - - - - - - - - - - - - - - L - - - - - - - - - - - - - - - -
5-HT1C.human           254  E E P - P G L S - - - - - - - - - - - - - - - - - - - - - - - L - - - - - - - - - - - - - - - -
5-HT1C.mouse           255  E E E L R N I S - - - - - - - - - - - - - - - - - - - - - - - L - - - - - - - - - - - - - - - -
5-HT1C.rat             255  E E E L A N M S - - - - - - - - - - - - - - - - - - - - - - - L - - - - - - - - - - - - - - - -
5-HT1D.canine          236  G K R F T T A Q - - - - - - - - - - - - - - - - - - - - - - - L - - - - - - - - - - - - - - - -
5-HT1D.human           236  G K R F T T A H - - - - - - - - - - - - - - - - - - - - - - - L - - - - - - - - - - - - - - - -
5-HT1D.rat             233  G K R F T T A Q - - - - - - - - - - - - - - - - - - - - - - - L - - - - - - - - - - - - - - - -
5-HT1E.human           222  L S N R S T D S - - - - - - - - - - - - - - - - - - - - - - - - - - - - - - - - - - - - - - - -
5-HT1F.human           221  A K E E V N G Q - - - - - - - - - - - - - - - - - - - - - - - V - - - - - - - - - - - - - - - -
5-HT1F.mouse           221  I K E E L N G Q - - - - - - - - - - - - - - - - - - - - - - - V - - - - - - - - - - - - - - - -
5-HT1F.rat             221  I K E E L N G Q - - - - - - - - - - - - - - - - - - - - - - - V - - - - - - - - - - - - - - - -
5-HT2.hamster          274  T R A - K L A S - - - - - - - - - - - - - - - - - - - - - - - F - - - - - - - - - - - - - - - -
5-HT2.human            274  T R A - K L A S - - - - - - - - - - - - - - - - - - - - - - - F - - - - - - - - - - - - - - - -
5-HT2.mouse            274  T R A - K L S S - - - - - - - - - - - - - - - - - - - - - - - F - - - - - - - - - - - - - - - -
5-HT2.rat              274  T R A - K L A S - - - - - - - - - - - - - - - - - - - - - - - F - - - - - - - - - - - - - - - -
5-HT2F.mouse           256  Q R L T R W T V - - - - - - - - - - - - - - - - - - - - - - - P - - - - - - - - - - - - - - - -
5-HT2F.rat             256  Q R L T R W T V - - - - - - - - - - - - - - - - - - - - - - - S - - - - - - - - - - - - - - - -
5-HT5.mouse            232  G S R K T N S - - - - - - - - - - - - - - - - - - - - - - - - - - - - - - - - - - - - - - - - -
5-HTdro1.drosophila    363  L Q Q A L N G T - - - - - - - - - - - - - - - - - - - - - - - G S P S A P Q A P P L G H T
5-HTdro2A.drosophila   578  G P Q P T T A T S A T T A A G T N E S E D Q C K A N G V E V L E D P Q L Q Q Q L E Q V Q Q L Q K S V
5-HTdro2B.drosophila   428  G T E L T T V S S - - - - - - - - D A D D Y R T S N A N E I I T - - - V S Q Q V A H A T Q - H H L I
```

FIGURE 1e.

```
5-HT1A.human          271  GGALCANGAVRQGDDGAALEVIEVHRVGNSKEHLPLPSEAGPTPCAP---
5-HT1A.rat1           272  VGTPCTNGAVRQGDDEATLEVIEVHRVGNSKEHLPLPSESGSNSYAP---
5-HT1A.rat2           272  VGTPCTNGAVRQGDDEATLEVIEVHRVGNSKEHLPLPSESGSNSYAP---
5-HT1B.human          257  ----ITDSPGSTSSVTSINSRVPDVP SES-GS-PVYVNQVKVRV SD---
5-HT1B.mouse          253  ----ITDSPGSTSSVTSINSRAPDVP SES-GS-PVYVNQVKVRV SD---
5-HT1B.rat            253  ----ITDSPGSTSSVTSINSRVPEVP SES-GS-PVYVNQVKVRV SD---
5-HT1C.human          262  ----DFLK-CCKRNTAEE----ENSANPNQDQNARR---RKKKER---
5-HT1C.mouse          264  -----NFLKCCCKK-GDEE----ENAPNPNPDQKP-R---RKKKEK---
5-HT1C.rat            264  -----NFLNCCCKKNGGEE----ENAPNPNPDQKP-R---RKKKEK---
5-HT1D.canine         245  -----ITGSAG--SSLCSLSPSLQEER SHAAGP-PLFFNHVQVKLAE---
5-HT1D.human          245  -----ITGSAG--SSLCSLNSSLHEQH SHSAGS-PLFFNHVKIKLAD---
5-HT1D.rat            242  -----ITGSAG--SSLCSLNPSLHESHTHTVGS-PLFFNQVKIKLAD---
5-HT1E.human          230  ------QNSFASCKLTQTFCVSDFST SDPTTEFEKFHASIRIPPFD---
5-HT1F.human          230  -----LLESGEKSTKSVSTSYVLEKSL SDPSTDFDKIHSTVRSLR SE---
5-HT1F.mouse          230  -----FLESGEKSIKLVSTSYMLEKSL SDPSTDFDRIHSTVKSPR SE---
5-HT1F.rat            230  -----LLESGEKSIKLVSTSYMLEKSL SDPSTDFDRIHSTVKSPR SE---
5-HT2.hamster         282  -----SFL--------PQ----- SSL SSEKLFQRSIH---REPG SY---
5-HT2.human           282  -----SFL--------PQ----- SSL SSEKLFQRSIH---REPG SY---
5-HT2.mouse           282  -----SFL--------PQ----- SSL SSEKLFQRSIH---RERG SY---
5-HT2.rat             282  -----SFL--------PQ----- SSL SSEKLFQRSIH---REPG SY---
5-HT2F.mouse          265  -----TVFLREDSSFSSPEKVAMLDGSHRDKILPNSSDETLMRRM SS---
5-HT2F.rat            265  -----TVLQREDSSFSSPEKMVMLDGSHKDKILPNSTDETLMRRM SS---
5-HT5.mouse           239  -----------VSPVPEAVEVKNRTQHPQMVFTVRHATVTFQTEG---
5-HTdro1.drosophila   385  ELASSGNGQRHSSVGNTSLTYSTCGGL SSGGGALAGHGSGGGVSG ST---
5-HTdro2A.drosophila  628  KSGGGGGASTSNATTITSISALSPQTPTSQGVGIAAAAAGPMTAKTSTLT
5-HTdro2B.drosophila  466  ASHLNAITPLAQSIAMGGVGCLTTTTP SEKALSGAGTVAGAVAGG SGS--


5-HT1A.human          318  ..................................................ASF
5-HT1A.rat1           319  ..................................................ACL
5-HT1A.rat2           319  ..................................................ACL
5-HT1B.human          297  ..................................................ALL
5-HT1B.mouse          293  ..................................................ALL
5-HT1B.rat            293  ..................................................ALL
5-HT1C.human          295  ..................................................RPR
5-HT1C.mouse          296  ..................................................RPR
5-HT1C.rat            297  ..................................................RPR
5-HT1D.canine         284  ..................................................GVL
5-HT1D.human          284  ..................................................SAL
5-HT1D.rat            281  ..................................................SIL
5-HT1E.human          270  ..................................................NDL
5-HT1F.human          272  ..................................................FKH
5-HT1F.mouse          272  ..................................................LKH
5-HT1F.rat            272  ..................................................LKH
5-HT2.hamster         307  ..................................................TGR
5-HT2.human           307  ..................................................TGR
5-HT2.mouse           307  ..................................................AGR
5-HT2.rat             307  ..................................................AGR
5-HT2F.mouse          307  ..................................................VGK
5-HT2F.rat            307  ..................................................AGK
5-HTdro1.drosophila   432  ..................................................GLL
5-HTdro2A.drosophila  678  SCNQSHPLCGTANESPSTPEPRSRQPTTPQQQPHQQAHQQQQQQQQLSSI
5-HTdro2B.drosophila  514  ..........GSGEEGAGT----------------EGKNAGVGLGGVLASI
```

FIGURE 1f.

```
5-HT1A.human          321  ERKNERNAEAKRKMALARERKTVKTLGIIMGTFILCWLPFFIVALVLPFC
5-HT1A.rat1           322  ERKNERNAEAKRKMALARERKTVKTLGIIMGTFILCWLPFFIVALVLPFC
5-HT1A.rat2           322  ERKNERNAEAKRKMALARERKTVKTLGIIMGTFILCWLPFFIVALVLPFC
5-HT1B.human          300  E----K-----KKLMAARERKATKTLGIILGAFIVCWLPFFIISLVMPIC
5-HT1B.mouse          296  E----K-----KKLMAARERKATKTLGIILGAFIVCWLPFFIISLVMPIC
5-HT1B.rat            296  E----K-----KKLMAARERKATKTLGIILGAFIVCWLPFFIISLVMPIC
5-HT1C.human          298  G---------TMQAINNERKASKVLGIVFFVFLIMWCPFFITNILSVLC
5-HT1C.mouse          299  G---------TMQAINNEKKASKVLGIVFFVFLIMWCPFFITNILSVLC
5-HT1C.rat            300  G---------TMQAINNEKKASKVLGIVFFVFLIMWCPFFITNILSVLC
5-HT1D.canine         287  E----R-----KRISAARERKATKTLGIILGAFIVCWLPFFVASLVLPIC
5-HT1D.human          287  E----R-----KRISAARERKATKILGIILGAFIVIWLPFFVVSLVLPIC
5-HT1D.rat            284  E----R-----KRISAARERKATKTLGIILGAFIICWLPFFVVSLVLPIC
5-HT1E.human          273  DHPGER-----QQISSTRERKAARILGLILGAFILSWLPFFIKELIVGL-
5-HT1F.human          275  EKSWRR-----QKISGTRERKAATTLGLILGAFIVCWLPFFVKELVVNVC
5-HT1F.mouse          275  EKSWRR-----QKISGTRERKAATTLGLILGAFIVCWLPFFVKELVVNVC
5-HT1F.rat            275  EKSWRR-----QKISGTRERKAATTLGLILGAFIVCWLPFFVKELVVNIC
5-HT2.hamster         310  R---------TMQSISNEQKACKVLGIVFFLFVVMWCPFFITNIMAVIC
5-HT2.human           310  R---------TMQSISNEQKACKVLGIVFFLFVVMWCPFFITNIMAVIC
5-HT2.mouse           310  R---------TMQSISNEQKACKGLGIVFFLFVVMWCPFFITNIMAVIC
5-HT2.rat             310  R---------TMQSISNEQKACKVLGIVFFLFVVMWCPFFITNIMAVIC
5-HT2F.mouse          310  R---------SAQTISNEQRASKALGVVFFLFLLMWCPFFITNLTLALC
5-HT2F.rat            310  K---------PAQTISNEQRASKVLGIVFLFFLLMWCPFFITNVTLALC
5-HT5.mouse           273  ----------DTWREQKEQRAALMVGILIGVFVLCWFPFFVTELISPLC
5-HTdro1.drosophila   435  GSPHHK----KLRFQLAKEKKASTTLGIIMSAFTVCWLPFFILALIRPF-
5-HTdro2A.drosophila  728  ANPMQKVNKRKETLEAKRERKAAKTLAIITGAFVVCWLPFFVMALTMPLC
5-HTdro2B.drosophila  539  ANPHQKLAKRRQLLEAKRERKAAQTLAIITGAFVICWLPFFVMALTMSLC


5-HT1A.human          371  ESSC--HMPTLLGAIINWLGYSNSLLNPVIYAYFNKDFQNAFKKIIKCNF
5-HT1A.rat1           372  ESSC--HMPALLGAIINWLGYSNSLLNPVIYAYFNKDFQNAFKKIIKCKF
5-HT1A.rat2           372  ENSC--HMPALLGAIINWLGYSNSLLNPVIYAYFNKDFQNAFKKIIKCKF
5-HT1B.human          341  KDAC--WFHLAIFDFFTWLGYLNSLINPIIYTMSNEDFKQAFHKLIRFKC
5-HT1B.mouse          337  KDAC--WFHMAIFDFFNWLGYLNSLINPIIYTMSNEDFKQAFHKLIRFKC
5-HT1B.rat            337  KDAC--WFHMAIFDFFNWLGYLNSLINPIIYTMSNEDFKQAFHKLIRFKC
5-HT1C.human          338  EKSCNQKLMEKLLNVFVWIGYVCSGINPLVYTLFNKIYRRAFSNYLRCNY
5-HT1C.mouse          339  GKACNQKLMEKLLNVFVWIGYVCSGINPLVYTLFNKIYRRAFSKYLRCDY
5-HT1C.rat            340  GKACNQKLMEKLLNVFVWIGYVCSGINPLVYTLFNKIYRRAFSKYLRCDY
5-HT1D.canine         328  RASC--WLHPALFDFFTWLGYLNSLINPIIYTVFNEEFRQAFQRVVHVRK
5-HT1D.human          328  RDSC--WIHPALFDFFTWLGYLNSLINPIIYTVFNEEFRQAFQKIVPFRK
5-HT1D.rat            325  RDSC--WIHPALFDFFTWLGYLNSLINPVIYTVFNEDFRQAFQRVVHFRK
5-HT1E.human          317  -SIY--TVSSEVADFLTWLGYVNSLINPLLYTSFNEDFKLAFKKLIRCRE
5-HT1F.human          320  -DKC--KISEEMSNFLAWLGYLNSLINPLIYTIFNEDFKKAFQKLVRCRC
5-HT1F.mouse          320  -EKC--KISEEMSNFLAWLGYLNSLINPLIYTIFNEDFKKAFQKLVRCRY
5-HT1F.rat            320  -EKC--KISEEMSNFLAWLGYLNSLINPLIYTIFNEDFKKAFQKLVRCRN
5-HT2.hamster         350  KESCNEHVIGALLNVFVWIGYLSSAVNPLVYTLFNKTYRSAFSRYIQCQY
5-HT2.human           350  KESCNEDVIGALLNVFVWIGYLSSAVNPLVYTLFNKTYRSAFSRYIQCQY
5-HT2.mouse           350  KESCNENVIGALLNVFVWIGYLSSAVNPLVYTLFNKTYRSAFSRYIQCQY
5-HT2.rat             350  KESCNENVIGALLNVFVWIGYLSSAVNPLVYTLFNKTYRSAFSRYIQCQY
5-HT2F.mouse          350  -DSCNQTTLKTLLEIFVWIGYVSSGVNPLIYTLFNKTFREAFGRYITCNY
5-HT2F.rat            350  -DSCNQTTLKTLLQIFVWVGYVSSGVNPLIYTLFNKTFREAFGRYITCNY
5-HT5.mouse           312  SWD----VPAIWKSIFLWLGYSNSFFNPLIYTAFNRSYSSAFKVFFSKQQ
5-HTdro1.drosophila   480  -ETM--HVPASLSSLFLWLGYANSLLNPIIYATLNRDFRKPFQEILYFRC
5-HTdro2A.drosophila  778  -AAC--QISDSVASLFLWLGYFNSTLNPVIYTIFSPEFRQAFKRILFGGH
5-HTdro2B.drosophila  589  -KEC--EIHTAVASLFLWLGYFNSTLNPVIYTIFNPEFRRAFKRILFGRK
```

FIGURE 1g.

```
5-HT1A.human          419  C R Q - - - - - - - - - - - - - - - - - - - - - - - - - - - - - - - - - - - - - - - - - - - - -
5-HT1A.rat1           420  C R R - - - - - - - - - - - - - - - - - - - - - - - - - - - - - - - - - - - - - - - - - - - - -
5-HT1A.rat2           420  C R R - - - - - - - - - - - - - - - - - - - - - - - - - - - - - - - - - - - - - - - - - - - - -
5-HT1B.human          389  T S - - - - - - - - - - - - - - - - - - - - - - - - - - - - - - - - - - - - - - - - - - - - - -
5-HT1B.mouse          385  A G - - - - - - - - - - - - - - - - - - - - - - - - - - - - - - - - - - - - - - - - - - - - - -
5-HT1B.rat            385  T G - - - - - - - - - - - - - - - - - - - - - - - - - - - - - - - - - - - - - - - - - - - - - -
5-HT1C.human          388  K V E K K P - P V R Q I P R V A A T A - - - - - - - - - L S G R E L N V N I Y R H T N E P V I E K A
5-HT1C.mouse          389  K P D K K P - P V R Q I P R V A A T A - - - - - - - - - L S G R E L N V N I Y R H T N E R V V R K A
5-HT1C.rat            390  K P D K K P - P V R Q I P R V A A T A - - - - - - - - - L S G R E L N V N I Y R H T N E R V A R K A
5-HT1D.canine         376  A S - - - - - - - - - - - - - - - - - - - - - - - - - - - - - - - - - - - - - - - - - - - - - -
5-HT1D.human          376  A S - - - - - - - - - - - - - - - - - - - - - - - - - - - - - - - - - - - - - - - - - - - - - -
5-HT1D.rat            373  A S - - - - - - - - - - - - - - - - - - - - - - - - - - - - - - - - - - - - - - - - - - - - - -
5-HT1E.human          364  H T - - - - - - - - - - - - - - - - - - - - - - - - - - - - - - - - - - - - - - - - - - - - - -
5-HT2.hamster         400  K E N R K P L Q L I L V N T I P A L A - - - - - - - - Y K S S Q L Q A G Q N K D S K E D A E P T D
5-HT2.human           400  K E N K K P L Q L I L V N T I P A L A - - - - - - - - Y K S S Q L Q M G Q K K N S K Q D A K T T D
5-HT2.mouse           400  K E N R K P L Q L I L V N T I P T L A - - - - - - - - Y K S S Q L Q V G Q K K N S Q E D A E P T A
5-HT2.rat             400  K E N R K P L Q L I L V N T I P A L A - - - - - - - - Y K S S Q L Q V G Q K K N S Q E D A E Q T V
5-HT2F.mouse          399  R A T K S V K A L R K F S S T L C F G N S M V E N S K F F T K H G I R N G I N P A M Y Q S P M R L R
5-HT2F.rat            399  Q A T K S V K V L R K C S S T L Y F G N S M V E N S K F F T K H G I R N G I N P A M Y Q S P V R L R
5-HTdro1.drosophila   527  S S L N - - - - - - - - - - - - - - - - - - - - - - - - - - - T M M R E N Y Y Q D Q Y G E
5-HTdro2A.drosophila  825  R P V H Y R S G K L - - - - - - - - - - - - - - - - - - - - - - - - - - - - - - - - - - - - - - -
5-HTdro2B.drosophila  636  A A A R A R S A K I - - - - - - - - - - - - - - - - - - - - - - - - - - - - - - - - - - - - - - -


5-HT1C.human          428  S D N E P G I E M Q V E N L E L P V N P S S V V S E R I S S V - - - - - - - - - - - - - - - - - -
5-HT1C.mouse          429  N D T E P G I E M Q V E N L E L P V N P S N V V S E R I S S V - - - - - - - - - - - - - - - - - -
5-HT1C.rat            430  N D P E P G I E M Q V E N L E L P V N P S N V V S E R I S S V - - - - - - - - - - - - - - - - - -
5-HT2.hamster         441  N D C S M V T L G K Q Q S E E T C T D N I N T V N E K V S C V - - - - - - - - - - - - - - - - - -
5-HT2.human           441  N D C S M V A L G K Q H S E E A S K D N S D G V N E K V S C V - - - - - - - - - - - - - - - - - -
5-HT2.mouse           441  N D C S M V T L G N Q H S E E M C T D N I E T V N E K V S C V - - - - - - - - - - - - - - - - - -
5-HT2.rat             441  D D C S M V T L G K Q Q S E E N C T D N I E T V N E K V S C V - - - - - - - - - - - - - - - - - -
5-HT2F.mouse          449  C S T I Q S S S I I L L D T L L T E N D G D K A E E Q V S Y I L Q E R A G L I L R E G D E Q D A R A
5-HT2F.rat            449  S S T I Q S S S I I L L N T F L T E N D G D K V E D Q V S Y I - - - - - - - - - - - - - - - - - -
5-HTdro1.drosophila   545  P P S Q R V M L G D E R H G A R E S F L - - - - - - - - - - - - - - - - - - - - - - - - - - - - -


5-HT2F.mouse          499  P W Q V Q E
```

FIGURE 1h.

1.16.11   5-HT$_{1A}$ Receptors

5-HT$_1$ binding sites labeled by ^{3}H-5-HT were first shown to be heterogeneous by David Nelson and colleagues in the early 1980s. Nonsigmoidal displacement of ^{3}H-5-HT by spiperone led to the suggestion that 5-HT$_1$ sites with high affinity for spiperone should be designated 5-HT$_{1A}$ sites while 5-HT$_1$ sites with relatively low affinity for spiperone should be designated 5-HT$_{1B}$ sites (Pedigo et al., 1981; Schnellmann et al., 1984). Eventually, the 5-HT$_{1A}$ receptor was more selectively labeled with a variety of radioligands, such as ^{3}H-8-hydroxy-2-(di-*n*-propylamino)-tetralin (8-OH-DPAT) (Gozlan et al., 1983; Hoyer et al., 1985). Other radioligands that have been used to label the 5-HT$_{1A}$ site are listed in Table 3.

In comparison to other 5-HT receptor subtypes, the 5-HT$_{1A}$ site displays high and selective affinity for 8-OH-DPAT, ipsapirone, 2-(2,6-dimethoxyphenoxyethyl)aminomethyl-1,4-benzodioxane (WB 4101) and 5-methoxydimethyltryptamine (5-MDMT) (Table 4). Compounds that bind selectively and with high affinity to the 5-HT$_{1A}$ receptor are aminotetralins (e.g., 8-OH-DPAT), pyrimidinylpiperazines (e.g., buspirone, ipsapirone), and benzodioxanes (e.g., WB 4101, spiroxatrine, 8-[4-[(1,4-benzodioxan-2-ylmethyl)amino]butyl]-8-azaspiro[4.5]decane-7,9-dione (MDL 72832)). Drugs that interact potently with the 5-HT$_{1A}$ receptor but are relatively nonselective are ergots (e.g., d-LSD, metergoline) and certain indoles (e.g., 5-HT, 5-CT, 5-methoxy-3-[1,2,3,6-tetrahydro-4-pyridinyl]-1H-indole (RU 24969)).

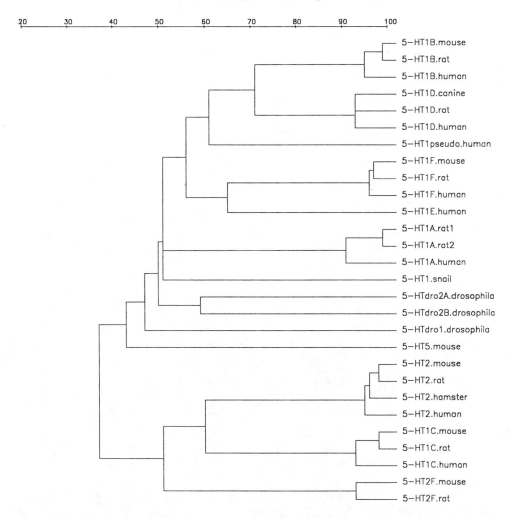

FIGURE 2.   Phylogenetic tree of 5-hydroxytryptamine receptors. The phylogenetic tree was constructed according to the method of Feng and Doolittle (1990). The length of each "branch" of the tree correlates with the evolutionary distance between receptor subpopulations.

1.16.11.1   Human 5-HT$_{1A}$ Receptors

The 5-HT$_{1A}$ receptor was the first 5-HT receptor that was successfully cloned (Kobilka et al., 1987) although its identity was not discovered until 1988 (Fargin et al., 1988). A DNA fragment in the human genome was cloned and sequenced after it was found to cross-hybridize with a full length β_2-adrenergic receptor cDNA at low stringency (Kobilka et al., 1987). The genomic clone was designated G-21 (the amino acid sequence is provided in Figure 1) and was found to contain an intronless gene which was located on human chromosome 5 (q11.2-q13). The tissue distribution of G21 was found to be unique compared to other identified receptors and was highest in lymphoid tissue.

The initial radioligand studies failed to detect any specific binding of ligands for the β_1-, α_1-, and α_2-adrenergic receptors as well for dopamine$_1$ and dopamine$_2$ receptors (Kobilka et al., 1987). However, subsequent radioligand binding studies using membranes from cells that had been transfected with G-21 DNA revealed

Table 3.   Radioligands Used to Label Mammalian G Protein-Coupled 5-HT Receptors

| | **5-HT$_1$ Receptors** | | | | |
	5-HT$_{1A}$	**5-HT$_{1B}$**	**5-HT$_{1D}$**	**5-HT$_{1E}$**	**5-HT$_{1F}$**
Diolabeled	^{3}H-5-HT	^{3}H-5-HT	^{3}H-5-HT	^{3}H-5-HT	^{125}I-LSD
	^{125}I-CYP	^{125}I-CYP*	^{3}H-5-CT		
	^{3}H-8-OH-DPAT	^{3}H-DHE	^{125}I-GTI		
	^{3}H-WB 4101	^{3}H-CP-93129			
	^{3}H-Buspirone	^{3}H-CP-96501			
	^{3}H-Ipsapirone				
	^{3}H-PAPP				
	^{3}H-p-azido-PAPP				
	^{3}H-Spiroxatrine				
	^{3}H-5-MeO-DPAC				
	^{3}H-Lisuride				
	^{125}I-PAPP				
	^{125}I-BH-8-MeO-N-PAT				
	^{125}I-N$_3$-NAP-spiperone				

| | **5-HT$_2$ Receptors** | | |
	5-HT$_2$	**5-HT$_{1C}$**	**5-HT$_{2F}$**
Diolabeled	^{3}H-5-HT	^{3}H-5-HT	
	^{125}I-LSD	^{125}I-LSD	^{125}I-LSD
	^{3}H-Mesulergine	^{3}H-Mesulergine	
	^{3}H-Mianserin	^{3}H-Mianserin	
	^{3}H-LSD	^{3}H-LSD	
	^{125}I-MIL	^{125}I-MIL	
	^{3}H-SCH 23390	^{3}H-SCH 23390	
	^{125}I-SCH 23892	^{125}I-SCH 23892	
	^{3}H-Spiperone		
	^{3}H-N-Me-spiperone		
	^{3}H-Ketanserin		
	^{3}H-DOB		
	^{77}Br-R(-)DOB		
	^{125}I-R(-)DOI		

* Certain species only.

Note: Data are derived from the literature including Hoyer and Fozard (1991).

specific ^{125}I-cyanopindolol binding, a radioligand which labels β-adrenergic, 5-HT$_{1A}$, and rodent 5-HT$_{1B}$ receptors (Fargin et al., 1988). The presence of 5-HT$_{1A}$ receptors was further substantiated by the specific binding of the selective 5-HT$_{1A}$ agonist ^{3}H-8-OH-DPAT. Scatchard analysis revealed a biphasic pattern with both a high (K_D = 0.06 nM) and a low (K_D = 14.5 nM) affinity component. 8-OH-DPAT, 5-HT, and ipsapirone potently inhibited ^{3}H-8-OH-DPAT binding to the transfected cells. The 5-HT$_{1A}$ receptor has seven membrane spanning units, with the amino acid sequence of these domains showing the least variability in comparison to other cloned biogenic amine receptors. The third cytoplasmic loop shows the greatest variability in its amino acid sequence and is believed to interact with the G-protein.

The effector coupling mechanisms of permanently expressed human 5-HT$_{1A}$ receptors have been evaluated in a variety of cell lines (Fargin et al., 1989; Raymond et al., 1989; Boddeke et al., 1992; Varrault et al., 1992) as well as in *E. coli* (Bertin et al., 1992). As predicted by biochemical studies in rat brain membranes, 5-HT potently inhibits forskolin-stimulated adenylate cyclase activity (EC_{50} = 20 nM). Moreover, this effect could be antagonized by both methiothepin and spiperone (Fargin et al., 1989). Interestingly, at micromolar concentrations of 5-HT, a stimu-

Table 4. Pharmacological Characteristics of Mammalian G-Protein-Coupled 5-HT Receptors

K_i values <10 nM 5-HT$_1$ Receptors				
5-HT$_{1A}$	**5-HT$_{1B}$**	**5-HT$_{1D}$**	**5-HT$_{1E}$**	**5-HT$_{1F}$**
5-HT		5-HT	5-HT	^{125}I-LSD
5-CT	5-CT	5-CT		
DHE	DHE			
RU 24969	RU 24969*			
I-CYP	I-CYP*			
5-MeOT		5-MeOT		
Bufotenine		Bufotenine		
Metergoline		Metergoline		
d-LSD		d-LSD		
5-MDMT	CP 93129			
8-OH-DPAT	CP 96501			
Ipsapirone				
Lisuride				
NAN 190				
WB 4101				

5-HT$_2$ Receptors		
5-HT$_2$	**5-HT$_{1C}$**	**5-HT$_{2F}$**
Cyproheptadine	Cyproheptadine	
d-LSD	d-LSD	
Mesulergine	Mesulergine	
Metergoline	Metergoline	Metergoline
Methiothepin	Methiothepin	
Methysergide	Methysergide	Methysergide
Mianserin	Mianserin	
Pizotifen	Pizotifen	
Ritanserin	Ritanserin	Ritanserin
(+)Butaclamol		
Chlorpromazine	5-HT	5-HT
Cinanserin		
DOI		DOI
DOB		
Ketanserin		
Pirenperone		
Spiperone		

Note: Data shown are derived from a variety of published reports including Hoyer et al. (1989), Peroutka (1990), and Hoyer and Fozard (1991).

lation of phospholipase C was also observed in HeLa cells with permanently expressed 5-HT$_{1A}$ receptors but not in COS-7 cells with transiently expressed 5-HT$_{1A}$ receptors. This effect did not appear to be secondary to an inhibition of adenylate cyclase and demonstrated that 5-HT$_{1A}$ receptors can stimulate phosphatidyl inositol hydrolysis (Raymond et al., 1989). A major unresolved issue is whether this pathway is functionally relevant in the central nervous system. Therefore, the 5-HT$_{1A}$ receptor appears capable of coupling to multiple G protein-associated effector systems in a single cell.

5-HT$_{1A}$ receptor-mediated release of Ca^{2+} was measured in 2 different HeLa cell lines expressing significantly different levels of human 5-HT$_{1A}$ receptors (Boddeke et al., 1992). Compounds such as ipsapirone and buspirone acted as agonists in the cell line with the greater number of receptors but acted as pure antagonists in the cell line with 6-fold fewer 5-HT_{1A} receptors. Similar data were also generated in NIH 3T3 cells transfected with human 5-HT_{1A} receptors where the efficacies of agonists increased with receptor density (Varrault et al., 1992). These data were interpreted to indicate that the intrinsic activity of drugs may vary depending upon the number of receptors expressed in the target tissue (Boddeke et al., 1992). This conclusion is significant since it implies that intrinsic agonist activity is not solely ligand dependent. This observation may also explain apparent contradictory conclusions in the literature concerning the intrinsic activity of pharmacological agents.

Site-specific and random mutagenesis studies have provided unique insights into receptor function. The alignment of G protein-coupled receptor sequences indicates that the hydrophilic components of the transmembrane segments are the most likely site of drug-receptor interactions. Mutation studies have suggested that have indicated that the VI and VII transmembrane domains of the adrenergic receptors are important in determining the specificity of antagonist binding. For example, certain β-adrenergic antagonists such as pindolol also bind to 5-HT_{1A} receptors with high affinity. Significant homologies exist between the VI and VII transmembrane segments of 5-HT_{1A} and β-adrenergic receptors.

In the first reported human 5-HT_{1A} mutagenesis study, a single amino acid residue, Asn_{385}, in the VII transmembrane segment of the human 5-HT_{1A} receptor was changed to Val (Guan et al., 1992). Both the wild type and mutant receptor genes were expressed in COS-7 cells. Radioligand binding studies were performed by using ^{3}H-5-HT, ^{3}H-8-OH-DPAT, and ^{125}I-CYP. The affinity of the mutant receptor for pindolol was significantly decreased by about 100-fold while its affinities for 5-HT and 8-OH-DPAT were essentially unchanged as compared to the wild-type receptor. These data suggest that Asn_{385} plays an important role in the specific interaction between 5-HT_{1A} receptors and pindolol. Ultimately, this approach should enhance the ability to design potent, selective, and novel 5-HT receptor active agents.

1.16.11.10   Rat 5-HT_{1A} Receptors

The rat 5-HT_{1A} receptor was first cloned, expressed, and characterized in 1990 (Albert et al., 1990). The rat 5-HT_{1A} receptor displays significant homologies to the human 5-HT_{1A} receptor in terms of its structure, pharmacology, biochemistry, and anatomical distribution in the brain. The receptor is derived from an intronless open reading frame encoding a 422 amino acid protein which is 89% identical with the human 5-HT_{1A} receptor (designated 5-HT1A.rat1 in Figure 1). A Scatchard analysis of ^{3}H-8-OH-DPAT binding to transfected cells revealed a biphasic pattern of binding ($K_H = 1.6$ nM; $K_L = 9.4$ nM) (Albert et al., 1990). The radioligand binding of the agonist was modulated by guanine nucleotides. Biochemical assays demonstrated that the rat 5-HT_{1A} receptor inhibited both basal and stimulated cAMP accumulation. Interestingly, 3 distinct RNA species (3.9, 3.6, and 3.3 kb) were detected for the rat 5-HT_{1A} receptor. The significance of this observation remains speculative.

These data have been largely confirmed by another laboratory (Fujiwara et al., 1990). However, Fujiwara et al. (1990) used a different rat genomic library than Albert et al. (1990) and reported that the rat 5-HT_{1A} receptor sequence was very similar, but not identical, to the sequence reported by Albert et al. (1990) (designated 5-HT1A.rat2 in Figure 1). Specifically, two codons were different but only one variation resulted in a different amino acid (i.e., an asparagine residue rather than a serine residue in the VI-VII cytoplasmic loop). The significance of this observation remains speculative.

1.16.11.11   Mouse 5-HT$_{1A}$ Receptors

Although no sequence data have been published, the mouse 5-HT$_{1A}$ receptor gene has been localized to distal mouse chromosome 13 (Sundaresan et al., 1989; Oakey et al., 1991).

1.16.12   5-HT$_{1B}$ Receptors

Contrary to recent theories based on species variations in pharmacological measurements (Heuring et al., 1987; Hoyer and Middlemiss, 1989), molecular biological data have demonstrated that 5-HT$_{1B}$ receptors exist in numerous species, including human (Jin et al., 1992), rat (Voigt et al., 1991), and mouse (Maroteaux et al., 1992). 5-HT$_{1B}$ receptors are pharmacologically distinct from 5-HT$_{1D}$ receptors in the rat and mouse but appear to be very similar, pharmacologically, in the human (Hamblin and Metcalf, 1991; Adham et al., 1992; Weinshank et al., 1992). For example, ^{125}I-cyanopindolol can be used to label 5-HT$_{1B}$ receptors in rodent but not human brain (Hoyer et al., 1985a). 5-HT$_{1B}$ sites display <10 nM affinity for RU 24969, 5-CT, and DHE and relatively low affinity for d-LSD and metergoline (compared to 5-HT$_{1A}$ and 5-HT$_{1D}$ receptors). In comparison to other 5-HT receptor subtypes, relatively few radioligands have been used to label 5-HT$_{1B}$ receptors (Table 3).

1.16.12.1   Human 5-HT$_{1B}$ Receptors

An intronless human gene encoding the 5-HT$_{1B}$ receptor was identified independently by a number of groups in early 1992. Although the amino acid sequence reported was identical in all publications, the receptor was designated the 5-HT$_{1B}$ (Jin et al., 1992; Hamblin et al., 1992; Mochizuki et al., 1992), 5-HT$_{1D\beta}$ (Adham et al., 1992; Weinshank et al., 1992; Demchyshyn et al., 1992), 5-HT$_{S12}$ (Levy et al., 1992), or 5-HT$_{1D\text{-like}}$ (Veldman and Bienkowski, 1992) receptor. The receptor is a 390 amino acid protein which is most abundantly expressed in the striatum but also in the hippocampus and frontal cortex. Two mRNA transcripts of different size were detected in all 3 brain regions. The gene has been localized to chromosome 6q13 (Jin et al., 1992). Scatchard analysis of the expressed receptor has been reported to display a monophasic binding pattern with a K$_D$ = 23 nM (Jin et al., 1992) in human embryonic kidney cells and an even lower affinity (K$_D$ = 50 to 200 nM) in Ltk- cells (Levy et al., 1992). High-affinity (i.e., nM) binding of ^{3}H-5-HT has been detected under certain assay conditions (Levy et al., 1992). The human 5-HT$_{1B}$ receptor (Jin et al., 1992) is 93% identical to the rat 5-HT$_{1B}$ receptor (Voigt et al., 1991) but only 61% identical to the dog (Libert et al., 1989) and human (Hamblin and Metcalf, 1991a) 5-HT$_{1D}$ receptors.

1.16.12.10   Rat 5-HT$_{1B}$ Receptors

The cloning and expression of a rat brain cDNA encoding the 5-HT$_{1B}$ receptor was first reported in 1991 (Voigt et al., 1991). A transient expression system was used to determine the pharmacological properties of the receptor. A two-site model was found to be most consistent with saturation data using ^{3}H-5-HT. The high-affinity component displayed a K$_D$ = 3 nM while the low-affinity component displayed a K$_D$ = 60 nM. The binding of ^{3}H-5-HT was inhibited by guanine nucleotides. However, no consistent effect of 5-HT could be detected on either basal or stimulated levels of cAMP levels in human embryonic kidney (HEK-293) cells expressing the rat 5-HT$_{1B}$ receptor.

These data have been confirmed and extended with the major exception that 5-HT was found to inhibit forskolin-stimulated cAMP levels in transfected Y-1 adrenal cells (Adham et al., 1992). A K_D value of 23 nM was reported in the Y-1 cell expression system (Adham et al., 1992).

1.16.12.11　Mouse 5-HT$_{1B}$ Receptors

A mouse genomic clone encoding the 5-HT$_{1B}$ receptor was isolated from a genomic library (Maroteaux et al., 1992). Membranes of transfected COS-7 cells revealed specific ^{3}H-5-HT binding with a $K_D = 48$ nM. As expected of a rodent 5-HT$_{1B}$ receptor, the most potent drugs analyzed were cyanopindolol, 5-CT, and RU 24969. 5-HT was found to inhibit adenylate cyclase activity in NIH 3T3 cells expressing the 5-HT$_{1B}$ receptor. The receptor was found to be expressed most abundantly in the striatum and in the Purkinje cells of the cerebellum. This pattern was felt to be most consistent with a postsynaptic localization of the 5-HT$_{1B}$ receptor.

1.16.13　5-HT$_{1D}$ Receptors

In 1987, another distinct subtype of the 5-HT$_1$ receptor family was identified in bovine brain membranes (Heuring and Peroutka, 1987). ^{3}H-5-HT binding in bovine caudate was examined in the presence of 100 nM 8-OH-DPAT and 100 nM mesulergine, a condition designed to displace over 90% of specific binding to 5-HT$_{1A}$ and 5-HT$_{1C}$ receptor sites (Heuring and Peroutka, 1987). However, a significant amount of ^{3}H-5-HT binding was still observed under such conditions. Moreover, the pharmacological characteristics of this apparently homogeneous group of receptors did not correlate with any previously defined 5-HT receptor binding site subtype. Therefore, this recognition site for ^{3}H-5-HT was designated the 5-HT$_{1D}$ binding site (Heuring and Peroutka, 1987).

To date, no potent (i.e., <10 nM) selective 5-HT$_{1D}$ agent has been described. 5-HT$_{1D}$ receptor binding sites display nanomolar affinity for 5-HT, 5-CT, 5-MeOT, and metergoline (Table 3). RU 24969 and (-)pindolol are approximately two orders of magnitude less potent at these sites than at 5-HT$_{1B}$ binding sites. Drugs selective for 5-HT$_{1A}$ receptors such as 8-OH-DPAT, ipsapirone and buspirone display micromolar affinities for 5-HT$_{1D}$ sites as do agents which display nanomolar potencies for 5-HT$_{1C}$, 5-HT$_2$, and 5-HT$_3$ sites such as mianserin, mesulergine, and ICS 205-930.

1.16.13.1　Human 5-HT$_{1D}$ Receptors

The sequence of the human 5-HT$_{1D}$ receptor was first reported by Hamblin and Metcalf (1991a). A human clone containing an intronless open reading frame was found to encode the 5-HT$_{1D}$ receptor and the gene has been localized to chromosome 1p34.3-36.3 (Libert et al., 1991). The 377 amino acid receptor was transfected into COS-7 cells and both pharmacological and biochemical studies were performed. Analysis of the transfected cell line revealed a pharmacological profile that was consistent with the presence of expressed 5-HT$_{1D}$ receptors. The saturation data using ^{3}H-5-HT indicated a single population of binding sites with a $K_D = 4.8$ nM. The rank order of potency of various drugs was 5-CT > 5-HT > RU 24969 > 8-OH-DPAT. In a permanent expression system, 5-HT was found to inhibit adenylate cyclase activity. These data have been confirmed and extended by other laboratories (Weinshank et al., 1992).

1.16.13.5   Canine 5-HT$_{1D}$ Receptors

An intronless member of the G protein-coupled receptor family was isolated from dog in 1989 (Libert et al., 1989). The receptor sequence showed structural homology (43% identity) with the 5-HT$_{1A}$ receptor and was designated the RDC4 receptor. However, neither radioligand nor functional assays were performed in the initial report. More recently, RDC4 has been determined to be the dog 5-HT$_{1D}$ receptor (Maenhaut et al., 1991; Zgombick et al., 1992). The receptor is a 377 amino acid protein which is highly homologous to the human 5-HT$_{1D}$ receptor (87% identity). Radioligand studies indicate that the expressed dog 5-HT$_{1D}$ receptor possesses a pharmacological profile that is similar to the human 5-HT$_{1D}$ receptor.

Transient expression of the dog 5-HT$_{1D}$ receptor in either COS-7 cells or *Xenopus* oocytes indicated that an increase in cAMP levels occurs with 5-HT stimulation (Maenhaut et al., 1991). However, using a LM(tk-) cell line stably expressing the dog 5-HT$_{1D}$ receptor, another laboratory reported that the receptor mediated the inhibition of adenylate cyclase (Zgombick et al., 1992). In this system, 5-HT alone had no effect on basal levels of cAMP. The basis of this apparent discrepancy has yet to be resolved.

1.16.13.9   Guinea Pig 5-HT$_{1D}$ Receptors

A partial sequence of the guinea pig 5-HT$_{1D}$ receptor has been reported (Weydert et al., 1992).

1.16.13.10   Rat 5-HT$_{1D}$ Receptors

The rat 5-HT$_{1D}$ receptor clone has been identified but only reported in abstract form (Hamblin and Metcalf, 1991). It was reported to display 85% identity with the human 5-HT$_{1D}$ receptor. The pharmacological profile of the rat 5-HT$_{1D}$ receptor was consistent with the properties of the human 5-HT$_{1D}$ receptor.

1.16.13.11   Mouse 5-HT$_{1D}$ Receptors

A partial sequence of the mouse 5-HT$_{1D}$ receptor has been reported (Weydert et al., 1992).

1.16.14   5-HT$_{1E}$ Receptors

1.16.14.1   Human 5-HT$_{1E}$ Receptors

A fourth member of the 5-HT$_1$ family of human receptors was reported in early 1992 and was designated the 5-HT$_{S31}$ receptor (Levy et al., 1992a) although it was subsequently characterized and defined as the 5-HT$_{1E}$ receptor (McAllister et al., 1992; Zgombick et al., 1992). The intronless gene encodes a 365 amino acid protein. Based on a phylogenetic tree analysis, 5-HT$_{1E}$ receptors branched from a receptor group which eventually evolved into 5-HT$_{1A}$, 5-HT$_{1B}$, and 5-HT$_{1D}$ receptors (Figure 1). When the 5-HT$_{1E}$ receptor was expressed in murine Ltk- cells, it was found to mediate the inhibition of adenylate cyclase. No radioligand binding studies were performed in this initial report.

However, McAllister et al. (1992) have reported on a detailed pharmacological characterization of the 5-HT$_{1E}$ receptor. Saturation analysis using ^{3}H-5-HT found a K$_D$ value of 15 nM. The most significant pharmacological characteristic of this site was the relatively low affinity of 5-CT (K$_i$ = 7100 nM). Sumatriptan (K$_i$ = 2300 nM) was also relatively weak at this receptor as compared to other 5-HT$_1$ receptor subtypes. In general, the pharmacological characteristics of this site correlated with the characteristics of the putative 5-HT$_{1E}$ receptor that had been identified previously in human brain membranes (Leonhardt et al., 1989). These initial observations have been confirmed by other laboratories (Zgombick et al., 1992).

1.16.15   5-HT$_{1F}$ Receptors

1.16.15.1   Human 5-HT$_{1F}$ Receptors

Although the sequence of the human 5-HT$_{1F}$ receptor has been reported, no pharmacological data are yet available (Lovenberg et al., 1992).

1.16.15.10   Rat 5-HT$_{1F}$ Receptors

The sequence and preliminary pharmacological characterization of a novel 5-HT receptor has been reported (Lovenberg et al., 1992). The intronless gene encodes for a 366 amino acid protein. This receptor was subsequently found to be highly homologous to the mouse 5-HT$_{1E\beta}$ receptor (Amlaiky et al., 1992). The receptor was labeled by ^{125}I-LSD. 5-HT displayed an IC$_{50}$ = 59 nM whereas sumatriptan displayed an IC$_{50}$ = 64 nM. 5-CT was reported to be inactive at this receptor. The receptor appeared to be inhibitory to adenylate cyclase. The authors concluded that this receptor was a novel 5-HT receptor in the 5-HT$_1$ receptor family.

1.16.15.11   Mouse 5-HT$_{1F}$ Receptors

A mouse 5-HT receptor that is most closely homologous to the human 5-HT$_{1E}$ receptor was isolated in 1992 (Amlaiky et al., 1992). The receptor consists of 367 amino acids and a poly (A) tail. The amino acid sequence was most homologous to the human 5-HT$_{1E}$ receptor (61% identity). Pharmacologically, the receptor was similar, but not identical, to the human 5-HT$_{1E}$ receptor. The receptor could be labeled by ^{125}I-LSD but not by ^{3}H-spiperone or ^{125}I-cyanopindolol (Amlaiky et al., 1992). 5-HT displayed approximately 100 nM affinity for the receptor whereas 5-CT was essentially inactive (i.e., K$_i$ > 1,000 nM). The 5-HT$_{1E\beta}$ receptor was found to be negatively coupled to adenylate cyclase. Therefore, due to the relatively low sequence homology between the human and mouse receptor, the authors were led to recommend that this receptor be designated the 5-HT$_{1E\beta}$ receptor. Given the fact that highly homologous clones have been identified in both humans and rat, the receptor might be more appropriately designated the 5-HT$_{1F}$ receptor.

1.16.21   5-HT$_2$ Receptors

Leysen and colleagues (Leysen et al., 1978) first identified and described the pharmacological characteristics of 5-HT$_2$ receptors. More recent molecular biological and pharmacological data suggest that subtypes exist within the 5-HT$_2$ receptor family. For example, both 5-HT$_{1C}$ and 5-HT$_{2F}$ receptors are included as members of the 5-HT$_2$ family since they share molecular biological, pharmacological, and biochemical characteristics with other members of the 5-HT$_2$ receptor family. Therefore, the 5-HT$_2$ family of receptors includes 5-HT$_2$, 5-HT$_{1C}$, and 5-HT$_{2F}$ receptors.

5-HT$_2$ receptors have been extensively analyzed due to the availability of a large number of potent and selective antagonists. Several radioligands such as ^{3}H-spiperone,

^{3}H-LSD, ^{3}H-mianserin, ^{3}H-ketanserin, ^{3}H-mesulergine, ^{125}I-LSD, and N$_1$methyl-2-^{125}I-LSD can be used to label 5-HT$_2$ receptors (Table 3). A number of classical serotonergic antagonists such as methysergide, cinanserin, metergoline, and methiothepin display high affinity for the 5-HT$_2$ binding site while 5-HT and related tryptamines display markedly lower affinity (Table 4). 8-OH-DPAT, RU 24969, and 5-HT$_3$ selective agents display low affinity for this site. Autoradiographic studies reveal high densities of 5-HT$_2$ receptor sites in the cerebral cortex and caudate, with all other brain regions having substantially fewer binding sites (Pazos et al., 1985).

1.16.21.1   Human 5-HT$_2$ Receptors

The human 5-HT$_2$ receptor consists of 471 amino acids and the gene is located on chromosome 13q14-q21 (Sparkes et al., 1991). The human 5-HT$_2$ receptor appears to exist in at least 2 allelic forms (Hartig et al., 1990). Of four human clones sequenced, two locations were identified that exhibited parallel nucleotide substitutions. The variations occurred at the third base position of a codon but did not result in a change in the amino acid sequence. Therefore, the allelic variants were silent (Hartig et al., 1990).

The cloned human 5-HT$_2$ receptor was expressed in a mouse fibroblast line and 5-HT was reported to inhibit the binding of ^{3}H-ketanserin with a K$_i$ value of 600 nM while both spiperone and (+)butaclamol displayed nanomolar affinity for the receptor (Hartig et al., 1990). A different pharmacological profile was observed when ^{3}H-DOB was used to label the human 5-HT$_2$ receptor clone. 5-HT, tryptamine, and ($\pm$)DOI were significantly more potent while spiperone and (+)butaclamol were less potent than when ^{3}H-ketanserin was used to label the human 5-HT$_2$ receptor. These data suggest that ^{3}H-ketanserin and ^{3}H-DOB label different, but potentially overlapping, regions of the human 5-HT$_2$ receptor.

The 471 amino acid sequence encoding the human 5-HT$_2$ receptor contains approximately 10% different amino acids than the amino acid sequence encoding the rat 5-HT$_2$ receptor (Saltzman et al., 1991). The pharmacological significance of these amino acid differences between species has recently been analyzed by mutagenesis studies (Kao et al., 1992). In the human 5-HT$_2$ receptor, a serine residue is present in position 242 (located in transmembrane segment V) whereas an alanine residue is present in the same location in rodents. Therefore, a mutant human 5-HT$_2$ receptor was created which replaced Ser 242 with an Ala residue (S242A mutation). Mesulergine displayed nanomolar affinity for rat and mutated human 5-HT$_2$ receptor but a greater than 50-fold lower affinity for the wild-type human 5-HT$_2$ receptor. These data indicate that a naturally occurring amino acid difference between the human and rat 5-HT$_2$ receptor can significantly affect the pharmacological properties of the receptor. This observation may have great significance to human drug development.

1.16.21.10   Rat 5-HT$_2$ Receptors

5-HT$_2$ receptor cDNA was isolated from a cloned cDNA library using probes generated from the 5-HT$_{1C}$ receptor amino acid sequence (Pritchett et al., 1988). The rat 5-HT$_2$ receptor is composed of 471 amino acids (Julius et al., 1990) (Table 2), as opposed to the incorrect total of 449 amino acids in the initial report (Pritchett et al., 1988). The rat 5-HT$_2$ receptor is 90% identical to the human 5-HT$_2$ but only 50% identical to the sequence encoding the rat 5-HT$_{1C}$ receptor. Radioligand binding on membranes from a transfected mammalian cell line reveal high affinity binding with ^{3}H-spiperone. 5-HT antagonists such as ketanserin and mianserin potently (K$_i$ values 1 to 2 nM) inhibited ^{3}H-spiperone binding to transfected cell membranes. By contrast, the 5-HT$_{1A}$ agonist, 8-OH-DPAT, and the 5-HT$_3$ antagonist, MDL 72222, displayed low affinity for ^{3}H-spiperone binding. 5-HT was reported to increase intracellular Ca^{2+} and activated phosphoinositide hydrolysis in the transfected cell line. These observations have been confirmed and extended by other laboratories (Julius et al., 1990; Apud et al., 1992).

Radioligand data from a cell line containing expressed rat 5-HT$_2$ receptors indicate that both ^{3}H-ketanserin and ^{3}H-DOB binding sites can be detected (Teitler et al., 1990; Branchek et al., 1990). The ^{3}H-DOB binding sites represented only approximately 30% of the density of the ^{3}H-ketanserin labeled binding sites (Branchek et al., 1990). These data seem to suggest 2 major possibilities: either the 5-HT$_2$ receptor in membrane preparations exists in at least 2 distinct, but noninterconverting, states which differ in their binding characteristics or posttranslational receptor changes may result in two slightly different forms of the 5-HT$_2$ receptor.

In either case, it appears that the actual binding site on the molecule labeled by ^{3}H-DOB must be distinct from the site(s) labeled by ^{3}H-ketanserin. Although some overlap of the sites may exist, the relative densities of the site (i.e., approximately 0 to 30% agonist binding sites depending on the receptor source) seems to suggest that conformationally distinct 5-HT$_2$ receptor subtypes may exist. Ongoing studies are likely to clarify these interesting issues in the near future. Pharmacologically, the 5-HT$_2$ agonist binding site displays nanomolar affinity for 5-HT. The affinity of 5-HT for this binding site is approximately 250 times more potent than its affinity for 5-HT$_2$ sites labeled by ^{3}H-ketanserin (McKenna and Peroutka, 1989). In addition, a series of phenalkylamines (e.g., DOB, DOI, DOM) have been reported to be approximately 100-fold more potent at agonist vs. antagonist radiolabled 5-HT$_2$ receptors.

1.16.21.11 Mouse 5-HT$_2$ Receptors

Yang et al. (1991) have reported that the mouse 5-HT$_2$ receptor consists of 471 amino acids. The gene is located on chromosome 14 (Sparkes et al., 1991; Liu et al., 1991) and consists of three exons separated by two introns (Nquyen et al., 1991). The pharmacological characteristics of the mouse 5-HT$_2$ receptor are similar to all the other known 5-HT$_2$ receptors (Yang et al., 1992).

1.16.21.12 Hamster 5-HT$_2$ Receptors

Chambard et al. (1990) have reported that the hamster 5-HT$_2$ receptor consists of 471 amino acids. The pharmacological characteristics of the hamster 5-HT$_2$ receptor have not yet been reported.

1.16.22 5-HT$_{1C}$ Receptors

The 5-HT$_{1C}$ binding site was discovered as a consequence of autoradiographic studies using ^{3}H-5-HT in 1984. Palacios and colleagues noted that certain serotonergic radioligands bound in extremely high density to choroid plexus (Pazos et al., 1984). When the pharmacological characteristics of this site were examined, it was found that the rank order of drug potencies did not correlate with drug affinities for 5-HT$_{1A}$, 5-HT$_{1B}$, or 5-HT$_2$ receptors. This site was designated the 5-HT$_{1C}$ site due to its high affinity for 5-HT. ^{3}H-5-HT, ^{3}H-mesulergine, ^{125}I-LSD, and a variety of other radioligands can be used to label the 5-HT$_{1C}$ receptor (Table 3).

To date, no selective 5-HT$_{1C}$ receptor agent has been described. In fact, the 5-HT$_{1C}$ binding shares a similar pharmacological profile to 5-HT$_2$ sites with certain important exceptions. First, agents such as cinanserin, ketanserin, and pirenperone are approximately 30 to 50 times less potent at 5-HT$_{1C}$ sites than at 5-HT$_2$ sites. Second, spiperone is a key agent in the differentiation of 5-HT$_{1C}$ from 5-HT$_2$ receptors since it is approximately 1000-fold weaker at 5-HT$_{1C}$ receptors. Third, both the 5-HT$_{1C}$ and 5-HT$_2$ receptors display high affinity for cyproheptadine, mesulergine, methysergide, and mianserin, but the 5-HT$_{1C}$ site displays relatively high affinity (i.e., 5 to 30 nM) for 5-HT (Table 4).

1.16.22.1   Human 5-HT$_{1C}$ Receptors

Saltzman et al. (1991) have reported that the human 5-HT$_{1C}$ receptor contains 458 amino acids. The pharmacological characteristics of the human 5-HT$_{1C}$ receptor have not yet been reported.

1.16.22.10   Rat 5-HT$_{1C}$ Receptors

A functional rat 5-HT$_{1C}$ receptor cDNA was isolated using methods involving RNA expression vectors combined with a sensitive electrophysiological assay (Julius et al., 1988). High affinity binding with ^{125}I-LSD to membranes transfected mouse fibroblast cells was demonstrated. Mianserin and 5-HT potently inhibited specific ^{125}I-LSD binding (K$_i$ values less than 20 nM), whereas spiperone was a relatively weak competitor. This pharmacological profile is consistent with the 5-HT$_{1C}$ receptor. The rat 5-HT$_{1C}$ receptor is a 460 amino acid protein. *In situ* hybridization studies show that, in addition to choroid plexus, 5-HT$_{1C}$ receptor mRNA is found in basal ganglia, hypothalamus, hippocampus, and spinal cord (Julius et al., 1988; Molineaux et al., 1989). Indeed, the distribution of the 5-HT$_{1C}$ receptor is considerably more widespread than that of the 5-HT$_2$ receptor (Molineaux et al., 1989).

In response to 5-HT, cells transfected with 5-HT$_{1C}$ receptors exhibit an increase in intracellular concentrations of Ca^{2+} (Julius et al., 1988). Neurophysiological analysis has shown that 5-HT elicits an inward current characteristic of the activation of the phosphatidylinositol signal transduction system in *Xenopus* oocytes transfected with 5-HT$_{1C}$ receptor cDNA (Briggs et al., 1991; Panicker et al., 1991). In addition, the rat 5-HT$_{1C}$ receptor has been shown to induce malignant transformation of transfected mouse fibroblast cells (Julius et al., 1989). These data indicate that the rat 5-HT$_{1C}$ receptor may function as a protooncogene under certain conditions.

1.16.22.11   Mouse 5-HT$_{1C}$ Receptors

Yu et al. (1991) have reported that the mouse 5-HT$_{1C}$ receptor consists of 459 amino acids. The gene is located on the X chromosome of the mouse (Yu et al., 1991) and consists of four exons and 3 introns (Nquyen et al., 1991). 5-HT elicits a depolarizing current in mRNA-injected *Xenopus* oocytes (Yu et al., 1991). Interestingly, a hydrophobicity analysis indicates that the mouse (as well as rat) 5-HT$_{1C}$ receptor may have 8 rather than the expected 7 transmembrane segments (Yu et al., 1991). The eighth hydrophobic region is located at the amino terminus of the receptor. The functional significance of this hydrophobic region, if any, remains unknown.

1.16.23   5-HT$_{2F}$ Receptors

1.16.23.10   Rat 5-HT$_{2F}$ Receptors

The 5-HT receptor which mediates contraction of the rat stomach fundus has been well characterized using traditional pharmacological methods. In 1992, the receptor was cloned and found to be structurally related to both 5-HT$_{1C}$ and 5-HT$_2$ receptors

(Kursar et al., 1992). The receptor consists of 479 amino acids with 7 hydrophobic domains and was designated the 5-HT$_{2F}$ receptor. The receptor was expressed stably and was found to couple to phosphatidyl inositol hydrolysis. ^{3}H-5-HT displayed high affinity (K$_D$ = 7.9 nM) for the 5-HT$_{2F}$ receptor. The affinities of a variety of tryptamine and piperazine derivatives for ^{3}H-5-HT-labeled 5-HT$_{2F}$ receptors correlated with their potencies in contracting the rat stomach fundus. Southern blot data combined with PCR data using 5-HT$_{2F}$ receptor-specific primers indicated the presence of introns (Kursar et al., 1992). Independently, Foguet et al. (1992) reported similar observations after cloning and then expressing the 5-HT$_{2F}$ receptor in COS cells. The receptor was also expressed in *Xenopus* oocytes where 5-HT was found to mediate chloride currents (presumably via activation of phospholipase C) via the 5-HT$_{2F}$ receptor (Foguet et al., 1992).

1.16.23.11 Mouse 5-HT$_{2F}$ Receptors

A DNA segment homologous to the third exons of the 5-HT$_{1C}$ and 5-HT$_2$ receptor genes was isolated from a mouse genomic library (Foguet et al., 1992). Using a quantitative PCR technique, the receptor was found to be highly expressed in the stomach fundus. Further characterization of this receptor was reported by Loric et al. (1992). The gene encodes for a 504 amino acid protein and includes 2 introns. ^{125}I-DOI was able to label the receptor (K$_D$= 25.8 nM). 5-HT was relatively weak at this receptor with an apparent affinity >1000 nM. Ritanserin displayed nanomolar affinity for the receptor while methysergide and cyproheptadine were slightly less potent antagonists. Biochemical assays indicated that the mouse 5-HT$_{2F}$ receptor modulate PI hydrolysis (Loric et al., 1992). *In situ* studies identified the receptor in brain, heart, and intestine.

1.16.31 Invertebrate$_1$ Receptors

1.16.31.21 Drosophila$_1$ Receptors

Molecular biological techniques have identified proteins in drosophila which appear to be members of the 5-HT$_1$ receptor family (Witz et al., 1990; Saudou et al., 1992). At least 3 members of this receptor family exist in drosophila and they have been designated 5-HT$_{dro1}$, 5-HT$_{dro2A}$, and 5-HT$_{dro2B}$ receptors. A sequence analysis has revealed that these receptors are most similar to mammalian 5-HT$_1$ receptors (Figure 1). The 5-HT$_{dro1}$ receptor consists of 564 amino acids with 8 putative hydrophobic domains. The first hydrophobic region observed is in the amino terminal end starting at amino acid position 28. The 5-HT$_{dro1}$ receptor, like other G protein receptors which activate adenylate cyclase, has a negatively charged amino acid (glutamate) in the C-terminal end at a position where receptors which inhibit adenylate cyclase have a positively charged amino acid. In keeping with this observation, the 5-HT$_{dro1}$ receptor was found to stimulate (rather than inhibit) cAMP levels in mammalian cells which expressed the cloned receptor. The EC$_{50}$ of 5-HT was 60 nM and the potent 5-HT$_{1A}$ selective agonist, 8-OH-DPAT, was inactive. The receptor was radiolabeled with nanomolar affinity by ^{125}I-LSD (K$_D$ = 0.64 nM). Dihydroergocryptine was found to compete for ^{125}I-LSD binding with nanomolar affinity as well as to antagonize the 5-HT-induced increase in cAMP activity. The 5-HT$_{dro1}$ receptor has been localized to the right arm of drosophila chromosome 3 at position 100 A.

1.16.41   Invertebrate$_{2A}$ Receptors

1.16.41.21   Drosophila$_{2A}$ Receptors

Two additional drosophila 5-HT receptors have been isolated and characterized (Saudou et al., 1992). The 5-HT$_{dro2A}$ and 5-HT$_{dro2B}$ receptors are highly homologous proteins that are encoded by a gene located on the right arm of the second chromosome in the region 56 A-B. The 5-HT$_{dro2A}$ receptors consists of 834 amino acids with a putative polyadenylation site. This receptor has 8 putative hydrophobic domains with the first domain located in the comparatively long amino terminal end between amino acid positions 106 and 120. The putative extracellular tail of the 5-HT$_{dro2A}$ receptor contains 8 sites for potential N-linked glycosylation.

1.16.42   Invertebrate$_{2B}$ Receptors

1.16.42.21   Drosophila$_{2B}$ Receptors

The 5-HT$_{dro2B}$ receptor consists of 645 amino acids and is encoded by a gene which contains 4 introns. A hydropathy analysis indicates that the receptor has only 7 putative transmembrane domains. Within the transmembrane domains, the receptor is 43% homologous to the human 5-HT$_{1A}$ receptor but only 30% homologous with the rat 5-HT$_{1C}$ receptor. This observation is consistent with the phylogenetic tree analysis described above (Figure 1) which indicates that the drosophila 5-HT receptors are derived from the same ancestral gene as the mammalian 5-HT$_1$ receptors.

All 3 drosophila receptors displayed high and similar affinity for ^{125}I-LSD binding with K$_D$ values ranging from 0.26 to 0.44 nM (Saudou et al., 1992). 5-HT displayed 2 to 16 μM affinity for the 3 receptors and dihydroergocryptine was the most potent antagonist with K$_i$ values ranging from 4 to13 nM. In contrast to the activation of adenylate cyclase observed with transfected 5-HT$_{dro1}$ receptors, both the 5-HT$_{dro2A}$ and 5-HT$_{dro2B}$ receptors inhibit adenylate cyclase and activate phospholipase C.

1.16.51   5-HT$_5$ Receptors

1.16.51.11   Mouse 5-HT$_5$ Receptors

Sequence homologies between other 5-HT receptors were used to isolate a novel 5-HT receptor from a mouse brain library (Plassat et al., 1992). Based on a phylogenetic tree analysis (Figure 1), this 357 amino acid receptor is distantly related to all other identified 5-HT receptor subtypes. Therefore, it has been initially designated the 5-HT$_5$ receptor. However, as shown in Figure 1, it is more closely related to the 5-HT$_1$ than the 5-HT$_2$ family of receptors.

Expressed 5-HT$_5$ receptors displayed high affinity for ^{125}I-LSD binding (K$_D$ = 0.34 nM) but did not bind ^{3}H-spiperone, ^{3}H-8-OH-DPAT, or ^{125}I-cyanopindolol. Both ergotamine and 2-bromo-LSD displayed <10 nM affinity for the receptor while 5-HT was less potent. 5-HT$_5$ receptor mRNA was localized to the cerebral cortex, hippocampus, olfactory bulb, and granular layer of the cerebellum. Biochemical studies showed that the 5-HT$_5$ receptor did not alter the levels of cAMP or inositol phosphates. Therefore, the second messenger system of the 5-HT$_5$ remains to be identified.

1.16.99 Other 5-HT Receptors

A variety of data indicate that multiple, as yet unidentified and/or uncharacterized, G protein-coupled 5-HT receptor subtypes may exist. It is highly likely that marked species exist such that novel 5-HT receptor subtypes will be defined in the years ahead. For example, Bockaert and colleagues have described a nonclassical $5-HT_4$ receptor which is positively coupled to adenylate cyclase (Dumuis et al., 1989). In cell cultures generated from mouse embryo colliculi, 5-HT stimulates cAMP production in a dose-dependent manner. 5-MeOT and 5-CT are full agonists whereas tryptamine, bufotenine, and 2-methyl-5-HT are partial agonists. In addition, further characterization of this site has revealed that $5-HT_3$ antagonists of the benzamide chemical class (i.e., zacopride, cisapride, BRL 24924, metoclopramide) act as agonists to stimulate cAMP levels in mouse embryo colliculi cell cultures (Dumuis et al., 1989). Spiperone, methiothepin, and ketanserin are ineffective in antagonizing 5-HT-induced increases in cAMP levels, whereas ICS 205-930, but not MDL 72222, weakly but competitively inhibited the effects of 5-HT (K_i = 997 nM). ICS 205-930 also competitively antagonizes 5-HT-induced increase in cAMP levels in adult guinea pig hippocampal membranes (K_i = 454 nM). This pharmacological profile is not consistent with any previously described 5-HT receptor family. Further studies are needed to determine whether the putative $5-HT_4$ receptor represents yet another major family of 5-HT receptors or, since the site is clearly linked to adenylate cyclase, whether it will eventually be classified (based on amino acid sequence homologies) as a member of the $5-HT_1$ family (Clarke et al., 1989).

Acknowledgments

I thank Jean M. Peroutka for excellent editorial assistance. This work was supported in part by the Kleiner Family Foundation and NIH Grant NS 25360-06.

REFERENCES

Adham, N., Romanienko, P., Hartig, P., Weinshank, R., and Branchek, T. (1992) The rat 5-hydroxytryptamine$_{1B}$ receptor is the species homolog of the human 5-hydroxytryptamine$_{1Dbeta}$ receptor. *Mol. Pharmacol.* 41, 1–7.

Albert, P. R., Zhou, Q.-Y., VanTol, H. H. M., Bunzow, J. R., and Civelli, O. (1990) Cloning, functional expression, and mRNA tissue distribution of the rat 5-hydroxytryptamine$_{1A}$ receptor gene. *J. Biol. Chem.* 265, 5825–5832.

Amlaiky, N., Ramboz, S., Boschert, U., Plassat, J., and Hen, R. (1992) Isolation of a mouse "5HT1E-like" serotonin receptor expressed predominatly in the hippocampus. *J. Biol. Chem.* (in press).

Apud, J. A., Grayson, D. R., DeErausquin, E., and Costa, E. (1992) Pharmacological characterization of regulation of phosphoinositide metabolism by recombinant $5-HT_2$ receptors of the rat. *Neuropharmacology* 31, 1–8.

Bertin, B., Freissmuth, M., Breyer, R. M., Schutz, W., Strosberg, A. D., and Marullo, S. (1992) Functional expression of the human serotonin 5-HT1A receptor in *Escherichia coli*. *J. Biol. Chem.* 267, 8200–8206.

Boddeke, H. W. G. M., Fargin, A., Raymond, J., Schoeffter, P., and Hoyer, D. (1992) Agonist/antagonist interactions with cloned human $5-HT_{1A}$ receptors: variations in intrinsic activity studied in transfected HeLa cells. *Naunyn-Schmiede. Arch. Pharmacol.* 345, 257–263.

Bradley, P. B., Engel, G., Feniuk, W., Fozard, J. R., Humphrey, P. P. A., Middlemiss, D. N., Mylecharane, E. J., Richardson, B. P., and Saxena, P. R. (1986) Proposals for the classification and nomenclature of functional receptors for 5-hydroxytryptamine. *Neuropharmacology* 25, 563–576.

Branchek, T., Adham, N., Macchi, M., Kao, H.-T., and Hartig, P. R. (1990) [3H]-DOB (4-Bromo-2,5-dimethoxyphenylisopropylamine) and [3H]ketanserin label two affinity states of the cloned human 5-hydroxytryptamine2 receptor. *Mol. Pharmacol.* 38, 604–609.

Briggs, C. A., Pollock, N. J., Frail, D. E., Paxson, C. L., Rakowski, R. F., Kang, C. H., and Kebabian, J. W. (1991) Activation of the 5-HT$_{1C}$ receptor expressed in *Xenopus* oocytes by the benzapines SCH 23390 and SKF 38393. *Br. J. Pharmacol.* 104, 1038–1044.

Chambard, J., VanObberghen-Schilling, E., Haslam, R. J., Vouret, V., and Pouyssegur, J. (1990) Chinese hamster serotonin (5-HT) type 2 receptor cDNA sequence. *Nucleic Acids Res.* 18, 5282.

Clarke, D. E., Craig, D. A., and Fozard, J. R. (1989) The 5-HT$_4$ receptor: Naughty, but nice. *Trends Pharmacol. Sci.* 10, 385–386.

Demchyshyn, L., Sunahara, R. K., Miller, K., Teitler, M., Hoffman, B., Kennedy, J. L., Seeman, P., VanTol, H. H. M., and Niznik, H. B. (1992) A human srotonin 1D receptor variant (5-HT1Dbeta) encoded by an intronless gene on chromosome 6. *Proc. Natl. Acad. Sci. U.S.A.* 89, 5522–5526.

Dumuis, A., Sebben, M., and Bockaert, J. (1989) The gastrointestinal prokinetic benzamide derivatives are agonists at the non-classical 5-HT receptor (5-HT$_4$) positively coupled to adenylate cyclase in neurons. *Naunyn-Schmiedeberg's Arch. Pharmacol.* 340, 403–410.

Fargin, A., Raymond, J. R., Lohse, M. J., Kobilka, B. K., Caron, M. G., and Lefkowitz, R. J. (1988) The genomic clone G-21 which resembles a beta-adrenergic receptor sequence encodes the 5-HT$_{1A}$ receptor. *Nature* 335, 358–360.

Fargin, A., Raymond, J. R., Regan, J. W., Cotecchia, S., Lefkowitz, R. J., and Caron, M. G. (1989) Effector coupling mechanisms of the cloned 5-HT$_{1A}$ receptor. *J. Biol. Chem.* 264, 4848–4852.

Feng, D. F. and Doolittle, R. F. (1990) Progressive alignment and phylogenetic tree construction of protein sequences. *Meth. Enzymol.* 183, 375–387.

Foguet, M., Hoyer, D., Pardo, L. A., Parekh, A., Kluxen, F. W., Kalkman, H. O., Stuhmer, W., and Lubbert, H. (1992) Cloning and functional characterization of the rat stomach fundus serotonin receptor. *EMBO J.* 11, 3481–3487.

Fujiwara, Y., Nelson, D. L., Kashihara, K., Varga, E., Roeske, W. R., and Yamamura, H. I. (1990) The cloning and sequence analysis of the rat serotonin-1A receptor gene. *Life Sci.* 47, 127–132.

Gozlan, H., Mestikawy, S. E., Pichat, L., Glowinski, J., and Hamon, M. (1983) Identification of presynaptic serotonin autoreceptors using a new ligand: ^{3}H-PAT. *Nature* 305, 140–142.

Guan, X. G., Peroutka, S. J., and Kobilka, B. K. (1992) Identification of a single amino acid residue responsible for the binding of a class of beta-receptor antagonists to 5-hydroxytryptamine1A receptors. *Mol. Pharmacol.* 41, 695–698.

Hamblin, M. W. and Metcalf, M. A. (1991) Cloning of a human 5-HT1D serotonin receptor gene and its rat homolog. *Soc. Neurosci. Abstr.* 17, 719.

Hamblin, M. W. and Metcalf, M. A. (1991a) Primary structure and functional characterization of a human 5-HT1D-type serotonin receptor. *Mol. Pharmacol.* 40, 143–148.

Hamblin, M. W., Metcalf, M. A., McGuffin, R. W., and Karpells, S. (1992) Molecular cloning and functional characterization of a human 5-HT$_{1B}$ serotonin receptor: A homologue of the rat 5-HT$_{1B}$ receptor with 5-HT$_{1D}$-like pharmacological specificity. *Biochem. Biophys. Res. Commun.* 184, 752–759.

Hartig, P., Kao, H.-T., Macchi, M., Adham, N., Zgombick, J., Weinshank, R., and Branchek, T. (1990) The molecular biology of serotonin receptors. *Neuropsychopharmacology* 3, 335–347.

Heuring, R. E. and Peroutka, S. J. (1987) Characterization of a novel ^{3}H-5-hydroxytryptamine binding site subtype in bovine brain membranes. *J. Neurosci.* 7, 894–903.

Heuring, R. E., Schlegel, J. R., and Peroutka, S. J. (1987) Species variations in RU 24969 interactions with non-5-HT$_{1A}$ binding sites. *Eur. J. Pharmacol.* 12, 279–282.

Hoyer, D., Engel, G., and Kalkman, H. O. (1985) Molecular pharmacology of 5-HT$_1$ and 5-HT$_2$ recognition sites in rat and pig brain membranes: Radioligand binding studies with [^{3}H]5-HT, [^{3}H]8-OH-DPAT, (-)[^{125}I]iodocyanopindolol, [^{3}H]mesulergine and [^{3}H]ketanserin. *Eur. J. Pharmacol.* 118, 13–23.

Hoyer, D., Engel, G., and Kalkman, H. O. (1985a) Characterization of the 5-HT$_{1B}$ recognition site in rat brain: Binding studies with (-)[^{125}I]iodocyanopindolol. *Eur. J. Pharmacol.* 118, 1–12.

Hoyer, D. and Fozard, J. R. (1991) 5-Hydroxytryptamine receptors. In *Receptor Data for Biological Experiments*, Doods, H. N. and VanMeel, J. C. A., Eds., Ellis Harwood, New York, pp. 35–37.

Hoyer, D. and Middlemiss, D. N. (1989) Species differences in the pharmacology of terminal 5-HT autoreceptors in mammalian brain. *TIPS* 10, 130–132.

Hoyer, D., Neijt, H. C., and Karpf, A. (1989) Competitive interaction of agonists and antagonists with 5-HT$_3$ recognition sites in membranes of neuroblastoma cells labelled with [^{3}H]ICS 205–930. *J. Recep. Res.* 9, 65–79.

Jin, H., Oksenberg, O., Ashkenazi, A., Peroutka, S. J., Duncan, M. V., Rozmahel, R., Yang, Y., Mengod, G., Palacios, J. M., and O'Dowd, B. F. (1992) Characterization of the human 5-hydroxytryptamine1B receptor. *J. Biol. Chem.* 267, 5735–5738.

Julius, D., Huang, K. N., Livelli, T. J., Axel, R., and Jessel, T. M. (1990) The 5HT2 receptor defines a family of structurally distinct but functionally conserved serotonin receptors. *Proc. Natl. Acad. Sci. U.S.A.* 87, 928–932.

Julius, D., Livelli, T. J., Jessel, T. M., and Axel, R. (1989) Ectopic expression of the serotonin 1c receptor and the triggering of malignant transformation. *Science* 244, 1057–1062.

Julius, D., MacDermott, A. B., Axel, R., and Jessel, T. (1988) Molecular characterization of a functional cDNA encoding the serotonin 1c receptor. *Science* 241, 558–564.

Kao, H., Adham, N., Olsen, M. A., Weinshank, R. L., Branchek, T. A., and Hartig, P. R. (1992) Site-directed mutagenesis of a single residue changes the binding properties of the serotonin 5-HT2 receptor from a human to a rat pharmacology. *FEBS Lett.* 307, 324–328.

Kobilka, B. K., Frielle, T., Collins, S., Yang-Feng, T., Kobilka, T. S., Francke, U., Lefkowitz, R. J., and Caron, M. G. (1987) An intronless gene encoding a potential member of the family of receptors coupled to guanine nucleotide regulatory proteins. *Nature* 329, 75–79.

Kursar, J. D., Nelson, D. L., Wainscott, D. B., Cohen, M. L. B., and M. (1992) Molecular cloning, functional expression and pharmacological characterization of a novel serotonin receptor (5-HT$_{2F}$) from rat stomach fundus. *Mol. Pharmacol.* (in press).

Leonhardt, S., Herrick-Davis, K., and Titeler, M. (1989) Detection of a novel serotonin receptor subtype (5-HT$_{1E}$) in human brain: Interaction with a GTP-binding protein. *J. Neurochem.* 53, 465-471.

Levy, F. O., Gudermann, T., Birnbaumer, M., Kaumann, A. J., and Birnbaumer, L. (1992a) Molecular cloning of a human gene (S31) encoding a novel serotonin receptor mediating inhibition of adenylate cyclase. *FEBS Lett.* 296, 201–206.

Levy, F. O., Gudermann, T., Perez-Reyes, E., Birnbaumer, M., Kaumann, A. J., and Birnbaumer, L. (1992) Molecular cloning of a human serotonin receptor (S12) with a pharmacological profile resembling that of the 5-HT$_{1D}$ subtype. *J. Biol. Chem.* 257, 7553–7562.

Leysen, J. E., Niemegeers, C. J. E., Tollenaere, J. P., and Laduron, P. M. (1978) Serotonergic component of neuroleptic receptors. *Nature* 272, 168–171.

Libert, F., Parmentier, M., Lefort, A., Dinsart, C., VanSande, J., Maenhaut, C., Simons, M. J., Dumont, J. E., and Vassart, G. (1989) Selective amplification and cloning of four new members of the G protein-coupled receptor family. *Science* 244, 569–572.

Libert, F., Passage, E., Parmentier, M., Simons, M., Vassart, G., and Mattei, M. (1991) Chromosomal mapping of A1 and A2 adenosine receptors, VIP receptor, and a new subtype of serotonin receptor. *Genomics* 11, 225–227.

Liu, J., Chen, Y., Kozak, C. A., and Yu, L. (1991) The 5-HT$_2$ serotonin receptor gene Htr-2 is tightly linked to Es-10 on mouse chromosome 14. *Genomics* 11, 231–234.

Loric, S., Launay, J., Colas, J., and Maroteaux, L. (1992) New mouse 5-HT2-like receptor. *FEBS Lett.* 312, 203–207.

Lovenberg, T. W., Erlander, M. G., Baron, B. M., Dudley, M. W., Danielson, P. E., Burns, J. E., Craft, C. M., and Sutcliffe, J. G. (1992) Cloning and functional expression of a novel rat 5-HT$_1$-like receptor. *Soc. Neurosci. Abstr.* 18, 464.

Maenhaut, C., VanSande, J., Massart, C., Dinsart, C., Libert, F., Monferini, E., Giraldo, E., Ladinsky, H., Vassart, G., and Dumont, J. E. (1991) The orphan receptor cDNA RDC4 encodes a 5-HT$_{1D}$ serotonin receptor. *Biochem. Biophys. Res. Commun.* 180, 1460–1468.

Maroteaux, L., Saudou, F., Amlaiky, N., Boschert, U., Plassat, J. L., and Hen, R. (1992) The mouse 5-HT1B serotonin receptor: Cloning, functional expression and localization in motor control centers. *Proc. Natl. Acad. Sci. U.S.A.* 89, 3020–3024.

McAllister, G., Charlesworth, A., Snodin, C., Beer, M. S., Noble, A. J., Middlemiss, D. N., Iverson, L. L., and Whiting, P. (1992) Molecular cloning of a serotonin receptor from human brain (5-HT1E): A fifth 5-HT1-like subtype. *Proc. Natl. Acad. Sci. U.S.A.* 89, 5517–5521.

McKenna, D. J. and Peroutka, S. J. (1989) Differentiation of 5-hydroxytryptamine$_2$ receptor subtypes using ^{125}I-R-(-)2,5-dimethoxy-4-iodo-phenylisopropylamine and ^{3}H-ketanserin. *J. Neurosci.* 9, 3482–3490.

Mochizuki, D., Yuyama, Y., Tsujita, R., Komaki, H., and Sagai, H. (1992) Cloning and expression of the human 5-HT1B-type receptor gene. *Biochem. Biophys. Res. Commun.* 185, 517–523.

Molineaux, S. M., Jessell, T. M., Axel, R., and Julius, D. (1989) 5-HT$_{1C}$ receptor is a prominent serotonin receptor subtype in the central nervous system. *Proc. Natl. Acad. Sci. U.S.A.* 86, 6793–6797.

Nquyen, H., Le, H. H., Lee, R. Y., Staufenbiel, M., Palacios, J. M., Mengod, G., and Lubbert, H. (1991) Does alternative exon usage contribute to serotonin receptor heterogeneity?. *Neurochem. Int.* 19, 433–436.

Oakey, R. J., Caron, M. G., Lefkowitz, R. J., and Seldin, M. F. (1991) Genomic organization of adrenergic and serotonin receptors in the mouse: Linkage mapping of sequence-related genes provides a method of examining mammalian chromosome evolution. *Genomics* 10, 338–344.

Panicker, M. M., Parker, I., and Miledi, R. (1991) Receptors of the serotonin 1C subtype expressed from cloned DNA mediate the closing of K+ membrane channels by brain mRNA. *Proc. Natl. Acad. Sci. U.S.A.* 88, 2560–2562.

Pazos, A., Cortes, R., and Palacios, J. M. (1985) Quantative autoradiography mapping of serotonin receptors in the rat brain. II. Serotonin-2 receptors. *Brain Res.* 346, 231–249.

Pazos, A., Hoyer, D., and Palacios, J. M. (1984) The binding of serotonergic ligands to the porcine choroid plexus: Characterization of a new type of serotonin recognition site. *Eur. J. Pharmacol.* 106, 539–546.

Pedigo, N. W., Yamamura, H. I., and Nelson, D. L. (1981) Discrimination of multiple ^{3}H-5-hydroxytryptamine binding site by the neuroleptic spiperone in the rat brain. *J. Neurochem.* 36, 220–226.

Peroutka, S. J. (1990) 5-hydroxytryptamine receptor subtypes. *Pharmacol. Toxicol.* 67, 373–383.

Peroutka, S. J. (1992) Phylogenetic tree analysis of G protein-coupled 5-HT receptors: Implications for receptor nomenclature. *Neuropharmacology* 31, 609–613.

Peroutka, S. J. and Snyder, S. H. (1979) Multiple serotonin receptors: Differential binding of ^{3}H-5-hydroxytryptamine, ^{3}H-lysergic acid diethylamide and ^{3}H-spiroperidol. *Mol. Pharmacol.* 16, 687–689.

Plassat, J., Boschert, U., Amlaiky, A., and Hen, R. (1992) The mouse 5-HT5 receptor reveals a remarkable heterogeneity within the 5-HT1D receptor family. *EMBO. J.* (in press).

Pritchett, D. B., Bach, A. W. J., Wozny, M., Taleb, O., DalToso, R., Shih, J. C., and Seeburg, P. H. (1988) Structure and functional expression of cloned rat serotonin 5-HT$_2$ receptor. *EMBO J.* 7, 4135–4140.

Raymond, J. R., Fargin, A., Middleton, J. P., Graff, J. M., Haupt, D. M., Caron, M. G., Lefkowitz, R. J., and Dennis, V. W. (1989) The human 5-HT$_{1A}$ receptor expressed in HeLa cells stimulates sodium-dependent phosphate uptake via protein kinase C. *J. Biol. Chem.* 264, 1943–1950.

Saltzman, A. G., Morse, B., Whitman, M. M., Ivanshchenko, Y., Jaye, M., and Felder, S. (1991) Cloning of the human serotonin 5-HT2 and 5-HT1C receptor subtypes. *Biochem. Biophys. Res. Commun.* 181, 1469–1478.

Saudou, F., Boschert, U., Amlaiky, N., Plassat, J., and Hen, R. (1992) A family of drosophila serotonin receptors with distinct intracellular signalling properties and expression patterns. *EMBO J.* 11, 7–17.

Schnellmann, R. G., Waters, S. J., and Nelson, D. L. (1984) [^{3}H]5-Hydroxytryptamine binding sites: Species and tissue variation. *J. Neurochem.* 42, 65–70.

Sparkes, R. S., Lan, N., Klisak, I., Mohandas, T., Diep, A., Kojis, T., Heinzmann, C., and Shih, J. (1991) Assignment of a serotonin 5HT-2 receptor gene (HTR2) to human chromosome 13q14-q21 and mouse chromosome 14. *Genomics* 9, 461–465.

Sundaresan, S., Yang-Feng, T. L., and Francke, U. (1989) Genes for HMG-CoA reductase and serotonin 1a receptor are on mouse chromosome 13. *Somatic Cell Mol. Genet.* 15, 465–469.

Teitler, M., Leonhardt, S., Weisberg, E. L., and Hoffman, B. J. (1990) 4-[125I]Iodo-(2,5-dimethoxy)phenylisoprobylamine and [3H]ketanserin labeling of 5-hydroxytryptamine2 (5-HT2) receptors in mammalian cells transfected with a rat 5HT2 cDNA: Evidence for multiple states and not multiple 5-HT2 receptor. *Mol. Pharmacol.* 38, 594–598.

Varrault, A., Journot, L., Audigier, Y., and Bockaert, J. (1992) Transfection of human 5-hydroxytryptamine$_{1A}$ receptors in NIH-3T$_3$ fibroblasts: effects of increasing receptor density on the coupling of 5-hydroxytryptamine$_{1A}$ receptors to adenylyl cyclase. *Mol. Pharmacol.* 41, 999–1007.

Veldman, S. A. and Bienkowski, M. J. (1992) Cloning and pharmacological characterization of a novel human 5-hydroxytryptamine1D receptor subtype. *Mol. Pharmacol.* 42, 439–444.

Voigt, M. M., Laurie, D. J., Seeburg, P. H., and Bach, A. (1991) Molecular cloning and characterization of a rat brain cDNA encoding a 5-hydroxytryptamine$_{1B}$ receptor. *EMBO J.* 10, 4017–4023.

Weinshank, R. L., Zgombick, J. M., Macchi, M. J., Branchek, T. A., and Hartig, P. R. (1992) The human serotonin 1D receptor is encoded by a subfamily of two distinct genes: 5-HT$_{1Dalpha}$ and 5-HT$_{1Dbeta}$. *Proc. Natl. Acad. Sci. U.S.A.* 89, 3630–3634.

Weydert, A., Cloez-Tayarani, I., Fillion, M., Simon-Chazottes, D., Guenet, J., and Fillion, G. (1992) Molecular cloning of two partial serotonin 5-HT$_{1D}$ receptor sequences in mouse and one in guinea pig. *C.R. Acad. Sci. Paris* 314, 429–435.

Witz, P., Amlaiky, N., Plassat, J. L., Maroteaux, L., Borreli, E., and Hen, R. (1990) Cloning and characterization of a Drosophila serotonin receptor that activates adenylate cyclase. *Proc. Natl. Acad. Sci. U.S.A.* 87, 8940–8944.

Yang, W., Chen, K., Grimsby, J., and Shih, J. C. (1991) Human 5-HT$_2$ receptor encoded by a multiple intron-exon containing gene. *Soc. Neurosci. Abstr.* 17, 405.

Yang, W., Chen, K., Lan, N. C., Gallher, T. K., and Shih, J. C. (1992) Gene structure and expression of the mouse 5-HT2 receptor. *J. Neurosci. Res.* 33, 196–204.

Yu, L., Nguyen, H., Le, H., Bloem, L. J., Kozak, C. A., Hoffman, B. J., Snutch, T. P., Lester, H. A., Davidson, N., and Lubbert, H. (1991) The mouse 5-HT$_{1C}$ receptor contains eight hydrophobic domains and is X-linked. *Mol. Brain Res.* 11, 143–149.

Zgombick, J. M., Weinshank, R. L., Macchi, M., Schechter, L. E., Branchek, T. A., and Hartig, P. R. (1992) Expression and pharmacological characterization of a canine 5-hydroxytryptamine$_{1D}$ receptor subtype. *Mol. Pharmacol.* 40, 1036–1042.

19

Olfactory Receptors

Marc Parmentier, M.D., Ph.D., Stéphane Schurmans, M.D., Frédéric Libert, Ph.D., Pierre Vanderhaeghen, M.D., and Gilbert Vassart, M.D., Ph.D.

1.19.0   Introduction

The olfactory receptor cells are bipolar neurons located in a specialized epithelium of the dorsal nasal cavity. They send axons toward the olfactory bulb through the cribriform plate of the ethmoid bone, and dendrites toward the nasal lumen. At the contact with the lumen, the dendrite ends in a swelling, the olfactory knob, from which 5 to 20 cilia project into the mucous layer coating the epithelium. The cilia are believed to be the main site of odor detection, while the soma contains the voltage-sensitive mechanisms responsible for the generation of action potentials. The olfactory receptor cells are unique among neurons in higher vertebrates for their continuous growth and ability to regenerate.

The olfactory system is able to discriminate among thousands of odorants. Odor recognition is thought to involve the interaction of the odorous compounds with specific receptors on the cilia of olfactory neurons. Exposure of the cilia to odorants leads to the rapid stimulation of either adenylate cyclase or phospholipase C, depending on the odorant (Pace et al., 1985; Sklar et al., 1986; Breer et al., 1990; Breer, 1991). Elevation of cAMP leads to the opening of a cyclic nucleotide-gated cation channel (Nakamura and Gold, 1987), and the subsequent depolarization initiates the generation of an action potential. Elevation of IP_3 levels in olfactory neurons is thought to gate plasma membrane Ca^{2+} channels (Restrepo et al., 1990).

Over the recent years, the understanding of the molecular mechanisms of olfactory perception (reviewed by Lancet, 1986 and Snyder et al., 1989) has made considerable progress, with the identification and subsequent molecular cloning of the main proteins involved in olfactory signal transduction, the GTP binding protein G_{olf} (Snyder et al., 1989), the olfactory-specific type III adenylyl cyclase (Bakalyar and Reed, 1990), and the nucleotide gated olfactory ion channel (Dhallan et al., 1990), and more recently with the cloning of a gene family encoding putative olfactory receptors, first in rat (Buck and Axel, 1991), and subsequently in human and dog (Parmentier et al., 1992; Selbie et al., 1992).

1.19.0.1   Size of the Olfactory Receptor Gene Family

The size of the multigene family encoding olfactory receptors was estimated from Southern blotting experiments and libraries screening using probes from the available

Table 1. Olfactory

Receptor	Second messenger	Species cloned	Chromosomal location	a.a. Sequence	Accession number	Primary reference
DTMT		Canine		313	S20571	Parmentier et al., 1992
OLFI15		Rat		314	P23274	Buck and Axel, 1991
OLFI9		Rat		314	P23272	Buck and Axel, 1991
HGMP07I		Human		314	S20572	Parmentier et al., 1992
OLFI14		Rat		312	P23273	Buck and Axel, 1991
OLfI3		Rat		310	P23269	Buck and Axel, 1991
OLFI8		Rat		312	P23271	Buck and Axel, 1991
OLFF5		Rat		313	P23266	Buck and Axel, 1991
OLFF12		Rat		317	P23268	Buck and Axel, 1991
OLFF3		Rat		333	P23265	Buck and Axel, 1991
OR3		Mouse		312	P23275	Nef et al., 1992
OLFF6		Rat		311	P23267	Buck and Axel, 1991
OLFI7		Rat		327	P23270	Buck and Axel, 1991
HGMP07J		Human		320	S20583	Parmentier et al., 1992

receptor fragments. In their original report, Buck and Axel (1991) estimated that 100 to 200 genes per haploid genome was a lower limit for rat. Our own estimate, for human and dog is 400 genes or more (Parmentier et al., 1992). Preliminary evidence suggests that members of this family of olfactory proteins are conserved in lower vertebrates as well as invertebrates, although the number of genes is lower in these species (Buck and Axel, 1991). From a structural viewpoint, the olfactory receptors can be divided into subfamilies (Figure 1e). Future studies will reveal if this "anatomic" classification can be correlated with the functional interaction of the receptors with specific classes of odorants and/or with different intracellular cascades.

We and others have isolated genomic clones containing tandemly arranged receptors. Clustering of related receptors is therefore a fair hypothesis. Evolution of this gene family could therefore involve unequal crossing-over and gene conversion, leading to expansion and further diversification of the family (Buck and Axel, 1991). A number of olfactory receptor pseudogenes have also been isolated from the human genome (Selbie et al. 1992; Parmentier et al., unpublished results). Given the limited role of olfaction for human, the evolutionary pressure to maintain a full set of functional molecules is probably low, and part of the receptor repertoire could have been mutated into pseudogenes. The hypothesis that some of these genes are inactive and that allelic variants are present in human populations is supported by the genetic determination of olfactory thresholds for odorants (Gross-Isseroff et al., 1992).

1.19.0.2 Cell and Tissue Specificity of Olfactory Receptor Gene Expression

From Northern blot analyses, Buck and Axel (1991) concluded that olfactory receptors were expressed specifically in olfactory receptor neurons. In the absence of functional data, this specific expression represented one of the main criteria in favor of the olfactory nature of these receptors. Immunohistochemistry using antibodies directed against a synthetic peptide corresponding to the second intracellular loop of the rat receptor I3 (Buck and Axel, 1991; Figure 1) indicates that olfactory receptor neurons expressing a given receptor are widely distributed in the olfactory epithelium, and that each cell expresses one or a small number of receptor types (Koshimoto et al., 1992).

Before the publication of rat olfactory receptors (Buck and Axel, 1991), our own distribution studies of transcripts of the orphan receptor HGMP07 indicated that the gene was expressed in dog testis (olfactory mucosa was not tested at that stage).

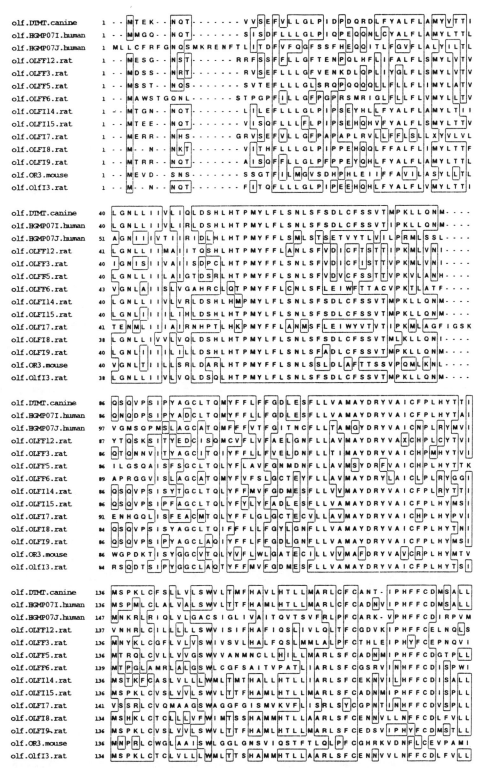

FIGURE 1. Amino acid sequences (single letter code) of putative olfactory receptors published to date. Panels a and b display the alignment of full size sequences (referenced in Table 1) in the format common to other chapters. Panels c and d display the alignment of full size (or almost full size) sequences. Panel e displays the alignment of partial sequences from 47 members of the olfactory receptor family. The human HGMP07E is from Schurmans et al. (1993). The human G1 (psuedogene), G3 and H8 are from Selbie et al. (1992). Other clones in panel e were obtained by low stringency PCR (Parmentier et al. 1992). HTPCR, HGPCR, DTPCR and DOPCR clones were isolated from testis cDNA, human genomic DNA, dog testis cDNA, and dog olfactory mucosa cDNA, respectively. The dendrogram represents the relative homology between the displayed sequences, using the Clustal software (Higgins and Sharp, 1988).

FIGURE 1b.

Consequently, testis cDNA libraries were screened (see below), and PCR amplification was made with human and dog testis cDNA as starting material (Parmentier et al., 1992). Multiple copies of one clone (DTMT) were obtained from the PCR reaction on dog testis mRNA, suggesting that it could represent the most abundant transcript of this receptor family. Northern blots and RNase protection assays performed with this probe confirmed this relative abundance. Expression in olfactory mucosa or testis cannot be correlated with the belonging of the receptors to specific subfamilies (Figure 1e). The clone with a relatively high level of expression in dog testis (DTMT) shares 82% identities with OLFI15, a rat cDNA cloned from the olfactory mucosa (Buck and Axel, 1991).

In order to explore in more detail the presence of candidate olfactory receptor transcripts in the testis, additional Northern blots prepared with RNA from testicular

```
                                                                         I                                 II
HGMP07I  (Hum)     1 MMGQNQT-SISDFLLLGLPIQPEQQNLCYALFLAMYLTTLLGNLLIIVLIRLDSHLHTPMYLFLSNLSFSDLCFSSVTIP  79
DTMT     (Dog)     1 MTEKNQT-VVSEFVLLGLPIDPDQRDLFYALFLAMYVTTILGNLLIIVLIQLDSHLHTPMYLFLSNLSFSDLCFSSVTMP  79
OLFI9    (Rat)     1 MTRRNQT-AISQFFLLGLPFPPEYQHLFYALFLAMYLTTLLGNLIIIILILLDSHLHTPMYLFLSNLSFADLCFSSVTMP  79
OLFI15   (Rat)     1 MTEENQT-VISQFLLLFLPIPSEHQHVFYALFLSMYLTTVLGNLIIIILIHLDSHLHTPMYLFLSNLSFSDLCFSSVTMP  79
OLFI3    (Rat)     1 M--NNQT-FITQFLLLGLPIPEEHQHLFYALFLVMYLTTILGNLLIIVLVQLDSQLHTPMYLFLSNLSFSDLCFSSVTMP  77
OLFI8    (Rat)     1 M--NNKT-VITHFLLLGLPIPPEHQQLFFALFLIMYLTTFLGNLLIIVLVQLDSHLHTPMYLFLSNLSFSDLCFSSVTML  77
OLFI14   (Rat)     1 MTGNNQT-LILEFLLLGLPIPSEYHLLFYALFLAMYLTIILGNLLIIVLVRLDSHLHMPMYLFLSNLSFSDLCFSSVTMP  79
OLFF5    (Rat)     1 MSSTNQS-SVTEFLLLGLSRQPQQQQLLFLLFLIMYLATVLGNLLIILAIGTDSRLHTPMYFFLSNLSFVDVCFSSTTVP  79
H8       (Hum)       ...........................................................PMYLFLSNLSFLDIGFISTIIP
OLFF3    (Rat)     1 MDSSNRT-RVSEFLLLGFVENKDLQPLIYGLFLSMYLVTVIGNISIIVAIISDPCLHTPMYFFLSNLSFVDICFISTTVP  79
OLFF12   (Rat)     1 MESGNSTRRFSSFFLLGFTENPQLHFLIFALFLSMYLVTVLGNLLIIMAIITQSHLHTPMYFFLANLSFVDICFTSTTIP  80
G3       (Hum)       .........EFLLLGLSEDPDLQPVLALLSLSLSMYLVMVLRNLLSILAVSSDSPLHTPMYFFLSNLCWPDIGFTSAMVP
HGMP07E  (Hum)     1 MDGGNQS-EGSEFLLLGMSESPEQQQILFWMFLSMYLVTVVGNVLIILAISSDSRLHTPVYFFLANLSFTDLFFVTNTIP  79
G1       (Hum)       ..............MAI*VTL*KSSILFWMFLSMYLVTVLGNVLIILAISSDSHLHTPMYFFLANLSFTDLFFVTNTIP
OLFF6    (Rat)   1 MAWSTGQNLS-TPGPFILLGFPGPRSMRIGLFLLFLVMYLLTVVGNLAIISLVGAHRCLQTPMYFFLCNLSFLEIWFTTACVP  82
OLFI7    (Rat)     1 MERRNHSGRVSEFVLLGFPAPAPLRVLLFFLSLLXYYLVLTENMLIIIAIRNHPTLHKPMYFFLANMSFLEIWYVTVTIP  80
HGMP07J  (Hum) 1 MLLCFRFGNQSMKRENFT-LITDFVFQGFSSFHEQQITLFGVFLALYILTLAGNIIIVTIIRIDLHLHTPMYFFLSMLSTSETVYTLVILP  90
OR3      (Mus)    1 MEVDSNSS--SGTFILMGVSDHPHLEIIFFAVILASYLLTLVGNLTIILLSRLDARLHTPMYFFLSNLSSLDLAFTTSSVP  79
                          Glyc

                                               III                                   IV
HGMP07I  (Hum)  KLLQNMQNQDPS----IPYADCLTQMYFFLLFGDLES-FLLVAMAYDRYVAICFPLHYTAIMSPMLCLALVALSWVLTTFHAMLHTLLMA 164
DTMT     (Dog)  KLLQNMQSQVPS----IPYAGCLTQMYFFLFFGDLES-FLLVAMAYDRYVAICFPLHYTTIMSPKLCFSLLVLSWVLTMFHAVLHTLLMA 164
OLFI9    (Rat)  KLLQNMQSQVPS----IPYAGCLAQIYFFLFFGDLGN-FLLVAMAYDRYVAICFPLHYMSIMSPKLCVSLVVLSWVLTTFHAMLHTLLMA 164
OLFI15   (Rat)  KLLQNMQSQVPS----IPFAGCLTQLYFYLYFADLES-FLLVAMAYDRYVAICFPLHYMSIMSPKLCVSLVVLSWVLTTFHAMLHTLLMA 164
OLFI3    (Rat)  KLLQNMRSQDTS----IPYGGCLAQTYFFMVFGDMES-FLLVAMAYDRYVAICFPLHYTSIMSPKLCTCLVLLLWMLTTSHAMMHTLLAA 162
OLFI8    (Rat)  KLLQNIQSQVPS----ISYAGCLTQIFFFLLFGYLGN-FLLVAMAYDRYVAICFPLHYTNIMSHKLCTCLLLVFWIMTSSHAMMHTLLAA 162
OLFI14   (Rat)  KLLQNMQSQVPS----ISYTGCLTQLYFFMVFGDMES-FLLVVMAYDRYVAICFPLRYTTIMSTKFCASLVLLLWMLTMTHALLHTLLIA 164
OLFF5    (Rat)  KVLANHILGSQA----ISFSGCLTQLYFLAVFGNMDN-FLLAVMSYDRFVAICHPLHYTTKMTRQLCVLLVVGSWVVANMNCLLHILLMA 164
H8       (Hum)  KMLDHISSGIKL----ISYGECLTQLYFSGLFADLDNNFLLAVMAYDRYVAISHPLHYALTMNSQRCVLLVAVSWVITILHALVHTLLVT
OLFF3    (Rat)  KMLVNIQTQNNV----ITYAGCITQIYFFLLFVELDN-FLLTIMAYDRYVAICHPMHYTVIMNYKLCGFLVLVSWIVSVLHALFQSLMML 164
OLFF12   (Rat)  KMLVNIYTQSKS----ITYEGCISQMCVFLVFAELGN-FLLAVMAYDRYVAXCHPLCYTVIVNHRLCILLLLLSWVISIFHAFIQSLIVL 165
G3       (Hum)  KMIVDTQSHSRV----ISHAGCLTQMSFLLLVACIEG-MLLTVMAYDCFVAICRPLHYPIIVNPHLCVFFVLVSFFLSLLDSQLHSWIVL
HGMP07E  (Hum)  KMLVNLQSHNKA----ISYAGCLTQLYFLVSLVALDN-LILAVMAYDRYVAICCPLHYTTAMSPKLCILLLSLCWVLSVLYGLIHTLLMT 164
G1       (Hum)  KMLVNFQSQNKA----ISYAGCLTQLYFLVSLVTLDN-LILAVMAYDRYVAICCPLHYVTAMSPGLCVLLLSLCWGLSVLYGLILTFLLT
OLFF6    (Rat)  KTLATFAPRGGV----ISLAGCATQMYFVFSLGCTEY-FLLAVMAYDRYLAICLPLRYGGIMTPGLAMRLALGSWLCGFSAITVPATLIA 167
OLFI7    (Rat)  KMLAGFIGSKENHGQLISFEACMTQLYFFLGLGCTEC-VLLAVMAYDRYVAICHPLHYPVIVSSRLCVQMAAGSWAGGFGISMVKVFLIS 169
HGMP07J  (Hum)  RMLSSLVGMSQP----MSLAGCATQMFFFVTFGITNC-FLLTAMGYDRYVAICNPLRYMVIMNKRLRIQLVLGACSIGLIVAITQVTSVF 175
OR3      (Mus)  QMLKNLWGPDKT----ISYGGCVTQLYVFLWLGATEC-ILLVVMAFDRYVAVCRPLHYMTVMNPRLCWGLAAISWLGGLGNSVIQSTFTL 164
                                Cys                                           PKC
```

FIGURE 1c.

```
                                    V                                                              VI

HGMP07I (Hum)  RLCFCADNVIPHFFCDMSALLKLAFSDTRVNEWVIFIMGGLILVIPFLLILGSYARIVSSILKVPSSKGICKAFSTCGSHLSVVSLFYGT  254
DTMT    (Dog)  RLCFCA-NTIPHFFCDMSALLKLACSDTQVNELVIFIMGGLILVIPFLLIITSYARIVSSILKVPSAIGICKVFSTCGSHLSVVSLFYGT  253
OLFI9   (Rat)  RLSFCEDSVIPHYFCDMSTILKVACSDTHDNELAIFILGGPIVVLPFLLIIVSYARIVSSIFKVPSSQSIHKAFSTCGSHLSVVSLFYGT  254
OLFI15  (Rat)  RLSFCADNMIPHFFCDISPLLKLACSDTHVNELVIFTVLIIVSYARVVASILKVPSTQGICKVFSTCGSHLSVVSLFYGT          254
OLFI3   (Rat)  RLSFCENNVLNFFCDLFVLLKLACSDTYINELMIFIMSTLLIIIPFFLIVMSYARIISSILKVPSTQSIHKVFSTCGSHLSVVSLFYGT  252
OLFI8   (Rat)  RLSFCENNVLHFFCDLFVLLKLACSDTYVNELMIHIMGVIIIVIPFVLIVISYAKIISSILKVPSTQSIHKVFSTCGSHLSVVSLFYGT  252
OLFI14  (Rat)  RLSFCEKNVILHFFCDISALLKLSCSDIYVNELMIYILGLIIIIPFLLIVMSYVRIFFSILKFPSIQDIYKVFSTCGSHLSVVTLFYGT  254
OLFF5   (Rat)  RKSFCADNMIPHFFCDGTPLLKLSCSDTHLNELMILTEGAVVMVTPFVCILISYIHTCAVLRVSSPRGGWKSFSTCGSHLAVVCLFYGT  254
H8      (Hum)  RLSFCGPNIIPHFFCDLIVPLLKLACSSTCVNDIVLILVPGTLLIAPFVCILMSYFYIALALAIRIDSPRGKQRAFSSCTSHLSVVSLFYST  254
OLFF3   (Rat)  ALPFCTHLEIPHYFCEPNQVIQLTCSDAFLNDLVIYFTLVLLATVPLAGIFYSYFKIVSSICAISSVRGKYKAFSTCASHLSVVSLFYST  254
OLFF12  (Rat)  QLTFCGDVKIPHFFCELNQLSQITCSDNFPSHLIMNLVPVMLAAISFSGILYSYFKIVSSIHSISTVQGKYKAFSTCASHLSIVSLFYST  255
G3      (Hum)  QLTIIKNVEISNLVCDPSQLLKLACSDSVLTNIFIYSIG·····················
HGMP07E (Hum)  RVTFCGSRKIHYIFCEMYLLRMACSNIQINHTVLIATGCFIFLIPFGVIISYVLIIRAILRIPSVSKKYKAFSTCASHLGAVSLFYGT  254
G1      (Hum)  RVTFCGPE-IHYLFCDMYILLWLACSNTHIIHTVLIATGCFIFLTPLGFMTTSYVRIVRTILQMPSASKKYKIFSTCASHLGVVSLFYGM  254
OLFF6   (Rat)  RLSYCGSRVINHFFCDISPWIVLSCTMDMSTAELTDFVLAIFILLGSCGITLVSYAYITTIIKIPSARGRHRAFSTCSSHLTVLIWYGS  257
OLFI7   (Rat)  RLPFCARK-VPHFFCDIRPVMKLSCIDTTVNEILTLIISVLVLVVPMGLVFISYVLIISTILKIASVEGRKKAFATCASHLTVVIVHYSC  259
HGMP07J (Hum)  QLPFCGHRKVDNFLCEVPAMIKLACGDTSLNEAVLNGVCTFFTVVPVSVILVSYCFIAQAVMKIRSVEGRRKAFNTCVSHLVVFLFYGS  274
OR3     (Mus)  QLPFCGHRKVDNFLCEVPAMIKLACGDTSLNEAVLNGVCTFFTVVPVSVILVSYCFIAQAVMKIRSVEGRRKAFNTCVSHLVVFLFYGS  254
                   Cys        Cys        Cys                                      **   **PKC
```

```
                                  VII

HGMP07I (Hum)  VIGLYLCSSANSSTLKDTVMAMMYTVVTPMLNPFIYSLRNRDMKGALSRVIHQKKTFFSL           314
DTMT    (Dog)  VIGLYLCPSANNSTVKETIMAMMYTVVTPMLNPFIYSLRNRDMKGALRRVICRKKITFSV           313
OLFI9   (Rat)  VIGLYLCPSANNSTVKETVMSLMYTMVTPMLNPFIYSLRNRDIKDALEKIMCKKQIPSFL           314
OLFI15  (Rat)  IIGLYLCPSANNSTVKETVMAMMYTVVTPMLNPFIYSLRNRDMKEALIRVLCKKKITFCL           314
OLFI3   (Rat)  IIGLYLCPAGNNSTVKEMVMAMMYTVVTPMLNPFIYSLRNRDMKRALIRVICSMKITL            310
OLFI8   (Rat)  IIGLYLCPSGDNFSLKGSAMAMMYTVVTPMLNPFIYSLRNRDMKQALIRVTCSKKISLPW          312
OLFI14  (Rat)  IFGIYLCPSGNNSTVKEIAMAMMYTVVTPMLNPFIYSLRNRDMKRALIRVICTKKISL            312
OLFF5   (Rat)  VIAVYFNPSSSHLAGRDMAAAVMYAVVTPMLNPFIYSLRNSDMKAALRKVLAMRFPSKQ           313
H8      (Hum)  AIGVYLCPPSSHSDGKDRVFSVMYTVMYTVVTPMLNPFIYSLRNRDMKGALGKLLGIKTS
OLFF3   (Rat)  GLGVYLSSAANNSSQASATASVMYTVVTPMVNPFIYSLRNKDVKSVLKKTLCEEVIRSPPSLLHFFLVLCHLPCFIFCY  333
OLFF12  (Rat)  GLGVYVSSAVVQSSHSAASASVMYTVVTPMLNPFIYSLRNKDVKRALERLLEGNCKVHHWTG  317
G3      (Hum)  ············
HGMP07E (Hum)  LCMVYLKPLHTYSVKDSVATV-MYAVVTPMNPFIYSLRNKDMHGALGRLLDKHFKRLT            313
G1      (Hum)  LAMVYLQPLHTYSMKDSVATV-MYAVLTPMNPFIYRLRNKDMHGAPGRVLWRPFQRPK
OLFF6   (Rat)  TIFLHVRTSVESSLDLTKAITVLNTIVTPVLNPFIYTLRNKDVKEALRRTVKGK                311
OLFI7   (Rat)  SIFIYARPKALSAFDTNKLVSVLYAVIVPLFNPIIYCLRNQDVKRALRRTLHLAQDQEANTNKGSKIG  327
HGMP07J (Hum)  ASIAYLKPKSENTREHDQLISVTYTVITPLLNPVVTLRNKEVKDALCRAVGGKFS              320
OR3     (Mus)  AIYGYLLPAKSSNQSQGKFISLFYSVVTPMVNPLIYTLRNKEVKGALGRLLGKGRGAS            312
                   Glyc                              PKC             CaPK
```

FIGURE 1d.

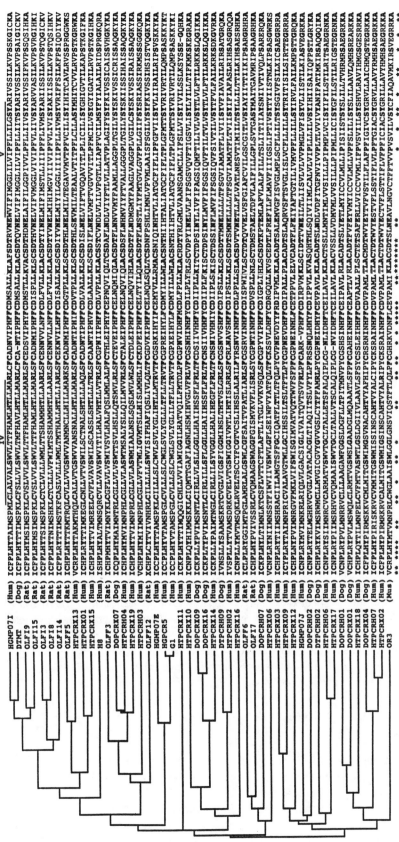

FIGURE 1e.

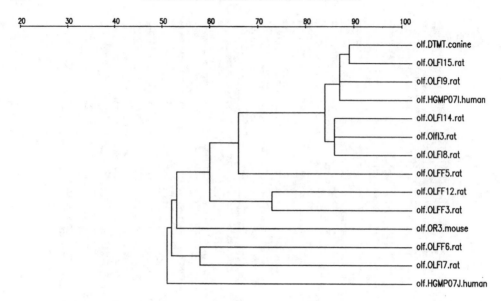

FIGURE 2. Dendrogram representing similarities between full size sequences displayed in fig. 1a, b, and referenced in Table 1.

fractions enriched in different cell types were probed with DTMT. The results demonstrate a high expression in fractions enriched in spermatocytes and spermatids. The presence of transcripts belonging to the putative olfactory receptor gene family in germ cells does not necessarily imply that the corresponding proteins are produced. However, it is tempting to speculate that they might encode receptors implicated in the chemotaxis of sperm cells during fertilization. Sperm chemotaxis, involving receptors of the guanylate cyclase family, has been clearly established in sea urchin (Garbers, 1989). Experimental evidence for the existence of a similar phenomenon in mammals has recently been provided (Ralt et al., 1991; Villanueva et al., 1990). If these receptors are effectively expressed and functional, it would mean that evolution has selected a common gene family to encode the sensors involved in the chemotaxis of the organism during both the diploid and haploid portions of its life cycle. It would also have obvious implications in the field of male and female infertility and may lead to the development of a new type of contraception.

Nef et al. (1992) analyzed the distribution of the mouse OR3 receptor transcripts by Northern blotting and *in situ* hybridization. From day 12 of embryonic life, the transcripts are located exclusively in subsets of cells of the olfactory mucosa (presumably olfactory neurons). At embryonic days 10 and 11, however, they describe OR3 transcripts in cells adjacent to the wall of the telencephalic vesicle, the primordium of the olfactory bulb. The authors suggest that the cells expressing the OR3 receptor could migrate from the central nervous system to the olfactory epithelium. This observation awaits confirmation for other receptors of the family.

1.19.0.3 Structural Features of Olfactory Receptors

All cloned putative olfactory receptors encode proteins of 310 to 333 amino acids with a calculated molecular weight ranging from 34,200 to 37,500 Da. They display common structural characteristics independently of the species of origin and will therefore be analyzed together. The identity between any two of the full size (or almost full size) sequences ranges from 38 to 85%, but partial sequences (such as the two dog clones DOPCRX09 and DOPCRX16) share more than 90% of identical residues. The identity between olfactory receptors and any other G protein-coupled receptor is not higher than 30%, showing that the cloned receptors constitute a well-defined subfamily of molecules. Within the family, the similarity is high within the second, sixth, and seventh transmembrane segments, particularly at the junction with the intracellular domains. The lowest similarity is found for the fifth transmembrane segment. Moderate conservation is found in the first, third, and fourth segments. The intracellular domains are generally more conserved than extracellular domains. All receptors contain a conserved hydrophobic stretch in their N-terminal extracellular segment. The third intracellular loop, which has been shown in other G protein-coupled receptors to be primarily involved in the coupling with G proteins, is not particularly well conserved compared to other parts of the molecule. Buck and Axel (1991) proposed that the relatively low conservation of the third, fourth, and fifth transmembrane segments (transmembrane segments are thought to mediate the ligand-binding specificity in most G protein-coupled receptors) could be correlated with the structural diversity of ligands that characterize this family of molecules. The regions conserved within the olfactory family (second, end of third, sixth, and seventh membrane spanning domains) are also those that show the highest similarity with, and between, other families of G protein-coupled receptors.

Given the limited number of available sequences, and the expected size of the gene family, it is at the moment difficult to determine if some of the cloned receptors are orthologous proteins (species homologs). The identity between mammalian orthologs in the family of G protein-coupled receptors ranges from 68 (dog and human C5a receptors) to 98% (rat and human cannabinoid receptors). The highest similarity between olfactory receptors cloned from different species is 85% (dog DTMT vs. human HGMP07I), which is well in the range for potential orthologous sequences. However, as stated above, identities of 90% can be found within a single species, indicating that either gene amplification occurred after the diversification of mammalian species, or that true orthologs will share more than 90% of their sequence.

Although there is little doubt that the subfamily of G protein-coupled receptor genes cloned in rat, human, and dog actually encodes (at least part of the) olfactory receptors, no functional evidence has been provided so far. To our knowledge, all attempts to associate a biological response (essentially the stimulation of adenylate cyclase or phospholipase C) to a cloned receptor expressed in different systems have failed up to now. Nevertheless, olfactory receptors were recently expressed in *Xenopus* oocytes by microinjection of mRNA extracted from rat and catfish olfactory mucosa, and activation of endogenous calcium-dependent chloride channels was recorded following stimulation by pyrazine (rat mRNA) or L-alanine (catfish mRNA) (Dahmen et al., 1992). The difficulty in obtaining a biological response with a cloned receptor can partially be explained by the extremely large number of potential odorous ligands, although olfactory receptors are expected to be sensitive to a rather large number of chemicals, with overlapping profiles between receptors (Sicard and Holley, 1984). Alternatively, additional protein factors could be needed for functional expression in heterologous expression systems. These putative factors could include extracellular proteins such as odorant-binding proteins (Pevsner et al., 1986), or specific intracellular proteins involved in signal transduction, as suggested for the FMLP and C5a receptors (Schultz et al., 1992a, b). The cloned receptors could also represent only one class of olfactory receptors that is not coupled to the two established cascades (cAMP and IP3), and other receptor families with distinct structural features could exist.

All olfactory receptor cloned to date share a common N-linked putative glycosylation site in their N terminal extracellular domain (Figure 1). Some receptors

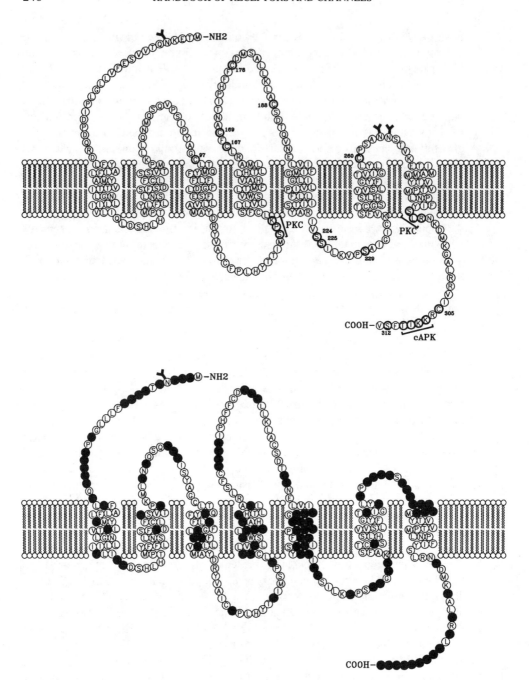

FIGURE 3. (A) Putative organization of an olfactory receptor (dog DTMT) in the plasma membrane, and localization of the possible regulatory elements. Cysteines potentially involved in disulfide bonds or palmitoylation, as well as potential sites for phosphorylation by protein kinase C (PKC), cAMP-dependent protein kinase (cAPK) and βARK-related kinases (serines in the third intracellular loop) are indicated. (B) Schematic representation of the amino acid conservation among putative olfactory receptors. White circles represent amino acids that are present in half or more of the sequences displayed in Figure 1. Black circles represent less conserved residues.

display additional glycosylation sites in the N terminal segment (HGMP07J), or the second (OLFI7, HGMP07E) or third (DTMT, OLFI9,...) extracellular loops. Glycosylation of olfactory receptors seem to be important for either ligand binding or signal transduction, since lectins, like concanavalin A or wheat germ agglutinin were shown to significantly reduce the cAMP or IP3 signals induced by odorants (Shirley et al., 1987; Breer, 1991).

Cysteines conserved in almost all sequences are found in the first and second extracellular loops of the receptors (Cys-97, -169, -178, and -188 in DTMT, Figure 3). Cysteine residues located at similar positions in the adrenergic receptors and in rhodopsin are thought to form disulfide bonds. A conserved cysteine is present in the C-terminal intracellular domain of some receptors (Cys-305 of DTMT, Figure 3) at a position identical to the Cys-341 of the β2-adrenergic receptor which has been shown to be palmitoylated.

As in visual and hormonal signal transduction, phosphorylation of receptors seems to be an important step in signal termination. Inhibition of cAMP-dependent kinase has been reported to increase the length of the cAMP response to odorants like citralva. Similarly, inhibitors of protein kinase C lead to sustained increases of IP3 levels in response to other odorants (Breer, 1991). Putative phosphorylation sites that could mediate this functional regulation are found in all cloned receptors (Figure 1). Almost all olfactory receptors contain in the C-terminal segment a potential site for protein kinase C (SXR). Some receptors reveal additional sites for PKC in the second and third intracellular loops and a potential site for the cAMP dependent protein kinase (KRXT). The phosphorylation of the serine- and threonine-rich C-terminal region of the β_2-adrenergic receptor is known to mediate homologous desensitization (Dohlman et al., 1991). The C-terminal region of olfactory receptors is not particularly rich in serines or threonines, but the third intracellular loop contains two to five residues that represent potential sites for phosphorylation by βARK or βARK-related kinases.

1.19.0.4 Perspectives

The cloning of the gene family encoding putative olfactory receptors is by itself a big step toward the understanding of the physiology of olfaction. Nevertheless, much still has to be learned about this gene family. Most importantly, it will be essential to confirm that the cloned family actually encodes the receptors that were characterized functionally in olfactory neurons. Refined estimations of the number of receptors and genomic organization of the clusters in different species will for sure be available shortly, as well as additional data concerning the distribution of specific receptors in olfactory mucosa. This will set the basis for studies dealing with the cell specificity of gene expression, the setting of olfactory neuron projections to the olfactory bulb during development and regeneration, and the evolution of the gene family. The potential relationship between olfactory receptors and sperm chemoattraction will await demonstration of the presence of the proteins in sperm cell membranes.

1.19.1   Olfactory Receptors

1.19.1.1   Human Olfactory Receptors

Homology cloning relying on sequence similarities between G protein-coupled receptors and low stringency polymerase chain reaction (PCR) allowed to isolate a series of genomic and cDNA clones encoding new putative members of this large gene family (Libert et al., 1989; Parmentier et al., 1989). Among the unidentified receptors (orphan receptors), a human genomic clone (HGMP07) was characterized by its expression in testis (see above), and its belonging to a large subfamily of genes sharing extensive sequence similarities. Human genomic and cDNA libraries were screened in order to obtain full-length clones corresponding to HGMP07. The result was the isolation of 16 different clones with strong similarities with HGMP07 (Parmentier et al., 1992). Genomic clones contained an open reading frame colinear with the cDNA sequences, demonstrating the absence of introns within the coding region. Sequence comparison further demonstrated that this gene subfamily is the human counterpart of the putative rat olfactory receptors cloned by Buck and Axel (1991). The full size clones are represented in Figure 1.

Degenerate primers specific for the subfamily were designed and used in PCR experiments (Parmentier et al., 1992). This led to the amplification of 21 clones from human testis cDNA (see above). The alignment of most of the available partial sequences extending between transmembranes III and VI is displayed in Figure 1e together with a dendrogram representing the structural relationships between the members of the subfamily.

Using a similar strategy, Selbie et al. (1992) isolated four clones encoding human olfactory receptors from genomic libraries. One of these clones (G1) has a coding sequence interrupted by stop codons and is therefore a pseudogene. A second clone (G5) is a fragment of the HGMP07E gene cloned by Schurmans et al. (1992). G1 and the other partial sequences (G3 and H8) are aligned with the other olfactory receptors in Figure 1.

We have started to delineate the genomic organization of this gene family by analyzing a number of genomic clones bearing one or several putative olfactory receptor genes. The localization in the human genome of one of these receptors (HGMP07E) was determined by *in situ* hybridization to metaphase chromosomes (Schurmans et al., 1992). A major peak of grains was obtained at the 17p13 band, but a second peak was also detected at the 3p13-3p21 segment. When used as a probe on a Southern blot, the HGMP07E probe hybridized to four related genes. These results allowed to map the HGMP07E probe to the 17p13 band of the human genome and suggest that other genes encoding receptors related to HGMP07E are located on the same chromosomal segment and on chromosome 3. A MspI restriction fragment length polymorphism was obtained and linkage analysis in CEPH families confirmed the localization on the polymorphic locus to the short arm of chromosome 17 (Schurmans et al., 1992). Preliminary evidence suggests that additional olfactory receptors would be located on chromosome 17 (Ben-Arie et al., 1992).

1.19.1.7 Dog Olfactory Receptors

Degenerate primers specific of the subfamily allowed to clone three partial clones from dog testis cDNA (see above) and eight from olfactory mucosa. The full coding region corresponding to one of the PCR clones (which was shown to be the most abundant transcript in testis) was subsequently isolated from a dog genomic library and named DTMT (Parmentier et al., 1992). The sequences are displayed in Figure 1.

1.19.1.10 Rat Olfactory Receptors

Rat olfactory receptors were cloned by Buck and Axel (1991) by low stringency polymerase chain reaction (Libert et al., 1989), starting from olfactory mucosa mRNA.

The selection of amplified products was based on the specific expression of the receptors in olfactory mucosa, and on the expected multiplicity of related sequences. Full open reading frames were further isolated from an olfactory mucosa cDNA library. Altogether, Buck and Axel describe ten full size putative olfactory receptors and eight partial sequences corresponding to PCR products. Using these data, Levy et al. (1991) reported the amplification of 9 partial sequences extending from transmembrane segments II to IV, and showing more diversity than the original 18 sequences from Buck and Axel (1991).

1.19.1.11 Mouse Olfactory Receptors

A mouse putative receptor clone (OR3) was isolated from olfactory mucosa by low stringency polymerase chain reaction (Nef et al., 1992). The OR3 receptor falls well within the putative olfactory receptor family (Figures 1 and 2), although the authors suggest that it could represent a different subfamily endowed with other functional properties.

Acknowledgments

We are grateful to J. E. Dumont for his continuous support and interest. This work was supported by the Belgian program on Interuniversity Poles of Attraction initiated by the Belgian State, Prime Minister's office, Science Policy programming. The Scientific responsibility is assumed by the authors. Also supported by grants from the Fonds de la Recherche Scientifique Médicale, and Boerhinger Ingelheim. M.P. and F.L. are respectively Chercheur Qualifié and Chargé de Recherche of the Fonds National de la Recherche Scientifique of Belgium.

REFERENCES

Bakalyar, H. A. and Reed, R. R. (1990) Identification of a specialized adenylyl cyclase that may mediate odorant detection. *Nature* 250, 1403–1406.

Ben-Arie, N., North, M., Khen, M., Margalit, T., Lehrach, H., and Lancet, D. (1992). Mapping the olfactory receptor "sub-genome": implications to human sensory polymorphisms. Proceedings of the Cold Spring Harbor meeting on "Genome Mapping and Sequencing", p. 280.

Breer, H. (1991) Molecular reaction cascades in olfactory signal transduction. *J. Steroid. Biochem. Molec. Biol.* 39, 621–625.

Breer, H., Boekhoff, I., and Tareilus, E. (1990) Rapid kinetics of second messenger formation in olfactory transduction. *Nature* 345, 65–68.

Buck, L. and Axel, R. (1991) A novel multigene family may encode odorant receptors: a molecular basis for odor recognition. *Cell* 65, 175–187.

Dahmen, N., Wang, H. L., and Margolis, F. L. (1992) Expression of olfactory receptors in xenopus oocytes. *J. Neurochem.* 58, 1176–1179.

Dhallan, R. S., Yau, K. W., Schrader, K. A., and Reed, R. R. (1990) Primary structure and functionnal expression of a cyclic nucleotide-activated channel from olfactory neurons. *Nature* 347, 184–187.

Dohlmann, H. G., Thorner, J., Caron, M. G., and Lefkowitz, R. J. (1991) Model systems for the study of seven-transmembrane-segment receptors. *Annu. Rev. Biochem.* 60, 653–688.

Garbers, D. L. (1989) Molecular basis for fertilization. *Annu. Rev. Biochem.* 58, 719–742.

Gross-Isseroff, R., Ophir, D., Bartana, A., Voet, H., and Lancet, D. (1992) Evidence for genetic determination in human twins of olfactory tresholds for a standart odorant. *Neurosci. Lett.* 141, 115–118.

Higgins, D. G. and Sharp, P. M. (1988) CLUSTAL: a package for perfoming multiple sequence aligment on a microcomputer. *Gene* 73, 237–244.

Jones, D. T. and Reed, R. R. (1989) Golf: an olfactory neuron specific-G protein involved in odorant signal transduction. *Science* 244, 790–795.

Koshimoto, H., Katoh, K., Yoshihara, Y., and Mori, K. (1992) Distribution of putative odour receptor proteins in olfactory epithelium. *NeuroReport* 3, 521–523.

Lancet, D. (1986) Vertebrate olfactory reception. *Annu. Rev Neurosci.* 9, 329–355.

Levy, N. S., Bakalyar, H. A., and Reed, R. R. (1991) Signal transduction in olfactory neurons. *J. Steroid. Molec. Biol.* 39, 633–637.

Libert, F., Parmentier, M., Lefort, A., Dinsart, C., Van Sande, J., Maenhaut, C., Simons, M. J., Dumont, J. E., and Vassart, G. (1989) Selective amplification and cloning of four new members of the G protein-coupled receptor family. *Science* 244, 569–572.

Nakamura, T. and Gold, G. H. (1987) A cyclic nucleotide-gated conductance in olfactory receptor cilia. *Nature* 325, 442–444.

Nef, P., Hermans-Borgmeyer, I., Artieres-Pin, H., Beasley, L., Dionne, V. E., and Heinemann, S. F. (1992) Spatial pattern of receptor expression in the olfactory epithelium. *Proc. Natl. Acad. Sci. U.S.A.* 89, 8948–8952.

Pace, U., Hanski, E., Salomon, Y., and Lancet, D. (1985) Odorant-sensitive adenylate cyclase of olfactory reception. *Nature* 316, 255–258.

Parmentier, M., Libert, F., Maenhaut, C., Lefort, A., Gerard, C., Perret, J., Van Sande, J., Dumont, J. E., and Vassart, G. (1989) Molecular cloning of the thyrotropin receptor. *Science* 246, 1620–1622.

Parmentier, M., Libert, F., Schurmans, S., Schiffmann, S., Lefort, A., Eggerickx, D., Ledent, C., Mollereau, C., Gérard, C., Perret, J., Grootegoed, A., and Vassart, G. (1992) Members of the olfactory receptor gene family are expressed in mammalian germ cells. *Nature* 355, 453–455.

Pevsner, J., Sklar, P. B., and Snyder, S. H. (1986) Odorant-binding protein localization to nasal glands and secretions. *Proc. Natl. Acad. Sci. U.S.A.* 83, 4942–4946.

Ralt, D., Goldenberg, M., Fetterolf, P., Thompson, D., Dor, J., Mashiach, S., Garbers, D. L., and Eisenbach, M. (1991) Sperm attraction to a follicular factor correlates with human egg fertilizability. *Proc. Natl. Acad. Sci. U.S.A.* 88, 2840–2844.

Restrepo, D., Miyamoto, T., Bryant, P. B., and Teeter, J. H. (1990) Odor stimuli trigger influx of calcium into olfactory neurons of the channel catfish. *Science* 249, 1166–1168.

Schultz, P., Stannek, P., Bischoff, S. C., Dahinden, C. A., and Gierschik, P. (1992a) Functional reconstitution of a receptor-activated signal transduction pathway in Xenopus laevis oocytes using the cloned human C5a receptor. *Cell. Signalling* 4, 153–161.

Schultz, P., Stannek, P., Voigt, M., Jakobs, K. H., and Gierschik, P. (1992b) Complementation of formyl peptide receptor-mediated signal transduction in Xenopus oocytes. *Biochem. J.* 284, 207–212.

Schurmans, S., Passage, E., Miot, F., Mattei, M. G., Vassart, G., and Parmentier, M. (1993) The HGMP07E gene encoding a putative olfactory receptor maps to the 17p12–17p13 region of the human genome and reveals a MspI restriction fragment length polymorphism. *Cytogenet. Cell Genet.* 63, 200–204.

Selbie, L. A., Townsend-Nicholson, A., Iismaa, T. P., and Shine, J. (1992) Novel G protein-coupled receptors: a gene family of putative human olfactory receptor sequences. *Mol. Brain Res.* 13, 159–163.

Shirley, S. G., Polak, E. H., Mather, R. A., and Dodd, G. H. (1987). The effect of concanavalin A on the rat electroolfactogram. *J. Biochem.* 245, 175–184.

Sicard, G. and Holley, A. (1984) Receptor cell responses to odorants: similarities and differences among odorants. *Brain Res.* 292, 283–296.

Sklar, P. B., Anholt, R. R. H., and Snyder, S. H. (1986). The odorant-sensitive adenylate cyclase of olfactory receptor cells: differential stimulation by distinct classes of odorants. *J. Biol. Chem.* 261, 15538–15543.

Snyder, S. H., Sklar, P. B., Hwang, P. M., and Pevsner, J. (1989) Molecular mechanisms of olfaction. *Trends Neurosci.* 12, 35–38.

Villanueva Diaz, C., Vadillo Ortega, F., and Kably Ambe, A. (1990) Evidence that human follicular fluid contains a chemoattractant for spermatozoa. *Fertility and Sterility* 54, 1180–1182.

20

Opioid And Opiate Receptors

Chris Evans, Ph.D

1.20.0   Introduction

The wealth of opioid peptide and opiate alkaloid agonists, partial agonists, and antagonists has provided excellent probes to study receptors of the endogenous opioid system. An early classification of opiate sites, based on pharmacological activity of a series of diverse opiate alkaloids, was proposed by Gilbert and Martin (1976). Three distinct sites were delineated: a mu receptor for which morphine was the prototypic ligand, a kappa receptor for which the prototypic ligand was ketocyclazocine, and a sigma receptor eliciting psychomimetic actions of opiates such as those induced by N-allylnorclazocine. With the discovery of enkephalins, Lord et al. (1977) reevaluated the opioid receptor classification and defined three classes of receptors — namely the mu, kappa, and delta. (The later definition of opioid effects as necessarily "naloxone reversible" subsequently excluded sigma receptors from the opioid receptor group.) The extensive utilization of this opioid receptor nomenclature is so solidly imbedded in the literature that an alternative naming system for cloned opioid receptors would be somewhat confusing at this juncture.

With the isolation of the enkephalins in the early 1970s, the opioid field blossomed with a myriad of endogenous opioid peptide ligands. Well over 20 different endogenous opioid peptides derived by highly specific proteolytic processing of four opioid precursors — proenkephalin (pEnk), prodynorphin (pDyn), proopiomelanocortin (pOMC), and the precursor for deltorphans (pDelt) — have been described (for reviews see Akil et al., 1984; Evans et al., 1988). Additionally, there are reports of morphinans in mammalian tissue (for review see Faull, 1993) and, thus, in contrast to many of the G-protein coupled receptors, the opioid receptors have a potential for interacting with a wide variety of endogenous ligands that differ in their affinities and selectivities for the various opioid receptor classes, their stability to extracellular enzymes, their neuroanatomical distribution and, perhaps, their mechanisms of receptor modulation (for review see Evans, 1993).

The evidence for delta, mu, and kappa opioid receptors eliciting their actions via G-proteins is convincing. All three classes of receptors have been demonstrated to couple via pertussis toxin-sensitive mechanisms to inhibit adenylate cyclase, activate potassium channels, and inhibit calcium channels. Furthermore, guanylnucleotides perturb agonist but not antagonist binding and activation of opioid receptors is accompanied by stimulated

GTP hydrolysis (for review of second messenger coupling see Loh and Smith, 1990; Childers, 1991; Di Chiara and North, 1992).

For reviews on the pharmacological features of the different opioid and opiate receptors and their selective ligands, see Weber et al., 1983; Pasternak 1988; Wollemann, 1990; Zimmerman and Leander, 1990; Rapaka and Porreca, 1991; Portoghese, 1992; Lutz and Pfister, 1992. For a review of the distribution of opioid receptor classes in brain, see Mansour et al., 1987.

The cloning of opioid receptors has been a particularly difficult endeavor. The instability of the receptor following solubilization has complicated purification of opiate receptor proteins. Expression cloning using frog oocytes and bacterial and mammalian expression systems has also been problematic. Possible reasons for these difficulties include low levels of mRNA encoding opioid receptors and highly folded mRNA species making cDNA copies difficult to generate.

Molecular approaches to characterizing the opioid receptors have resulted in the isolation of a number of different cDNA clones. The cDNA encoding a protein purified by morphine affinity chromatography (OBCAM) shows homology to the phosphatidyl-inositol-linked members of the immunoglobulin superfamily but does not contain seven transmembrane domains or a G-protein binding motif (e.g., as implicated in GAP 43) and thus does not show the anticipated homology to G protein-coupled receptors (Schofield et al., 1989). There are no data demonstrating functional opiate-mediated activity via OBCAM in an expression system. Currently, there is insufficient evidence to determine whether OBCAM can be classified as an opiate receptor.

1.20.1   Mu Opioid Receptors

Mu receptor binding is broadly distributed throughout the peripheral and central nervous systems and mu receptor-mediated biological effects are exhibited by cells of the immune system. In addition to morphine, a number of the endogenous opioid peptides are high-affinity agonists at mu receptors including processing products derived from pEnk (e.g., Met and Leu enkephalin, metorphamide, and BAM 18), Leu enkephalin from pDyn, and b-endorphin from pOMC. The opiate antagonist naloxone has high affinity for mu receptors although beta-funaltrexamine has higher selectivity. There are a number of selective synthetic peptide agonists including morphiceptin (Tyr-Pro-Phe-Pro amide) and DAMGO (D-Ala-2, Me-Phe4, Gly(ol)5 enkephalin). In addition, somatostatin ana-logues are high-affinity antagonists at mu receptors (Pelton et al., 1985) which raises the possibility that somatostatin and mu receptors might be closely related with regard to their binding pockets. Based upon binding and pharmacologic analyses, multiple mu receptors (mu1 and mu2) have been postulated. Mu 1 receptors are specifically antago-nized by naloxazine and participate in morphine supraspinal analgesia, whereas mu2 receptors are implicated in spinal analgesia (Paul et al., 1989). The discovery of morphinans in mammals has implicated alkaloids as potential endogenous mu ligands. However, the significance of these alkaloid opiates in the CNS remains unclear. There are data demonstrating that mu receptors can be coupled to adenylate cyclase in cell lines and a number of earlier electrophysiological studies suggested coupling to both calcium and potassium channels (see Note for recent advances).

1.20.2   Delta Opioid Receptors

The distribution of delta opioid receptors in the CNS is more discrete than that of other classes of opioid receptors. Delta receptors bind with high affinity to the enkephalins.

Deltorphans, a family of opioid peptides recently isolated from frog skin (Erspamer et al., 1989), also have high affinity for delta receptors; however, they are considerably more selective than enkephalins. The deltorphans are extraordinary in that they have specifically posttranslationally modified amino acids within the structure corresponding to the d isomers — a rare event in nature. Selective synthetic peptide agonists for delta receptors include DPDPE ([D-Pen^2, D-Pen^5] enkephalin) and DTLET ([D-Thr2,Leu5,Thr6] enkephalin). High-affinity delta selective peptide and alkaloid antagonists have been developed, including TIPP (Tyr-tetrahydroisoquinoline-3-carboxylic acid-Phe-Phe) and the alkaloid Naltrindole. There is growing evidence in the literature to suggest multiple delta opioid receptors. The evidence stems from observations on the ability for various delta-selective antagonists [D-Ala2,Leu5,Cys6]enkephalin (DALCE), N,N-diallyl-Try-Aib-Aib-Phe-Leu-OH (ICI 174,864) and naltrindole-5′-isothiocyanate (5′-NTII), to differentially antagonize the actions of the delta selective agonists [D-Pen2,D-Pen5]enkephalin (DPDPE) and [D-Ala2]deltorphin II (DELT). The lack of cross tolerance between DPDPE and deltorphin II lends support to this premise although a clear differentiation using binding has not unequivocally separated the two sites (Mattia et al., 1991). From antibody blocking studies there is evidence that in the hybrid cell line NG108-15 delta opioid receptors couple to adenylate cyclase via G_i2 (McKenzie and Milligan, 1990). However, the coupling of delta opioid receptors to Ca^{++} and K^+ channels appears to involve G_{oA} (Taussig at al., 1990). The participation of G_{oA} is implied by the retention of delta receptor activity following pertussis toxin treatment of cells transfected with a series of mutant G-proteins genetically manipulated to be resistant to ADP ribosylation by pertussis toxin. It is important to note that somatostatin receptors, like delta opioid receptors, also appear to couple to channels via G_{oA} (Taussig et al. 1990).

1.20.2.11 Mouse Delta Receptors

Utilizing transient expression screening of cDNA libraries constructed from the delta opioid receptor-rich mouse/rat hybrid cell line, NG108-15, two groups have independently identified murine clones encoding a delta opioid receptor (Evans et al., 1992; Kieffer et al., 1992). On expression in COS cells both clones exhibit the anticipated selectivity for a delta opioid receptor with regard to binding of a series of peptide and alkaloid ligands. In addition one of these clones [mouse Delta Opioid Receptor (mDOR$_A$)] confers naloxone reversible opiate inhibition of forskolin-stimulated adenylate cyclase on transfection of COS cells (Evans et al., 1992).

The nucleotide sequence of mDOR$_A$ predicts a protein of 372 amino acids. The slight sequence discrepancy in the protein-coding region of the two published clones was technical and the correct sequence is as reported by Evans and colleagues. Hydropathy analysis indicates seven transmembrane domains and mimics hydropathy plots of many other members of the G-protein receptor family. The third intracellular loop is small consisting of approximately 23 amino acids and is almost identical in both length and hydrophilicity to the somatostatin 1 receptor (Keith et al., 1993). There are two consensus sites for N-linked glycosylation sites at the N-terminus of the opioid receptor (asn[18] and asn[33]) and a pair of cysteine residues proposed to form a disulfide bond occur in the first two extracellular loops (cys[121] and cys[198]). Aspartate residues appear in the second and third proposed transmembrane domains (asp[95] and asp[128]) and analogous residues have been implicated in ligand binding by other receptors that couple to G-proteins. Predicted amphiphilic domains also occur at two sites; one in the third cytoplasmic loop and the other at the C-terminus of the protein. Studies of related receptors have shown amphiphilic domains to be important for G-protein coupling. The C-terminus of the protein contains several cysteines (cys[328], cys[333], cys[337]) that may undergo palmitoylation; although, based on sequence homology and positioning following the seventh transmembrane domain, cys[333] appears the likely site for this posttranslational modification. Intracellular consensus sites are numerous for phosphorylation by cAMP/cGMP dependent protein kinases (Thr[82, 84, 260, 352], Ser[242,247,255]) and protein kinase C (Thr[352], Ser[242, 255, 344]).

mDOR$_A$ shows highest homology to the receptors for somatostatin (especially the somatostatin 1 receptor) and the receptors for angiotensin and the chemotactic factors interleukin-8 and N-formyl peptide. mDOR$_A$ shows substantially less homology to the tachykinin receptors (including the closely related putative kappa receptor).

Northern analysis shows multiple bands but at present there is no conclusive data concerning possible alternative splice products. Initial sequence and PCR analysis of both the human and mouse mDOR$_A$ genes clearly shows introns in the coding region and the potential for alternative splicing. Southern blot analysis of mouse DNA shows that the mDOR$_A$ sequence is encoded in a rather small segment of genomic material (single Bam H1 band of approx 5KB) (see Note for recent advances).

1.20.3   Kappa Opioid Receptors

The kappa receptors, like mu receptors, are widely distributed throughout the central and peripheral nervous systems. The neoendorphins and the dynorphins have high affinity for kappa opioid receptors. However, the distribution of kappa receptors does not always reflect the localization of dynorphins and fragments of proenkephalin such as BAM18 and metorphamide have high affinity for kappa receptors and mimic some of the biological effects of dynorphins. There is evidence that the kappa receptors (like mu and delta receptors) can couple to both ion channels and adenylate cyclase. Pharmacological evidence is accumulating for a complex multiplicity of kappa binding sites (for overviews see Leslie, 1987; Clark et al., 1989; Traynor, 1989).

1.20.3.1   Putative Human Kappa Receptors

Recent efforts to identify a kappa opioid receptor by interaction with an analog of dynorphin have led to the isolation of a placental cDNA encoding a G protein-coupled receptor with very high homology to the neurokinin B receptor (Xie et al., 1992). Expressed in COS cells, this cDNA confers binding to the alkaloid opiate ligand bremazocine that can be displaced with a variety of opioid and opiate ligands. However, this cDNA confers low affinity for bremazocine and analysis of the binding sites shows lack of appropriate ligand selectivity expected for either kappa, mu, or delta opioid receptors. Although this receptor binds opioid and opiate ligands there remains significant doubt as to its actual endogenous ligand and in lieu of further data this receptor is presently categorized in the neurokinin section (see Note for recent advances).

1.20.99   Other Opioid Receptors

The multiple mDOR$_A$ transcripts and introns in the mDOR$_A$ protein coding region of the gene provide a distinct possibility that different primary amino acid sequences may be derived from the mDOR$_A$ gene. This may shed light on the pharmacological differences of opioid receptors that have been reported. However, the presence of the multiple transcripts in the NG108-15 cells which express only delta opioid receptors makes it unlikely that any of these transcripts encode different receptor classes. The

Table 1.

Receptor	Second messenger	Species cloned	Chromosomal location	a.a. Sequence	GenBank accession	Primary reference
Kappa		Human		440	M84605	Goldstein et al., 1992
Delta		Mouse		372	L07271	Evans et al., 1992
Delta		Mouse		372	L06322	Kieffer et al., 1992
Delta		Mouse		372	L11064	Yasuda et al., 1993
Delta		Rat		372	D16348	Fakuda et al., 1993
Mu		Rat		398	L13069	Chen et al., 1993
Mu		Rat		398	D16349	Fakuda et al., 1993
Kappa		Mouse		380	L11065	Yasuda et al., 1993

most plausible hypothesis is that genes closely related to $mDOR_A$ will be responsible for the diversity of opioid receptors described in the literature using classical pharmacology.

Note Added in Proof

Since the initial writing of this review a number of opioid receptors (including a mu and kappa opioid receptor) have been cloned (Chen et al., 1993; Fukuda et al., 1993; Yasuda et al., 1993). All are closely homologous to the initial delta opioid receptor clone. Constraints on the page proofs do not permit detailed analysis of these additional receptors in this edition however the GenBank accession numbers are included in Table 1.

REFERENCES

Akil, H., Watson, S., Young, E., Lewis, M., Khachaturian, H., and Walker, J. (1984) Endogenous opioids: biology and function. *Annu. Rev. Neurosci.* 7, 223–255.

Childers, S. R. (1991) Opioid receptor-coupled second messenger systems. *Life Sci.* 48, 1991–2003.

Clark, J. A, Liu, L., Price, M., Hersh, B., Edelson, M., and Pasternak, G. W. (1989) Kappa opiate receptor multiplicity: evidence for two U50,488-sensitive kappa 1 subtypes and a novel kappa 3 subtype. *J. Pharmacol. Exp. Ther.* 251, 461–468.

Di Chiara, G. and North, R. A. (1992) Neurobiology of opiate abuse. *Trends Pharmacol. Sci.* 13, 185–193.

Erspamer, V., Melchiorri, P., Falconieri-Erspamer, G., Negri, L., Corsi, R., Severini, C., Barra, D., Simmaco, M., and Kreil, G. (1989). Deltorphins: a family of naturally occurring peptides with high affinity and selectivity for delta opioid binding sites. *Proc. Natl. Acad. Sci. U.S.A.* 86, 5188–5192.

Evans, C. J., Fredrickson, R., and Hammond, D. (1988) Processing of endogenous opioid precursors. In *The Opiate Receptors,* Pasternak, G. W., Ed., Humana Press, Clifton, N.J., pp. 23–71.

Evans, C. J., Keith, D. E., Jr., Morrison, H., Magendzo, K., and Edwards, R. H. (1992) Cloning of a delta opioid receptor by functional expression. *Science* 258, 1952–1955.

Evans. C. J. (1993) Diversity among the opioid receptors. In *Biological Basis of Substance Abuse,* Korenman, S. G. and Barchas, J. D., Eds., Oxford University Press, New York, pp. 31–48.

Faull, K. F. (1993) Mammalian morphine alkaloids: Identification and possible *in vivo* biosynthesis. In *Biological Basis of Substance Abuse,* Korenman, S. G. and Barchas, J. D., Eds., Oxford University Press, New York, pp. 267–283.

Gilbert, P. E. and Martin, W. R. (1976) The effects of morphine and nalorphine-like drugs in the nondependent, morphine-dependent and cyclazocine-dependent chronic spinal dog. *J. Pharmacol. Exp. Ther.* 198, 66–82.

Keith, D. E., Jr., Anton, B., and Evans, C. J. (1993) Characterization and mapping of a delta opioid receptor clone from NG108-15 cells. *Proc. Western Pharmacol. Soc.* 36, 299–306.

Kieffer, B. L., Befort, K., Gaveriaux-Ruff, C., and Hirth, C. G. (1992) The delta-opioid receptor: Isolation of a cDNA by expression cloning and pharmacological characterization. *Proc. Natl. Acad. Sci. U.S.A.* 89, 12048–12052.

Leslie, F. (1987) Methods used for the study of opioid receptors. *Pharmacol. Rev.* 39, 197–249.

Loh, H. H. and Smith, A. P. (1990) Molecular characterization of opioid receptors. *Annu. Rev. Pharmacol. Toxicol.* 30, 123–147.

Lord, J. A., Waterfield, A. A., Hughes, J., and Kosterlitz, H. W. (1977) Endogenous opioid peptides: multiple agonists and receptors. *Nature* 267, 495–499.

Lutz, R. A. and Pfister, H. P. (1992) Opioid receptors and their pharmacological profiles. *J. Receptor Res.* 12, 267–286.

Mansour, A., Khachaturian, H., Lewis, M. E., Akil, H. and Watson, S. J. (1987) Autoradiographic differentiation of mu, delta, and kappa opioid receptors in the rat forebrain and midbrain. *J. Neurosci.* 7, 2445–2464.

Mattia, A., Vanderah, T., Mosberg, H. I., and Porreca, F. (1991) Lack of antinociceptive cross-tolerance between [D-Pen2,D-Pen5]enkephalin and [D-Ala2]deltorphin II in mice: evidence for delta receptor subtypes. *J. Pharmacol. Exp. Ther.* 258, 583–587.

McKenzie, F. R. and Milligan, G. (1990) Delta-opioid-receptor-mediated inhibition of adenylate cyclase is transduced specifically by the guanine-nucleotide-binding protein Gi2. *Biochem. J.* 267, 391–398.

Pasternak, G. W. (1988) Multiple morphine and enkephalin receptors and the relief of pain. *JAMA* 259, 1362–1367.

Paul, D., Bodnar, R. J., Gistrak, M. A., and Pasternak, G. W. (1989) Different mu receptor subtypes mediate spinal and supraspinal analgesia in mice. *Eur. J. Pharmacol.* 168, 307–314.

Pelton, J. T., Gulya, K., Hruby, V. J., Duckles, S. P., and Yamamura, H. I. (1985) Conformationally restricted analogs of somatostatin with high mu-opiate receptor specificity. *Proc. Natl. Acad. Sci. U.S.A.* 82, 236–239.

Portoghese, P. S. (1992) The role of concepts in structue-activity relationship studies of opioid ligands. *J. Med. Chem.* 35, 1927–1937.

Rapaka, R. S. and Porreca, F. (1991) Development of delta opioid peptides as nonaddicting analgesics. *Pharmaceutical Res.* 8, 1–8.

Schofield, P. R., McFarland, K. C., Hayflick, J. S., Wilcox, J. N., Cho, T. M., Roy, S., Lee, N. M., Loh, H. H., and Seeburg, P. H. (1989) Molecular characterization of a new immunoglobulin superfamily protein with potential roles in opioid binding and cell contact. *EMBO J.* 8, 489–495.

Taussig, R., Sanchez, S., Rifo, M., Golman, A. G., and Belardetti, F. (1992) Inhibition of the omega-conotoxin-sensitive calcium current by distinct G proteins. *Neuron* 8, 799–809.

Traynor, J. (1989) Subtypes of the kappa-opioid receptor: fact or fiction? *Trends Pharmacol. Sci.* 10, 52–53.

Weber, E., Evans, C. J., and Barchas, J. D. (1983) Multiple endogenous ligands for opioid receptors. *Trends Neurosci.* 6, 333–336.

Wollemann, M. (1990) Recent developments in the research of opioid receptor subtype molecularcharacterization. *J. Neurochem.* 54, 1095–1101.

Xie, G. X., Miyajima, A., and Goldstein, A. (1992) Expression cloning of cDNA encoding a seven-helix receptor from human placenta with affinity for opioid ligands. *Proc. Natl. Acad. Sci. U.S.A.* 89, 4124–4128.

Zimmerman, D. M. and Leander, J. D. (1990) Selective opioid receptor agonists and antagonists: research tools and potential therapeutic agents. *J. Med. Chem.* 33, 895–902.

Chen, Y., Mestek, A., Liu, J., Hurley, J. A. and Yu, L. (1993) Molecular cloning and functional expression of a mu-opioid receptor from rat brain. *Mol. Pharmacol.* 44, 8–12.

Fukuda, K., Kato, S., Mori, K., Iwabe, N., Miyata, T., Nishi, M. and Takeshima, H. (1993) Primary structures and expression from cDNAs of rat opioid receptor delta- and mu-subtypes. *FEBS Lett.* 327, 311–314.

Yasuda, K., Raynor, K., Kong, H., Breder, C. D., Takeda, J., Reisine, T. and Bell, G. I. (1993) Cloning and functional comparison of kappa and delta opioid receptors from mouse brain *Proc. Natl. Acad. Sci. U.S.A.* 90, 6736–6740.

22

Opsins

Thomas P. Sakmar, M.D.

1.22.0 Introduction

Visual pigments comprise a large family of G protein-coupled receptors. They share homology with other G protein-coupled receptor types; however, there is significant specialization in visual pigments not found in other receptor families (Dohlman et al., 1991; Fryxell and Meyerowitz, 1991; Hargrave, 1991; Hargrave and McDowell, 1992). The visual pigments of many species of vertebrates and invertebrates have been studied by absorption spectroscopy or microspectrophotometry of visual organs. Therefore, historically vertebrate visual pigments have been classified on the basis of photoreceptor cell type of the retina in which they were detected (Bowmaker and Dartnall, 1980). Rod cells, responsible for dim-light vision, contain rhodopsin ("red" opsin). Cone cells, responsible for bright-light and color vision, contain iodopsins ("violet" opsin), also known as cone pigments or color vision pigments. Cloning of opsins from a variety of species has allowed more detailed comparisons and phylogenetic classifications based on structural, spectral, and biochemical properties of visual pigments (Applebury and Hargrave, 1986; Findlay and Pappin, 1986; Nathans, 1987).

Pigments are made up of opsin apoprotein plus chromophore. The chromophore is a cofactor and not a ligand as in other seven helix receptors. It is linked covalently via a Schiff base bond to a specific lysine residue. The chromophore-binding pocket resides in the membrane-embedded domain of the protein. One of two retinoids serves as the chromophore for all visual pigments. The chromophore in most vertebrate pigments is the aldehyde of vitamin A, 11-*cis*-retinal. The chromophore in some fishes and amphibians is the aldehyde of vitamin A_2, 11-*cis*-3-dehydroretinal, which contains an additional carbon-carbon double bond in the β-ionone ring (Figure 1).

Photoisomerization of the 11-*cis* to all-*trans* form of the chromophore is the primary event in visual signal transduction, and it is the only light-dependent step (Schoenlein et al., 1991; Wald, 1968). Retinal isomerization activates the pigment, allowing it to interact with a specific heterotrimeric G protein. In the case of the vertebrate visual system, G protein activation leads to the activation of a cyclic-GMP phosphodiesterase (Fung et al., 1981), and the closing of a cyclic-GMP-gated cation channel (Kaupp et al., 1989; Kaupp, 1991). In short, light causes a graded hyperpolarization of the photoreceptor cell. The amplification, modulation, and regulation of the light response is of great physiological importance and has been discussed in detail (Chabre, 1985; Koch, 1992; Stryer, 1991).

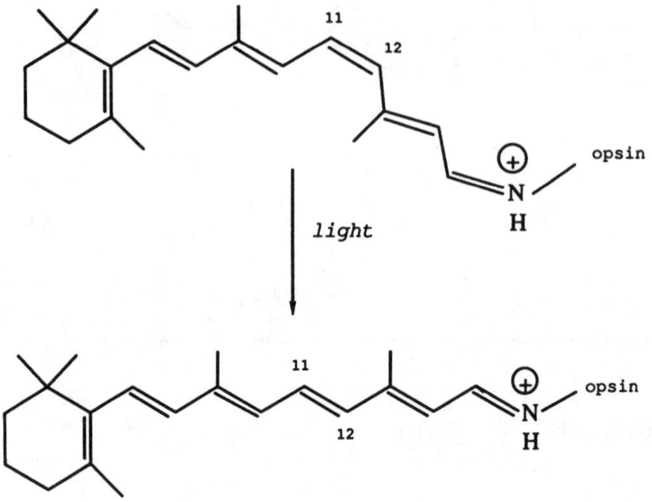

FIGURE 1. Photoisomerization of 11-*cis*-retinal to all-*trans*-retinal is the only light-dependent event in vision. All vertebrate visual pigments contain one of two chromophores, vitamin A aldehyde (11-*cis*-retinal), or vitamin A_2 aldehyde (11-*cis*-3-dehydroretinal), which contains an additional carbon-carbon double bond in the β-ionone ring. The chromophore is covalently linked as a cofactor to a specific opsin lysine residue via a Schiff base bond.

All pigments are tuned to a characteristic wavelength of maximal absorption (λ_{max}). Despite the fact that retinal is the universal chromophore, the λ_{max} values of visual pigments span the visible spectrum, i.e., from near ultraviolet at about 400 nm to far visible red at about 600 nm. Distinct chromophore-protein interactions are responsible directly or indirectly for spectral tuning in visual pigments. Thus, differences in primary structure result in differences in spectral properties.

As pigment genes from a variety of species have been cloned and characterized, models of pigment evolution have been proposed (Okano et al., 1989; Wang, S. et al., 1992; Yokoyama and Yokoyama, 1989, 1990a). The homology in the opsin family of genes indicates that divergent evolution occurred from a single precursor retinal-binding protein to form long- and short-wavelength absorbing prototypes. The long-wavelength prototype diverged to form red and green pigments. The short-wavelength prototype then diverged to form a blue pigment and the family of rhodopsins and rhodopsin-like green pigments. Details of sequence comparisons are discussed below.

It is interesting to note that retinal-based pigments are found in unicellular algae and prokaryotes. It is possible that these pigments are the precursors of the super-family of G protein-coupled receptors (Yokoyama and Yokoyama, 1989). One of the most widely studied membrane proteins has been bacteriorhodopsin (br), the light-driven proton pump of *Halobacterium halobium*. Detailed spectroscopic studies of mutants and a high resolution structure for br have led to a basic understanding of the mechanism of light-driven proton pumping (Henderson et al., 1990; Khorana, 1988). Some structural aspects of br are relevant to understanding the structures of visual pigments (Henderson and Schertler, 1990).

A number of structural features are shared by visual pigments. Like all G protein-coupled receptors, they consist of seven hydrophobic domains (Dratz and Hargrave, 1982). A Glu or Asp/Arg/Tyr tripeptide sequence is found at the cytoplasmic border of the third transmembrane domain. This domain is conserved in most G protein-coupled receptors and has been shown to be involved in G protein interaction (Franke et al., 1990; Sakmar et al., 1989). A lysine residue that acts as the linkage site for the chromophore is conserved within the seventh transmembrane segment in all pigments. A pair of highly conserved cysteine residues is found on the extracelluar surface and may form a disulfide bond. In many pigments, a carboxylic acid residue that acts as the counterion to the protonated, positively charged Schiff base is conserved within the third transmembrane segment (Nathans, 1990; Sakmar et al., 1989; Zhukovsky and Oprian, 1989). Sites of light-dependent phosphorylation (serine and threonine residues) are found at the carboxyl-terminal of most visual pigments. These sites may be analogous to phosphorylation sites found on the carboxyl-terminal domains of other G protein-coupled receptors (Benovic et al., 1986).

1.22.1   Rhodopsins

1.22.1.1   Human Rhodopsin and Cone Pigments

Rhodopsin is the pigment of the retinal photoreceptor rod cell that is responsible for dim-light vision. The gene for human rhodopsin was cloned from a genomic library and sequenced (Nathans and Hogness, 1984). The gene is located on chromosome 3. It codes for a protein with 348 amino acid residues and contains four introns. The primary structure is 93.4% homologous to that of bovine rhodopsin. The key structural, spectral, and biochemical features of rhodopsins in general are discussed in Section 1.22.1.4.

Retinitis pigmentosa is a group of hereditary progressive blinding diseases with varible clinical presentations. One form of the disease, autosomal dominant retinitis pigmentosa (ADRP) was linked to a mutation in the gene for rhodopsin (Dryja et al., 1990; Farrar, 1990). About 30 different rhodopsin gene mutations have been reported in ADRP patients (Dryja et al., 1990; Dryja et al., 1991; Gal et al., 1991; Humphries et al., 1992; Inglehearn et al., 1991; Keen et al., 1991; Sung et al., 1991a). The mutations reported would result in alterations in all domains of rhodopsin; extracellular, membrane-embedded, and cytoplasmic. A study was carried out in which site-directed mutant opsin genes corresponding to ADRP genotypes were prepared (Sung et al., 1991b). When expressed in 293S cells in tissue culture, the mutant opsins displayed a heterogeneity of spectral properties and cellular transport behavior. Some mutants were defective in chromophore binding, others in cellular transport and insertion into the plasma membrane. However, some mutations had no apparent effect. One mutation linked to ADRP is a replacement of the Schiff base lysine by glutamic acid. This mutation should prevent chromophore Schiff base formation. Interestingly, a similar mutant of bovine rhodopsin was shown to have constitutive activity without chromophore addition in *in vitro* transducin activation assays (Robinson et al., 1992). The molecular pathophysiology of ADRP remains to be fully elucidated.

Human trichromatic color vision, at the level of the photoreceptor, requires the presence of three cone pigments with broad overlapping spectral absorption. Three genomic and cDNA clones encoding the opsin apoproteins of these pigments were cloned and characterized (Nathans et al., 1986b). The amino acid sequences of these opsins are about 41% identical to that of human rhodopsin. The green and red opsins are about 96% identical to each other and about 43% identical to the blue opsin. The spectral properties of human cone pigments have been studied by a variety of techniques ranging from psychophysical color matching to microspectrophotometry (Bowmaker and Dartnall, 1980; Winderickx et al., 1992). Recently, however, the human cone pigment genes were expressed in tissue culture cells, reconstituted with 11-*cis*-retinal, and studied by ultra-violet-visible spectroscopy (Merbs and Nathans, 1992a; Oprian et al., 1991). The λ_{max} values reported in the two studies were as follows: blue, 426 nm; green, 530 nm; red, 552 and 557 nm for polymorphic variants (Merbs and Nathans, 1992a); and blue, 424 nm, green, 530 nm, and red, 560 nm (Oprian et al., 1991). These studies confirmed the assignments based on genetic analysis of the cloned pigment genes.

Analysis of the arrangement of the cone opsin genes on the X chromosome has led to a detailed understanding of the molecular genetics of inherited variations in color vision (Nathans et al., 1986a). In males with normal color vision, a single red opsin gene resides with one or more green opsin genes in a head-to-tail tandem array. In one type of color vision defect, anomalous trichromacy, unequal intragenic recombination can result in an opsin gene that is a hybrid between green and red opsin genes. It was proposed that these hybrids would have anomalous spectral properties (Nathans et al., 1986a; Neitz et al., 1989). Recently, this genetic hypothesis was confirmed experimentally by obtaining absorption spectra for heterologously expressed hybrid pigments responsible for anomalous trichromacy (Merbs and Nathans, 1992b). The molecular genetics of blue cone monocromacy has also been elucidated (Nathans et al., 1989). A genetic model to account for the absence of the green and red genes has been tested in transgenic mice (Wang et al., 1992). The results suggest that a conserved 5′ region interacts with the green or red gene promoter to determine which gene is expressed in a given cone cell.

Comparisons of amino acid residues in the chromophore binding pocket of rhodopsin, cone pigments, and hybrid cone pigments has led to a number of proposals regarding the specific amino acid residues responsible for spectral tuning in the visual pigments (Kosower, 1988; Nathans et al., 1986b; Neitz and Jacobs, 1986; Neitz et al., 1989). Of the 15 amino acid differences between green and red pigments, three hydroxyl-bearing amino acid residues may be predominantly responsible for the spectral shift: Ser-180, Tyr-277, and Thr-285 (Chan et al., 1992; Merbs and Nathans, 1992b; Neitz et al., 1991).

1.22.1.3 Monkey Rhodopsin

Partial nucleotide sequence information from polymerase chain reaction nucleotide sequencing of monkey visual pigment exons has led to models of spectral tuning and to predictions about the identity of specific amino acids involved in human red-green color vision (Neitz et al., 1991; Williams et al., 1992). New world monkeys, such as marmosets (*Callithrix jacchus jacchus*), tamarins (e.g., *Saguinus fuscicollis*), and squirrel monkeys (e.g., *Saimiri sciureus*) have dicromatic vision with a blue pigment and a single long-wavelength pigment. However, there is a striking polymorphism in the long-wavelength pigment, such that females with two X-linked opsin genes are effectively trichromatic. Using pairwise comparisons of opsin gene sequences from a number of individual male monkeys with a range of pigment λ_{max} values, a model was proposed in which amino acid residues at three sites account for the spectral variation between human green and red cone pigments: postions 180, 277, and 285 (Neitz et al., 1991). This model was tested by introducing mutations at these sites in bovine rhodopsin (Chan et al., 1992) and by expression of hybrid human pigment genes (Merbs and Nathans, 1992b).

1.22.1.4 Bovine Rhodopsin

Bovine rhodopsin is the most extensively studied G protein-coupled receptor. A large amount of pigment (as much as 0.4 mg) can be obtained from a single bovine retina by a sucrose density gradient centrifugation preparation of rod outer segment disk membranes. The pigment can be further purified by lectin affinity chromatography on conconavalin A Sepharose resin. Rhodopsin is stable in solubilized form in a variety of detergents, including digitonin, dodecylmaltoside, and octyl glucoside. Bovine rhodopsin was the first G protein-coupled receptor to be sequenced by amino acid sequencing (Hargrave et al., 1983; Ovchinnikov, 1982) and the first to be cloned (Nathans and Hogness, 1983). The cloning of a β-adrenergic receptor (Dixon et al., 1986) led to the identification of the structural homologies that now define the super-family of G protein-coupled receptors.

Bovine rhodopsin is 348 amino acid residues in length, and is identical to human rhodopsin at all but 23 positions. Hydrophobicity profiles are consistent with seven transmembrane segments (Dratz and Hargrave, 1983). The amino-terminal domain is remarkable for two N-linked glycosylation sites. The carboxyl-terminal domain is rich in serine and threonine residues, which are phosphorylated in a light-dependent manner by rhodopsin kinase. The retinal Schiff base linkage is at Lys-296.

Bovine rhodopsin has a broad absorption maximum (λ_{max}) at about 500 nm (Figure 2). Upon photoisomerization of the chromophore, the pigment is converted to metarhodopsin II (MII) with an λ_{max} value of 380 nm. The MII intermediate is characterized by a deprotonated Schiff base chromophore linkage. MII is the active form of the receptor that catalyzes guanine nucleotide exchange by transducin (Ganter et al., 1989; Longstaff et al., 1986).

Recent work on the structure and function of bovine rhodopsin focusing on molecular biology has been reviewed (Nathans, 1992; Khorana, 1992). These studies have elucidated the role of the extracellular domain in structure, key retinal-protein interactions, and the domains involved in transducin activation. The extracellular loops of bovine rhodopsin have been shown in a deletion analysis to be important for proper folding of the receptor that allows cellular processing and chromophore binding (Doi et al., 1990). In addition, two conserved cysteine residues on the extracelluar domain, Cys-110 and Cys-187, were shown to be essential for proper folding of opsin (Karnik et al., 1988). These two residues were shown to form a disulfide linkage (Karnik and Khorana, 1990).

A number of studies have been carried out to investigate the retinal-binding domain of bovine rhodopsin. Several spectroscopic methods such as resonance Raman spectroscopy, Fourier-transform infrared difference spectroscopy, and nmr spectros-copy have been employed (Birge, 1990; Sawatzki et al., 1990; Siebert, 1990). Other approaches have included reconstitution of opsin apoprotein with synthetic retinal analogs (Honig et al., 1980) and photochemical cross-linking (Nakayama and Khorana, 1990). Recently, site-directed mutagenesis in combination with ultraviolet-visible spectroscopy and other spectroscopic and biochemical techniques has provided useful information about the retinal binding pocket (Bhattacharya et al., 1992; Nakayama and Khorana, 1991). One important result from the study of mutant bovine rhodopsin pigments was the identification of the retinylidene Schiff base counterion as Glu-113 (Nathans, 1990; Sakmar et al., 1989; Zhukovsky and Oprian, 1989). A model for the retinal binding pocket in bovine rhodopsin has been proposed based on a study of mutant pigments by microprobe resonance Raman spectroscopy (Figure 3) (Lin et al., 1992).

The domains of rhodopsin that interact with transducin have also been studied by site-directed mutagenesis of bovine rhodopsin. Relatively large segments of cytoplas-mic loops are required for proper rhodopsin-transducin interaction (Franke et al., 1992; König et al., 1989). However, single amino acid substitutions within these domains can have dramatic effects on transducin activation (Franke et al., 1988; Gurevich et al.,

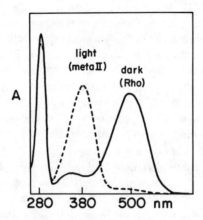

FIGURE 2. An ultraviolet-visible absorption spectrum of purified bovine rhodopsin shows a characteristic broad visible absorbance with a λ_{max} value of 500 nm. The 280-nm peak represents the protein component. After exposure to light, the pigment is converted to a peak with a λ_{max} value of 380 nm characteristic of metarhodopsin II. This is the active form of the receptor that interacts with the rod cell G protein, transducin.

1990; Sakmar et al., 1989). In addition, transducin binding, and activation of bound transducin were shown to be discrete steps involving different surface domains of the receptor (Franke et al., 1990).

Recently, bovine rhodopsin mutants with substitutions at the site of the Schiff base linkage (Lys-296) and the counterion (Glu-113) were employed to show that a covalent bond to the 11-*cis*-retinal chromophore is not required for light-dependent activation of transducin (Zhukovsky et al., 1991). Furthermore, rhodopsin mutants were reported that have constitutive activity. They activate transducin in the absence of chromophore (Robinson et al., 1992). In this context, the main physiological reason for a covalent linkage between opsin and chromophore may be to reduce dark noise and to provide an extremely rapid light-dependent activation. These results emphasize

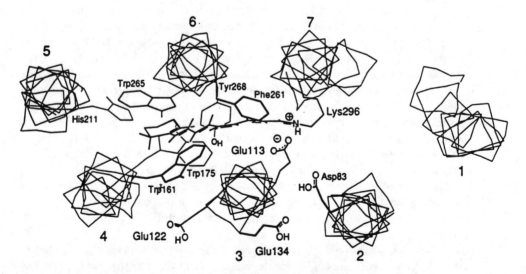

FIGURE 3. A schematic representation of the retinal-binding pocket of bovine rhodopsin (Lin et al., 1992). The pigment is viewed from above the plane of the membrane bilayer. Selected amino acid side chains are displayed and numbered.

similarities among the opsins and other receptors in terms of the mechanism of G protein activation.

1.22.1.11 Mouse Rhodopsin

The complete nucleotide sequence of the mouse opsin gene including introns and flanking sequences was reported (Al-Urbaidi et al., 1990). Evidence was also presented that the heterogeneity of mouse opsin transcript was due to the use of multiple polyadenylation sites in the 3′-flanking region. The existence of this clone is especially important because of the potential for using transgenic mice in studies of vertebrate visual development and of the molecular pathophysiology of human retinitis pigmentosa. The gene for rhodopsin is located on chromosome 6 (Elliott et al., 1990).

Ultraviolet retinal sensitivity has been identified in rodent retinas (Jacobs et al., 1991), but the cloning of a vertebrate ultraviolet pigment gene has not been reported.

1.22.1.15 Chicken Rhodopsin

Chicken retinas contain an abundance of cone cells compared with other vertebrate retinas, and the chicken visual system has been studied in detail using a number of approaches in addition to molecular biology. Five chicken visual pigments have been partially purified and characterized spectrally (Chen et al., 1989; Okano et al., 1989). The cloned DNA sequences of three chicken opsins have been reported: chicken rhodopsin (Takao et al., 1988), chicken red iodopsin (Kuwata et al., 1990; Tokunaga et al., 1990; Yoshizawa and Kuwata, 1991), and a chicken opsin that is likely to correspond to the chicken green iodopsin (Wang et al., 1992). The latter pigment is of particular interest because it displays an λ_{max} value (495 nm) similar to that of rhodopsin (500 nm), yet it is sensitive to hydroxylamine bleaching. This opsin was reported to define a new branch of the visual pigment gene family that is now known to contain the goldfish green pigment and the gecko rhodopsin (see below).

1.22.1.18 Octopus and Squid Rhodopsin

The primary structures of octopus (*Paroctopus defleini*) rhodopsin (Dergachev et al., 1989; Ovchinnikov et al., 1988) and squid (*Logligo forbesi*) rhodopsin (Hall et al., 1991) have been reported. The amino acid sequences of the octopus and squid rhodopsin are homologous and are most similar to the *Drosophila* opsins (see below). Several features of these invertebrate pigments distinguish them from vertebrate pigments. For example, in place of a glutamic acid identified to serve as the Schiff base counterion at position 113 in bovine rhodopsin, both of these pigments have a tyrosine residue. Some features of invertebrate rhodopsins such as the stability and acidity constant (pK_a) values of their photoproducts are suggestive of bovine rhodopsin mutants with replacements of Glu-113. However, replacement of Glu-113 in bovine rhodopsin by Tyr does not result in a mutant pigment with properties similar to those of invertebrate rhodopsins (Sakmar, unpublished results).

The carboxyl-terminal region of octopus rhodopsin and squid rhodopsin consists of two domains. The first domain is a charged region that may contain a calcium binding site in addition to multiple threonine and serine residues. The second domain is a proline-rich region that consists of a multiple (8 to 10 times) repeat of a pentapeptide with the following consensus sequence: Pro-Pro-Gln-Gly-Tyr. This structure may be involved in regulating rhodopsin-rhodopsin and rhodopsin-cytoskeleton interactions within the visual microvilli of cephalopods (Hall et al., 1991).

1.22.1.21 *Drosophila* Rhodopsin

The invertebrate visual transduction system has been elucidated by biochemical, electrophysiological, and genetic studies. There are fundamental differences between vertebrate and invertebrate signaling. For example, at the level of the photoreceptor, the chromophore in some higher orders of insects is likely to be 11-*cis*-3-hydroxyretinal, rather than 11-*cis*-retinal (Goldsmith et al., 1986). Photoactivation of the pigments results in a stable meta (M) state that displays a red-shifted absorption maximum relative to that of the dark form of the pigment. The heterotrimeric G protein activated by the pigment in turn activates a visual phopholipase C, resulting in increases in cellular inositol-3-phosphate and diacylglycerol. Calcium is mobilized and ultimately sodium channels are opened resulting in a depolarization of the photoreceptor cell (Smith et al., 1991a; Smith et al., 1991b).

The visual system of the fruit fly (*Drosophila melanogaster*) has proven especially valuable for the study of invertebrate visual pigments and visual system development. The *Drosophila* compound eye is made up of about 800 ommatidia. Each ommatidium contains 8 photoreceptor cells: 6 outer cells (R1–R6) and 2 central cells (R7 and R8). In addition, simple eyes called ocelli are found on the top of the fly head. The cells fall into one of 3 spectral classes: R1–R6 are blue-sensitive photoreceptors with absorption peaks at about 470 nm, R7 and R8 are ultraviolet receptors, and the ocelli are violet sensitive with absorption peaks at about 420 nm (Feiler et al., 1988; Harris et al., 1976).

Four opsin genes (Rh1–Rh4) have been cloned and characterized. The *ninaE* gene was shown to be the structural gene for Rh1, which is found in the R1–R6 cells (O'Tousa et al., 1985; Zuker et al., 1985). Although the Rh1 rhodopsin shares only about 22% homology at the amino acid level with bovine rhodopsin, the seven transmembrane segment motif and a number of other structual features are conserved. The Rh2 gene encodes the ocelli opsin (Cowman et al., 1987; Feiler et al., 1988; Fryxell and Meyerowitz, 1987). The Rh3 and Rh4 opsins are expressed in the pair of central ommatidial cells (Montell et al., 1987; Zuker et al., 1987). The Rh2 gene is 67% homologous, and the Rh3 and Rh4 genes are only about 38% homologous to the Rh1 opsin.

An opsin of the larger fly species (*Calliphora erythrocephala*) was also cloned and displayed an 86% amino acid identity with the *Drosophila* Rh1 opsin (Huber et al., 1990).

1.22.1.22 Fish Rhodopsin and Iodopsins

The visual systems of fish have evolved to adapt to a variety of specific underwater environments (Loew and Lythgoe, 1978). For example, the visual spectrum in fish is somewhat shifted to longer wavelengths from that of other vertebrates because of the use of the chromophore 11-*cis*-3-dehydroretinal, which contains one additonal carbon-carbon double bond. Some species of fish exhibit ultraviolet spectral sensitivity, as well as sensitivity to polarized light (Bowmaker et al., 1991; Hárosi and Hashimoto, 1983). Recently, a number of fish opsins have been cloned and characterized. Of particular interest has been the cloning of long-wavelength sensitive visual pigments from fish. Comparison of fish long-wavelength pigments with mammalian long-wavelength pigments allows insights regarding the evolutionary mechanism of visual development.

The Mexican blind cave fish (*Astyanax fasciatus*) shows rhodopsin and green-spectral responses in its pineal organ. Three opsin genes have been cloned from a genomic library that have been tentatively assigned to be a red-like opsin and two related green-like opsins (Yokoyama and Yokoyama, 1990a, 1990b). A rhodopsin has also been cloned from lamphrey (*Lampetra japonica*) (Histomi et al., 1991).

Table 1. Opsin

Receptor	Second messenger	Species cloned	Chromosomal location	a.a. Sequence	GenBank Accession	Primary reference
Rhodopsin		Human	3	348	P08100	Nathans and Hogness, 1984
		Bovine		348	P02699	Ovchinnikov, 1982
		Sheep		348	P02700	Pappin et al., 1984
		Hamster		348	P28681	Gale et al., 1992
		Mouse	6	348	P15409	Baehr et al., 1988
		Chicken		351	P22328	Takao et al., 1988
		Xenopus		354	P29403	Saha et al., 1992
		Calliphora fly		371	P22269	Huber et al., 1990
		Lamprey		353	P22671	Hisatomi et al., 1991
		Octopus		455	P09241	Ovchinnikov et al., 1988
		Squid		452	P24603	Hall et al., 1991
		Drosophilia1		381	P06002	O'Tousa et al., 1985
		Drosophilia2		381	P08099	Cowman et al., 1986
		Drosophilia3		383	P04950	Fryxell et al., 1987
		Drosophilia4		378	P08255	Montel et al., 1987
Opsin (blue)		Human		348	P03999	Nathans et al., 1986
		Chicken		361	P28682	Okano et al., 1992
Opsin (red)		Human1		364	P04000	Nathans et al., 1986
		Human2				
		Chicken		362	P28683	Okano et al., 1992
		Astyanax		342	P22332	Yokoyama et al., 1990
Opsin (green)		Human		364	P04001	Nathans et al., 1986
		Chicken		355	P28683	Okano et al., 1992
		Astyanax1		355	P22330	Yokoyama et al., 1990
		Astyanax2		353	P22331	Yokoyama et al., 1990
Opsin (Violet)		Chicken		347	P28684	Okano et al., 1992

Goldfish (*Carassius auratus*) show rod and cone spectral sensitivities similar to those of humans. Five opsin genes from goldfish have been cloned and expressed (Johnson et al., 1992). The clones were assigned to a rod opsin, a red cone opsin that may be a polymorphic variant, a blue cone opsin that shows homology to both human blue and human rod opsins, and two green opsins that share some biochemical properties with chicken green opsins.

An evolutionary model has been described that takes into account the sequence information from human, chicken, fish, and gecko opsin sequences (Johnson et al., 1992; Yokoyama and Yokoyama, 1989). In summary, short- and long-wavelength sensitve opsins diverged from a common ancestor. The short-wavelength opsin precursor then diverged into blue opsins and a family of rhodopsin. The rhodopsin family includes goldfish green, chicken green, and gecko rod opsins that share biochemical properties with rhodopsins such as resistance to hydroxylamine reaction. The physiological roles of the rhodopsin-like green opsins is not understood. The long-wavelength opsin precursor went on to form human red and green iodopsins, cave fish red and green opsins, gecko green opsin, and chicken and goldfish red opsins.

1.22.4   Other Retinal-Based Pigments

Bacteriorhodopsin (br) is a light-driven proton pump found in *Halobacterium halobium* (Stoeckenius and Bogomolni, 1982). Like visual pigments, br is an integral membrane protein and employs a Schiff base-linked retinal (all-*trans*-retinal) as a chromophore. Br has been studied extensively by spectroscopic, biochemical, and genetic methods (Khorana,

1988). In addition, a high resolution structure from cryoelectron microscopy has been reported for br (Henderson et al., 1990). This structure confirmed that seven segments transverse the membrane bilayer and that they are predominantly α-helical in secondary structure. A number of molecular models of visual pigments and other G protein-coupled receptors have been based on the structure of br and the relevance of structural comparisons has been discussed (Birge, 1990; Henderson and Schertler, 1990).

Other retinal-based pigments have been identified in prokaryotes (Duschil et al., 1990; Scharf et al., 1992; Stoeckenius and Bogomolni, 1982) and in unicellular algae (Beckmann and Hegemann, 1991; Foster et al., 1991; Gualtieri et al., 1992). These pigments are involved in phototactic and photophobic responses, and are thus a part of primitive sensory systems. For example, rhodopsin has been shown to regulate calcium currents in *Chlamydomonas* (Harz and Hegemann, 1991). The primary structures of prokaryotic sensory rhodopsins have been reported (Blanck et al., 1989; Lanyi et al., 1990).

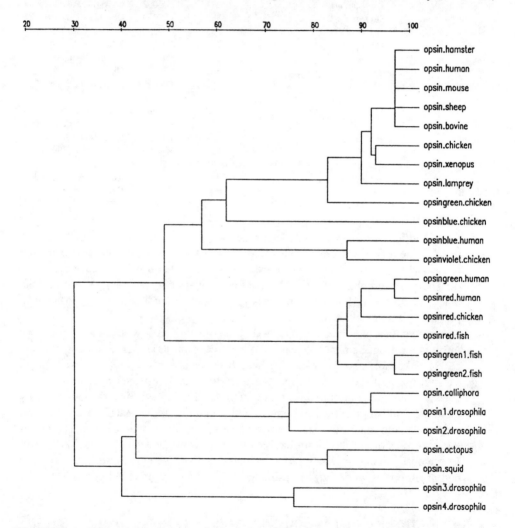

FIGURE 4. Primary structures of cloned opsins. Primary structures of prokaryotic sensory rhodopsins have been reported (Blanck et al., 1989; Lanyi et al., 1990).

```
opsin.calliphora     1 MERY--ST-----PLIGPSFAALT--NGSV--TD-KVTPDMAHLVHPYWN
opsin.chicken        1 MN----G------TE-GQDFY----------VPMS-NKTGVVR-SPFEYP
opsin.cow            1 MN----G------TE-GPNFY----------VPFS-NKTGVVR-SPFEAP
opsin.hamster        1 MN----G------TE-GPNFY----------VPFS-NATGVVR-SPFEYP
opsin.human          1 MN----G------TE-GPNFY----------VPFS-NATGVVR-SPFEYP
opsin.lamprey        1 MN----G------TE-GDNFY----------VPFS-NKTGLAR-SPYEYP
opsin.mouse          1 MN----G------TE-GPNFY----------VPFS-NVTGVGR-SPFEQP
opsin.octopus        1 --------------MVESTTLVNQT-----------WWYNPTVD--IHPHWA
opsin.sheep          1 MN----G------TE-GPNFY----------VPFS-NKTGVVR-SPFEAP
opsin.squid          1 ----------MGRDIP-DNET---------WWYNPYMD--IHPHWK
opsin1.drosophila    1 MESFAVAA-----AQLGPHFAPLS--NGSV--VD-KVTPDMAHLISPYWN
opsin2.drosophila    1 MERSHLPETPFDLAHSGPRFQAQSSGNGSV--LD-NVLPDMAHLVNPYWS
opsin3.drosophila    1 MESGNVSSSLFGNVSTALRPEARLSAETRL--LGWNVPPEELRHIPEHWL
opsin4.drosophila    1 ME------PLCNASEPPLRPEARSSGNGDLQFLGWNVPPDQIQYIPEHWL
opsinblue.chicken    1 MHPPRPT------TDLPEDFY----------IPMALDAPNITALSPFLVP
opsinblue.human      1 MR---------KMSEEEFY----------L-F--KNISSV--GPWDGP
opsingreen.chicken   1 MN----G------TE-GINFY----------VPMS-NKTGVVR-SPFEYP
opsingreen.human     1 MAQQWSL------QRLAGRHPQDSYEDSTQSSIFTYTNSNSTR-GPFEGP
opsingreen1.fish     1 MAAHADE------PVFAARRYN---EETTRESAFVYTNANNTR-DPFEGP
opsingreen2.fish     1 MAAH--E------PVFAARRHN---EDTTRESAFVYTNANNTR-DPFEGP
opsinred.chicken     1 MA-AWE------AAFAARRRHEE-EDTTRDSVFTYTNSNNTR-GPFEGP
opsinred.fish        1 MGDQWGD------AVFAARRRG---DDTTREAAFTYTNSNNTK-DPFEGP
opsinred.human       1 MAQQWSL------QRLAGRHPQDSYEDSTQSSIFTYTNSNSTR-GPFEGP
opsinviolet.chicken  1 MS-----------SDDDFY----------L-F--TNGSVP--GPWDGP


opsin.calliphora    39 QFPAMEPKWAKFLAAYMVLIATISWCGNGVVIYIFSTTKSLRTPANLLVI
opsin.chicken       28 QYYLAEPWKFSALAAYMFMLILLGFPVNFLTLYVTIQHKKLRTPLNYILL
opsin.cow           28 QYYLAEPWQFSMLAAYMFLLIMLGFPINFLTLYVTVQHKKLRTPLNYILL
opsin.hamster       28 QYYLAEPWQFSMLAAYMFLLIVLGFPINFLTLYVTVQHKKLRTPLNYILL
opsin.human         28 QYYLAEPWQFSMLAAYMFLLIVLGFPINFLTLYVTVQHKKLRTPLNYILL
opsin.lamprey       28 QYYLAEPWKYSALAAYMFFLILVGFPINFLTLFVTVQHKKLRTPLNYILL
opsin.mouse         28 QYYLAEPWQFSMLAAYMFLLIVLGFPINFLTLYVTVQHKKLRTPLNYILL
opsin.octopus       26 KFDPIPDAVYYSVGIFIGVVGIIGILGNGVVIYLFSKTKSLQTPANMFII
opsin.sheep         28 QYYLAEPWQFSMLAAYMFLLIVLGFPINFLTLYVTVQHKKLRTPLNYILL
opsin.squid         25 QFDQVPAAVYYSLGIFIAICGIIGCVGNGVVIYLFTKTKSLQTPANMFII
opsin1.drosophila   41 QFPAMDPIWAKILTAYMIMIGMISWCGNGVVIYIFATTKSLRTPANLLVI
opsin2.drosophila   48 RFAPMDPMMSKILGLFTLAIMIISCCGNGVVVYIFGGTKSLRTPANLLVL
opsin3.drosophila   49 TYPEPPESMNYLLGTLYIFFTLMSMLGNGLVIWVFSAAKSLRTPSNILVI
opsin4.drosophila   45 TQLEPPASMHYMLGVFYIFLFCASTVGNGMVIWIFSTSKSLRTPSNMFVL
opsinblue.chicken   35 QTHLGSPGLFRAMAAFMFLLIALGVPINTLTIFCTARFRKLRSHLNYILV
opsinblue.human     25 QYHIAPVWAFYLQAAFMGTVFLIGFPLNAMVLVATLRYKKLRQPLNYILV
opsingreen.chicken  28 QYYLAEPWKYRLVCCYIFFLISTGLPINLLTLLVTFKHKKLRQPLNYILV
opsingreen.human    44 NYHIAPRWVYHLTSVWMIFVVIASVFTNGLVLAATMKFKKLRHPLNWILV
opsingreen1.fish    41 NYHIAPRWVYNLASLWMIIVVIASIFTNSLVIVATAKFKKLRHPLNWILV
opsingreen2.fish    39 NYHIAPRWVYNVSSLWMIFVVIASVFTNGLVIVATAKFKKLRHPLNWILV
opsinred.chicken    41 NYHIAPRWVYNLTSVWMIFVVAASVFTNGLVLVATWKFKKLRHPLNWILV
opsinred.fish       41 NYHIAPRWVYNLATCWMFFVVVASTVTNGLVLVASAKFKKLRHPLNWILV
opsinred.human      44 NYHIAPRWVYHLTSVWMIFVVTASVFTNGLVLAATMKFKKLRHPLNWILV
opsinviolet.chicken 23 QYHIAPPWAFYLQTAFMGIVFAVGTPLNAVVLWVTVRYKRLRQPLNYILV
```

FIGURE 4b.

```
opsin.calliphora     89  NLAISDFGIMITNT-PMMGINLFYETWV--LGPLMCDIYGGLGSAFGCSS
opsin.chicken        78  NLVVADL-FMVFGGFTTTMYTSMNGYFV--FGVTGCYIEGFFATLGGEIA
opsin.cow            78  NLAVADL-FMVFGGFTTTLYTSLHGYFV--FGPTGCNLEGFFATLGGEIA
opsin.hamster        78  NLAVADL-FMVFGGFTTTLYTSLHGYFV--FGPTGCNLEGFFATLGGEIA
opsin.human          78  NLAVADL-FMVLGGFTSTLYTSLHGYFV--FGPTGCNLEGFFATLGGEIA
opsin.lamprey        78  NLAMANL-FMVLFGFTVTMYTSMNGYFV--FGPTMCSIEGFFATLGGEVA
opsin.mouse          78  NLAVADL-FMVFGGFTTTLYTSLHGYFV--FGPTGCNLEGFFATLGGEIA
opsin.octopus        76  NLAMSDLSFSAINGFPLKTISAFMKKWI--FGKVACQLYGLLGGIFGFMS
opsin.sheep          78  NLAVADL-FMVFGGFTTTLYTSLHGYFV--FGPTGCNLEGFFATLGGEIA
opsin.squid          75  NLAFSDFTFSLVNGFPLMTISCFMKYWV--FGNAACKVYGLIGGIFGLMS
opsin1.drosophila    91  NLAISDFGIMITNT-PMMGINLYFETWV--LGPMMCDIYAGLGSAFGCSS
opsin2.drosophila    98  NLAFSDFCMMASQS-PVMIINFYYETWV--LGPLWCDIYAGCGSLFGCVS
opsin3.drosophila    99  NLAFCDF-MMMVKT-PIFIYNSFHQGYA--LGHLGCQIFGIIGSYTGIAA
opsin4.drosophila    95  NLAVFDL-IMCLKA-PIF--NSFHRGFAIYLGNTWCQIFASIGSYSGIGA
opsinblue.chicken    85  NLALANL-LVILVGSTTACYSFSQMYFA--LGPTACKIEGFAATLGGMVS
opsinblue.human      75  NVSFGGF-LLCIFSVFPVFVASCNGYFV--FGRHVCALEGFLGTVAGLVT
opsingreen.chicken   78  NLAVADL-FMACFGFTVTFYTAWNGYFV--FGPVGCAVEGFFATLGGQVA
opsingreen.human     94  NLAVADL-AETVIASTISVVNQVYGYFV--LGHPMCVLEGYTVSLCGITG
opsingreen1.fish     91  NLAIADL-GETVLASTISVFNQVFGYFV--LGHPMCIFEGWTVSVCGITA
opsingreen2.fish     89  NLAIADL-GETVLASTISVINQIFGYFI--LGHPMCVFEGWTVSVCGITA
opsinred.chicken     91  NLAVADL-GETVIASTISVINQISGYFI--LGHPMCVVEGYTVSACGITA
opsinred.fish        91  NLAIADL-LETLLASTISVCNQFFGYFI--LGHPMCVFEGFTVATCGIAG
opsinred.human       94  NLAVADL-AETVIASTISIVNQVSGYFV--LGHPMCVLEGYTVSLCGITG
opsinviolet.chicken  73  NISASGF-VSCVLSVFVVFVASARGYFV--FGKRVCELEAFVGTHGGLVT

opsin.calliphora    136  ILSMCMISLDRYNVIVKGM-AGQPMTIKLAIMKIALIWFMASIWTLAPVF
opsin.chicken       125  LWSLVVLAVERYVVVCKPM-SNFRFGENHAIMGVAFSWIMAMACAAPPLF
opsin.cow           125  LWSLVVLAIERYVVVCKPM-SNFRFGENHAIMGVAFTWVMALACAAPPLV
opsin.hamster       125  LWSLVVLAIERYVVICKPM-SNFRFGENHAIMGVVFTWIMALACAAPPLV
opsin.human         125  LWSLVVLAIERYVVVCKPM-SNFRFGENHAIMGVAFTWVMALACAAPPLA
opsin.lamprey       125  LWSLVVLAIERYIVICKPM-GNFRFGNTHAIMGVAFTWIMALACAAPPLV
opsin.mouse         125  LWSLVVLAIERYVVVCKPM-SNFRFGENHAIMGVVFTWIMALACAAPPLV
opsin.octopus       124  INTMAMISIDRYNVIGRP-MAASKKMSHRRAFLMIIFVWMWSIVWSVGPVF
opsin.sheep         125  LWSLVVLAIERYVVVCKPM-SNFRFGENHAIMGVAFTWVMALACAAPPLV
opsin.squid         123  IMTMTMISIDRYNVIGRPMSASKKMSHRKAFIMIIFVWIWSTIWAIGPIF
opsin1.drosophila   138  IWSMCMISLDRYQVIVKGM-AGRPMTIPLALGKIAYIWFMSSIWCLAPAF
opsin2.drosophila   145  IWSMCMIAFDRYNVIVKGI-NGTPMTIKTSIMKILFIWMMAVFWTVMPLI
opsin3.drosophila   145  GATNAFIAYDRFNVITRPM-EGK-MTHGKAIAMIIFIYMYATPWVVACYT
opsin4.drosophila   141  GMTNAAIGYDRYNVITKPM-NRN-MTFTKAVIMNIIIWLYCTPWVVLPLT
opsinblue.chicken   132  LWSLAVVAFERFLVICKPL-GNFTFRGSHAVLGCVATWVLGFVASAPPLF
opsinblue.human     122  GWSLAFLAFERYIVICKPF-GNFRFSSKHALTVVLATWTIGIGVSIPPFF
opsingreen.chicken  125  LWSLVVLAIERYIVVCKPM-GNFRFSATHAMMGIAFTWVMAFSCAAPPLF
opsingreen.human    141  LWSLAIISWERWMVVCKPF-GNVRFDAKLAIVGIAFSWIWAAVWTAPPIF
opsingreen1.fish    138  LWSLTIISWERWVVVCKPF-GNVKFDGKWAAGGIIFAWTWAIIWCTPPIF
opsingreen2.fish    136  LWSLTIISWERWVVVCKPF-GNVKFDGKWAAGGIIFSWVWAIIWCTPPIF
opsinred.chicken    138  LWSLAIISWERWFVVCKPF-GNIKFDGKLAVAGILFSWLWSCAWTAPPIF
opsinred.fish       138  LWSLTVISWERWVVVCKPF-GNVKFDGKMATAGIVFTWVWSAVWCAPPIF
opsinred.human      141  LWSLAIISWERWLVVCKPF-GNVRFDAKLAIVGIAFSWIWSAVWTAPPIF
opsinviolet.chicken 120  GWSLAFLAFERYIVICKPF-GNFRFSSRHALLVVVATWLIGVGVGLPPFF
```

FIGURE 4c.

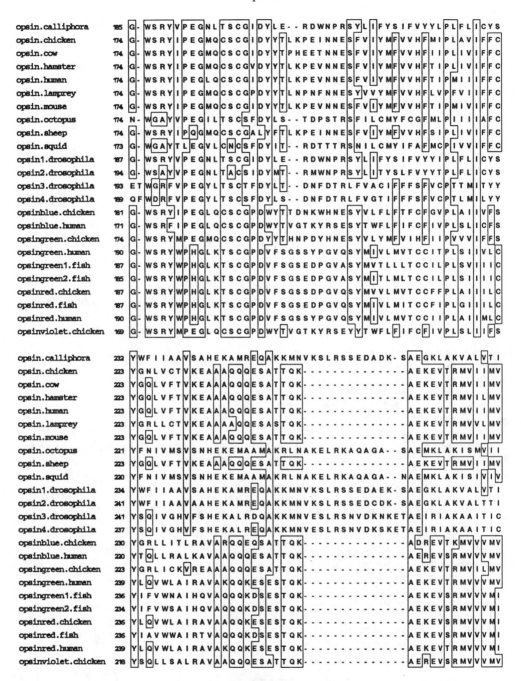

FIGURE 4d.

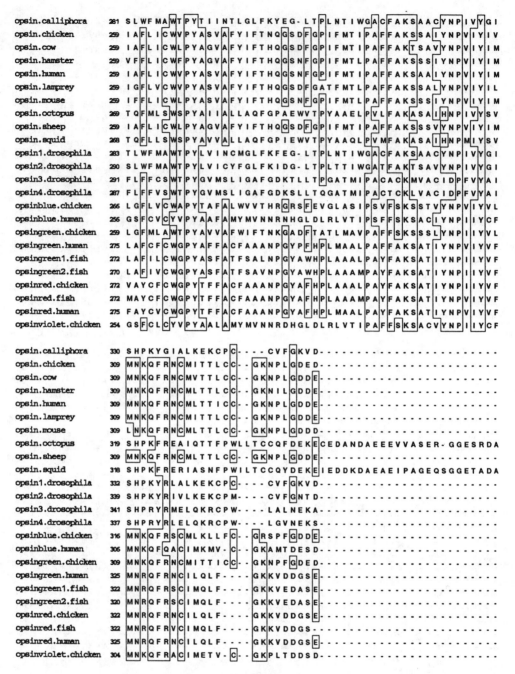

opsin.calliphora	281	S L W F M A W T P Y T I I N T L G L F K Y E G - L T P L N T I W G A C F A K S A A C Y N P I V Y G I
opsin.chicken	259	I A F L I C W V P Y A S V A F Y I F T N Q G S D F G P I F M T I P A F F A K S S A I Y N P V I Y I V
opsin.cow	259	I A F L I C W L P Y A G V A F Y I F T H Q G S D F G P I F M T I P A F F A K T S A V Y N P V I Y I M
opsin.hamster	259	V F F L I C W F P Y A G V A F Y I F T H Q G S N F G P I F M T L P A F F A K S S S I Y N P V I Y I M
opsin.human	259	I A F L I C W V P Y A S V A F Y I F T H Q G S N F G P I F M T I P A F F A K S A A I Y N P V I Y I M
opsin.lamprey	259	I G F L V C W V P Y A S V A F Y I F T H Q G S D F G A T F M T L P A F F A K S S A L Y N P V I Y I L
opsin.mouse	259	I F F L I C W L P Y A S V A F Y I F T H Q G S N F G P I F M T L P A F F A K S S S I Y N P V I Y I M
opsin.octopus	269	T Q F M L S W S P Y A I I A L L A Q F G P A E W V T P Y A A E L P V L F A K S A I H N P I V Y S V
opsin.sheep	259	I A F L I C W L P Y A G V A F Y I F T H Q G S D F G P I F M T I P A F F A K S S S V Y N P V I Y I M
opsin.squid	268	T Q F L L S W S P Y A V V A L L A Q F G P I E W V T P Y A A Q L P V M F A K S A I H N P M I Y S V
opsin1.drosophila	283	T L W F M A W T P Y L V I N C M G L F K F E G - L T P L N T I W G A C F A K S A A C Y N P I V Y G I
opsin2.drosophila	290	S L W F M A W T P Y L V I C Y F G L F K I D G - L T P L T T I W G A T F A K T S A V Y N P I V Y G I
opsin3.drosophila	291	F L F F C S W T P Y G V M S L I G A F G D K T L L T P G A T M I P A C A C K M V A C I D P F V Y A I
opsin4.drosophila	287	F L F F V S W T P Y G V M S L I G A F G D K S L L T Q G A T M I P A C T C K L V A C I D P F V Y A I
opsinblue.chicken	266	L G F L V C W A P Y T A F A L W V V T H R G R S F E V G L A S I P S V F S K S S T V Y N P V I Y V L
opsinblue.human	256	G S F C V C Y V P Y A A F A M Y M V N N R N H G L D L R L V T I P S F F S K S A C I Y N P I I Y C F
opsingreen.chicken	259	L G F M L A W T P Y A V F W I F T N K G A D F T A T L M A V P A F F S K S S L Y N P I I Y V L
opsingreen.human	275	L A F C F C W G P Y A F F A C F A A A N P G Y P F H P L M A A L P A F F A K S A T I Y N P V I Y V F
opsingreen1.fish	272	L A F I L C W G P Y A S F A T F S A L N P G Y A W H P L A A A L P A Y F A K S A T I Y N P I I Y V F
opsingreen2.fish	270	L A F I V C W G P Y A S F A T F S A V N P G Y A W H P L A A A M P A Y F A K S A T I Y N P I I Y V F
opsinred.chicken	272	V A Y C F C W G P Y T F F A C F A A A N P G Y A F H P L A A A L P A Y F A K S A T I Y N P I I Y V F
opsinred.fish	272	M A Y C F C W G P Y T F F A C F A A A N P G Y A F H P L A A A M P A Y F A K S A T I Y N P V I Y V F
opsinred.human	275	F A Y C V C W G P Y T F F A C F A A A N P G Y A F H P L M A A L P A Y F A K S A T I Y N P V I Y V F
opsinviolet.chicken	254	G S F C L C Y V P Y A A L A M Y M V N N R D H G L D L R L V T I P A F F S K S A C V Y N P I I Y C F

opsin.calliphora	330	S H P K Y G I A L K E K C P C - - - - C V F G K V D - - - - - - - - - - - - - - - - -
opsin.chicken	309	M N K Q F R N C M I T T L C C - - G K N P L G D E D - - - - - - - - - - - - - - - - -
opsin.cow	309	M N K Q F R N C M V T T L C C - - G K N P L G D D E - - - - - - - - - - - - - - - - -
opsin.hamster	309	M N K Q F R N C M L T T L C C - - G K N I L G D D E - - - - - - - - - - - - - - - - -
opsin.human	309	M N K Q F R N C M L T T I C C - - G K N P L G D D E - - - - - - - - - - - - - - - - -
opsin.lamprey	309	M N K Q F R N C M I T T L C C - - G K N P L G D D E - - - - - - - - - - - - - - - - -
opsin.mouse	309	L N K Q F R N C M L T T L C C - - G K N P L G D D D - - - - - - - - - - - - - - - - -
opsin.octopus	319	S H P K F R E A I Q T T F P W L L T C C Q F D E K E C E D A N D A E E E V V A S E R - G G E S R D A
opsin.sheep	309	M N K Q F R N C M L T T L C C - - G K N P L G D D E - - - - - - - - - - - - - - - - -
opsin.squid	318	S H P K F R E R I A S N F P W I L T C C Q Y D E K E I E D D K D A E A E I P A G E Q S G G E T A D A
opsin1.drosophila	332	S H P K Y R L A L K E K C P C - - - - C V F G K V D - - - - - - - - - - - - - - - - -
opsin2.drosophila	339	S H P K Y R I V L K E K C P M - - - - C V F G N T D - - - - - - - - - - - - - - - - -
opsin3.drosophila	341	S H P R Y R M E L Q K R C P W - - - - L A L N E K A - - - - - - - - - - - - - - - - -
opsin4.drosophila	337	S H P R Y R L E L Q K R C P W - - - - L G V N E K S - - - - - - - - - - - - - - - - -
opsinblue.chicken	316	M N K Q F R S C M L K L L F C - - G R S P F G D D E - - - - - - - - - - - - - - - - -
opsinblue.human	306	M N K Q F Q A C I M K M V - C - - G K A M T D E S D - - - - - - - - - - - - - - - - -
opsingreen.chicken	309	M N K Q F R N C M I T T I C C - - G K N P F G D E D - - - - - - - - - - - - - - - - -
opsingreen.human	325	M N R Q F R N C I L Q L F - - - - G K K V D D G S E - - - - - - - - - - - - - - - - -
opsingreen1.fish	322	M N R Q F R S C I M Q L F - - - - G K K V E D A S E - - - - - - - - - - - - - - - - -
opsingreen2.fish	320	M N R Q F R S C I M Q L F - - - - G K K V E D A S E - - - - - - - - - - - - - - - - -
opsinred.chicken	322	M N R Q F R N C I L Q L F - - - - G K K V D D G S E - - - - - - - - - - - - - - - - -
opsinred.fish	322	M N R Q F R V C I M Q L F - - - - G K K V D D G S - - - - - - - - - - - - - - - - - -
opsinred.human	325	M N R Q F R N C I L Q L F - - - - G K K V D D G S E - - - - - - - - - - - - - - - - -
opsinviolet.chicken	304	M N K Q F R A C I M E T V - C - - G K P L T D D S D - - - - - - - - - - - - - - - - -

FIGURE 4e.

```
opsin.octopus    368  A Q M K E M M A M M Q K M Q A Q Q A A Y Q P P P P P Q G Y P P Q G Y P P Q G A Y P P P Q G Y P P Q G
opsin.squid      368  A Q M K E M M A M M Q K M Q A Q Q Q Q - Q P A Y P P Q G Y P P Q G Y P P - - - - P P P Q G Y P P Q G

opsin.calliphora  352  - - - - - - - - - - - - - - - D G K A S D A T S Q A - T N N E S E T K A - - -
opsin.chicken     333  - - - - - - - - - - - - - - T - - S A G K T E T S S V S T S Q V S P A - -
opsin.cow         333  - - - - - - - - - - - - - - A - - S - - - T T V S K T E T S Q V A P A - -
opsin.hamster     333  - - - - - - - - - - - - - - A - - S - - - A T A S K T E T S Q V A P A - -
opsin.human       333  - - - - - - - - - - - - - - A - - S - - - A T V S K T E T S Q V A P A - -
opsin.lamprey     333  - - - - - - - - - - - - - S G A - - S T S K T E V S S V S T S P V S P A - -
opsin.mouse       333  - - - - - - - - - - - - - - A - - S - - - A T A S K T E T S Q V A P A - -
opsin.octopus     418  Y P P Q G Y P P Q G Y - - P P Q G A P P Q V E A P Q G A P P Q G V D N Q A Y Q A
opsin.sheep       333  - - - - - - - - - - - - - - A - - S - - - T T V S K T E T S Q V A P A - -
opsin.squid       413  Y P P Q G Y P P Q G Y P P P P Q G P P P Q G P P P Q A A P P Q G V D N Q A Y Q A
opsin1.drosophila 354  - - - - - - - - - - - - - D G K S S D A Q S Q A - T A S E A E S K A - - -
opsin2.drosophila 361  - - - - - - - - - - - - - E P K P D A P A S D T E T T S E A D S K A - - -
opsin3.drosophila 363  - - - - - - - - - - - - P E S S A V A S T S T T Q E P Q Q T T A A - - -
opsin4.drosophila 359  - - - - - - - - - - - - G E I S S A Q S T - T T Q E Q Q Q T T A A - - -
opsinblue.chicken 340  - - - - - - - - - - - - D V S - G S S Q A T Q V S S V S S S H V A P A - - -
opsinblue.human   329  - - - - - - - - - - - T C - S S Q K T E V S T V S S T Q V G P N - - -
opsingreen.chicken 333 - - - - - - - - - - V S S T V S Q S K T E V S S V S S S Q V S P A - - -
opsingreen.human  347  - - - - - - - - - L - - - S S A S K T E V S S V - - S S V S P A - - -
opsingreen1.fish  344  - - - - - - - - - V - - - - S G S T T E V S T A S - - - - - - - -
opsingreen2.fish  342  - - - - - - - - - V - - - - S G S T T E V S T A S - - - - - - - -
opsinred.chicken  344  - - - - - - - - - V - - - - S T S R T E V S S V S N S S V S P A - - -
opsinred.human    347  - - - - - - - - - L - - - S S A S K T E V S S V - - S S V S P A - - -
opsinviolet.chicken 327 - - - - - - - - - - - - - A S T S A Q R T E V S S V S S S Q V G P T - -
```

FIGURE 4f.

REFERENCES

Al-Ubaidi, M. R., Pittler, S. J., Champagne, M. S., Triantafyllos, J. T., McGinnis, J. F., and Baehr, W. (1990) Mouse opsin: gene structure and molecular basis of multiple transcripts. *J. Biol. Chem.* 265, 20563–20569.

Applebury, M. L. and Hargrave, P. A. (1986) Molecular biology of the visual pigments. *Vision Res.* 26, 1881–1895.

Beckmann, M. and Hegemann, P. (1991) In vitro identification of rhodopsin in the green algae *Chlamydomonas. Biochemistry,* 30, 3692–3697.

Benovic, J. L., Mayor, F. J., Somers, R. L., Caron, M. G., and Lefkowitz, R. J. (1986) Light-dependent phosphorylation of rhodopsin by β-adrenergic receptor kinase. *Nature* 321, 869–872.

Bhattacharya, S., Ridge, K. D., Knox, B. E., and Khorana, H. G. (1992) Light-stable rhodopsin. I. A rhodopsin analog reconstituted with a nonisomerizable 11-*cis* retinal derivative. *J. Biol. Chem.* 267, 6763–6779.

Birge, R. R. (1990) Nature of the primary photochemical events in rhodopsin and bacteriorhodopsin. *Biochim. Biophys. Acta* 1016, 293–327.

Blanck, A., Oesterhelt, D., Ferrando, E., Schegk, E. S., and Lottspeich, F. (1989) Primary structure of sensory rhodopsin, a prokaryotic photoreceptor. *EMBO J.* 8, 3963–3971.

Bowmaker, J. K. and Dartnall, H. J. A. (1980) Visual pigments of rod and cones in a human retina. *J. Physiol.* 298, 501–511.

Bowmaker, J. K., Thorpe, A., and Douglas, R. H. (1991) Ultraviolet-sensitive cones in the goldfish. *Vision Res.* 31, 349–352.

Chabre, M. (1985) Trigger and amplification mechanisms in visual phototransduction. *Annu. Rev. Biophys. Chem.* 14, 331–360.

Chan, T., Lee, M., and Sakmar, T. P. (1992) Introduction of hydroxyl-bearing amino acids causes bathochromic spectral shifts in rhodopsin. Amino acid substitutions responsible for red-green color pigment spectral tuning. *J. Biol. Chem.* 267, 9478–9480.

Chen, J. G., Nakamura, T., Ebrey, T. G., Ok, H., Konno, K., Derguini, F., Nakanishi, K., and Honig, B. (1989) Wavelength regulation in iodopsin, a cone pigment. *Biophys. J.* 55, 725–729.

Cowman, A. F., Zuker, C. S., and Rubin, G. M. (1986) An opsin gene expressed in only one photoreceptor cell type of the *Drosophila* eye. *Cell* 44, 705–710.

Dergachev, A. E., Artamov, I. D., Bespalov, I. A., Zolotarev, A. S., and Abdulaev, N. G. (1989) Octopus rhodopsin: unusual C-terminal fragment. *J. Protein Chem.* 8, 382–384.

Dixon, R. A. F., Kobilka, B. K., Strader, D. J., Benovic, J. L., Kohlman, H. G., Frielle, T., Bolanowski, M. A., Bennett, C. D., Rands, E., Diehle, R. E., Mumford, R. A., Slater, E. E., Sigal, I. S., Caron, M. G., Lefkowitz, R. J., and Strader, C. D. (1986) Cloning of the gene and cDNA for mammalian β-adrenergic receptor and homology with rhodopsin. *Nature* 321, 75–79.

Dohlman, H. G., Thorner, J., Caron, M. G., and Lefkowitz, R. J. (1991) Model systems for the study of seven-transmembrane-segment receptors. *Annu. Rev. Biochem.* 60, 653–688.

Doi, T., Molday, R. S., and Khorana, H. G. (1990) Role of the intradiscal domain in rhodopsin assembly and function. *Proc. Natl. Acad. Sci. U.S.A.* 87, 4991–4995.

Dratz, E. A. and Hargrave, P. A. (1983) The structure of rhodopsin and the rod outer segment disk membrane. *Trends Biochem. Sci.* 8, 128–131.

Dryja, T. P., McGee, T. L., Hahn, L. B., Cowley, G. S., Olson, J. E., Reichel, E., Sandberg, M. A., and Berson, E. L. (1990) Mutations within the rhodopsin gene in patients with autosomal dominant retinitis pigmentosa. *N. Engl. J. Med.* 323, 1302–1307.

Dryja, T. P., McGee, T. L., Reichel, E., Hahn, L. B., Cowley, G. S., Yandell, D. W., Sandberg, M. A., and Berson, E. L. (1990) A point mutation of the rhodopsin gene in one form of retinitis pigmentosa. *Nature* 343, 364–366.

Dryja, T. P., Hahn, L. B., Cowley, G. S., McGee, T. L., and Berson, E. L. (1991) Mutation spectrum of the rhodopsin gene among patients with autosomal dominant retinitis pigmentosa. *Proc. Natl. Acad. Sci. U.S.A.* 88, 9370–9374.

Duschil, A., Lanyi, J. K., and Zimányi, L. (1990) Properties and phototchemistry of a halorhodopsin from the haloalkalophile, *Natronobacterium pharaonis. J. Biol. Chem.* 265, 1261–1267.

Elliott, R. W., Sparkes, R. S., Mohandas, T., Grant, S. G., and McGinnis, J. F. (1990) Localization of the rhodopsin gene to the distal half of mouse chromosome 6. *Genomics* 6, 635–644.

Farrar, G. J., McWilliam, P., Bradley, D. G., Kenna, P., Lawler, M., Sharp, E. M., Humphries, M. M., Eiberg, H., Conneally, P. M., Trofatter, J. A., and Humphries, P. (1990) Autosomal dominant retinitis pigmentosa: linkage to rhodopsin and evidence for genetic heterogeneity. *Genomics* 8, 35–40.

Feiler, R., Harris, W. A., Kirschfeld, K., Wehrhahn, C., and Zuker, C. S. (1988) Targeted misexpression of a *Drosophila* opsin gene leads to altered visual function. *Nature* 333, 737–741.

Findlay, J. B. C. and Pappin, D. J. C. (1986) The opsin family of proteins. *Biochem. J.* 238, 625–642.

Foster, K. W., Saranak, J., and Dowben, P. A. (1991) Spectral sensitivity, structure and activation of eukaryotic rhodopsins: activation spectroscopy of rhodopsin analogs in Chlamydomonas. *J. Photochem. Photobiol.* 8, 385–408.

Franke, R. R., König, B., Sakmar, T. P., Khorana, H. G., and Hofmann, K. P. (1990) Rhodopsin mutants that bind but fail to activate transducin. *Science* 250, 123–125.

Franke, R. R., Sakmar, T. P., Oprian, D. D., and Khorana, H. G. (1988) A single amino acid substitution in rhodopsin (lysine 248 to leucine) prevents activation of transducin. *J. Biol. Chem.* 263, 2119–2122.

Franke, R. R., Sakmar, T. P., Graham, R. M., and Khorana, H. G. (1992) Structure and function in rhodopsin. Studies of the interaction between the rhodopsin cytoplasmic domain and transducin. *J. Biol. Chem.* 267, 14767–14774.

Fryxell, K. J. and Meyerowitz, E. M. (1991) The evolution of rhodopsins and neurotransmitter receptors. *J. Mol. Evol.* 33, 367–378.

Fryxell, K. J. and Meyerowitz, E. M. (1987) An opsin gene that is expressed only in the R7 photoreceptor cell of *Drosophila. EMBO J.* 6, 443–451.

Fung, B. K.-K., Hurley, J. B., and Stryer, L. (1981) Flow of information in the light-triggered cyclic nucleotide cascade of vision. *Proc. Natl. Acad. Sci. U.S.A.* 78, 152–156.

Gal, A., Artlich, A., Ludwig, M., Niemeyer, G., Olek, K., Schwinger, E., and Schinzel, A. (1991) Pro-347-Arg mutation of the rhodopsin gene in autosomal dominant retinitis pigmentosa. *Genomics* 11, 468–470.

Ganter, U. M., Schmid, E. D., Perez-Sala, D., Rando, R. R., and Siebert, F. (1989) Removal of the 9-methyl group of retinal inhibits signal transduction in the visual process. A Fourier transform infrared and biochemical investigation. *Biochemistry* 28, 5954–5962.

Goldsmith, T. H., Marks, B. C., and Bernard, G. D. (1986) Separation and identification of geometric isomers of 3-hydroxyretinoids and occurrence in the eyes of insects. *Vision Res.* 26, 1763–1769.

Gualtieri, P., Pelosi, P., Passarelli, V., and Barsanti, L. (1992) Identification of a rhodopsin photoreceptor in *Euglena gracilis*. *Biochim. Biophys. Acta* 1117, 55–59.

Gurevich, V. V., Zozulya, S. A., Zerf, E. P., Pokrovskaya, I. D., and Obukhova, T. A. (1990) Bovine rhodopsin: amino acid substitutions Asp-83 to Asn and Glu-134 to Gln prevent activation of cyclic GMP phosphodiesterase. *Biomed. Sci.* 1, 527–530.

Hall, M. D., Hoon, M. A., Ryba, N. J. P., Pottinger, J. D. D., Keen, J. N., Saibil, H. R., and Findlay, J. B. C. (1991) Molecular cloning and primary structure of squid (*Loligo forbesi*) rhodopsin, a phospholipase C-directed G-protein-linked receptor. *Biochem. J.* 274, 35–40.

Hargrave, P. A., McDowell, J. H., Curtis, D. R., Wang, J. K., Jusczak, E., Fong, S. L., Mohanna Rao, J. K., and Argos, P. (1983) The structure of bovine rhodopsin. *Biophys. Struct. Mech.* 9, 235–244.

Hargrave, P. A. and McDowell, J. H. (1992) Rhodopsin and phototransduction: a model system for G protein-linked receptors. *FASEB J.* 6, 2323–2331.

Hargrave, P. A. (1991) Seven-helix receptors. *Curr. Op. Struc. Biol.* 1, 575–581.

Hárosi, F. I. and Hashimoto, Y. (1983) Ultraviolet visual pigment in a vertebrate: a tetrachromatic cone system in the dace. *Science* 222, 1021–1023.

Harris, W. A., Stark, W. S., and Walker, J. A. (1976) Genetic dissection of the photoreceptor systme in the compound eye of *Drosophila melanogaster*. *J. Physiol.* 256, 415–439.

Harz, H. and Hegemann, P. (1991) Rhodopsin-regulated calcium currents in *Chlamydomonas*. *Nature* 351, 489–491.

Henderson, R., Baldwin, J. M., Ceska, T. A., Zemlin, F., Beckmann, E., and Downing, K. H. (1990) Model for the structure of bacteriorhodopsin based on high-resolution electron cryo-microscopy. *J. Mol. Biol.* 213, 899–929.

Henderson, R. and Schertler, G. F. (1990) The structure of bacteriorhodopsin and its relevance to the visual opsins and other seven-helix G-protein coupled receptors. *Philos. Trans. R. Soc. Lond. [Biol.]* 326, 379–389.

Hisatomi, O., Iwasa, T., Tokunaga, F., and Yasui, A. (1991) Isolation and characterization of lamprey rhodopsin cDNA. *Biochem. Biophys. Res. Commun.* 174, 1125–1132.

Honig, B., Dinur, U., Nakanishi, K., Balogh-Nair, V., Gawinowica, M. A., Arnaboldi, M., and Motto, M. G. (1980) An external point-charge model for wavelength regulation in visual pigments. *J. Am. Chem. Soc.* 101, 7084–7086.

Huber, A., Smith, D., Zuker, C. S., and Paulsen, R. (1990) Opsin of *Calliphora* peripheral photoreceptors R1–6. Homology with *Drosophila* Rh1 and posttranslational processing. *J. Biol. Chem.* 265, 17906–17910.

Humphries, P., Kenna, P., and Farrar, G. J. (1992) On the molecular genetics of retinitis pigmentosa. *Science* 256, 804–808.

Inglehearn, C. F., Bashir, R., Lester, D. H., Jay, M., Bird, A. C., and Bhattacharya, S. S. (1991) A 3-bp deletion in the rhodopsin gene in a gamily with autosomal dominant retinitis pigmentosa. *Am. J. Hum. Genet.* 48, 26–30.

Jacobs, G. H., Neitz, J., and Deegan II, J. F. (1991) Retinal receptors in rodents maximally sensitive to ultraviolet light. *Nature* 353, 655–657.

Johnson, R. L., Grant, K. B., Zanke. T. C., Boehm, M. F., Merbs, S. L., Nathans, J., and Nakanishi, K. (1993) Cloning and expression of goldfish opsin sequences. *Biochemistry*, 32, 208–214.

Karnik, S. S. and Khorana, H. G. (1990) Assembly of functional rhodopsin requires a disulfide bond between cysteine residues 110 and 187. *J. Biol. Chem.* 265, 17520–17524.

Karnik, S. S., Sakmar, T. P., Chen, H.-B., and Khorana, H. G. (1988) Cysteine residues 110 and 187 are essential for the formation of correct structure in bovine rhodopsin. *Proc. Natl. Acad. Sci. U.S.A.* 85, 8459–8463.

Kaupp, U. B. (1991) The cyclic nucleotide-gated channels of vertebrate photoreceptors and olfactory epithelium. *Trends Neurosci.* 14, 150–157.

Kaupp, U. B., Niidome, T., Tanabe, T., Terada, S., Bönigk, W., Stühmer, W., Cook, N. J., Kangawa, K., Matsuo, H., Hirose, T., Miyata, T., and Numa, S. (1989) Primary structure and functional expression for complementary DNA of the rod photoreceptor cyclic GMP-gated channel. *Nature* 342, 762–766.

Keen, T. J., Inglehearn, C. F., Lester, D. H., Bashir, R., Jay, M., Bird, A. C., Jay, B., and Bhattacharya, S. S. (1991) Autosomal dominant retinitis pigmentosa: four new mutations in rhodopsin, one of them in the retinal attachment site. *Genomics* 11, 199–205.

Khorana, H. G. (1988) Bacteriorhodopsin, a membrane-protein that uses light to translocate protons. *J. Biol. Chem.* 263, 7439–7442.

Khorana, H. G. (1992) Rhodopsin, photoreceptor of the rod cell. An emerging pattern for structure and function. *J. Biol. Chem.* 267, 1–4.

Koch, K.-W. (1992) Biochemical mechanism of light adaptation in vertebrate photoreceptors. *Trends Biochem. Sci.* 17, 307–311.

König, B., Arendt, A., McDowell, J. H., Kahlert, M., Hargrave, P. A., and Hofmann, K. P. (1989) Three cytoplasmic loops of rhodopsin interact with transducin. *Proc. Natl. Acad. Sci. U.S.A.* 86, 6878–6882.

Kosower, E. M. (1988) Assignment of groups responsible for the "opsin shift" and light absorptions of rhodopsin and red, green, and blue iodopsins (cone pigments). *Proc. Natl. Acad. Sci. U.S.A.* 85, 1076–1080.

Kuwata, O., Imamoto, Y., Okano, T., Kokame, K., Kojima, D., Matsumoto, H., Morodome, A., Fukada, Y., Shichida, Y., Yasuda, K., Shimura, Y., and Yoshizawa, T. (1990) The primary structure of iodopsin, a chicken red-sensitive cone pigment. *FEBS Lett.* 272, 128–132.

Lanyi, J. K., Duschl, A., Hatfield, G. W., May, K., and Oesterhelt, D. (1990) The primary structure of a halorhodopsin from *Natronobacterium pharaonis*. Structural, functional and evolutionary implications for bacterial rhodopsins and halorhodopsins. *J. Biol. Chem.* 265, 1253–1260.

Lin, S. W., Sakmar, T. P., Franke, R. R., Khorana, H. G., and Mathies, R. A. (1992) Resonance Raman microprobe spectroscopy of rhodopsin mutants: effect of substitutions in the third transmembrane helix. *Biochemistry* 31, 5105–5111.

Loew, E. R. and Lythgoe, J. N. (1978) The ecology of cone pigments in teleost fishes. *Vision Res.* 18, 715–722.

Longstaff, C., Calhoon, R. D., and Rando, R. R. (1986) Deprotonation of the Schiff base of rhodopsin is obligate in the activation of the G protein. *Proc. Natl. Acad. Sci. U.S.A.* 83, 4209–4213.

Merbs, S. L. and Nathans, J. (1992a) Absorption spectra of human cone pigments. *Nature* 356, 433–435.

Merbs, S. L. and Nathans, J. (1992b) Absorption spectra of the hybrid pigments responsible for anomalous color vision. *Science* 258, 464–466.

Montell, C., Jones, K., Zuker, C., and Rubin, G. (1987) A second opsin gene expressed in the ultraviolet-sensitive R7 photoreceptor cells of *Drosophila melanogaster*. *J. Neurosci.* 7, 1558–1566.

Nakayama, T. A. and Khorana, H. G. (1990) Orientation of retinal in bovine rhodopsin determined by cross-linking using a photoactivatable analog of 11-*cis*-retinal. *J. Biol. Chem.* 265, 15762–15769.

Nakayama, T. A. and Khorana, H. G. (1991) Mapping of the amino acids in membrane-embedded helices that interact with the retinal chromophore in bovine rhodopsin. *J. Biol. Chem.* 266, 4269–4275.

Nathans, J. (1990) Determinants of visual pigment absorbance: identification of the retinylidene Schiff's base counterion in bovine rhodopsin. *Biochemistry* 29, 9746–9752.

Nathans, J. (1987) Molecular biology of visual pigments. *Annu. Rev. Neurosci.* 10, 163–194.

Nathans, J. (1992) Rhodopsin: structure, function, and genetics. *Biochemistry* 31, 4923–4930.

Nathans, J., Davenport, C. M., Maumenee, I. H., Lewis, R. A., Hejtmancik, J. F., Litt, M., Lovrien, R., Weleber, R., Bachynski, B., Zwas, F., Klingaman, R., and Fishman, G. (1989) Molecular genetics of human blue cone monochromacy. *Science* 245, 831–845.

Nathans, J. and Hogness, D. S. (1983) Isolation, sequence analysis, and intron-exon arrangement of the gene encoding bovine rhodopsin. *Cell* 34, 807–814.

Nathans, J. and Hogness, D. S. (1984) Isolation and nucleotide sequence of the gene encoding human rhodopsin. *Proc. Natl. Acad. Sci. U.S.A.* 81, 4851–4855.

Nathans, J., Piantanida, T. P., Eddy, R. L., Shows, T. B., and Hogness, D. S. (1986a) Molecular genetics of inherited variation in human color vision. *Science* 232, 203–210.

Nathans, J., Thomas, D., and Hogness, D. S. (1986b) Molecular genetics of human color vision: the genes encoding blue, green, and red pigments. *Science* 232, 193–202.

Neitz, J. and Jacobs, G. H. (1986) Polymorphism of the long-wavelength cone in normal human colour vision. *Nature* 323, 623–625.

Neitz, J., Neitz, M., and Jacobs, G. H. (1989) Analysis of fusion gene and encoded photopigment of colour-blind humans. *Nature* 342, 679–682.

Neitz, M., Neitz, J., and Jacobs, G. H. (1991) Spectral tuning of pigments underlying red-green color vision. *Science* 252, 971–974.

Okano, T., Fukada, Y., Artamonov, I. D., and Yoshizawa, T. (1989) Purification of cone visual pigments from chicken retina. *Biochemistry* 28, 8848–8856.

Okano, T., Kojima, D., Fukada, Y., Shichida, Y., and Yoshizawa, T. (1992) Primary structures of chicken cone visual pigments: Vertebrate rhodopsins have evolved out of cone visual pigments. *Proc. Natl. Acad. Sci. U.S.A.* 89, 5932–5936.

Oprian, D. D., Asenjo, A. B., Lee, N., and Pelletier, S. L. (1991) Design, chemical synthesis, and expression of genes for the three human color vision pigments. *Biochemistry* 30, 11367–11372.

O'Tousa, J. E., Baehr, W., Martin, R. L., Hirsh, J., Pak, W. L., and Applebury, M. L. (1985) The Drosophila ninaE gene encodes an opsin. *Cell* 40, 839–850.

Ovchinnikov, Y. A. (1982) Rhodopsin and bacteriorhodopsin: structure-function relationships. *FEBS Lett.* 148, 179–191.

Robinson, P. R., Cohen, G. B., Zhukovsky, E. A., and Oprian, D. D. (1992) Constitutively active mutants of rhodopsin. *Neuron* 9, 719–725.

Sakmar, T. P., Franke, R. R., and Khorana, H. G. (1989) Glutamic acid 113 serves as the retinylidene schiff base counterion in bovine rhodopsin. *Proc. Natl. Acad. Sci. U.S.A.* 86, 8309–8313.

Sawatzki, J., Fischer, R., Scheer, H., and Siebert, F. (1990) Fourier-transform Raman spectroscopy applied to photobiological systems. *Proc. Natl. Acad. Sci. U.S.A.* 87, 5903–5906.

Schoenlein, R. W., Peteanu, L. A., Mathies, R. A., and Shank, C. V. (1991) The first step in vision: femtosecond isomerization of rhodopsin. *Science* 254, 412–415.

Scharf, B., Pevec, B., Hess, B., and Engelhard, M. (1992) Biochemical and photochemical properties of the photophobic receptors from *Halobacterium halobium* and *Natronobacterium pharaonis*. *Eur. J. Biochem.* 206, 359–366.

Siebert, F. (1990) Resonance Raman and infrared difference spectroscopy of retinal proteins. *Methods Enzymol.* 189, 123–136.

Smith, D. P., Stamnes, M. A., and Zuker, C. S. (1991a) Signal transduction in the visual system of Drosophila. *Annu. Rev. Cell Biol.* 7, 161–190.

Smith, D. P., Ranganathan, R., Hardy, R. W., Marx, J., Tsuchida, T., and Zuker, C. S. (1991b) Photoreceptor deactivation and retinal degeneration mediated by a photoreceptor-specific protein kinase C. *Science* 254, 1478–1484.

Stoeckenius, W. and Bogomolni, R. A. (1982) Bacteriorhodopsin and related pigments of halobacteria. *Annu. Rev. Biochem.* 52, 587–616.

Stryer, L. (1991) Visual excitation and recovery. *J. Biol. Chem.* 266, 10711–10714.

Sung, C.-H., Davenport, C. M., Hennessey, J. C., Maumenee, I. H., Jacobson, S. G., Heckenlively, J. R., Howakowski, R., Fishman, G., Gouras, P., and Nathans, J. (1991a) Rhodopsin mutations in autosomal dominant retinitis pigmentosa. *Proc. Natl. Acad. Sci. U.S.A.* 88, 6481–6485.

Sung, C.-H., Schneider, B. G., Agarwal, N., Papermaster, D. S., and Nathans, J. (1991b) Functional heterogeneity of mutant rhodopsins responsible for autosomal dominant retinitis pigmentosa. *Proc. Natl. Acad. Sci. U.S.A.* 88, 8840–8844.

Takao, M., Yasui, A., and Tokunaga, F. (1988) Isolation and sequence determination of the chicken rhodopsin gene. *Vision Res.* 28, 471–480.

Tokunaga, F., Iwasa, T., Miyagishi, M., and Kayada, S. (1990) Cloning of cDNA and amino acid sequence of one of chicken cone visual pigments. *Biochem. Biophys. Res. Commun.* 173, 1212–1217.

Wald, G. (1968) The Molecular basis of visual excitation. *Nature* 219, 800–807.

Wang, S.-Z., Adler, R., and Nathans, J. (1992) A visual pigment from chicken that resembles rhodopsin: amino acid sequence, gene structure, and functional expression. *Biochemistry* 31, 3309–3314.

Wang, Y., Macke, J. P., Merbs, S. L., Zack, D. J., Klaunberg, B., Bennet, J., Gearhart, J., and Nathans, J. (1992) A locus control region adjacent to the human red and green visual pigment genes. *Neuron* 9, 429–440.

Williams, A. J., Hunt, D. M., Bowmaker, J. K., and Mollon, J. D. (1992) The polymorphic photopigments of the marmoset: spectral tuning and genetic basis. *EMBO J.* 11, 2039–2045.

Winderickx, J., Lindsey, D. T., Sanocki, E., Teller, D. Y., Motulsky, A. G., and Deeb, S. S. (1992) Polymorphism in red photopigment underlies variation in colour matching. *Nature* 356, 431–433.

Yokoyama, R. and Yokoyama, S. (1990a) Convergent evolution of the red- and green-like visual pigment genes in fish, *Astyanax fasciatus*, and human. *Proc. Natl. Acad. Sci. U.S.A.* 87, 9315–9318.

Yokoyama, R. and Yokoyama, S. (1990b) Isolation, DNA sequence and evolution of a color visual pigment gene of the blind cave fish *Astyanax fasciatus*. *Vision Res.* 30, 807–816.

Yokoyama, S. and Yokoyama, R. (1989) Molecular evolution of human visual pigment genes. *Mol. Biol. Evol.* 6, 186–197.

Yoshizawa, T. and Kuwata, O. (1991) Iodopsin, a red-sensitive cone visual pigment in the chicken retina. *Photochem. Photobiol.* 54, 1061–1070.

Zhukovsky, E. A. and Oprian, D. D. (1989) Effect of carboxylic acid side chains on the absorption maximum of visual pigments. *Science* 246, 928–930.

Zhukovsky, E. A., Robinson, P. R., and Oprian, D. D. (1991) Transducin activation by rhodopsin without a covalent bond to the 11-*cis*-retinal chromophore. *Science* 251, 558–560.

Zuker, C. S., Cowman, A. F., and Rubin, G. M. (1985) Isolation and structure of a rhodopsin gene from *D. melanogaster*. *Cell* 40, 851–858.

Zuker, C. S., Montell, C., Jones, K., Laverty, T., and Rubin, G. M. (1987) A rhodopsin gene expressed in photoreceptor cell R7 of the *Drosophila* eye: homologies with other signal-transducing molecules. *J. Neurosci.* 7, 1550–1557.

24

Tachykinin Receptors

James E. Krause, Bruce S. Sachais, and Paul Blount

1.24.0   Introduction

The tachykinin peptides comprise a family of structurally related peptides that were originally discovered on the basis of their atropine-resistant rapid stimulation of smooth-muscle contraction. These peptides have been isolated from vertebrate and nonvertebrate species, and are structurally characterized by the conserved carboxyl terminal region of Phe-X-Gly-Leu-Met-NH$_2$ where the X residue usually is an aromatic or aliphatic amino acid. There exists at least five mammalian tachykinin peptides, including Substance P (SP), neurokinin A* (NKA), neurokinin B (NKB), neuropeptide K (NPK), and neuropeptide gamma (NPγ), that are synthesized, processed, and secreted from tachykinin-secreting neurons. NPK and NPgamma are amino terminal extended forms of NKA. The primary structures of these and some other relevant peptide agonists and antagonists are displayed in Figure 1. Mammalian tachykinin receptors consist of three types, based upon pharmacological, biochemical, and molecular characterization. These receptors have been named neurokinin-1 (NK-1) — the receptor with highest affinity for substance P, neurokinin-2 (NK-2) — the receptor with highest affinity for neurokinin A, and neurokinin-3 (NK-3), the receptor with highest affinity for neurokinin B. Prior to the use of molecular biological techniques, the classification of mammalian tachykinin receptors was based predominantly on various pharmacological properties of the mammalian receptors expressed in various tissue contexts. This classification was based on the potency of selective agonists and selective peptidic antagonists, and also on different patterns of desensitization of receptor responses in the continued presence of agonist stimulation. For example, "SP receptors" were shown in membrane preparations to display high affinity (subnanomolar to nanomolar) for SP and were blocked by relatively weak peptide antagonists, and in tissue preparations they displayed desensitization of responses. Consequently, the early classification systems were based on the availability of selective agonists and partially selective but weak antagonists.

In the past few years, molecular biological studies coupled with expression analysis has confirmed and further clarified the existence of multiple tachykinin receptors. These studies, along with the recent discovery and development of selective, high affinity nonpeptidic antagonists for some of the tachykinin receptors, now conclusively documents the existence of multiple tachykinin receptors both within and between species. The extent of receptor diversity based on molecular studies to date is essentially that previously suggested based on the pharmacological studies performed previously (Table 1). That is,

* Neurokinin A is also known as substance K, and neurokinin B is known as neuromedin K.

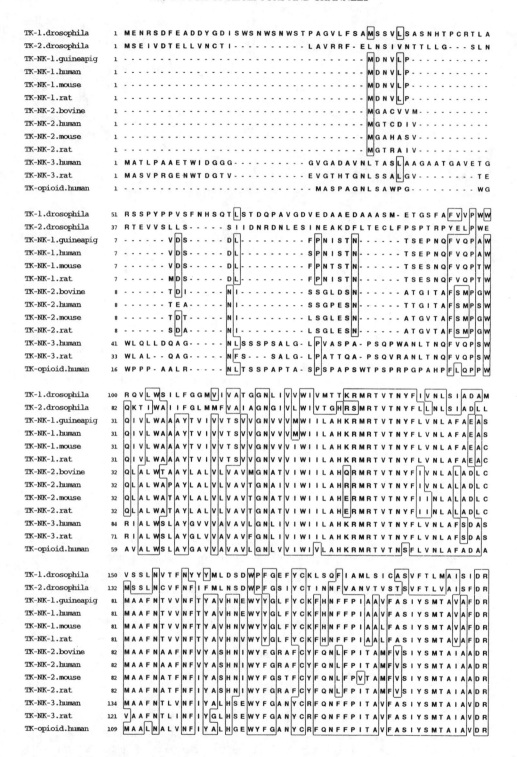

FIGURE 1. Alignment of the amino acid sequences of tachykinin receptors. The sequences were obtained from commercial protein databases such as GenBank and EMBL. The references for each sequence are cited in the text.

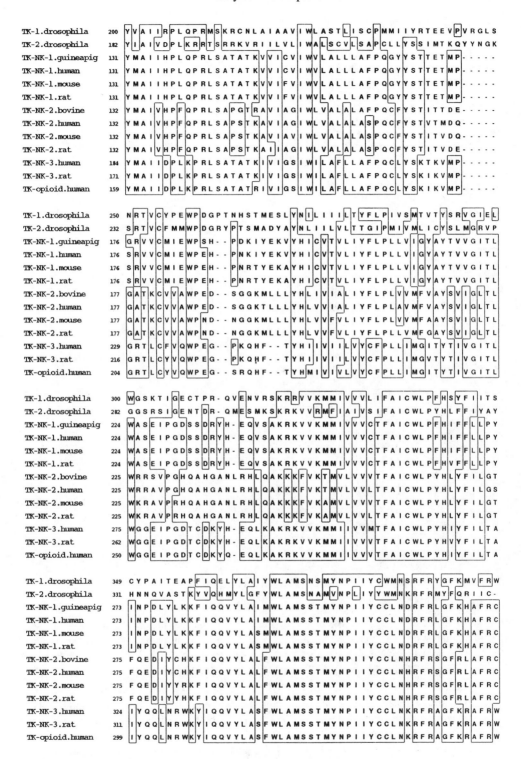

FIGURE 1b.

HANDBOOK OF RECEPTORS AND CHANNELS

FIGURE 1c.

there seems to be only a single receptor species that has nanomolar affinity for each naturally occurring peptide agonist, in contrast to the multiplicity of G-protein coupled receptors displaying high affinity for 5-hydroxytryptamine, dopamine, or acetylcholine, for example. Within the group of G-protein-coupled tachykinin receptors, evolutionary relationships between the established tachykinin receptor types were determined by a phylogenetic tree analysis developed by Feng and Doolittle (1990), shown in Figure 2. The length of each branch of the tree correlates with the evolutionary distance between receptors. Consequently, mammalian G-protein coupled tachykinin receptors have differentiated into three distinct major branches, each branch corresponding to a type of receptor (NK-1, NK-2, and NK-3) displaying nanomolar affinity for its cognate ligand. For all mammalian tachykinin receptor types identified, the interspecies variation is low such that there is greater than 90% identity between species homologs. In addition, two recently described receptors from Drosophila, called TK-1 and TK-2, appear to be tachykinin receptors based on the fact that mammalian tachykinin peptides can activate responses

Table 1. Molecular Biological Characteristics of G-Protein Coupled Tachykinin Receptors

Receptor	Second messenger	Species cloned	Chromosomal location	a.a. Sequence	Accession number	Primary references
NK-1	IP_3-Ca^{++};cAMP*	Human	2	407	P25103	Takeda, et al., 1991a; Girard et al., 1991; Hopkins et al., 1991
	IP_3-Ca^{++};cAMP*	Guinea pig		407	X64323	Yokota et al., 1989; Hershey and Krause, 1990
		Rat		407	P14600	
		Mouse		407	S20304	Sundelin et al., 1992
NK-2	IP_3-Ca^{++};cAMP*	Human	10	398	P21452	Girard et al., 1990; Kris et al., 1991
	IP_3-Ca^{++};cAMP*	Cow		384	P90551	Masu et al., 1987
	IP_3-Ca^{++};cAMP*	Rat		390	P16610	Sasai and Nakanishi, 1989
	IP_3-Ca^{++}	Mouse		384	S20303	Sundelin et al., 1992
NK-3	IP_3-Ca^{++}	Human		465	P29371	Takahashi et al., 1992; Buell et al., 1992
	IP_3-Ca^{++};cAMP*	Rat		452	P16177	Shigemoto et al., 1990
DTK-1	IP_3-Ca^{++}	Drosophila	99D1-2 on the right arm of chromosome 3	519	X62711	Li et al., 1991
DTK-2	IP_3-Ca^{++}	Drosophila	86C on the right arm of chromosome 3	504	M77168	Monnier et al., 1992

* Ligand-stimulated cAMP responses seem to be cell context dependent; see text for details.

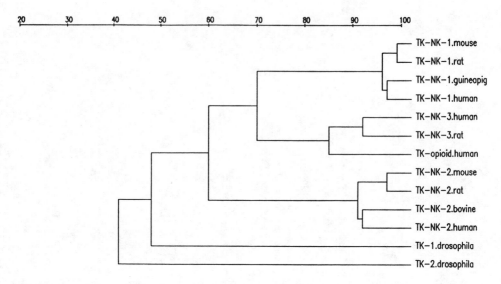

FIGURE 2. Phylogenetic tree of tachykinin receptors. The phylogenetic tree was constructed according to the method of Feng and Doolittle (1990). The length of each "branch" of the tree correlates with the evolutionary distance between receptor subpopulations.

mediated by these receptors. However, the natural Drosophila agonist(s) for these receptors have not been defined, and whether there exists tachykinin peptides structurally related to their mammalian counterparts remains uncertain. These receptors are distantly related to each other (approximately 40% identity), while the TK-1 receptor is more closely related to the mammalian tachykinin receptors than is the TK-2 receptor.

1.24.1   NK-1 or Substance P Receptors

The NK-1 receptor is the receptor type to which substance P is the most potent naturally occurring agonist (and therefore was originally named the substance P receptor). Several radioligands have been used to identify and biochemically characterize the NK-1 receptor (Table 2). Since substance P binds to all three types of neurokinin receptors, the NK-1 receptor is identified pharmacologically by the rank order of potency of various tachykinin peptides. For the naturally occurring mammalian tachykinins this order is SP > NPγ > NKA > NKB; for nonmammalian tachykinin peptides, this order physalaemin > eledoisin > kassinin.

Until the recent cloning of the NK-1 receptor in several species (Figures 2 and 3) the receptor was studied in membrane preparations from various tissues including rat brain and dog carotid artery. Early studies had suggested that this receptor was a member of the G protein-coupled superfamily because responses elicited in the oocyte expression system were slow in onset, consistent with the activation of second messengers (Parker et al., 1986), and that radioligand binding was modulated by GTP and nonhydrolyzable GTP analogs (Lee et al., 1983; MacDonald and Boyd, 1989). Now, with the cloning of this receptor, studies to determine the structural basis of NK-1 receptor function are greatly facilitated.

All endogenous mammalian tachykinin peptides share the same carboxyl-terminal sequence of Phe-X-Gly-Leu-Met-NH$_2$, where X is one of several aliphatic or aromatic amino acids. In substance P this residue is Phe[8]. This conserved carboxyl

Table 2.   Radioligands Used to Label Mammalian G-Protein-Coupled Tachykinin Receptors

NK-1 Receptor	NK-2 Receptor	NK-3 Receptor
^{125}I-BH-Substance P	^{3}H-Neurokinin A	^{3}H-Neurokinin B
^{3}H-Substance P	^{125}I-Neurokinin A	^{3}H-Eledoisin
^{125}I-Tyr^8-Substance P	^{125}I-Neuropeptide γ	^{125}I-BH-Eledoisin
^{125}I-Tyr^{-1}-Substance P		^{125}I-Scyliorhinin II
^{3}H-Pro^9-Substance P		^{3}H-Senktide
^{125}I-Physalaemin		
^{3}H-Physalaemin		
^{125}I-BH-[Sar^9,Met(O$_2$)^{11}]-Substance P		
^{3}H-[Sar^9,Met(O$_2$)^{11}]-Substance P		

terminus is recognized by the neurokinin receptors and is likely responsible for the cross reactivity of tachykinins among tachykinin receptors. For the NK-1 receptor, the importance of this region has been shown directly in that the carboxyl terminal hexapeptide Gln^6-Phe^7-Phe^8-Gly^9-Leu^{10}-Met^{11}-NH$_2$ is approximately equipotent to substance P in tissue response assays (Lee et al., 1982; for reviews see Sandberg and Iversen, 1982; Sandberg, 1985). It has been shown by alanine substitution that the invariant Phe at position seven of substance P plays a dominant role in the activity of substance P (Couture et al., 1979). The free acid form of the substance P is on the order of 1000-fold less potent than the amidated peptide (Lee et al., 1982; Cascieri and Liang, 1983), although the methyl ester derivative of substance P has been shown to be between equipotent (in physiological assays) (Watson et al., 1983) and 50-fold less potent (displacement assays) (Cascieri and Liang, 1983; Sachais and Krause, unpublished) than the native peptide. These studies indicate that Phe^7 and the carboxyamide moiety are important peptide regions in receptor recognition.

The NK-1 receptor is a member of the G protein-coupled superfamily of receptors. Several radioligands have been used to identify this receptor site (Table 2). High-affinity binding of radiolabeled substance P is sensitive to GTP and nonhydrolyzable analogs of GTP (Lee et al., 1983; MacDonald and Boyd, 1989). This decreased binding has been attributed to the conversion of high-affinity binding sites to a lower affinity state which is minimally 20-fold lower in affinity that the high affinity state (Luber-Narod et al., 1990). This effect is due to an increase in the dissociation rate for the ligand and is reversible upon removal of the added guanine nucleotide (Luber-Narod et al., 1990). A loss of high affinity, GTP sensitive, binding is also seen upon alkaline treatment of NK-1 receptor containing membranes (a treatment which dissociates G protein complexes from receptors). This binding is restored by addition of purified G$_i$/G$_o$ (MacDonald and Boyd, 1989). Although G$_i$/G$_o$ can act to restore the high-affinity binding of substance P, these G proteins may not be candidates for the *in vivo* G proteins since G protein modulation of substance P binding is generally insensitive to treatment with pertussis toxin.

The initial design of peptide antagonists of the tachykinins was based on modifications of the primary structure of SP. Two successful approaches have been used, one using a linear peptide approach where a variety of amino acids (including natural L amino acids, unnatural L amino acids, and unnatural D amino acids) replace natural residues, and the other approach being that based on the identification of key conformational requirements (i.e., constraints). The former approach has resulted in the generation of spantide (Folkers et al., 1984) and subsequently spantide II (Folkers et al., 1990, see Figure 1). Spantide II, compared to spantide has a higher pA$_2$ value (negative logarithm of antagonist concentration producing a twofold shift of the agonist concentration-activity curve) of 7.7 at the NK-1 receptor and does not possess significant mast cell releasing activity. The latter approach culminated in the development of a spirolactam substituted compound called GR 71251 (Ward et al., 1990; Figure 4A), which has a pA_2 value of 7.7 at the NK-1 receptor. The peptidic structures of these and other related antagonists may be expected to limit oral bioavailability and consequently their testing to evaluate their clinical potential.

Recently several potent nonpeptide NK-1 receptor antagonists have been reported (Figure 4A). The first of these antagonists was CP-96,345, a quinuclidine based compound which has subnanomolar affinity for the NK-1 receptor of most species tested (Snider et al., 1991). CP-96,3435 shows high selectivity for the NK-1 receptor over NK-2 and NK-3 receptors, and is greater than 100-fold more potent than the peptide antagonists described above (Rouissi et al., 1991). Other recently discovered nonpeptide SP antagonists include WIN 51708, a steroid-based molecule with a nitrogen-containing heterocycle attached to the A ring (Venepalli et al., see Figure 4A), and RP 67580, a substituted perhydroisoindole (Garret et al., 1991, see Figure 4A). Finally, a fermentation product isolated from *Stroptomyces violaceoniger* was catalytically hydrogenated and shown to be an antagonist at NK-1 and NK-2 sites (Morimoto et al., 1992). The structure of this novel cyclic peptide derivative (FK 224) is shown in Figure 4C.

Interestingly, the nonpeptide antagonist CP-96,345 has an affinity for rat and mouse NK-1 receptors which is 50- to 100-fold lower than other species (Snider et al., 1991; Takeda et al., unpublished). This species difference is of great interest from a protein biochemical and drug discovery point of view, as the NK-1 receptor from all species from which it has been cloned show very high identity (Figure 3). Recently, the molecular basis of this species selectivity has been elucidated based on a mutational analysis of rat and human NK-1 receptors (Fong et al., 1993; Sachais et al., 1993).

The nonpeptide antagonist WIN 51708, which has been recently described by investigators at Sterling Winthrop, also interacts with the human NK-1 receptor differently than the rat receptor. In stably transfected cell lines expressing the rat or human NK-1 receptor, WIN 51708 has an IC_{50} value of 25 nM, while for the human receptor its IC_{50} is approximately 15 µM (Sachais and Krause, unpublished). Further studies should establish the molecular basis of this dramatic species selectivity.

The NK-1 receptor has been shown to activate the hydrolysis of inositol phospholipids in many systems including central (Mantyh et al., 1984; Torrens et al., 1989) and peripheral (Merritt and Rink, 1987; Tachado et al., 1991) tissues. In the rat CNS, this activation has been shown to correlate with substance P binding sites (Mantyh et al., 1984). Activation of the NK-1 receptor can also stimulate cAMP accumulation. Substance P-mediated increases in cAMP levels have been shown in some systems (Yamachita et al., 1983; Narumi and Maki, 1978; Hanley et al., 1980), but not others (Watson, 1984; Hunter et al., 1985). Substance P activates either phosphatidyl inositol hydrolysis or cAMP production, but not both in iris sphincter, and is species dependent (Tachado et al., 1991). Heterologous expression of NK-1 receptors in several cell lines has established that the NK-1 receptor is capable of coupling to both of these second messenger systems (Nakajima et al., 1992; Takeda et al., 1992; Mitsuhashi et al., 1992).

In rat parotid acinar cells it has been shown that substance P causes a rapid but transient increase in inositol phospholipid hydrolysis, while muscarinic agonists cause a more prolonged response (Sugiya et al., 1987; Merritt and Rink, 1987). Reapplication of substance P does not induce a second response, eliminating peptide degradation by proteases as the mechanism underlying this transient response. Application of substance P after stimulation with the muscarinic agonist does, however, induce a response similar to that seen with substance P on naïve cells. Application of muscarinic agonist after substance P stimulation gives a prolonged response. These studies show

AGONISTS

Substance P   Arg-Pro-Lys-Pro-Gln-Gln-Phe-Phe-Gly-Leu-Met-NH$_2$

Neurokinin A   His-Lys-Thr-Asp-Ser-Phe-Val-Gly-Leu-Met-NH$_2$

Neuropeptide K   Asp-Ala-
                               -Asp-Ser-Ser-Ile-Glu-Lys-Gln-Val-Ala-Leu-Leu-
                               -Lys-Ala-Leu-Tyr-Gly-His-Gly-Gln-Ile-Ser-His-
                               -Lys-Arg-His-Lys-Thr-Asp-Ser-Phe-Val-Gly-Leu-Met-NH$_2$

Neuropeptide γ   Asp-Ala-Gly-His-Gly-Gln-Ile-Ser-His-
                               -Lys-Arg-His-Lys-Thr-Asp-Ser-Phe-Val-Gly-Leu-Met-NH$_2$

Physalaemin   pGlu-Ala-Asp-Pro-Asn-Lys-Phe-Tyr-Gly-Leu-Met-NH$_2$

Eledoisin   pGlu-Pro-Ser-Lys-Asp-Ala-Phe-Ile-Gly-Leu-Met-NH$_2$

Kassinin   Asp-Val-Pro-Lys-Ser-Asp-Gln-Phe-Val-Gly-Leu-Met-NH$_2$

Neurokinin B   Asp-Met-His-Asp-Phe-Phe-Val-Gly-Leu-Met-NH$_2$

Locustatachykinin I   Gly-Pro-Ser-Gly-Phe-Tyr-Gly-Val-Arg-NH$_2$

Locustatachykinin II   Ala-Pro-Leu-Ser-Gly-Phe-Tyr-Gly-Val-Arg-NH$_2$

Cys^{3,6},Tyr^8,Pro^9-Substance P   Arg-Pro-Cys-Pro-Gln-Cys-Phe-Tyr-Pro-Leu-Met-NH$_2$
                                                    |_____|

SELECTIVE AGONISTS

<u>NK-1</u>
[Sar^9,Met(O$_2$)^{11}]-SP   Arg-Pro-Lys-Pro-Gln-Gln-Phe-Phe-Sar-Leu-Met(O$_2$)-NH$_2$

<u>NK-2</u>
[Nle^{10}]-NKA(4-10)   Asp-Ser-Phe-Val-Gly-Leu-Nle-NH$_2$
[Lys^5,MeLeu^{9}Nle^{10}]NKA(4-10)   Asp-Lys-Phe-Val-Gly-MePhe-Nle-NH$_2$

<u>NK-3</u>
senktide   succinyl-Asp-Phe-MePhe-Gly-Leu-Met-NH$_2$
[MePhe^7]-NKB(4-10)   Asp-Phe-Phe-MePhe-Gly-Leu-Met-NH$_2$

NK-1 RECEPTOR ANTAGONISTS
Spantide   DArg-Pro-Lys-Pro-Gln-Gln-DTrp-Phe-DTrp-Leu-Leu-NH$_2$
Spantide II*   DLys(Nic)-Pro-Pal(3)-Pro-DPhe(Cl$_2$)-Asn-DTrp-Phe-DTrp-Leu-Nle-NH$_2$

SELECTIVE NK-2 RECEPTOR ANTAGONISTS

L-659877   cyclo(Gln-Trp-Phe-Gly-Leu-Met)
MDL 29,913   cyclo(Gln-Trp-Phe-Gly-LeuΨ(CH$_2$NCH$_3$)Leu)
MEN 10,207   Asp-Tyr-DTrp-Val-DTrp-DTrp-Arg-NH$_2$

FIGURE 3. Primary structures of mammalian tachykinin peptides and other relevant tachykinin receptor peptide agonists and peptide antagonists.

A

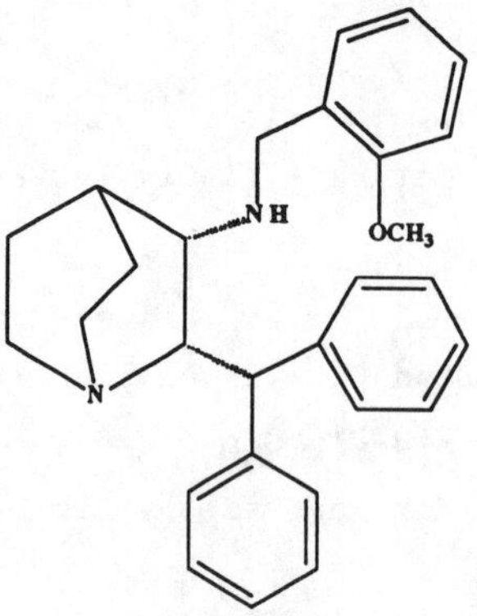

CP-96,345

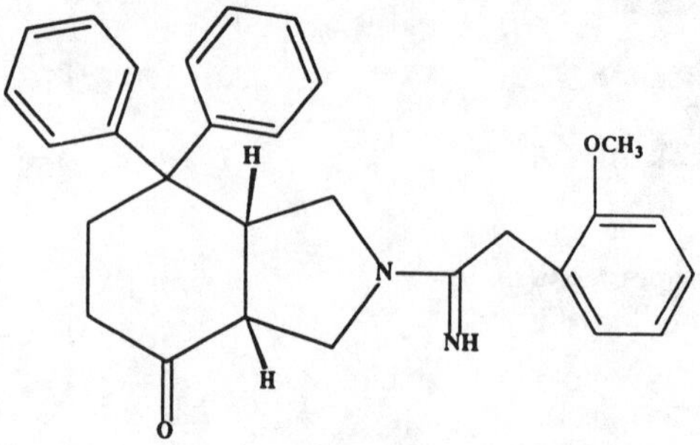

RP 67580

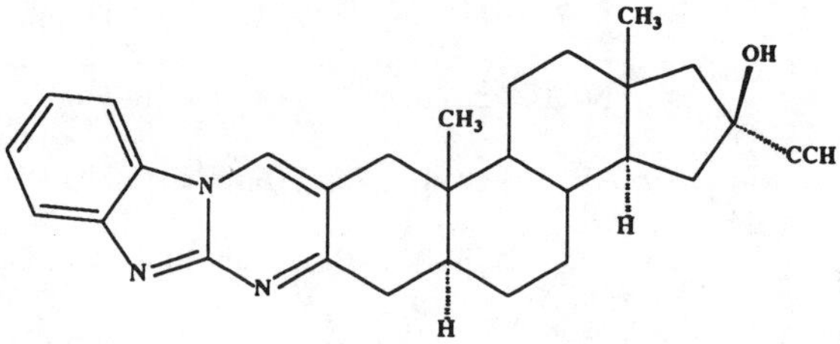

WIN 51708

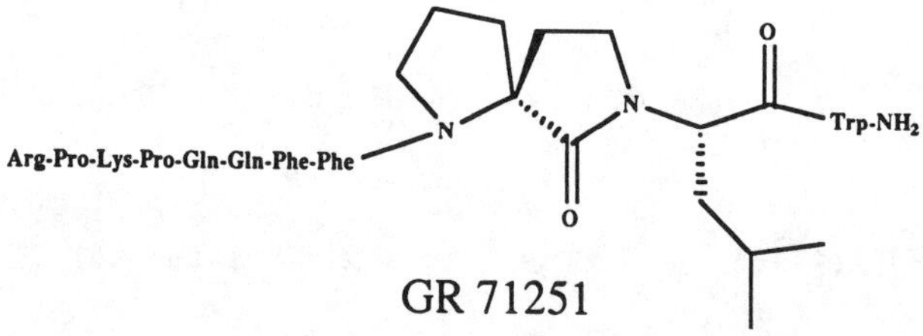

GR 71251

FIGURE 4. Recently discovered nonpeptide and modified peptide antagonists of mammalian NK-1 and NK-2 receptors. (A) NK-1 receptor antagonists. (B) NK-2 receptor antagonist. (C) Combined NK-1 and NK-2 receptor antagonist.

that substance P loses its ability to activate PI-hydrolysis under conditions where muscarinic agonists are still able to do so. This phenomenon has been referred to as homologous desensitization; however, it is not entirely clear from these studies whether the origin of this phenomenon is at the level of the NK-1 receptor or some postreceptor event in the signal transduction system.

1.24.1.1 Human NK-1 Receptors

The human NK-1 receptor cDNA has been cloned and functionally expressed (Takeda et al., 1991; Gerard et al., 1991; Hopkins et al., 1991). The primary sequences of the

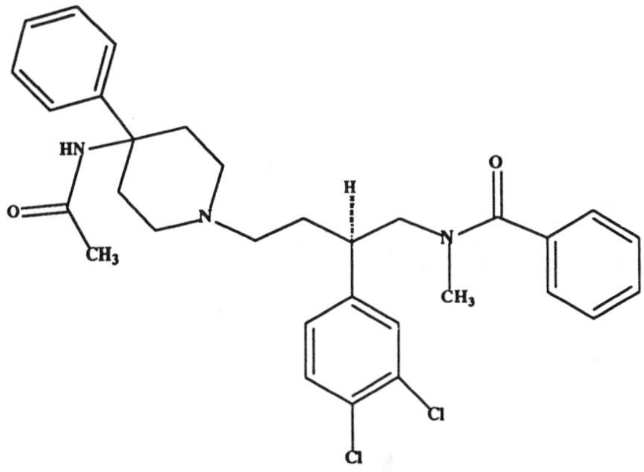

SR 48968

FIGURE 4b.

human NK-1 receptor is 94.5% identical to the rat NK-1 receptor (see below), differing by only 22 out of 407 residues (Takeda et al., 1991) (Figure 3). Of these differences, six occur in putative membrane spanning regions, eight in extracellular regions, and eight in intracellular regions. mRNA transcripts are present throughout the gastrointestinal system and in lung (Joslin et al., 1991). In this study, central nervous system was not examined.

The gene for the human NK-1 receptor has also been isolated and partially characterized and found to have a similar structural organization to the rat gene (see below, Takeda et al., 1991; Gerard et al., 1991). This gene has been localized to human chromosome 2 as a single copy gene based upon analysis of somatic cell hybrids (Gerard et al., 1991; Hopkins et al., 1991).

1.24.1.9   Guinea Pig NK-1 Receptors

A cDNA for the guinea-pig NK-1 receptor has been isolated from guinea-pig uterus and functionally expressed in COS-7 cells (Gorbulev et al., 1992). The relative affinities of mammalian tachykinins for this receptor was shown to be SP >> NKA > NKB, which is consistent with NK-1 receptors of other species.

1.24.1.10   Rat NK-1 Receptors

The rat NK-1 receptor cDNA was the first NK-1 receptor to be cloned and functionally expressed (Yokota et al., 1989; Hershey and Krause, 1990). Two quantitative studies of the tissue distribution of the NK-1 receptor mRNA in the rat have been performed (Hershey et al., 1991; Tschuda et al., 1991). NK-1 receptor transcripts are widely distributed in the central nervous system and many peripheral tissues. Relatively high levels are present in the striatum, hypothalamus, hippocampus, olfactory bulb, and medulla, and in the gastrointestinal tract, salivary glands, and urinary bladder.

The structure of the gene encoding the rat NK-1 receptor has been reported (Hershey et al., 1991). This gene is greater that 45 kb in length and contains 5 exons. The splice sites of these exons occur at the borders of sequences encoding putative membrane spanning domains of the receptor and the distinct exonic regions have been suggested to represent evolutionarily important functional units of the receptor. The 5'-

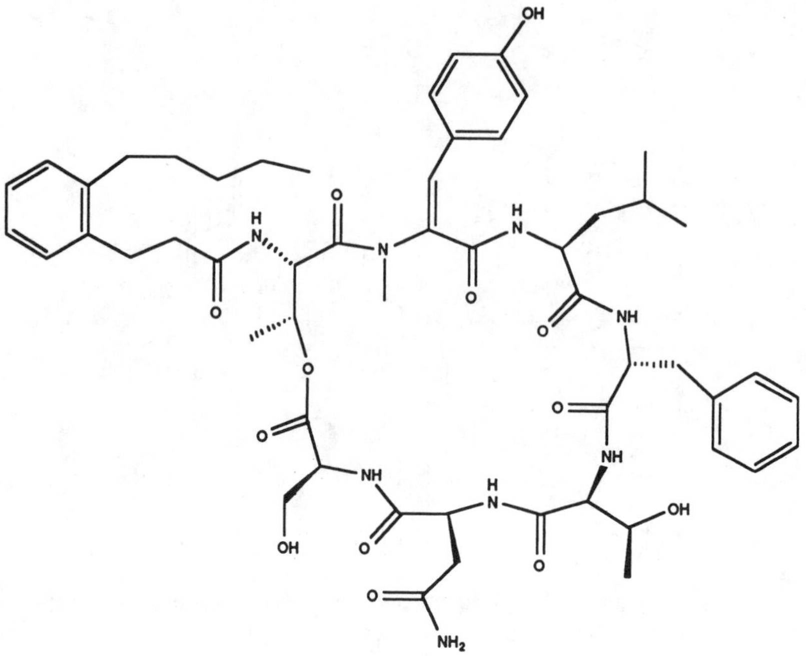

FK224

FIGURE 4c.

flanking region contains the consensus sequence for the cAMP response element as well as a sequence which resembles the Ca^{2+} inducibility sequence of c-fos. These and other elements are conserved between rat and human genes and may play a role in the transcriptional regulation of the NK-1 receptor. Since the NK-1 receptor can activate both of these second messenger systems (see above), it has been suggested that stimulation of the NK-1 receptor directly influences NK-1 receptor expression at the transcriptional level, perhaps as part of a receptor resensitization process (Hershey et al., 1991).

1.24.1.11   Mouse NK-1 Receptors

The murine NK-1 receptor cDNA has been cloned and functionally expressed (Sundellin et al., 1992). In an oocyte expression system the rank order of potency for this receptor is SP > NKA > NKB, consistent with NK-1 receptors of other species. Its primary structure is 99% identical to rat and 97% identical to human NK-1 receptors. Some exons of the gene for this receptor have been isolated and the gene appears to be greater than 30 kb in size. The gene organization appears to be similar to the rat and human NK-1 receptor genes (see above).

1.24.2   NK-2 or Substance K or Neurokinin A Receptors

Historically, the existence of multiple receptors for SP were proposed by Lee et al. (1982). This hypothesis was supported by work determining the rank order of potency of SP, its C-terminal fragments, analogs, and related tachykinins in systems including contracting guinea pig ileum, potentiating electrically evoked contractions of rat vas deferens preparations, and in displacement of ^{3}H-SP binding to rat brain membranes. Subsequent isolation of the endogenous mammalian tachykinins NKA and NKB supported the multiple receptor hypothesis, and further studies led to the characterization of the NK-2 receptor.

Table 2 lists several radioligands useful to identify NK-2 receptor sites. The binding properties of the NK-2 receptor are distinct from those of NK-1 receptor. The rank order of potency of endogenous mammalian ligands is NPγ = NPKγ ≥ NKA > NKB > SP, with NKB and SP having lesser affinity than the other ligands by at least one to two orders of magnitude. Similar to the NK-1 receptor, several synthesized and constrained peptide agonists and antagonists have been developed and utilized for studies to determine the requirements for receptor activation in model tissue preparations (Buck and Shatzer, 1988; Regoli et al., 1988; Dion et al., 1990; Lavielle et al., 1990; Saviano et al., 1991; Morimoto et al., 1992; Murai et al., 1992). More recently, a higher affinity nonpeptide antagonist (SR 48968) for the NK-2 receptor has been discovered (Emonds-Alt et al., 1992; see Figure 4B). Using peptide antagonists, Maggi et al., (1990) found heterogeneity between NK-2 receptors expressed in rabbit pulmonary artery and hamster trachea. In addition, two other peptidic antagonists including a cyclic peptide L-659,877 (McKnight et al., 1991) and a linear hexapeptide analog R-396 (Dion et al., 1990) have been developed and these compounds also differentiate between NK-2 receptor types present in different species. The NK-2 receptors present in the rabbit pulmonary artery, when induced to contract by the NK-2 selective agonist [β-A1a^8]-NKA (4-10), showed a rank order of antagonist potencies of MEN 10,376 > L-659,877 > R 396. By contrast, the rank order of potency of these antagonists on [β-A1a^8]-NKA (4-10)-induced contractions of hamster trachea was L-659,877 > R 396 > MEN 10,376 (Maggi et al., 1992). Heterogeneity in NK-2 receptor antagonist binding, and agonist-binding sensitivity to GTP analogu and divalent cations, has also been observed between the hamster urinary bladder and bovine stomach (van Giersbergen et al., 1991). The mechanisms of such heterogeneity are currently unknown, but may be attributed to either species differences, to receptor subtypes, or to coupling with multiple G-proteins.

Binding studies, and, more recently, analysis of RNA expression, has demonstrated that NK-2 receptors have a pattern of expression distinct from the NK-1 receptor. Investigators generally agree that the NK-2 receptor is found in several peripheral tissues including duodenum, large and small intestine, urinary bladder, adrenal gland, and pulmonary artery. Several of these tissues, including rat vas deferens and rabbit pulmonary artery, have been used as model systems for the characterization of the NK-2 receptor. Several investigators have also reported the absence of detectable receptors with NK-2 binding properties in the central nervous system (Buck et al., 1986; Saffroy et al., 1988; Dietl and Palacios, 1990). However, one report by Dam et al. (1990), reported ^{125}I-NPγ binding sites in rat brain. In this study, the profile of ^{125}I-NPγ binding in this study was similar to ^{125}I-NKA binding, but different from ^{125}I-BHSP binding supporting the hypothesis that Npγ is detecting the NK-2 receptor. RNA blot hybridization has thus far been unsuccessful at detecting NK-2 message in whole brain or brain regions in rat (Tsuchida et al., 1990); however, the more sensitive technique of nuclease protection has detected low levels in hippocampus, striatum, and spinal cord at levels 500-fold lower than urinary bladder (Takeda and Krause, 1991). Hence, the NK-2 receptor is observed in high concentrations in several peripheral tissues, and is expressed in low concentrations in brain.

The NK-2 receptor has the distinction among the tachykinin receptors of having been the first to be cloned. The isolation of the bovine NK-2 receptor cDNA (below) has allowed for the subsequent isolation and characterization of other tachykinin receptors and NK-2 receptors from other species by homologous hybridization, or by PCR using oligonucleotide primers of conserved regions. The NK-2 receptors are less conserved between species than are the NK-1 receptors, often having less than 90% identity (see Figures 2 and 3). Some of this divergence occurs in the carboxyl terminal tail; each species cloned thus far have different sized "tails" with the human having a tail 14, 14, and 9 amino acids longer than the bovine, mouse, and rat predicted proteins, respectively.

1.24.2.1 Human NK-2 Receptors

The cDNA of the human NK-2 receptor has been cloned (Gerard, et al., 1990; Kris et al., 1991), and expressed in 3T3 cells where a mitogenic function has been proposed (Kris et al., 1991); stimulation of cell growth by NKA had been previously observed in cultured arterial smooth muscle cells and human skin fibroblasts in the absence of serum (Nilsson et al., 1985). The cloning and characterization of the human gene demonstrated that, similar to the NK-1 receptor, the message is composed of five exons, suggesting a common evolutionary origin for the NK-1 and NK-2 receptors (Gerard, et al., 1990). The gene encoding the NK-2 receptor has been located on the tenth chromosome (Gerard, et al., 1990). In contrast to the wealth of potential regulatory elements 5′ of the rat NK-1 receptor, other than a putative TATA and GC-like box, no other known transcriptional control signals were observed within 500 bp of the presumed initiating methionine of the human NK-2 receptor gene (Gerard, et al., 1990).

1.24.2.4 Bovine NK-2 Receptors

The bovine NK-2 receptor was the first tachykinin receptor to be cloned. The oocyte expression system, was used in a novel way to isolate the cDNA clone by expressing pools of cDNA and then subdividing the pools with subsequent expression to ultimately isolate the cDNA encoding this receptor (Masu et al., 1987). This expressed cDNA was most potently activated by neurokinin A compared to neurokinin B or substance P. The analysis of the predicted protein structure of the NK-2 receptor (Masu et al., 1987) demonstrated that this receptor contained many features found in the family of G-protein coupled receptors including 7 putative transmembrane domains.

In addition to expression in oocytes (Masu et al., 1987), the NK-2 receptor from different species has been expressed in several cell types via transfection. Such experiments have allowed for detailed binding studies and the direct monitoring of second messengers activated by receptor stimulation. In transfected CHO cells, Eistetter et al. (1991) performed a series of experiments suggesting that in this system, the stimulation of bovine NK-2 receptor led to the formation of IP_3, cAMP, and output of arachidonic acid and prostaglandin E_2. In this study, a correlation of inhibition of arachidonic acid and prostaglandin E_2 release and the inhibition of cAMP production suggested a causative relationship between these responses. In contrast, expression of bovine NK-2 receptor in a rat glioma cell line, C6-2B, revealed that in this system, receptor stimulation led to an increase of intracellular Ca^{++} (presumably from IP_3 synthesis), that apparently led to a decrease in cAMP production when other receptors coupled to Gs were stimulated (DeBernardi et al., 1991). Hence, the response of the NK-2 receptor may be dependent upon the cell-type within which it is expressed.

1.24.2.10 Rat NK-2 Receptors

The cDNA for the rat NK-2 receptor has also been cloned and expressed in transfected tissue culture cells. A detailed binding study of cloned rat tachykinin receptors expressed in transfected monkey kidney COS-7 cells demonstrated that the putative NK-2 receptor had many of the binding properties found in some tissues (Ingi et al., 1991). In transfected CHO cells, Nakajima et al. (1992) demonstrated that the rat NK-1 and NK-2 receptors, when stimulated, led to the stimulation in both PI hydrolysis and cAMP formation. In this system, both responses could be generated in membrane preparations, suggesting that both second messengers systems were directly linked to receptor activation. In addition, a difference in the magnitude of the cAMP, but not the IP's response has been observed between the NK-1 and NK-2 receptors (Takeda et al., 1992). More recently, the study of chimeras of rat NK-1 and NK-2 receptors expressed in stably transfected CHO cells has led to the finding that, although the NK-1 and NK-2 receptors have high homology (see Figures 2 and 3), the homologous structural domains of the third cytoplasmic loop and the carboxyl tail of these receptors may not have functionally equivalent roles (Blount and Krause, 1992). These differences in the way that NK-1 and NK-2 receptors encode functional units, combined with differences in protein expression in host cells, may explain the differences observed in duration of SP and NKA responses in tissues.

1.24.2.11   Mouse NK-2 Receptors

The murine NK-2 receptor genomic and cDNA have been cloned, and the cDNA has been expressed in *Xenopus* oocytes (Sundelin et al., 1992). The rank order of agonist potency for this cloned mouse NK-2 receptor expressed in oocytes was consistent with what has been found in tissues, with NKA being most potent, followed by NKB and SP. Similar to the NK-1 receptor, and NK-2 receptors from other species, analysis of the genomic organization demonstrated that the receptor is encoded in five exons.

1.24.3   NK-3 Receptors

After the discovery of NKA and NKB, NK-3 receptors were initially suggested based upon ligand binding studies performed with rat central nervous system and in other tissue membrane preparations (Torrens et al., 1984; Buck et al., 1984; Cascieri et al., 1985). In these studies, radiolabeled agonists ([125I]-Bolton-Hunter-substance P; [125I]-Bolton-Hunter-eledoisin) were displaced by natural agonists with substance P most potent with the former radiolabel, and neurokinin B being most potent with the latter radiolabel. Research from Selinger's laboratory substantially clarified this issue with a combination of functional and ligand binding studies. Initially, they performed functional studies in guinea pig ileum in which substance P and neurokinin B elicit contraction both directly through activation of a muscle cell receptor and indirectly through activation of a neuronal receptor, leading to acetylcholine release which stimulates muscle contraction via muscarinic receptors (Laufer et al., 1985). In these studies, the muscle receptor was clearly distinct from the neuronal receptor, with neurokinin B being 50 times more potent than substance P or neurokinin A at the neuronal receptor. At the muscle receptor, substance P was most potent. These investigators further developed a highly selective radioligand, N--([125I]desamino-3-iodotyrosyl)-[Asp^{5,6},N-methylPhe^8]substance P(5-11)heptapeptide, abbreviated [125I]-Bolton-Hunter-NH-Senktide for selective neurokinin B receptor peptide, which selectively binds to this receptor type. Consequently, in rat cerebral cortex membranes, this ligand binds to the NK-3 receptor with high affinity ($K_d = 0.9$ nM) and binding is reversible and saturable. A linear Scatchard plot and a Hill coefficient close to unity were obtained which further lends credence to the notion that this radioligand labels a single class of binding sites. The use of this radioligand also provided further support to the suggestion that the rat brain binding site was similar or identical to the receptor site in guinea pig ileum myenteric plexus. This and other NK-2 receptor radioligands listed in Table 2.

1.24.3.1   Human NK-3 Receptors

Takahashi and co-workers have isolated genomic clones of a putative NK-3 receptor by cross-hybridization with the cognate rat probe (Takahashi et al., 1992). These workers isolated and identified a genomic sequence believed to be NK-3 based upon a high degree of homology to the rat sequence. The deduced amino acid sequence indicated that the putative NK-3 receptor protein was 88.2% identical to the rat and consists of 465 residues, and the amino terminal domain is 13 amino acids longer than that of the rat NK-3 receptor. Expression studies have recently been using a PCR generated cDNA with a similar sequence used to establish that this receptor sequence does indeed encode a functional NK-3 receptor (Buell et al., 1982; Huang et al., 1992). The agonist potency profile in ligand binding studies was NKB > NKA > SP.

Recently, Xie and co-workers (Xie et al., 1992) isolated a cDNA (called HKIR) from a human placental library by an expression cloning strategy in which a putative G-protein coupled receptor was found to have an amino acid sequence with 79% identity to the rat NK-3 receptor, with three gaps of unrelated sequence. However, the expressed receptor binds opioid ligands, albeit weakly, but not tachykinin ligands. Under the same conditions, the cloned rat NK-3 receptor bound ^{3}H-eledoisin but not opioids. These results suggest that putative receptors with relative high sequence identity do not necessarily encode a receptor with similar agonist activities. Thus, one must be cautioned about the identification of receptors based on homology alone with no functional expression data at hand.

1.24.3.10   Rat NK-3 Receptors

In 1990, Nakanishi and co-workers cloned and functionally expressed the rat NK-3 receptor (Shigemoto et al., 1990). By using cDNA fragments of the bovine NK-2 receptor, reduced stringency RNA blot hybridization demonstrated multiple hybridizing cDNAs of which one corresponded to the NK-3 receptor. This was based on electrophysiological analysis of RNA transcribed from a clone isolated from a minilibrary (which was prepared from a sucrose density fraction that contained NK-3 receptor mRNA activity based on oocyte expression). The cloned receptor was found to be activated by tachykinin peptides, and most potently by NKB. The NK-3 receptor protein consists of 452 residues, and by hydropathy analysis was shown to have seven hydrophobic domains. Electrophysiological assays of peptide-elicited responses were performed in *Xenopus* oocytes previously injected with RNA synthesized from the NK-3 receptor cDNA, and NKB was much more potent than either SP or NKA. Transient transfection studies were performed in COS cells, and ligand binding studies further documented that this cDNA encodes the NK-3 receptor. ^{125}I-Bolton-Hunter-eledoisin binding was saturable with a K_d of 0.64 nM. Displacement studies of radiolabeled eledoisin binding were also performed and K_i values for NKB, SP, and NKA were 4.2×10^{-10} M, 2.9×10^{-8} M, and 1.9×10^{-7} M, respectively. mRNA expression studies have demonstrated that NK-3 receptor transcripts are expressed relatively abundantly in several brain regions including cerebellum, hypothalamus, and cortex, and also at lower levels in the gastrointestinal system.

1.24.4   Putative Opioid Receptors

1.24.4.1   Human Putative Opioid Receptors

A putative opioid receptor was identified by Xie et al. (1992). However, this receptor is most homologous to the tachykinin receptors, as opposed to the opioid receptor identified by Evans et al. (1992). This issue is discussed in greater detail in the Opioid Receptor chapter by Evans and colleagues (this volume).

1.24.5   Invertebrate TK-1 Receptors

1.24.5.21   Drosophila TK-1 Receptors

The availability of DNA sequences and probes for mammalian tachykinin receptors has resulted in homology screening of cDNA and genomic libraries to further identify and characterize tachykinin-like receptor sequences. Consequently, Forte and colleagues (1991) screened a *Drosophila melanogaster* embryo cDNA library with a bovine stomach NK-2 receptor cDNA and isolated overlapping cDNAs with 40 to 48% amino acid identity to mammalian tachykinin receptors within putative transmembrane domains. Within the core region of receptor sequences (from transmembrane I through VII domains), there is approximately a 25 to 30% level of identity. The cDNA, called DTK-1, had an open reading frame resulting in a protein of 519 residues. When the transcribed mRNA was injected into *Xenopus* oocytes, a response to substance P was induced that was sensitive to pertussis toxin. Unlike responses elicited by mammalian tachykinin receptor RNAs injected into oocytes, TK-1 receptor responses did not appear to desensitize. Current-response relationships were determined for substance P, physalaemin, and Cys[3,6], Tyr[8], Pro[9]-substance P, and the former two peptides were equipotent while with the latter peptide tenfold lower concentrations were required to activate the chloride conductance response. The latter peptide was previously shown by Lavielle et al. (1986) to be a more potent analog of substance P active at NK-1 receptors. On the other hand, two substance P antagonists, spantide and [D-pro[4],D-trp[7,9]]-substance P 4-11, at micromolar concentrations were unable to block or attenuate responses of induced oocytes to the above-mentioned tachykinin agonists. As no tachykinin peptides have been isolated or identified from Drosophila, these workers were not able to test natural ligand candidates for this receptor. The TK-1 mRNA is developmentally regulated, with highest levels at late embryonic stages and by first instar larvae. In adult animals, expression is predominantly in head.

1.24.5   Inverterbrate TK-2 Receptors

1.24.5.21   Drosophila TK-2 Receptors

Monnier et al. (1992) used a polymerase chain reaction strategy with oligonucleotides corresponding to transmembrane III and VI regions to isolate from a *Drosophila melanogaster* genomic library a partial sequence encoding a putative G-protein-coupled receptor. This sequence was then used to screen a Drosophila head cDNA library, in which overlapping clones were used to identify an open reading frame coding for a protein of 504 residues. Within putative transmembrane domains, this sequence displays 38, 35, and 32% identify with rat NK-3, NK-1, and NK-2 sequences, respectively. The cDNA was functionally expressed in NIH-3T3 cells by assessment of the stimulation of inositol phosphate production by agonists. Little effect was observed with substance P, physalaemin, or other mammalian or insect peptides as agonists; however, the locust tachykinin peptide, locustatachykinin II, stimulated inositol phosphate metabolism. Dose-response analysis demonstrated an EC_{50} of ~$2.5 \times 10^{-9} M$, whereas locustatachykinin I was ineffective even at micromolar concentrations. The broad spectrum substance P antagonist, spantide, blocks the response elicited by locustatachykinin II. Once again, as for the DTK-1 receptor, endogenous Drosophila tachykinin peptides are not structurally characterized and as such putative endogenous ligands have not been examined. DTK-2 receptor mRNA is expressed in a developmentally regulated manner with highest expression at hours 12 to 16 of embryogenesis, a time period of major neuronal development. In the adult, DTK-2 receptor expression occurs predominantly in head.

REFERENCES

Blount, P. and Krause, J. E. (1992) The roles of the putative third cytoplasmic loop and cytoplasmic carboxyl tail of the NK-1 and NK-2 receptors in agonist induced cAMP responses in stably transfected CHO cells. *Reg. Peptides* 51, 538.

Buck, S. H. and Shatzer, S. A. (1988) Agonist and antagonist binding to tachykinin peptide NK-2 receptors. *Life Sci.* 42, 2701–2708.

Buck, S. H. and Krstenansky, J. L. (1987) The dogfish peptides scyliorhinin I and scyliorhinin II bind with differential selectivity to mammalian tachykinin receptors. *Eur. J. Pharmacol.* 144, 109–111.

Buck, S. H., Helke, C. J., Burcher, E., Shults, C. W., and O'Donohue, T. L. (1986) Pharmacologic characterization and autoradiographic distribution of binding sites for iodinated tachykinins in the rat central nervous system. *Peptides* 7, 1109–1120.

Buck, S. H., Burcher, E., Shults, C. W., Lovenberg, W., and O'Donohue, T. L. (1984) Novel pharmacology of substance K-binding sites: a third type of tachykinin receptor. *Science* 266, 987–989.

Buell, G., Schulz, M. F., Arkinstall, S. J., Maury, K., Missotten, M., Adami, N., Talabot, F., and Kawashima, E. (1992) Molecular characterisation, expression and localisation of human neurokinin-3 receptor. *FEBS Lett.* 299, 90–95.

Cascieri, M. A., Chicchi, G. G., and Liang, T. (1985) Demonstration of two distinct tachykinin receptors in rat brain cortex. *J. Biol. Chem.* 260, 1501–1507.

Cascieri, M. A. and Liang, T. (1983) Characterization of the substance P receptor in rat brain cortex membranes and the inhibition of radioligand binding by guanine nucleotides. *J. Biol. Chem.* 258, 5158–5164.

Couture, R., Fournier, A., Magnan, J., St-Pierre, S., and Regoli, D. (1979) Structure-activity studies on substance P. *Can. J. Pysiol.* 57, 1427–1436.

Dam, T.-V., Takeda, Y., Krause, J. E., Escher, E., and Quirion, R. (1990) g-Preprotachykinin-(72–92)-peptide amide: An endogenous preprotachykinin I gene-derived peptide that preferentially binds to neurokinin-2 receptors. *Proc. Natl. Acad. Sci. U.S.A.* 87, 246–250.

DeBernardi, M. A., Seki, T., and Brooker, G. (1991) Inhibition of cAMP accumulation by intracellular calcium mobilization in C6-2B cells stably transfected with substance K receptor cDNA. *Proc. Natl. Acad. Sci. U.S.A.* 88, 9257–9261.

Dietl, M. M. and Palacios, J. M. (1991) Phylogeny of tachykinin receptor localization in the vertebrate central nervous system: apparent absence of neurokinin-2 and neurokinin-3 binding sites in the human brain. *Brain Res.* 539, 211–222.

Dion, S., Nantel, R. F., Jukic, D., Rhaleb, N. E., Tousignant, S., Telemasque, S., Drapeau,

G., Regoli, D., Naline, C., Advenier, C., Rovero, P., and Maggi, C. A. (1990) Structure-activity study of neurokinins: Antagonists for the Neurokinin-2 receptor. *Pharm.* 41, 184–194.

Eistetter, H. R., Church, D. J., Mills, A., et al. (1991) Recombinant bovine neurokinin-2 receptor stably expressed in Chinese hamster ovary cells couples to multiple signal transduction pathways. *Cell Regulation* 2, 767–779.

Emonds-Alt, X., Vilain, P., Goulaouic, P., Proietto, D., Van Broeck, D., Advenier, C., Naline, E., Neliat, G., LeFur, G., and Breliere, J. C. (1992) A potent and selective nonpeptide antagonist of the neurokinin A (NK-2) receptor. *Life Sci.* 50, 101–106.

Folkers, K., Hakanson, R., Horig, J., Jie-Cheng, X., and Leander, S. (1984) Biological evaluation of substance P antagonists. *Br. J. Pharmacol.* 83, 449–456.

Fong, T. M., Yu, H., and Strader, C. D. (1992) Molecular basis for the species selectivity of the neurokinin-1 receptor antagonists CP-96,345 and RP67580. *J. Biol. Chem.* 267, 25668–25671.

Garret, C., Carruette, A., Fardin, V., Moussaoui, S., Peyronei, J., Blanchard, L., and Laduron, P. M. (1991) Pharmacological propertied of a potent and selective nonpeptide substance P antagonist. *Proc. Natl. Acad. Sci. U.S.A.* 88, 10208–10212.

Gerard, N. P., Garraway, L. A., Eddy, R. L., Shows, T. B., Iijima, H., Paquet, J., and Gerard, C. (1991) Human substance P receptor (NK-1): Organization of the gene, chromosome localization, and functional expression of cDNA clones. *Biochemistry* 30, 10640–10646.

Gerard, N. P., Eddy, R. L., Jr., Shows, T. B., and Gerard, C. (1990) The human neurokinin A (substance K) receptor. *J. Biol. Chem.* 265, 20455–20462.

Gorbulev, V., Akhundova, A., Luzius, H., and Fahrenholz, F. (1992) Molecular cloning of substance P receptor cDNA from guinea-pig uterus. *Biochem. Biophys. Acta* 1131, 99–102.

Guard, S. and Watson, S. P. (1991) Tachykinin receptor types: classification and membrane signalling mechanisms. *Neurochem. Int.* 18, 149–169.

Guard, S., Dam, T.-V., Watson, S. P., Martinelli, B., Watling, K. J., and Quirion, R. (1991) [^{3}H]Sar^9, Met(O$_2$)^{11}SP and [^{3}H]Succinyl-[Asp^6, MePhe^8]-SP(6–11) (Senktide): new selective radioligands for NK-1 and NK-3 tachykinin receptors. *Du Pont Biotech Update* February, 4–7.

Hanley, M. R., Lee, C.-M., Jones, L. M., and Michell, R. H. (1980) Similar effects of substance P and related peptides on salivation and phosphatidylinositol turnover in rat salivary glands. *Mol. Pharmacol.* 18, 78–83.

Hershey, A. D., Dykema, P. E., and Krause, J. E. (1991) Organization, structure, and expression of the gene encoding the rat substance P receptor. *J. Biol. Chem.* 266, 4366–4374.

Hershey, A. D. and Krause, J. E. (1990) Molecular characterization of a functional cDNA encoding the rat substance P receptor. *Science* 247, 958–961.

Hopkins, B., Powell, S. J., Dankal, P., Briggs, I., and Graham, A. (1991) Isolation and characterization of the human jung NK-1 receptor cDNA. *Biochem. Biophys. Res. Commun.* 180, 1110–1117.

Huang, R.-R. C., Cheung, A. H., Mazina, K. E., Strader, C. D., and Fong, T. M. (1992) cDNA sequence and heterologous expression of the human neurokinin-3 receptor. *Biochem. Biophys. Res. Commun.* 184, 966–972.

Hunter, J. C., Goedert, M., and Pinnock, R. D. (1985) Mammalian tachykinin-induced hydrolysis of inositol phospholipids in rat brain slices. *Biochem. Biophys. Res. Commun.* 127, 616–622.

Ingi, T., Kitajima, Y., Minamitake, Y., and Nakanishi, S. (1991) Characterization of ligand-binding properties and selectivities of three rat tachykinin receptors by transfection and functional expression of their cloned cDNAs in mammalian cells. *J. Pharmacol. Exp. Ther.* 259, 968–975.

Joslin, G., Krause, J. E., Hershey, A. D., Adams, S. P., Fallon, R. J., and Perlmutter, D. H. (1991) Amyloid-β peptide, substance P, and bonbesin bind to the serpin-enzyme complex receptor. *J. Biol. Chem.* 266, 21897–21902.

Kris, R. M., South, V., Saltzman, A., Felder, S., Ricca, G. A., Jaye, M., Huebner, K., Kagan, J., Croce, C. M., and Schlessinger, J. (1991) Cloning and expression of the human substance K receptor and analysis of its role in mitogenesis. *Cell Growth Differentiation* 2, 15–22.

Laufer, R., Gilon, C., Chorev, M., and Selinger, Z. (1988) Desensitization with a selective agonist discriminates between multiple tachykinin receptors. *J. Pharmacol. Exp. Ther.* 245, 639–643.

Laufer, R., Gilon, C., Chorev, M., and Selinger, Z. (1986) Characterization of a neurokinin B receptor site in rat brain using a highly selective radioligand. *J. Biol. Chem.* 261, 10257–10263.

Laufer, R., Wormser, U., Friedman, Z. Y., Gilon, C., Chorev, M., and Selinger, Z. (1985) Neurokinin B is a preferred agonist for a neuronal substance P receptor and its action is antagonized by enkephalin. *Proc. Natl. Acad. Sci. U.S.A.* 82, 7444–7448.

Lavielle, S., Chassaing, G., Ploux, O., Louillet, D., Besseyre, J., Julien, S., Marquet, A., Convert, O., Beaujouan, J.-C., Torrens, Y., Bergstrom, L., Saffroy, M., and Glowinski, J. (1988) Analysis of tachykinin binding site interactions using constrained analogues of tachykinins. *Biochem. Pharmacol.* 37, 41–49.

Lee, C.-M., Javitch, J. A., and Snyder, S. H. (1983) ^{3}H-Substance P binding to salvary gland membranes: regulation by guanyl nucleotides and divalent cations. *Mol. Pharmacol.* 23, 563–569.

Lee, C.-M., Iversen, L. L., Hanley, M. R., and Sandberg, B. E. B. (1982) The possible existence of multiple receptors for substance P. *Naunyn Schmiedeberg Arch. Pharmacol.* 318, 281–287.

Li, X. J., Wolfgang, W., Wu, Y.-N., North, R. A., and Forte, M. (1991) Cloning, heterlogous expression and developmental regulation of a Drosophila receptor for tachykinin-like peptides. *EMBO J.* 10, 3221–3229.

Luber-Narod, J., Boyd, N. D., and Leeman, S. E. (1990) Guanine nucleotides decrease the affinity of substance P binding to its receptor. *Eur. J. Pharmacol.* 188, 185–191.

MacDonald, S. G. and Boyd, N. D. (1989) Regulation of Substance P receptor affinity by guanine nucleotide-binding proteins. *J. Neurochem.* 53, 264–272.

Maggi, C. A., Eglezos, A., Quartara, R., Patacchini, R., and Giachetti, A. (1992) Heterogeneity of NK-2 tachykinin receptors in hamster and rabbit smooth muscles. *Reg. Peptides* 37, 85–93.

Maggi, C. A., Patacchini, R., Giuliani, S., Rovero, P., Dion, S., Regoli, D., Giachetti, A., and Meli, A. (1990) Competitive antagonists discriminate between NK_2 tachykinin receptor subtypes. *Br. J. Pharmacol.* 100, 588–592.

Mantyh, P. W., Pinnock, R. D., Downes, C. P., Goedert, M., and Hunt, S. P. (1984) Correlation between inositol phospholipid hydrolysis and substance P receptors in rat CNS. *Nature* 309, 795–797.

Masu, Y., Nakayama, K., Tamaki, H., Harada, Y., Kuno, M., and Nakanishi, S. (1987) cDNA cloning of bovine Substance-K receptor through oocyte expression system. *Nature* 329, 836–838.

Merritt, J. E. and Rink, T. J. (1987) The effects of substance P and carbachol on inositol tris- and tetrakisphosphate formation and cytosolic free calcium in rat parotid acinar cells. *J. Biol. Chem.* 262, 14912–14916.

Mitsuhashi, M., Ohashi, Y., Scichijo, S., Shristian, C., Sudduth-Klinger, J., Harrowe, G., and Payan, D. G. (1992) Multiple intracellular signaling pathways of the neuropeptide substance P receptor. *J. Neurosci. Res.* 32, 437–443.

Monnier, D., Colas, J.-F., Rosay, P., Hen, R., Borrelli, E., and Maroteaux, L. (1992) NKD, a developmentally regulated tachykinin receptor in Drosophila. *J. Biol. Chem.* 267, 1298–1302.

Morimoto, H., Murai, M., Maeda, Y., Yamaoka, M., Nishikana, M., Kiyotoh, S., and Fujii, T. (1992) FR 224, a novel cyclopeptide substance P antagonist with NK-1 and NK-2 selectivity. *J. Pharmacol. Exp. Ther.* 262, 398–402.

Murai, M., Morimoto, H., Maeda, Y., Kiyotoh, S., Nishikawa, M., and Fujii, T. (1992) Effects of FK 224, a novel compound NK-1 and NK-2 receptor antagonist, on airway constriction and airway edema induced by neurokinins and sensory nerve stimulation in guinea pigs. *J. Pharmacol. Exp. Ther.* 262, 403–408.

Nakajima, Y., Tsuchida, K., Negishi, M., Ito, S., and Nakanishi, S. (1992) Direct linkage of three tachykinin receptors to stimulation of both phosphotydylinositol hydrolysis and cyclic AMP cascades in transfected Chinese hamster ovary cells. *J. Biol. Chem.* 267, 2437–2442.

Narumi, S. and Maki, Y. (1978) Stimulatory effects of substance P on neurite extension and cyclic AMP levels in cultured neuroblastoma cells. *J. Neurochem.* 30, 1321–1326.

Nilsson, J., von Euler, A. M., and Dalsgaard, C.-J. (1985) Stimulation of connective tissue cell growth by substance P and substance K. *Nature* 315, 61–63.

Parker, I., Sumikawa, K., and Miledi, R. (1986) Neurotensin and substance P receptors expressed in Xenopus oocytes by messenger RNA from rat brain. *Proc. R. Soc. Lond.* 229, 151–159.

Petitet, F., Beaujouan, J.-C., Saffroy, M., Torrens, Y., Chassaing, G., Lavielle, S., Besseyre, J., Garret, C., Carruette, A., and Glowinski, J. (1991) Further demonstration that [Pro⁹]-substance P is a potent and selective ligand of NK-1 tachykinin receptors. *J. Neurochem.* 56, 879–889.

Regoli, D., Drapeau, G., Dion, S., and Couture, R. (1988) New selective agonists for neurokinin receptors: pharmacological tools for receptor characterization. *Trends Pharmacol Sci.* 9, 290–295.

Regoli, D., Elscher, E., Drapeau, G., D'Orleans-Juste, P., and Mizrahi, J. (1984) Receptors for substance P. III. classification by competitive antagonists. *Eur. J. Pharmacol.* 97, 179–189.

Rosell, S., Bjorkroth, U., Xu, J., and Folkers, K. (1983) The pharmacological profile of a substance P (SP) antagonist. Evidence for the existence of subpopulations of SP receptors. *Acta Physiol. Scand.* 117, 445–449.

Rouissi, N., Gitter, B. D., Waters, D. C., Howbert, J. J., Nixon, J. A,. and Regoli, D. (1991) Selectivity and specificity of new, non-peptide, quinuclidine antagonists of substance P. *Biochem. Biophys. Res. Commun.* 176, 894–901.

Sachais, B. S., Snider, R. M., Lowe, J. A., III, and Krause, J. E. (1993) Molecular basis for the species selectivity of the substance P antagonist CP-96,345. *J. Biol. Chem.* (in press).

Saffroy, M., Beaujouan, J.-C., Torrens, Y., Besseyre, J., Bergstrom, L., and Glowinski, J. (1988) Localization of tachykinin binding sites (NK1, NK2, NK3 Ligands) in the rat brain. *Peptides* 9, 227–241.

Sandberg, B. E. B. (1985) Structure-activity relationships for substance P: a review. In *Substance P: Metabolism and Biological Actions*, Jordan, C. C. and Oehme, P., Eds., Taylor and Francis, Philadelphia, pp. 65–81.

Sandberg, B. E. B. and Iversen, L. L. (1982) Substance P. *J. Med. Chem.* 25, 1009–1015.

Saviano, G., Temussi, P. A., Motta, A., Maggi, C. A., and Rovero, P. (1991) Conformation-activity relationship of tachykinin Neurokinin A (4–10) and of some [Xaa⁸] analogs. *Biochemistry* 30, 10175–10181.

Shigemoto, R., Yokota, Y., Tsuchida, K., and Nakanishi, S. (1990) Cloning and expression of a rat neuromedin K receptor cDNA. *J. Biol. Chem.* 265, 623–628.

Snider, R. M., Constantine, J. W., Lowe, J. A., et al. (1991) A potent nonpeptide antagonist of the substance P (NK1) receptor. *Science* 251, 435–437.

Sugiya, H., Tennes, K. A., and Putney, J. W., Jr. (1987) Homologous desensitization of substance P induced inosotol polyphosphate formation in rat parotid acinar cells. *Biochem. J.* 244, 647–653.

Sundellin, J. B., Provvendini, D. M., Wahlestedt, C. R., Pohl, J. S., and Peterson, P. A. (1992) Molecular cloning of the murine substance K and substance P receptor genes. *Eur. J. Biochem.* 203, 625–631.

Tachado, S. D., Akhtar, R. A., Yousufazai, S. Y. K., and Abdel-Latif, A. A. A. (1991) Species differences in the effects of substance P on inositol trisphosphate accumulation and cyclic AMP formation, and on contraction in isolated iris sphincter of the mammalian eye: Differences in receptor density. *Exp. Eye Res.* 53, 729–739.

Takahashi, K., Tanaka, A., Hara, M., and Nakanishi, S. (1992) The primary structure and gene organization of human substance P and neuromedin K receptors. *Eur. J. Biochem.* 204, 1025–1033.

Takeda, Y., Blount, P., Sachais, B. S., Hershey, A. D., Raddatz, R., and Krause, J. E. (1992) Ligand binding and kinetics of substance P and neurokinin A receptors stably expressed in Chinese hamster ovary cells and evidence for differential stimulation of inositol 1,4,5-trisphosphate and cyclic AMP second messenger responses. *J. Neurochem.* 59, 740–745.

Takeda, Y. and Krause, J. E. (1991) Pharmacological and molecular biological studies on the diversity of rat tachykinin NK-2 receptor subtypes in rat CNS, duodenum, vas deferens, and urinary bladder. In *Substance P and Related Peptides*, Leeman, S. E., Krause, J. E., and Lembeck, F., Eds., Annals of the New York Academy of Sciences, New York, 632, 479–482.

Takeda, Y., Chou, K. B., Takeda, J., Sachais, B. S., and Krause, J. E. (1991) Molecular cloning, structural characterization and functional expression of the human substance P receptor. *Biochem. Biophys. Res. Commun.* 179, 1232–1240.

Torrens, Y., Daguet de Montety, M. C., El Etr, M., Beaujouan, J.-C., and Glowinski, J. (1989) Tachykinin receptors of the NK-1 type (substance P) coupled positively to phospholipase C on cortical astrocytes from the newborn mouse in primary culture. *J. Neurochem.* 52, 1913–1918.

Torrens, Y., Lavielle, S., Chassaing, G., Marquet, A., Glowinski, J., and Beaujouan, J. C. (1984) Neuromedin K, a tool to further distinguish two central tachykinin binding sites. *Eur. J. Pharmacol.* 102, 381–382.

Tsuchida, K., Shigemoto, R., Yokota, Y., and Nakanishi, S. (1990) Tissue distribution and quantitation of the mRNA for three rat tachykinin receptors. *Eur. J. Biochem.* 193, 751–757.

van Giersbergen, P. L. M., Shatzer, S. A., Henderson, A. K., Lai, J., Nakanishi, S., Yamamura, H. I., and Buck, S. H. (1991) Characterization of a tachykinin peptide NK-2 receptor transfected into murine fibroblast B82 cells. *Proc. Natl. Acad. Sci. U.S.A.* 88, 1661–1665.

Venepalli, B. R., Aimone, L. D., Appel, K. C., Bell, M. R., Dority, J. A., Goswami, R., Hall, P. L., Kumar, V., Lawrence, K. B., Logan, M. E., Scensny, P. M., Seelye, J. A., Tomczuk, B. E., and Yanni, J. M. (1992) Synehesis and subatance P receptor binding activity of Androstano[3,2-b]pyrymido[1,2-a]benzimidizoles. *J. Med. Chem.* 35, 374–378.

Watson, S. (1984) The action of substance P on contraction, inositol phospholipids, and adenylate cyclase in rat small intestine. *Biochem. Pharmacol.* 33, 3733–3737.

Watson, S. P., Sandberg, B. E. B., Hanley, M. R., and Iversen, L. L. (1983) Tissue selectivity of substance P alkyl esters: suggestive of multiple receptors. *Eur. J. Pharmacol.* 87, 77–84.

Wormser, U., Laufer, R., Hart, Y., Chorev, M., Gilon, C., and Selinger, Z. (1986) Highly selective agonists for substance P receptor subtypes. *EMBO J.* 5, 2805–2808.

Xie, G.-X., Miyajima, A., and Goldstein, A. (1992) Expression cloning of cDNA encoding a seven-helix receptor from human placenta with affinity for opioid ligands. *Proc. Natl. Acad. Sci. U.S.A.* 89, 4124–4128.

Yamachita, K., Koide, Y., and Aiyochi, Y. (1983) Effects of substance P on thyroidal cyclic AMP levels and thyroid hormone from canine thyroid slices. *Life Sci.* 32, 2163–2166.

Yokota, Y., Sasai, Y., Tanaka, K., Fijiwana, T., Tsuchida, K., Shigemoto, R., Kakizuka, A., Ohkubo, H., and Nakanishi, S. (1989) Molecular characterization of a functional cDNA for rat substance P receptor. *J. Biol. Chem.* 264, 17649–17652.

99

Miscellaneous Receptors

Xiao-Ming Guan, Ph.D.

1.99.0 Introduction

The G protein-coupled receptors (GCRs) have been shown to be the essential elements in transmembrane signaling mechanisms in a broad range of biological systems including mammalian cells as well as several lower organisms such as yeast, slime mold, and viruses. As is evident from previous chapters in this volume, our knowledge on GCRs has expanded tremendously over the past few years as a consequence, at least in part, of the rapid advancement in molecular cloning techniques. To date, the genes of many GCRs have been cloned and analyzed, and the number of new clones is still increasing rapidly. Such studies will greatly facilitate our understanding of the structure and function of GCRs. This chapter is intended to provide a synopsis as well as some highlights on the current knowledge on several cloned GCRs not covered in the previous chapters. They include the receptors for cAMP in dictyostelium, yeast mating factors, N-formyl peptide, C5a, platelet-activating factors, thrombin, Thromboxane A2, neuropeptide Y, and neurotensin. Two classes of gene products that resemble structurally the GCRs, i.e., HCMV-encoded GCR homologues and endothelial differentiation gene are also included. The amino acid sequences of these receptors are listed in Figure 1.

1.99.1 C5a Analphylatoxin Receptors

1.99.1.1 Human C5a Receptors

C5a is a glycoprotein of 74-amino acid residues derived from the 5th component of complement and, like fMLP (see below), C5a is involved in the chemotactic responses in leukocytes through interactions with the specific cell surface receptors (Goldstein, 1988; Krych et al., 1992). Earlier biochemical and pharmacological characterizations of the C5a receptor indicated that the putative receptor was a polypeptide with an apparent molecular mass of about 40 to 55 kDa (Rollins and Springer, 1985; Johnson and Chenoweth, 1985; Gerard et al., 1989), and capable of binding to [^{125}I]C5a with a single high affinity (Kd = 1 to 7 n*M)* in human polymorphonuclear leukocytes (Chenoweth and Hugli, 1978). Further studies demonstrated that C5a was able to stimulate GTPase activity (Feltner et al., 1986), and the binding

```
MiscC5a.canine        1 . . . . . . . . . . . . . . . . . . . . M A S M N F S P P E Y P D Y . . . . . G T A T L D P N I . . .
MiscC5a.human         1 . . . . . . . . . . . . . . . . . . . . M N S F N Y T T P D Y G H Y . . . . D D K D T L D L N T . . .
MiscCMV-UL27.viral    1 . . . . . . . . . . . . . . . . . . . . M T T S T N N Q T . . . . . . . . . L T Q V S N M T N . . .
MiscCMV-UL28.viral    1 . . . . . . . . . . . . . . . . . . . . M T P T T T T A E . . . . . . . . . L T T E F D Y D E . . .
MiscCMV-UL33.viral    1 . . . . . . . . . . . . . . . . . . . . M T G P L F A I R T T E A V . . . . L N T F I I F V G G . . .
MiscFML1.human        1 . . . . . . . . . . . . . . . . . . . . M E T . N F S I P . . . . . . . . . . . . L N E . . .
MiscFMLP.human        1 . . . . . . . . . . . . . . . . . . . . M E T . N S S L P . . . . . . . . . . . . . T N I . . .
MiscFMLP2.human       1 . . . . . . . . . . . . . . . . . . . . M E T . N F S T P . . . . . . . . . . . . L N E . . .
MiscNPY.bovine        1 . . . . . . . . . . . . . . . . . . . . M E G I R I F T S D N Y T E . . . . D D L G S G D Y D S . . .
MiscNPY.drosophila    1 . . M Y Y I A H Q Q P M L R N E D D N Y Q E G Y F I R P D P A S L . . . . I Y N T T A L P A D D E G
MiscNPY.human1        1 . . . . . . . . . . . . . . . . . . . . M N S T L F S Q V E N H S V . . . . H S N F S E K N A Q . . .
MiscNPY.human2        1 . . . . . . . . . . . . . . . . . . . . M E G I S I Y T S D N Y T E . . . . E M G S G D Y D S . . .
MiscNPY.mouse         1 . . . . . . . . . . . . . . . . . . . . M N S T L F S K V E N H S I . . . . H Y N A S E . N S P . . .
MiscNPY.rat           1 . . . . . . . . . . . . . . . . . . . . M N S T L F S R V E N Y S V . . . . H Y N V S E . N S P . . .
MiscNT.rat            1 . . . . . M H L N S S V P Q G T P G E P D A Q P F S G P Q S E M E . . . . A T F L A L S L S N G S G
MiscPAFR.quineapig    1 . . . . . . . . . . . . . . . . . . . . M E . . . . . . . . . . . . . . . . . . . . . . . . . . .
MiscPAFR.human        1 . . . . . . . . . . . . . . . . . . . . M E . . . . . . . . . . . . . . . . . . . . . . . . . . .
MiscSTE2.yeast        1 . . . . . . . . . . . . . . . . . . . . M S D A A P S L S N L F Y D . . . . P T Y N P G . Q S T . . .
MiscSTE3.yeast        1 . . . . . . . . . . . . M S Y K S A I I G L C L L A V I L L A P P L A W H S H T K N I P A I I L . .
MiscTXA2.human        1 . . . . . . . . . . . . . . . . . . . . M W P N G S S L G P C . . . . . . . . . . . . . . . . . . .
MiscTXA2.mouse        1 . . . . . . . . . . . . . . . . . . . . M W P N G T S L G A C . . . . . . . . . . . . . . . . . . .
MiscThrombin.hamster  1 M G P Q R L L L V A A G L S L C G P L L S S R V P V R Q P E S E M . . . . T D A T V N P R S F F L R
MiscThrombin.human    1 M G P R R L L L V A A C F S L C G P L L S A R T R A R R P E S K A . . . . T N A T L D P R S F L L R
MiscThrombin.rat      1 M G P R R L L L V A V G L S L C G P L L S S R V P M R Q P E S E R M Y A T P Y A T P N P R S F F L R
MisccAR1.slimemold    1 . . . . . . . . . . . M G L L D G N P A N E T S L V L L L F A D F . . . S S M L G C M A V L I . .
Miscedg-1.human       1 . . . . . . . . . . . . . . . . . . . . M G P T S V P L V K A H R S . . . . S V S D Y V N Y D I . . .
MisckSTE2.yeast       1 . . . . . . . . . . . . . . . . . . . . M S . G K Q D L S P L G L Y . . . . S S Y D P T . K G L . . .
Miscmam2.yeast        1 . . . . . . . . . . . . . . . . . . . . M R Q P W W K D F T I . . . . . . . P D A S A I I H Q N . . .
```

```
MiscC5a.canine        24 . . . . . . . . . . . . . . . . . . . . . . F V D E S L N T P K L . . . . . . . . . S V . . . . . .
MiscC5a.human         25 . . . . . . . . . . . . . . . . . . . . . . P V D K T S N T . . L . . . . . . . . . R V . . . . . .
MiscCMV-UL27.viral    19 . . . . . . . . . . . . . . . . . . . . . . H T L N S T E I Y Q L . . . . . . . . . F E . . . . . .
MiscCMV-UL28.viral    19 . . . . . . . . . . . . . . . . . . . . . . D A T P C V F T D V L . . . . . . . . . N Q . . . . . .
MiscCMV-UL33.viral    25 . . . . . . . . . . . . . . . . . . . . . . P L N A I V L I T Q L L . . . . . . . . T N R V . . . . .
MiscFML1.human        12 . . . . . . . . . . . . . . . . . . . . . . T E E V L P E P A G H . . . . . . . . . T V . . . . . .
MiscFMLP.human        12 . . . . . . . . . . . . . . . . . . . . . . S G G T P A V S A G Y . . . . . . . . . L F . . . . . .
MiscFMLP2.human       12 . . . . . . . . . . . . . . . . . . . . . . Y E E V S Y E S A G Y . . . . . . . . . T V . . . . . .
MiscNPY.bovine        25 . . . . . . . . . . . . . . . . . . . . . . M K E P C F R E E N A . . . . . . . . . H F . . . . . .
MiscNPY.drosophila    45 S N Y G Y G S T T T L S G L Q F E T Y N I T V M M N F S C D D Y D L L S E D M W S S A Y . . . . . .
MiscNPY.human1        25 . . . . . . . . . . . . . . . . . . . . . . L L A F E N D D C H L . . . P L . . A M I . . . . .
MiscNPY.human2        24 . . . . . . . . . . . . . . . . . . . . . . M K E P C F R E E N A . . . . . . . . . N F . . . . .
MiscNPY.mouse         24 . . . . . . . . . . . . . . . . . . . . . . L L A F E N D D C H L . . . P L . . A V I . . . . .
MiscNPY.rat           24 . . . . . . . . . . . . . . . . . . . . . . F L A F E N D D C H L . . . P L . . A V I . . . . .
MiscNT.rat            42 . . . . . . . . . . . . . . . . . . . . . . N T S E S D T A G P N S D L D V N T D I Y . . . . .
MiscPAFR.quineapig    3 . . . . . . . . . . . . . . . . . . . . . . L N S S S R V D S E F . . . . . . . R Y . . . . .
MiscPAFR.human        3 . . . . . . . . . . . . . . . . . . . . . . P H D S S H M D S E F . . . . . . . R Y . . . . .
MiscSTE2.yeast        24 . . . . . . . . . . . . . . . . . . . . . . I N Y T S I Y G N G S T I . . . . . . . T F D E . . .
MiscSTE3.yeast        37 . . . . . . . . . . . . . . . . . . . . . . I T W L L T M N L T C I V D A A I W S D D D F . . . . . .
MiscTXA2.human        12 . . . . . . . . . . . . . . . . . . . . . . F R P T N I T L E E R . . . . . . . . . . . . . . .
MiscTXA2.mouse        12 . . . . . . . . . . . . . . . . . . . . . . F R P V N I T L Q E R . . . . . . . . . . . . . . .
MiscThrombin.hamster  47 N P G E N T F E L I P L G D E E E K N E S T L P E G R A I Y L N K S H . S P A P L A P F I S E D A S
MiscThrombin.human    47 N P . N D K Y E . . P F W E D E E K N E S G L T E Y R L V S I N K S S P L Q K Q L P A F I S E D A S
MiscThrombin.rat      51 N P S E D T F E Q F P L G D E E E K N E S I P L E G R A V Y L N K S R F P P M P P P P F I S E D A S
MisccAR1.slimemold    34 . . . . . . . . . . . . . . . . . . . . . . G F W R L K L . L R N H V T K V I . . . A C F . . . . . .
Miscedg-1.human       25 . . . . . . . . . . . . . . . . . . . . . . I V R H Y N Y T G K L . . . . . . . . N I S A . . . .
MisckSTE2.yeast       23 . . . . . . . . . . . . . . . . . . . . . . I S Y T S L Y G S G T T V . . . . . . . T F E E . . . .
Miscmam2.yeast        22 . . . . . . . . . . . . . . . . . . . . . . I T I V S I V G . E I E . . . . . . . . . . . . . . .
```

FIGURE 1. Alignment of the amino acid sequences of miscellaneous G protein-coupled receptors. The sequences were obtained from commercial protein databases such as GenBank and EMBL. The references for each sequence are cited in the text.

```
MiscC5a.canine        37  - - - - - - - P D M I A L V I F V M V F - L V G V P G N F L V V W V T G - - F E V R - R T I N A I W
MiscC5a.human         36  - - - - - - - P D I L A L V I F A V V F - L V G V L G N A L V V W V T A - - F E A K - R T I N A I W
MiscCMV-UL27.viral    32  - - - - - - - Y T R L G V W L M C I V G - T F L N V - - L V I T T I L Y - - Y R R K K K S P S D T Y
MiscCMV-UL28.viral    32  - - - - - - - S K P V T L F L Y G V V F - L F G S I G N F L V I F T I T - - W R R R I Q C S G D V Y
MiscCMV-UL33.viral    41  - - - - - - - L G Y S T P T I Y M T N L Y S T N F L T L T V L P F I V L S N Q W L L P A G V A S - -
MiscFML1.human        25  - - - - - - L W I F S L L V H G V T F - V F G V L G N G L V I W V A G - - F R M T - R T V N T I C
MiscFMLP.human        25  - - - - - - L D I I T Y L V F A V T F - V L G V L G N G L V I W V A G - - F R M T - H T V T T I S
MiscFMLP2.human       25  - - - - - - L R I L P L V V L G V T F - V L G V L G N G L V I W V A G - - F R M T - R T V T T I C
MiscNPY.bovine        38  - - - - - - N R I F L P T V Y S I I F - L T G I V G N G L V I L V M G - - Y Q K K L R S M T D K Y
MiscNPY.drosophila    89  - - - - - F K I I V Y M L Y I P I F - I F A L I G N G T V C Y I V Y - - S T P R M R T V T N Y F
MiscNPY.human1        41  - - - - - - F T L A - - L A Y G A V I - I L G V S G N L A L I I I I L - - K Q K E M R N V T N I L
MiscNPY.human2        37  - - - - - - N K I F L P T I Y S I I F - L T G I V G N G L V I L V M G - - Y Q K K L R S M T D K Y
MiscNPY.mouse         40  - - - - - - F T L A - - L A Y G A V I - I L G V S G N L A L I I I I L - - K Q K E M R N V T N I L
MiscNPY.rat           40  - - - - - - F T L A - - L A Y G A V I - I L G V S G N L A L I I I I L - - K Q K E M R N V T N I L
MiscNT.rat            63  - - - - - - S K V L V T A I Y L A L F - V V G T V G N S V T A F T L A R K K S L Q S L Q S T V H Y
MiscPAFR.guineapig    16  - - - - - - - - - T L F P I V Y S I I F - V L G I I A N G Y V L W V F A R L Y P S K K L N E I K I F
MiscPAFR.human        16  - - - - - - - - - T L F P I V Y S I I F - V L G V I A N G Y V L W V F A R L Y P C K K F N E I K I F
MiscSTE2.yeast        41  - - - - - L Q G L V N S T V T Q A I - M F G V R C G A A A L T L I V - - M W M T S R S R K T P I
MiscSTE3.yeast        60  - - - - - L T R W D G K G W C D I V I K L - Q V G A N I G I S C A V T N I I Y N L H T I L K A D
MiscTXA2.human        23  - - - - - - R L I A S P W F A A S F C V V G L A S N L L A L S V L A G A R Q G G S H T R S S F L
MiscTXA2.mouse        23  - - - - - - R A I A S P W F A A S F C A L G L G S N L L A L S V L A G A R P G A G - P R S S F L
MiscThrombin.hamster  96  G Y L T S P W L R L F I P S V Y T F V F - V V S L P L N I L A I A V F V - - L K M K V K K P A V V Y
MiscThrombin.human    94  G Y L T S S W L T L F V P S V Y T G V F - V V S L P L N I M A I V V F I - - L K M K V K K P A V V Y
MiscThrombin.rat      101 G Y L T S P W L T L F I P S V Y T F V F - I V S L P L N I L A I A V F V - - F R M K V K K P A V V Y
MisccAR1.slimemold    53  - - - - - - - C A T S F C K D F P S T I L T L T N T A V N G G F P C - - - - - - Y - L Y A I V - - -
Miscedg-1.human       40  - - - - - - - D K E N S I K L T S V V F I L I C C F I I L E N I F V L L T I W K T K K F H R P M Y Y
MisckSTE2.yeast       40  - - - - - - - L Q I F V N K K I T Q G I - L F G T R I G A A G L A I I V - - L W M V S K N R K T P I
Miscmam2.yeast        33  - - - - - - - V P V S T I D A Y E R D R L L T G M T L S A Q L A L G V L T I L M V C L L S S S E K R


MiscC5a.canine        76  - F L - N L A V - - - - A D L L S - C L A L P I L F S S I V - Q Q G Y W P F G N A - - - A C - - - R
MiscC5a.human         75  - F L - N L A V - - - - A D F L S - C L A L P I L F T S I V - Q H H H W P F G G A - - - A C - - - S
MiscCMV-UL27.viral    70  - I C - N L A V - - - - A D L L I - V V G L P F F L E Y A K - H H P K L S - R E V - - - V C - - - S
MiscCMV-UL28.viral    72  - F I - N L A A - - - - A D L L F - V C T L P L W M Q Y L L - D H N S L A - - S V - - - P C - - - T
MiscCMV-UL33.viral    82  C K F L S V I Y - - - Y S S C T V G F A T V A L I A A D R Y R V L H K R T Y A R Q - - - S Y - - - R
MiscFML1.human        64  - Y L - N L A L - - - - A D F S F - S A I L P F R M V S V A - M R E K W P F A S F - - - L C - - - K
MiscFMLP.human        64  - Y L - N L A V - - - - A D F C F - T S T L P F F M V R K A - M G G H W P F G W F - - - L C - - - K
MiscFMLP2.human       64  - Y L - N L A L - - - - A D F S F - T A T L P F L I V S M A - M G E K W P F G W F - - - L C - - - K
MiscNPY.bovine        78  - R L - H L S V - - - - A D L L F - V L T L P F W A V D A V - - - A N W Y F G K F - - - L C - - - K
MiscNPY.drosophila    129 - I A - S L A I - - - - G D I L M S F F C E P S S F I S L F - I L N Y W P F G L A - - - L C - - - H
MiscNPY.human1        79  - I V - N L S F - - - - S D L L V A I M C L P F T F V - Y T - L M D H W V F G E A - - - M C - - - K
MiscNPY.human2        77  - R L - H L S V - - - - A D L L F - V I T L P F W A V D A V - - - A N W Y F G N F - - - L C - - - K
MiscNPY.mouse         78  - I V - N L S F - - - - S D L L V A V M C L P F T F V - Y T - L M D H W V F G E T - - - M C - - - K
MiscNPY.rat           78  - I V - N L S F - - - - S D L L V A V M C L P F T F V - Y T - L M D H W V F G E T - - - M C - - - K
MiscNT.rat            105 - H L G S L A L - - - - S D L L I L L L A M P V E L Y N F I W V H H P W A F G D A - - - G C - - - R
MiscPAFR.guineapig    56  - M V - N L T V - - - - A D L L F - L I T L P L W I V Y Y S - N Q G N W F L P K F - - - L C - - - N
MiscPAFR.human        56  - M V - N L T M - - - - A D M L F - L I T L P L W I V Y Y Q - N Q G N W I L P K F - - - L C - - - N
MiscSTE2.yeast        81  F I I N Q V S L - - - - F - L I I L H S A L Y F K Y L L S N Y S S V T Y A L T G F - - - P - - - - Q
MiscSTE3.yeast        102 S V L P D L S S W T K I V K D L V I S L F T P V M V M G F S Y L L Q V F R Y G I A R Y N G C Q N L -
MiscTXA2.human        65  T F L C G L V L - - - - T D F L G L L V T G T I V V S Q H A - A L F E W H A V D P - - - G C - - - R
MiscTXA2.mouse        64  A L L C G L V L - - - - T D F L G L L V T G A I V A S Q H A - A L L D W R A T D P - - - S C - - - R
MiscThrombin.hamster  143 - M L - H L A M - - - - A D V L F - V S V L P L K I S Y Y F - S G S D W Q F G S G - - - M C - - - R
MiscThrombin.human    141 - M L - H L A T - - - - A D V L F - V S V L P F K I S Y Y F - S G S D W Q F G S E - - - L C - - - R
MiscThrombin.rat      148 - M L - H L A M - - - - A D V L F - V S V L P F K I S Y Y F - S G T D W Q F G S G - - - M C - - - R
MisccAR1.slimemold    86  I T Y G S F A C W L W T L - C L A I S I Y M L I V K R E - - - - P E P E R F E K Y Y Y L C W G L P
Miscedg-1.human       83  - F I G N L A L - - - - S D L L - - A G V A Y T A N L L L S G A T T Y K L T P A - - - Q W - - - F
MisckSTE2.yeast       80  F I I N Q I S L - - - F L I L L H S S L F L R Y L L G D Y A - - S V V F N F - T L - - - F S - - - Q
Miscmam2.yeast        76  - - - - K H P V - - F V F N S A S I V A M C L R A I L N I V T I C S N S Y S I L - - - V N - - - Y
```

FIGURE 1b.

```
MiscC5a.canine        112  ILPSLILLNMY----ASIL-LLTTISADRFVLVFNP-----I---WCQNYR
MiscC5a.human         111  ILPSLILLNMY----ASIL-LLATISADRFLLVFKP-----I---WCQNFR
MiscCMV-UL27.viral    105  GLNACFYICLF----AGVC-FLINLSMDRYCVIVWG-----V---ELNRVR
MiscCMV-UL28.viral    106  LLTACFYVAMF----ASLC-FITEIALDRYYAIVY----------MRYR
MiscCMV-UL33.viral    123  STYMILLLTWL----AGLI-FSVPAAVYTTVVMHHD-----A---NDTNNT
MiscFML1.human        100  LVHVMIDINLF----VSVY-LITIIALDRCICVLHP-----A---WAQNHR
MiscFMLP.human        100  FLFTIVDINLF----GSVF-LIALIALDRCVCVLHP-----V---WTQNHR
MiscFMLP2.human       100  LIHIVVDINLF----GSVF-LIGFIALDRCICVLHP-----V---WAQNHR
MiscNPY.bovine        112  AVHVIYTVNLY----SSVL-ILAFISLDRYLAIVHA-----T---NSQKPR
MiscNPY.drosophila    166  FVNYSQAVSVL----VSAY-TLVAISIDRYIAIMWP---------LKPR
MiscNPY.human1        115  LNPFVQCVSIT----VSIF-SLVLIAVERHQLIINP---------RGWR
MiscNPY.human2        111  AVHVIYTVNLY----SSVL-ILAFISLDRYLAIVHA-----T---NSQRPR
MiscNPY.mouse         114  LNPFVQCVSIT----VSIF-SLVLIAVERHQLIINP---------RGWR
MiscNPY.rat           114  LNPFVQCVSIT----VSIF-SLVLIAVERHQLIINP---------RGWR
MiscNT.rat            144  GYYFLRDACTY----ATAL-NVASLSVERYLAICHP-----F---KAKTLM
MiscPAFR.guineapig    92   LAGCLFFINTY----CSVA-FLGVITYNRFQAVKYP-----I---KTAQAT
MiscPAFR.human        92   VAGCLFFINTY----CSVA-FLGVITYNRFQAVTRP-----I---KTAQAN
MiscSTE2.yeast        119  FISRG-DVHVY-GA-TNIIQVLLVASIETSLVFQIK---------VIFTGD
MiscSTE3.yeast        151  LSPTWITTVLYTM-WMLIWSFVGAVYATLVLFVFYKKRKDVRDILHCTNS
MiscTXA2.human        104  LCRFMGVVMIF-FGLSPLL-LGAAMASERYLGITRP-----F----SRPAV
MiscTXA2.mouse        103  LCYFMGVAMVF-FGLCPLL-LGAAMASERFVGITRP-----F----SRPTA
MiscThrombin.hamster  179  FATAAFYCNMY----ASIM-LMTVISIDRFLAVVYP-----I---QSLSWR
MiscThrombin.human    177  FVTAAFYCNMY----ASIL-LMTVISIDRFLAVVYP-----M---QSLSWR
MiscThrombin.rat      184  FATAACYCNMY----ASIM-LMTVISIDRFLAVVYP-----I---QSLSWR
MisccAR1.slimemold    131  LISTIVMLAKNTVQFVGNWCWIGVSFTGYRFGLFY---------
Miscedg-1.human       119  LREGSMFVALS----ASVF-SLLAIAIERYITMLKM---------KLHNG
MisckSTE2.yeast       118  SISRN-DVHVY-GA-TNMIQVLLVAAVEISLIFQVR---------VIFKGD
Miscmam2.yeast        113  -GFILNMVHMY----VHVFNILILLLAPVIIFTAEM----S---MMIQVR

MiscC5a.canine        150  GPQLAWAACSVA-WAVALLL-TV----PSF-IFRGVHT-----
MiscC5a.human         149  GAGLAWIACAVA-WGLALLL-TI----PSF-LYRVVRE-----
MiscCMV-UL27.viral    143  NNKRATCWVVIF-WILAVLM-GM----PHY-LMYSHTN-----
MiscCMV-UL28.viral    140  PVKQA-CLFSIFWWIFAVII-AI----PHF-MVVTKKD-----
MiscCMV-UL33.viral    161  NGHATCVLYFVAEEVHTVLLSWK----VLLTMVWGAAPVIMMTW---F-Y
MiscFML1.human        138  TMSLAKRVMTGL-WIFTIVL-TL----PNF-IFWTTISTTNGDT---Y-C
MiscFMLP.human        138  TVSLAKKVIIGP-WVMALLL-TL----PVI-IRVTTVPGKTGTV---A-C
MiscFMLP2.human       138  TVSLAMKVIVGP-WILALVL-TL----PVF-LFLTTVTIPNGDT---Y-C
MiscNPY.bovine        150  KLLAEKVVYVGV-WLPAVLL-TI----PDL-IF-ADIKEVDERY---I-C
MiscNPY.drosophila    201  ITKRYATFIIAGVWFIALAT-AL----PI---PIVSGLD-------I-P
MiscNPY.human1        150  PNNRHAYVGIAVIWVLAVAS-SL----PFL-IYQVMTDEP-----F-Q
MiscNPY.human2        149  KLLAEKVVYVGV-WIPALLL-TI----PDF-IF-ANVSEADDRY---I-C
MiscNPY.mouse         149  PNNRHAYIGITVIWVLAVAS-SL----PFV-IYQILTDEP-----F-Q
MiscNPY.rat           149  PNNRHAYIGITVIWVLAVAS-SL----PFV-IYQILTDEP-----F-Q
MiscNT.rat            182  SRSRTKKFISAI-WLASALL-AI----PML-FTMGLQNRSGDGT---HPG
MiscPAFR.guineapig    130  TRKRGIALSLVI-WV-AIVA-AA----SYF-LVMDSTNVVSNKA---G-S
MiscPAFR.human        130  TRKRGISLSLVI-WV-AIVG-AA----SYF-LILDSTNTVPDSA---G-S
MiscSTE2.yeast        158  NFKRIGLMLTSISFTLGIATVTMYFVSAVKGMIVTYNDVSATQDKY-FNA
MiscSTE3.yeast        200  GLNLTRFARLLI-FCFIIILVMFPF--SVYTFVQDLQQVEGHYT---FKN
MiscTXA2.human        144  ASQRRAWATVGLVWAAALALGLL----PLLGVGRYTVQYPG--------
MiscTXA2.mouse        143  TS-RRAWATVGLVWVAAGALGLL----PLLGLGRYSVQYPG--------
MiscThrombin.hamster  217  TLGRANFTCLVI-WVMAIMG-VV----PLL-LKEQTTRVPG-----
MiscThrombin.human    215  TLGRASFTCLAI-WALAIAG-VV----PLV-LKEQTIQVPG-----
MiscThrombin.rat      222  TLGRANFTCVVI-WVMAIMG-VV----PLL-LKEQTTQVPG-----
MisccAR1.slimemold    166  ----GPF--LFI-WAISAVL-VGLT---SRYTYVVIHNGVSDN-----KE
Miscedg-1.human       155  SNNFRLFLLISACWVISLIL-GGL----PIM-GWNCISALSS--------C
MisckSTE2.yeast       157  SYKGVGRILTSISAVLGFTTVVMYFITAVKSMTSVYSDLTKTSDRYFFNI
Miscmam2.yeast        151  -IICAHD--RKTQRIMTVISACL----TVLVLAFWITNMCQ---------
```

FIGURE 1c.

```
MiscC5a.canine        181  E Y - F P F W M T C G V D Y S G V G V L V E R G V A I L R L L M - G F L G P L V - I L S I C Y T F L
MiscC5a.human         180  E Y - F P P K V L C G V D Y S - H D K R R E R A V A I V R L V L - G F L W P L L - T L T I C Y T F I
MiscCMV-UL27.viral    174  - - - - - - N E C V G E F A N E T S G W F P V F L N T K V N I C G Y L A P I A - L M A Y T Y N R M
MiscCMV-UL28.viral    171  - - - - - - N Q C M T D Y D Y L E V S - Y P I I L N V E L M L G A F V I P L S - V I S Y C Y Y R I
MiscCMV-UL33.viral    203  A F F Y S T V Q R T S Q K Q R S R T L T - - - F V S V L L I S F V A L Q T P Y V S L M I F N - S Y A
MiscFML1.human        177  I F N F A F W G D T A V E R L N V F I T M A K V F L I L H F I I - G F T V P M S - I I T V C Y G I I
MiscFMLP.human        177  T F N F S P W T N D P K E R I N V A V A M L T V R G I I R F I I - G F S A P M S - I V A V S Y G L I
MiscFMLP2.human       177  T F N F A S W G G T P E E R L K V A I T M L T A R G I I R F V I - G F S L P M S - I V A I C Y G L I
MiscNPY.bovine        188  D - - - R F Y P S - - - - - - D L W L V - - - V F Q F Q H I V V - G L L L P G I - V I L S C Y C I I
MiscNPY.drosophila    234  M S P W H T K C E K Y I C R E M W P S R S Q E Y Y Y T L S L F A L Q F V V P L G - V L I F T Y A R I
MiscNPY.human1        186  N V T L D A Y K D K Y V C F D Q F P S D S H R L S Y T T L L L V L Q Y F G P L C - F I F I C Y F K I
MiscNPY.human2        187  D - - - R F Y P N - - - - - - D L W V V - - - V F Q F Q H I M V - G L I L P G I - V I L S C Y C I I
MiscNPY.mouse         185  N V S L A A F K D K Y V C F D K F P S D S H R L S Y T T L L L V L Q Y F G P L C - F I F I C Y F K I
MiscNPY.rat           185  N V S L A A F K D K Y V C F D K F P S D S H R L S Y T T L L L V L Q Y F G P L C - F I F I C Y F K I
MiscNT.rat            222  G L V C T P I V D T A T V K V V I Q V N T F M S F L F P M L V I S I L N T V I A N K L T V M V H Q A
MiscPAFR.guineapig    168  G N I T R C F E H - - Y E K G S K P V L I I H I C I V L G F F I - V F L L I L F C N L V I I H T L L
MiscPAFR.human        168  G N V T R C F E H - - Y E K G S V P V L I I H I F I V F S F F L - V F L I I L F C N L V I I R T L L
MiscSTE2.yeast        207  S T I L L A S S I N F M S F V L V V K L I L A I R S R R F L G L K Q F D S F H I L L M S C Q S L L
MiscSTE3.yeast        244  T H S S T I W N T I I K F D P G R P I Y N I W L Y V L M S Y L V F L I F G L G S D A L H M Y S K F L
MiscTXA2.human        181  - - - S W C F L T L G A E S G D V A F G L L F S M L G G L S V G L S F L L N T V S V A T L C H V Y H
MiscTXA2.mouse        179  - - - S W C F L T L G T Q R G D V V F G L I F A L L G S A S V G L S L L L N T V S V A T L C R V Y H
MiscThrombin.hamster  251  - - - - L N I T T C H D V L N E T L L Q G F Y S Y Y F S A F S A V F F L V P L I - I S T I C Y M S I
MiscThrombin.human    249  - - - - L N I T T C H D V L N E T L L E G Y Y A Y Y F S A F S A V F F F V P L I - I S T V C Y V S I
MiscThrombin.rat      256  - - - - L N I T T C H D V L N E T L L H G F Y S Y Y F S A F S A I F F L V P L I - I S T V C Y T S I
MisccAR1.slimemold    200  K H L T Y Q F K - L I N Y I I V F L V C - - W V F A V V N R I V - N G L N M F P P A L N I L H T Y L
Miscedg-1.human       192  S T V L P L Y H K H Y I L F C T T V F T L L L L S I V I L Y C R - I Y S L V R T R S R R L T F R K N
MisckSTE2.yeast       207  A S I L L S S S V N F M T L L L T V K L I L A V R S R R F L G L K Q F D S F H V - L L I M S F Q T L
Miscmam2.yeast        185  - - - Q I Q Y L L W L T P L S S K T I V G Y S W P Y F I A K I L F A F S I I F H - S G V F S Y K L F


MiscC5a.canine        228  L I - - R T - - - - - - - - - - - - - W S R K A T R S T K T L K - - - - V V - - V A V V V S F F V
MiscC5a.human         226  L L - - R T - - - - - - - - - - - - - W S R R A T R S T K T L K - - - - V V - - V A V V A S F F I
MiscCMV-UL27.viral    216  V R - - F I - - - - - - - - - - - I N Y V G K W H M Q T L H - - - - V L - - L V V V V S F A S
MiscCMV-UL28.viral    212  S R - - I V - - - - - - - - - - - A V S Q S R H K G R I V R - - - - V L - - I A V V L V F I I
MiscCMV-UL33.viral    249  T T A W P M - - - - - - - - - - - Q C E H L T L R R T I G T L A R V V P H L - - H C L I N P I L Y
MiscFML1.human        225  A A - - K I - - - - - - - - - - - H R N H M I K S S R P L R - - - - V F - - A A V V A S F F I
MiscFMLP.human        225  A T - - K I - - - - - - - - - - - H K Q G L I K S S R P L R - - - - V L - - S F V A A A F F L
MiscFMLP2.human       225  A A - - K I - - - - - - - - - - - H K K G M I K S S R P L R - - - - V L - - T A V V A S F F I
MiscNPY.bovine        224  I S - - K L - - - - - - - - - - - - S H S K G Y Q K R K A L K - - - - T T - - V I L I L T F F A
MiscNPY.drosophila    283  T I - - R V W A K R P P G E A E T N R D Q R M A R S K R K M V K - - - - - M M - - L T V V I V F T C
MiscNPY.human1        235  Y I - - R L - - K R R N N M M D K M R D N K Y R S S E T K R I N I - - - - M L - - L S I V V A F A V
MiscNPY.human2        223  I S - - K L - - - - - - - - - - - - S H S K G H Q K R K A L K - - - - T T - - V I L I L A F F A
MiscNPY.mouse         234  Y I - - R L - - K R R N N M M D K I R D S K Y R S S E T K R I N I - - - - M L - - L S I V V A F A V
MiscNPY.rat           234  Y I - - R L - - K R R N N M M D K I R D S K Y R S S E T K R I N V - - - - M L - - L S I V V A F A V
MiscNT.rat            272  A E Q G R V C T V G T H N G L E H S T F N M T I E P G R V Q A L R H G V L V L - - R A V V I A F V V
MiscPAFR.guineapig    215  R Q - - P V - - - - - - - - - - - K Q Q R N A E V R R R A L W - - - - M V - - C T V L A V F V I
MiscPAFR.human        215  M Q - - P V - - - - - - - - - - - Q Q Q R N A E V K R R A L W - - - - M V - - C T V L A V F I I
MiscSTE2.yeast        257  V P S I I F - - - - - - - - - - - I L A Y S L K P N Q G T D V L T T V A T L L - - A V L S L P L S S
MiscSTE3.yeast        294  R S I K L G - - - - - F V L D M W K R F I D K N K E K R V G I L L - - - - N K L S S R K E S R N P F
MiscTXA2.human        228  G Q - - E A - - - - - - - - - - - A Q Q R P R D S E V E M M A - - - - Q L - - L G I M V V A S V
MiscTXA2.mouse        226  T R - - E A - - - - - - - - - - - T - Q R P R D C E V E M M V - - - - Q L - - V G I M V V A T V
MiscThrombin.hamster  296  I R C L S S - - - - - - - - - - - S S V A N R S K K S R A L F - - - - L S - - A A V F C V F I V
MiscThrombin.human    294  I R C L S S - - - - - - - - - - - S A V A N R S K K S R A L F - - - - L S - - A A V F C I F I I
MiscThrombin.rat      301  I R C L S S - - - - - - - - - - - S A V A N R S K K S R A L F - - - - L S - - A A V F C I F I V
MisccAR1.slimemold    246  - S V S H G - - - - - F - - - - W A S V T F I Y N N - - - P L M W - - - R Y F G A K I L T V F T F
Miscedg-1.human       241  I S - - - - - - - - - - - - - - K A S R S S E N V A L L K - - - - T V - - I I V L S V F I A
MisckSTE2.yeast       256  I F P S I L - - - - - - - - - F I L A Y A L N P N Q G T D T L T S I A T L L - - V T L S L P L S S
Miscmam2.yeast        231  R A - - - - - - - - - - - - - I L I R K K I G Q F P F G - - - - P M - - Q C I L V I S C Q
```

FIGURE 1d.

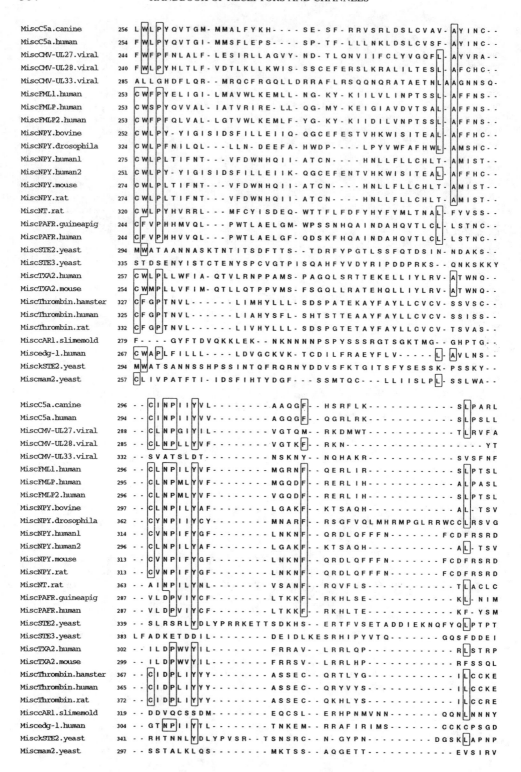

FIGURE 1e.

```
MiscC5a.canine        322  R Q - - - - - - - - - V L A E - E S V G R D S K S - - - - - - - - - - - - - - I T L S T V D T P A
MiscC5a.human         320  R N - - - - - - - - - V L T E - E S V V R E S K S - - - - - - - - - - - - - - F T R S T V D T M A
MiscCMV-UL27.viral    314  C C - - - - - - - - - C V K Q E I P Y Q D I D I E L Q K D I Q R R A K H T K R T H Y D R K N A P M E
MiscCMV-UL28.viral    304  V C - - - - - - - - - - - - - W P S F A S D - - - - - - - - - - - - - - - S F - - - - P A M Y
MiscCMV-UL33.viral    357  P S - - - - - - - - G T W- K G G G Q K T A S N D - - - - - - - - - - - - T S T K I P H R L S Q S H
MiscFML1.human        322  E R - - - - - - - - A L T E V P D S A Q T S N T - - - - - - - - - - - - - H T T S A S P P E E
MiscFMLP.human        321  E R - - - - - - - - A L T E - - D S T Q T S D T - - - - - - - - - - - - - A T N S T L P S A E
MiscFMLP2.human       322  E R - - - - - - - - A L S E - - D S A P T N D T - - - - - - - - - - - - - A A N C A S P P A E
MiscNPY.bovine        322  S R - - - - - - - - G S S L K I L S K G K R G G - - - - - - - - - - - - - H S S V S T E S E S
MiscNPY.drosophila    400  D R M N A T S G T G P A L P L N R M N T S T T Y I - - - - - - - - - S A R R K P R A T S L R A N P L
MiscNPY.human1        345  D D Y E - - - - - - - T I A M S T M H T D V S K T - - - - - - - - - S L K Q A S P V A F K K I N N N
MiscNPY.human2        321  S R - - - - - - - - G S S L K I L S K G K R G G - - - - - - - - - - - - - H S S V S T E S E S
MiscNPY.mouse         344  D D Y E - - - - - - - T I A M S T M H T D V S K T - - - - - - - - - S L K Q A S P V A F K K I - S M
MiscNPY.rat           344  D D Y E - - - - - - - T I A M S T M H T D V S K T - - - - - - - - - S L K Q A S P V A F K K I - S M
MiscNT.rat            389  P G W R - - - - - - H R R K K R P T F S R K P N S - - - - - - - - - - - - - M S S N H A F S T S
MiscPAFR.guineapig    312  R S - - - - - - - - S Q K C S R V T T D T G T E - - - - - - - - - - - - - M A I P I N H T P V
MiscPAFR.human        312  R S - - - - - - - - S R K C S R A T T D T V T E - - - - - - - - - - - - - V V V P F N Q I P G
MiscSTE2.yeast        385  S S K N T R I G P F A D A S Y K E G E V E P V D M - - - - - - - - Y T P D T A A D E E A R K F W T
MiscSTE3.yeast        418  S L G G F S K V T L D Y S E K L H N S A S - S N F - - - - - - - - - - E G E S L C Y S P A S K E E
MiscTXA2.human        328  R S - - - - - - - - - - - - - - - - - - - - - - - - - - - - - - - - - - - L S L Q P Q L T Q
MiscTXA2.mouse        325  Q A - - - - - - - - - - - - - - - - - - - - - - - - - - - - - - - - - - - V S L R R P P A Q
MiscThrombin.hamster  393  S S - - - - - - - - D P N S Y N S T G Q L M P S K - - - - - - - - - - - - - M D T C S S H L N N
MiscThrombin.human    391  S S - - - - - - - - D P S S Y N S S G Q L M A S K - - - - - - - - - - - - - M D T C S S N L N N
MiscThrombin.rat      398  S S - - - - - - - - D S N S C N S T G Q L M P S K - - - - - - - - - - - - : - - M D T C S S H L N N
MisccAR1.slimemold    350  G L Q Q - - - - - - N Y N D E G S S S S S L S S S - - - - - - - - - - D E E K Q T V E M Q N I Q I
Miscedg-1.human       335  S A G K F K R P I I A G M E F S R S K S D N S S H - - - - - - - - - P Q K D E G D N P E T I M S S
MisckSTE2.yeast       375  N C V G H N G S T M S V N D K N G A H A T C V Q N - - - - - - N V T L N T D S T L N Y S N V D T Q
Miscmam2.yeast        323  - D - - - - - - - - - R T F D I K H T P S D D Y - - - - - - - - - - - - - - S I S D E S E T


MiscC5a.canine        347  Q K - - - S Q G V - - - - - -
MiscC5a.human         345  Q K - - - T Q A V - - - - - -
MiscCMV-UL27.viral    355  S G - - - E E E F L L - - - -
MiscCMV-UL28.viral    319  P G - - - T T A - - - - - -
MiscCMV-UL33.viral    385  H N - - - L S G V - - - - - -
MiscFML1.human        348  T E - - - L Q A M - - - - - -
MiscFMLP.human        345  V A - - - L Q A K - - - - - -
MiscFMLP2.human       346  T E - - - L Q A M - - - - - -
MiscNPY.bovine        348  S S - - - F H S S - - - - - -
MiscNPY.drosophila    441  S C - - - G E T S P L R - - -
MiscNPY.human1        379  D D - - - N E K I - - - - - -
MiscNPY.human2        347  S S - - - F H S S - - - - - -
MiscNPY.mouse         377  N D - - - N E K V - - - - - -
MiscNPY.rat           377  N D - - - N E K I - - - - - -
MiscNT.rat            418  A T - - - R E T L Y - - - - -
MiscPAFR.guineapig    338  N P - - - I K N - - - - - - -
MiscPAFR.human        338  N S - - - L K N - - - - - - -
MiscSTE2.yeast        426  E D - - - N N N L - - - - - -
MiscSTE3.yeast        456  N S S S N E H S S E N T A G P
MiscTXA2.human        339  R S - - - G L Q - - - - - - -
MiscTXA2.mouse        336  A M - - - L S G P - - - - - -
MiscThrombin.hamster  420  S I - - - Y K K L L A - - - -
MiscThrombin.human    418  S I - - - Y K K L L T - - - -
MiscThrombin.rat      425  S I - - - Y K K L L A - - - -
MisccAR1.slimemold    383  S T S T N G Q G N N - - - - -
Miscedg-1.human       375  G N - - - V N S S S - - - - -
MisckSTE2.yeast       418  D T - - - S K I L M T T - - -
Miscmam2.yeast        345  K K - - - W T - - - - - - - -
```

FIGURE 1f.

of C5a to its receptor was inhibited by pertussis toxin as well as various guanine nucleotides (Siciliano et al., 1990), thereby indicating the importance of G protein in the signal transduction mechanism. It has been proposed that the activation of phospholipase C is responsible for mediating the function of C5a receptor (Snyderman and Uhing, 1988).

The gene for human C5a receptor was cloned from cDNA libraries prepared from HL-60 and U937 cells (Boulay et al., 1991; Gerard and Gerard, 1991) and mapped to chromosome 19 (Bao et al., 1992). The open reading frame encodes a 350 amino acid protein, and shares 34% sequence homology with fMLP receptor (Boulay et al., 1991; Gerard and Gerard, 1991). When the cDNA was expressed in cell lines, the transfected cells were able to bind to C5a ligands (Boulay et al., 1991; Gerard and Gerard, 1991), and to evoke functional responses by activating phosphatidylinositol turnover (Gerard and Gerard, 1991) and mobilizing the intracellular calcium (Didsbury et al., 1992). Southern analysis suggests that C5a receptor is encoded by a single intronless gene (Gerard and Gerard, 1991).

1.99.1.5 Canine C5a Receptors

The canine C5a receptor was cloned by Perret et al. (1992).

1.99.2 HCMV-Encoded G Protein-Coupled Receptor Homologues

1.99.2.25 Human CMV-Encoded Receptors

Human cytomegalovirus (HCMV) is a herpesvirus with a genome that encodes approximately 200 genes (Chee et al., 1990; Bankier et al., 1991). Recently, a family of three HCMV genes were identified to encode putative polypeptides structurally resembling the cellular GCRs (Chee et al., 1990). The DNA sequence analysis indicates that these three HCMV open reading frames, named UL33, US27, and US28, encode putative proteins of 390, 362, and 323 amino acid residues, respectively. They all contain seven putative transmembrane segments and share 24 to 40% sequence identity within themselves and about 20 to 30% sequence identity with other mammalian GCRs such as $beta_2$ adrenergic, 5-hydroxytryptamine$_{1A}$, muscarinic cholinergic and substance K receptors. Several amino acid residues believed to be structurally and functionally important for GCRs are also conserved in the HCMV homologues. Northern analysis of mRNA in HCMV infected cells demonstrated that all three HCMV homologue genes were transcribed during the viral infection as two sets of 3′-coterminal mRNAs (Welch et al., 1991).

More recently another homologous gene of GCR family has been identified from *Herpesvirus saimiri* genome (Nicholas et al., 1992). This gene, ECRF3, encodes a putative polypeptide of 321 amino acids, and shares approximately 20% sequence identity with the HCMV homologues. ECRF3 also possesses other features characteristic of GCRs.

The endogenous ligand(s) and the functional significance of these virus-encoded putative GCRs are currently unknown. They may play a role in the interactions between the virus and its host during and after the infections (Chee et al., 1990; Nicholas et al., 1992).

1.99.3 Endothelial Differentiation Gene (edg-1) Product

1.99.3.1 Human edg-1 Receptors

Endothelial differentiation gene (edg-1) is a cDNA clone isolated from human umbilical vein endothelial cells (Hla and Maciag, 1990). edg-1 is an immediate-early gene, which is induced by the tumor promoter phorbol 12-myristate 13-acetate. The principal open reading frame of edg-1 encodes a putative protein of 380 amino acids, which resembles structurally the superfamily of GCRs. In comparison to other members of the GCR family, edg-1 shares significant degree of sequence similarity within the carboxyl-terminal half of the receptor. The edg-1 translational product also contains a unique structural feather: a leucine-zipper motif from amino acids 47 to 68. The transcripts of edg-1 are very abundant in endothelial cells, but they are also detected at lower levels in other tissues including vascular smooth muscle, fibroblasts, melanocytes, and human brain. Although the ligand for edg-1 has not yet been identified, edg-1 has been proposed to be involved in the regulation of differentiation of endothelial cells (Hla and Maciag, 1990).

1.99.4 fMLP Receptors

The interaction of the proinflammatory bacterial peptide formyl-methionine-leucine-phenylalanine (fMLP) with its receptor on the surface of neutrophils elicits a series of biological activities, including chemotaxis, degranulation, and production of superoxide (Snyderman and Goetzel, 1981; Snyderman and Pike, 1984; Snyderman and Uhing, 1988). Radioligand binding using fMet-Leu-[^{3}H]Phe to the membranes of neutrophils or differentiated human myeloid HL-60 cells revealed a high- and a low-affinity binding sites with the Kd values of about 1 nM and 20 to 40 nM, respectively (Koo et al., 1982; Snyderman et al., 1984; Gierschick et al., 1989). The high-affinity sites can be converted to the low-affinity ones by treating the membranes with GppNHp or GTPgammaS (Snyderman et al., 1984; Snyderman and Uhing, 1988). When similar binding studies were performed on whole cell preparation from neutrophils, only one class of binding site (Kd = 22.3 nM) was observed (Koo et al., 1984). The absence of high-affinity sites on intact cells might be the consequence of high intracellular levels of GTP (Snyderman et al., 1984). Biochemical studies with photoaffinity labeling and cross-linking techniques indicate that fMLP receptor in human neutrophils or HL-60 cells is a glycoprotein with a molecular mass of about 50,000 to 70,000 Da, which can be separated into two isoforms based on their isoelectric points (Dolmatch and Niedel, 1983; Malech et al., 1985).

The regulation of fMLP receptor appears to be quite complex. Upon exposure to its ligand, the fMLP receptor undergoes both up (Fletcher et al., 1982; Snyderman and Pike, 1984) and down regulation (Vitkauskas et al., 1980). In addition, the function of fMLP receptor is influenced by a broad range of factors such as tumor necrosis factor (McLeish et al., 1991), granulocyte-macrophage colony-stimulating factor (Weisbart et al., 1986), methylation (Pike and Snyderman, 1982), and lipopolysaccharide (Goldman et al., 1986).

Numerous studies have indicated that the fMLP receptor is mainly coupled to phospholipase C via pertussis toxin- and cholera toxin-sensitive G protein(s) (Verghese et al., 1986; Bradford and Rubin, 1986; Smith et al., 1987; Snyderman and Uhing,

1988; Bommakanti et al., 1992). The activation of phosphalipase C stimulates the generation of inositol triphosphate which in turn causes release of intracellular Ca++ (Volpi et al., 1984; Bradford and Rubin, 1986; Korchak et al., 1984). In addition, activation of protein kinase C (DiVirgilio et al., 1984; Snyderman and Uhing, 1988), phosphalipase D (Pai et al., 1988), and phosphalipase A2 (Nielson et al., 1991) has also been proposed to play a role in mediating the function of fMLP receptor.

1.99.4.1 Human fMLP Receptors

cDNA for fMLP receptor was first cloned from a library prepared from differentiated human myeloid HL-60 cells (Boulay et al., 1990). The clone contains a 1050 base pair open reading frame, encoding a 350 amino acid residue protein with the seven hydrophobic segments. Expression of the clone in COS cells resulted in specific binding of fMLP with high- and low-affinity sites (Kd = 0.5 to 1 nM and 5 to 10 nM, respectively). The gene for fMLP receptor has been localized to chromosome 19 (Murphy et al., 1992).

Although there is no conclusive pharmacological evidence for the existence of fMLP receptor subtypes, the biochemical results suggest that more than one form of receptor may be present for fMLP (Dolmatch and Niedel, 1983; Malech et al., 1985). Recent molecular cloning studies provide further support for this notion. For example, a variant of the original human fMLP receptor clone was isolated with nearly identical coding sequence (with only two residue changes at positions 101 and 346), but distinct 5'- and 3'-untranslated regions (Boulay et al., 1990a). More recently, another cDNA clone was isolated from a HL-60 granulocyte library with fMLP receptor cDNA probes (Ye et al., 1992; Murphy et al., 1992). This clone, also located on chromosome 19, encodes a protein of 351 amino acid residues which shares 69% sequence identity to the human fMLP receptor. The expression of the cDNA in fibroblast cells confers calcium mobilization activity upon exposure to micromolar concentration of fMLP although the transfected cells exhibited little binding activity to formyl-peptide ligand (Ye et al., 1992; Murphy et al., 1992). The precise relationship between these clones and the fMLP receptor remains to be established.

1.99.5   Neuropeptide Y (NPY) Receptors

Neuropeptide Y (NPY), a 36-amino acid polypeptide, is widely distributed in the brain and peripheral systems, and involved in diverse biological functions as a neurotransmitter or neuromodulator (Sheikh, 1991; Dumont et al., 1992). A structurally related peptide, peptide YY (PYY), is a hormone present in endocrine cells in the gastrointestinal tract, and appears to share the same or related receptors with NPY to exert its functions (Sheikh, 1991; Dumont et al., 1992). Earlier binding studies using radiolabeled NPY, PYY, and various analogs have demonstrated the presence of specific binding sites in CNS and peripheral tissues with Kd values ranging from about 0.05 to 3 nM (Dumont et al., 1992). It is now believed that multiple subtypes of NPY receptors exist, and each subtype may be involved in distinct biological activities (Sheike, 1991). To date, at least three subtypes of NPY receptor have been proposed primarily based on the functional and binding data: Y_1-like receptor has a similar high affinity for NPY, PYY, and [Leu^{31},Pro^{34}]NPY (a NPY analog), but very low affinity for C-terminal fragments of NPY such as NPY$_{18-36}$; Y_2-like receptor displays high affinity for NPY, PYY, and C-terminal fragment NPY$_{18-36}$, but low affinity for [Leu^{31},Pro^{34}]NPY; Y_3-like receptor is characterized by high affinity for NPY, but low affinity for PYY (Michel, 1991; Sheike, 1991). The existence of heterogeneous receptors for NPY is also supported biochemically by affinity labeling studies, which indicate that Y_1 and Y_2 receptors are structurally different glycoproteins with apparent molecular mass of 70 and 50 kDa, respectively (Sheikh and Williams, 1990). In addition, NPY receptors also seem to be coupled to multiple signal transduction pathways via the pertussis toxin-sensitive G protein(s). The activation of all three subtypes of NPY receptors has been shown to have an inhibitory effect on adenylyl cyclase, and in some cases such as Y_1, NPY-stimulated mobilization of intracellular calcium has also been reported (Michel, 1991; Sheikh, 1991).

1.99.5.1   Human NPY Receptors

Human Y_1 cDNA clone has been isolated recently (Larhammar et al., 1992; Herzog et al., 1992). The human Y_1 clone encodes a 384 amino acid protein, which shares 93% sequence homology to the rat Y_1 clone. Southern analysis of genomic DNA suggests that human genome contains a single Y_1 gene (Larhammar et al., 1992). Comparison of the amino acid sequence to other GCRs indicates that Y_1 resembles substance K receptor most closely with about 30% sequence identity. Characterization of cell lines transfected with human Y_1 receptor exhibited pharmacological properties typical of Y_1 subtype (Larhammar et al., 1992; Herzog et al., 1992). It is of interest to note that there seems to be a cell type specific coupling between Y_1 receptor and second messengers. For example, the pertussis toxin-sensitive mobilization of intracellular calcium was observed in transfected CHO cells, but not in transfected 293 cells. By contrast, NPY-mediated inhibition of adenylyl cyclase was apparent in the 293 cells, but not in the CHO cells (Herzog et al., 1992).

1.99.5.4   Bovine NPY Receptors

A bovine cDNA clone for NPY (LCR1) has been reported (Rimland et al., 1991). LCR1 contains an open reading frame of 353 amino acids, displaying only 21% sequence homology in the putative transmembrane regions to the human Y_1 receptor. Expression of LCR1 in COS and CHO cells displayed specific $[^{125}I]$-NPY binding with a Kd of 1.3 nM. Moreover, the binding of $[^{125}I]$-NPY was decreased by Gpp(NH)p in a dose-dependent manner, indicating the interaction with G proteins. Northern blot analysis revealed that the mRNA of LCR1 was widely distributed in the brain and peripheral tissues. Preliminary binding experiment showed that $NPY_{13\text{-}16}$ (the Y_2 selective analog) had a slightly high affinity to displace $[^{125}I]$-NPY binding to LCR1 than $[Leu^{31},Pro^{34}]NPY$ (the Y_1 selective analog). On the other hand, LCR1 has a very low affinity to PYY (Ki > 1 μM). Therefore, LCR1 may encode Y_3 subtype of NPY receptor (Rimland et al., 1991).

1.99.5.10   Rat NPY Receptors

The first clone for NYP receptor was isolated from a rat forebrain cDNA library in 1990 (Eva et al., 1990) although its identity was not known until recently (Krause et al., 1992). The cDNA contains an open reading frame of 349 amino acid residues and displays the putative seven hydrophobic transmembrane segments (Eva et al., 1990). Binding of $[^{125}I]$-NPY to 293 cells transfected with the cloned receptor revealed a single, high-affinity binding site (Kd = 0.7 nM), and a characteristic pharmacological profile of Y_1 subtype of NYP receptors (Krase et al., 1992). Furthermore, a NYP-elicited inhibition of adenylyl cyclase and a stimulated mobilization of intracellular calcium was detected in 293 cells transiently expressing the cDNA clone (Krase et al., 1992).

1.99.5.21 Drosophila NPY Receptors

Recently, another cDNA clone of NPY was isolated from drosophila (Li et al., 1992). This clone, named PR4, encodes a protein of 449 amino acid residues, and shares 23% homology to human Y_1 receptor. When expressed in *Xenopus* oocytes, PR4 exhibited characteristic electrophysiological response upon NPY and PYY stimulation. The preliminary data suggest that the protein encoded by PR4 may be the Y_2 subtype of NPY receptor (Li et al., 1992).

1.99.6 Neurotensin (NT) Receptors

1.99.6.10 Rat NT Receptors

Originally isolated from bovine hypothalamic tissues, the tridecapeptide neurotensin (NT) is present in both central and peripheral systems, and exhibits multiple biological activities as a neurotransmitter/neuromodulator (Leeman and Carraway, 1982; Kasckow and Nemeroff, 1991). The saturable, specific and high-affinity binding of [^{3}H]NT or [^{125}I]NT has been detected in a number of tissues and cultured cells of many species (Kitabgi et al., 1985; Kanba et al., 1986; Dana et al., 1991). Two types of NT binding sites have been described in rat brain: sites 1 have high affinity (Kd = 0.05 to 0.2 n*M),* but low binding capacity, and their binding is inhibited by GTP and Na^+ ions; sites 2, by contrast, display relatively low affinity (Kd = 3 to 10 n*M),* but higher capacity. The binding of NT to these sites is not sensitive to GTP and cations, but can be selectively displaced by levocabastine, an antihistamine-1 drug bearing no structural similarity to NT (Kitabgi et al., 1987; Mazella et al., 1987, Kitabgi and Vincent, 1986). Anatomically, a clearly distinct regional distribution of the two binding sites has been demonstrated in rat brain by the receptor radioautographic studies (Kitabgi et al., 1987). Besides rats, the levocabastine-sensitive, low affinity NT binding sites were also observed in mouse brain, but were absent from the samples prepared from rabbit brain, human brain, mouse neuroblastoma N1E115 cells, and human colonic adenocarcinoma HT29 cells, all of which possess the high affinity, levocabastine-insensitive binding sites (Kitabgi et al., 1987). The activation of the high-affinity NT receptor has been shown to stimulate inositol phospholipid hydrolysis, intracellular calcium mobilization, cGMP production, and to decrease cAMP formation (Goedert et al., 1984, Mazella et al., 1987; Turner et al., 1990). The occupancy of the receptor also induces homologous and heterologous desensitization (Turner et al., 1990). However, the second messenger and functional significance of the low affinity, levocabastine-sensitive binding site remain unknown. Recently, the purification of NT receptors from bovine, mouse, and rat brains has been reported by using affinity chromatography (Mills, et al., 1988; Mazella et al., 1989; Miyamoto et al., 1991). The purified receptors have the apparent molecular mass ranging from 55 to 100 kDa, and retain the ability to bind NT although the affinity varies from 0.26 to 5.5 n*M*. Whether such differences represent the species or procedural variation, or the distinct subtypes of binding sites as mentioned above is to be elucidated.

    The molecular cloning and expression of rat NT receptor has been recently reported (Tanaka et al., 1990). The cDNA clone encodes a 424-amino acid protein with a calculated molecular weight of 47,052. The sequence analysis indicates that the receptor protein contains seven putative transmembrane domains and shares about 18 to 24% sequence homology to other cloned GCRs. Northern blot analysis shows that the messenger RNA for cloned receptor can be detected both in the brain and peripheral tissues. Expression of the cloned receptor in *Xenopus* oocyte and COS cells results in a characteristic pharmacological response to NT with a single high-affinity binding site (Kd = 0.16 n*M*). In addition, the radioligand binding to the cloned receptor is not affected by levocabastine. Thus, the cloned receptor is consistent with the profile of the sites 1. The molecular identity of the low affinity, levocabastine-sensitive binding sites 2 for NT may be unveiled by future cloning studies.

1.99.7   Platelet-Activating Factor (PAF) Receptors

Platelet-activating factor (PAF) is a potent phospholipid inflammatory mediator involved in a wide range of biological activities in various cells, tissues, and organs (Pinckard et al., 1988; Prescott et al., 1990; Shukla, 1992). Pharmacologic properties of PAF receptor has been extensively characterized. Using [^{3}H]PAF or its analogs, the specific binding sites for PAF have been demonstrated to exist in a number of cells and tissues with the Kd values in the ranges of 10^{-9} to 10^{-10} M (Hwang, 1990). In most cells, a single class of binding site was observed although high- and low-affinity binding sites were also reported in some cases (Hwang, 1990). In resemblance to other GCRs, PAF receptor undergoes rapid desensitization after receptor occupancy (Schwertschlag and Whorton, 1988; Chao et al., 1989). One major effort in PAF receptor research is to develop effective PAF antagonists. Up to now, a variety of compounds with diverse chemical structures have been identified to specifically inhibit PAF receptor (Handley, 1990; Hwang, 1990; Sunkel et al., 1990; Herbert et al., 1991; Crowley et al., 1991; Underwood, 1992).

Several lines of evidence suggest that PAF receptor-mediated signal transduction process is modulated by G proteins: (1) GTP modulated the binding of PAF (Ng and Wong, 1986; Hwang, 1990); (2) PAF stimulated GTPase activity (Hwang et al., 1989; Avdonin et al., 1985); and (3) pertussis toxin inhibited PAF-stimulated activities in some, but not all, systems (Shukla, 1992). Although the exact type(s) of G protein involved has not been identified, PAF receptor appears to couple to inositol phospholipid turnover through multiple mechanisms. PAF has been shown to activate phospholipase C (Shukla, 1991), which in turn generates inositol triphosphate and diglyceride as second messengers to elevate intracellular calcium and activate protein kinase C, respectively. PAF also activates phospholipase A2 to release arachidonic acid and stimulates phospholipase D to produce phosphatidic acid (Shukla, 1992; Kanaho et al., 1991). The precise role of the latter metabolic products in transmembrane signaling remains to be established. Finally, protein kinase C may also play a role in the signaling pathway (Ong et al., 1991).

1.99.7.1   Human PAF Receptors

A cDNA clone for the human PAF receptor has been reported by numerous laboratories (Nakamura et al., 1991; Ye et al., 1991; Kunz et al., 1992). The clone encodes a 342 amino acid protein, which shares 83% sequence identity to the guinea pig lung PAF receptor and possesses certain structural features common to GCRs such as seven hydrophobic putative membrane spanning segment, except that the human receptor does not contain the potential N-linked glycosylation sites at the N-terminus. Functionality of the clone was manifested by the detection of the electrophysiological response to PAF in *Xenopus* oocytes injected with the transcript of the cDNA (Nakamura et al.,

1991). When expressed in COS cells, the encoded receptor bound to [^{3}H]PAF with similar pharmacological properties to human platelet, and underwent desensitization upon incubation of the PAF (Nakamura et al., 1991; Kunz et al., 1992). Furthermore, the activation of the expressed receptors led to the production of inositol trisphosphate (Nakamura et al., 1991) and calcium mobilization (Ye et al., 1991). Southern analysis suggests that only a single copy of PAF receptor gene is present in the human genome (Kunz et al., 1992). However, many other experiments suggest that there may exist different PAF receptors in different cells and tissues (Stewart and Dusting, 1988; Hwang, 1988; Paulson et al., 1990; Hwang, 1991) and in different species (Ostermann et al., 1991; Hwang, 1991). Indeed, in *Xenopus* oocyte expression system, transcripts of different size (from 3.5 to 6 kb) have been shown to possess PAF receptor activity (Murphy et al., 1990). The definitive evidence for PAF receptor subtypes may await for future molecular cloning studies.

1.99.7.9   Guinea Pig PAF Receptors

The gene for PAF receptor was first cloned from guinea pig lung by functional expression. The cDNA clone encodes a 342-amino acid protein, which displays 7 putative hydrophobic transmembrane segments (Honda et al., 1991). When expressed in COS cells, the expressed receptor was able to bind to PAF antagonists WEB 2086 with a Kd value of 6.4 nM (Honda et al., 1991).

1.99.8   Dictyostelium (Slime Mold) cAMP Receptors

1.99.8.24   Slime Mold cAMP Receptors

The cell surface cAMP receptor plays a central role in chemotaxis and development of the eukaryotic microorganism *Dictyostelium discoideum*. cAMP binds to cAMP receptor in a highly specific manner (Van Haastert, 1983) and triggers developmentally important responses via a number of second messenger systems including cAMP, cGMP, IP$_3$, and Ca^{++}. These responses are transient in nature since the cell adapts quickly to the sustained cAMP stimulation (Janssens, 1987; Gerisch, 1987; and Firtel, 1989).

The binding of cAMP to its receptor appears to be heterogeneous. Two major binding sites have been defined based on the kinetic studies: the fast dissociating site (A site) and the slow dissociating site (B site) (Van Haastert and De Wit, 1984; Van Haastert et al., 1986). A site can be further divided into high- (A^H) and low-affinity (A^L) forms with Kd values of 60 and 450 nM, respectively, for [^{3}H]cAMP binding. B site can also be divided into B^S and B^{SS} forms. They have a similar affinity for [^{3}H]cAMP (Kd = 12.5 nM), but a tenfold difference in their dissociation rate constants (Van Haastert and De Wit, 1984; Van Haastert et al., 1986). During the binding reaction, A^H site is rapidly converted to A^L site with a reduction of the total number of A site. On the other hand, the number of B site remains unchanged, but its affinity is decreased following receptor occupancy (Van Haastert and De Wit, 1984; Kesbeke and Van Haastert, 1985). Biochemical characterization of cAMP receptor revealed that the receptor protein migrated as a doublet (Mr = 40,000 and 43,000 Da) on a SDS-PAGE gel as identified by photoaffinity labeling with 8-azido-[^{32}P]cAMP (Theibert et al., 1984). These two forms of the receptor are believed to represent the same protein with a different degree of phosphorylation, which is regulated by cAMP (Klein et al., 1987). Stimulation of cAMP induces a rapid and reversible conversion of the 40-kDa form to the 43-kDa form, and such modification by phosphorylation, catalyzed presumably by protein kinase A (Luderus et al., 1986) and protein kinase C (Van Haastert, et al., 1985), has been proposed to cause adaptation of the receptor function (Deveotes and Sherring, 1985; Klein et al., 1987; Vaughan and Devreotes, 1988). The function of cAMP receptor can also be antagonized pharmacologically by several cAMP and adenosine derivatives in either competitive or noncompetitive manners (Van Haastert, 1983).

The involvement of G protein in cAMP receptor-mediated signal transduction pathway was first implicated from several pharmacological studies. [^{3}H]cAMP binding to the receptor was affected by guanosine di- and triphosphates (Van Haastert, 1984; Van Haastert et al., 1986). Moreover, cAMP increased the binding of [^{3}H]GTP (De Wit and Snaar-Jagalska, 1985), stimulated GTPase activity (Snaar-Jagalska et al., 1988a), and modulated the activity of adenylate cyclase in *Dictyostelium discoideum* (Theibert and Devreotes, 1986; Van Haastert et al., 1987). Recently, the genes for two Gα subunits (Gα1 and Gα2) were cloned from *Dictyostelium discoideum* (Pupillo et al. 1989). Gα2 is believed to be coupled to the cAMP receptor, and mutations in this gene disrupts various known cAMP receptor-mediated processes (Kumagai et al., 1989). As mentioned above, the activation of the surface cAMP receptor elicits multiple responses, and these responses may be mediated by different forms of receptor through different G proteins and second messenger systems (Snaar-Jagalska et al., 1988b). It has been proposed that A site is most likely coupled to adenylate cyclase via an undefined G protein(s) (Van Haastert, 1985; Snaar-Jagalska et al., 1988b; Kumagai et al., 1989). The activation of this receptor induces an elevation of the intracellular level of cAMP, which relays the signal and can be released into the extracellular medium, where it activates the surface receptor as a positive feedback (Firtel et al., 1989). In contrast, B site is likely coupled to phospholipase C via Gα2 protein and the activation of this receptor is associated with the chemotaxis responses and the production of IP$_3$, Ca^{++} and cGMP (Snaar-Jagalska, et al., 1988b; Firtel et al., 1989; Kumagai et al., 1989). The elevation of cGMP and cytosolic Ca^{++} is probably a secondary response to the production of inositol 1,4,5-triphosphate because a similar result can be obtained following the addition of IP$_3$ to permeabilized cells (Europe-Finner and Newell, 1985; Small et al., 1986; Newell et al., 1988; Ginsburg and Kimmel, 1989).

The cDNA for cAMP receptor was first cloned by Klein et al. in 1988. It encodes a polypeptide of 392 amino acid residues with a calculated molecular weight of about 44 kDa. The hydropathy analysis of cAMP receptor reveals seven hydrophobic, putative transmembrane segments, a common feature shared by the cloned GCRs. There is an asparagine-linked glycosylation site near the N-terminus, and there are 18 serine and 6 threnine residues within the 132 amino acid residues of the C-terminal sequence. This cytoplasmic tail has been proposed to be the site for cAMP-stimulated phosphorylation (Klein et al., 1988; Vaughan and Devreotes, 1988). Using cAMP receptor cDNA as a probe, a genomic gene for cAMP receptor (cAR1) was also cloned (Saxe et al., 1991a). cAR1 is present as a single copy in the genome of *Dictyostelium*, and contains two introns located at 38 base pair upstream from the start codon and within the putative 3rd transmembrane segment, respectively. Expression of cloned cAR1 in cells produced a 40-kDa protein, which was capable of binding to [^{3}H]cAMP (Klein et al., 1988; Johnson et al., 1991). However, cAR1 expressed in growing cells did not appear to couple to its normal effectors although cAR1 did undergo phosphorylation upon cAMP stimulation (Johnson et al., 1991). The genetic study suggested that the structural domain for G protein interaction and crucial phosphorylation may reside within the amino terminal four fifths of the cAR1 protein whereas the C-terminal one-fifth sequence may not be critical for development of *Dictyostelium* (Sun and Devreotes, 1991).

More recently, the genomic clones for at least two other cAMP receptor subtypes (cAR2 and cAR3) were identified using low stringency hybridization (Saxe et al., 1991a,b). Like cAR1, cAR2 and cAR3 also contain an intervening sequence within the putative transmembrane segment 3, and these 3 receptors share about 60% identical (or 70% similar) amino acid sequences within the transmembrane and loop domains. The C-terminal sequences of these receptors, however, are very different. cAR2 and cAR3 do not have serine/threonine clusters as is present in cAR1 (Saxe et al., 1991b). Expression of cAR2 and cAR3 in cells yielded the receptor proteins of an apparent molecular mass of 39 and 62 kDa, respectively (Johnson et al., 1992). Radioligand binding using [^{3}H]cAMP demonstrated multiple binding states for different subtypes, and the binding parameters can be influenced by the salt used in the assay buffers. For example, in phosphate buffer, there were two affinity states for cAR1 (30 and 300 nM) and cAR3 (20 and 500 nM), but the affinity for cAR2 was nondetectable. When 3-M ammonium sulfate was added in the buffer, the affinity states were changed to 4 nM for cAR1, 11 nM for cAR2, 4 nM, and 200 nM for cAR3 (Johnson et al., 1992). Furthermore, the three cAMP receptor subtypes also exhibited a subtle difference in the cyclic nucleotide specificity (Johnson et al., 1992). The functional significance of each subtype of cAMP receptor is not clear at present. It has been shown that the temporal and spatial pattern of the expression of cAR1, cAR2, and cAR3 are quite distinct from each other. The expression pattern of cAR1 was suggestive of its role in the cAMP signal relay response whereas cAR3 was found to be expressed during the maximal chemotaxis and, therefore, could be coupled to phospholipase C (Saxe et al., 1991b). However, a recent study in which the cell aggregation and development in *Dictyostelium* was blocked by disrupting the cAR1 gene suggested that cAR1 may be linked to Gα2 and mediating the chemotaxis responses (Sun and Devreotes, 1991). Hence, the precise relationship between cAR1, cAR2, cAR3, and different kinetic forms of the receptor (i.e., A^H, A^L, B^S, and B^{SS}) remains to be established.

1.99.9   Thrombin Receptors

In addition to the principal role of converting fibrinogen to fibrin in blood coagulation cascade, thrombin displays a broad range of activities involving cellular activation of both platelet and nonplatelet origins (Shuman, 1986; Coughlin et al., 1992). For many years, despite great efforts by many investigators, the identity of the functional thrombin receptor was not clear. At least part of the difficulties is due to the fact that multiple mechanisms may be involved in thrombin-mediated activations and that there is often a lack of consistency between the binding and function of thrombin. As a serine protease, thrombin appears to exert its effects via both a proteolytic process and a receptor-related mechanism (Shuman, 1986; Jamieson, 1988; McNicol et al., 1989). Initial specific binding of [^{125}I]thrombin was demonstrated in platelets (Tollenfsen et al., 1974; Ganguly, 1974). More detailed analysis revealed a high (Kd = 0.3 nM) and a moderate (Kd = 11 nM) affinity binding sites, each of which was coupled to different processes in human platelets (Jamieson, 1988). The high-affinity site resembles the conventional receptor, and requires continuing occupancy by thrombin to mediate responses. The activation of this receptor is coupled to the inhibition of adenylyl cyclase and the stimulation of phospholipase A2. In contrast, the function mediated by the moderate affinity site requires only transient exposure to thrombin, and is linked to phospholipase C and therefore the production of inositol trisphosphate and diacylgelcerol, which in turn elevate intracellular calcium concentration and protein kinase C activity, respectively. The activation of these moderate-affinity sites is characteristic for a protease-catalyzed reaction (Jamieson, 1988; Lapetina, 1990; Colman, 1991). It is now believed that G proteins may be involved in thrombin receptor-mediated responses. Although the exact identity of the G protein(s) remains to be determined, a pertussis toxin-sensitive G protein(s) was shown to couple the receptor to phospholipase C (Brass et al., 1991) and adenylyl cyclase (Brass, 1992), mobilization of intracellular calcium (Meylon et al.,1992), and phospholipase A2 (Gupta et al., 1990). Like other GCRs, thrombin receptor-mediated responses also exhibit homologous desensitization upon stimulation by thrombin (Paris et al., 1988; Halldorsson et al., 1991), and such phenomenon has been proposed to result from distinct mechanisms such as protein phosphorylation and proteolysis (Levin and Santell, 1991; Brass, 1992).

1.99.9.1   Human Thrombin Receptors

Over the years, many attempts have been made to identify the receptor molecule for thrombin. A number of cellular proteins were shown to bind to thrombin, but whether any of them represents the functional thrombin receptor is unclear (Shuman, 1986; Jamieson, 1988). Recent molecular cloning of a functional human thrombin receptor cDNA represented a breakthrough in understanding both structure and function of thrombin receptor (Vu et al., 1991). The cDNA clone encodes a protein of 425-amino acid residues. The hydropathy plot reveals that it belongs to the superfamily of GCRs with seven putative transmembrane segments. Unlike most other receptors in this family, however, it appears to contain a cleavable signal sequence at the N-terminus of the protein. The expression of the cloned receptor was detected in thrombin-responsive megakaryocyte-like cell lines as well as in human platelets and vascular endothelial cells. The identity of the cloned receptor was evidenced by the expression experiments. Microinjection of the transcript of thrombin receptor into *Xenopus* oocyte conferred thrombin-mediated responses with EC_{50} of 50 pM and the effect can be blocked by thrombin antagonist hirudin (Vu et al., 1991). Thrombin-induced phosphoinositide hydrolysis was also observed in CV-1 cells transfected with the cloned cDNA (Hung et al., 1992a). Furthermore, the cloned receptor has been demonstrated to be necessary for thrombin-induced platelet activation (Hung et al., 1992b).

Analysis of amino acid sequence of N-terminus of the receptor reveals the presence of a thrombin cleavage site at position 41 (arginine) and a highly acidic sequence, 13 amino acid distal to the cleavage site, which resembles the C-terminal tail of a thrombin binding molecule, hirudin. Alteration of the putative cleavage site by mutagenesis resulted in a mutant receptor that failed to respond to thrombin. On the other hand, a synthetic peptide mimicking the new amino terminus following thrombin cleavage proved to be a potent agonist at both wild-type and noncleavable mutant thrombin receptors (Vu et al., 1991). Further studies indicate that all information necessary for receptor activation is accommodated by proteolysis of the receptor after arginine 41, and that the highly acidic domain downstream from the thrombin cleavage site plays an important role in the interaction with thrombin (Vu, 1991a). Based on above observations, a novel model of thrombin receptor activation was put forward: thrombin interacts with its receptor through the cleavage recognition site and downstream anion-binding exosite binding domain, and then cleaves the receptor after arginine 41, thereby unmasking a new amino terminus. This newly created amino terminus then acts as a tethered agonist which binds to and activates the receptor (Vu et al., 1991, 1991a). In support of this model, the "tethered peptide ligand" has been shown to be able to elicit various thrombin-stimulated activities, including activation of phospholipase C (Huang et al., 1991; Brass, 1992), inhibition of adenylyl cyclase

(Seiler et al., 1992; Brass, 1992). Recent studies indicate that the critical structural requirement for receptor activation resides within the first 6 amino acid residues (SFLLRN) of the "tethered-ligand", especially Phe2, Arg5, and the free ammonium group of Ser1 (Vassallo et al., 1992; Hui et al., 1992; Scarborough et al., 1992).

1.99.9.10   Rat Thrombin Receptors

The rat thrombin receptor has been cloned by Zhong et al. (1992).

1.99.9.12   Hamster Thrombin Receptors

A hamster cDNA clone of thrombin receptor has also been reported (Rasmussen et al., 1991). The hamster receptor contains 427 amino acid residues, which shares 79% sequence identity to the human clone and exhibits a similar putative membrane topology. Like the human receptor, there is a thrombin cleavage site at arginine 41 along with a cluster of negatively charged residues about 20 amino acids C-terminal to the cleavage site, although the overall sequence homology in the N-terminal extracellular domain is quite low (55%) between the two species (Rasmussen et al., 1991). Whether such differences represent merely the species variation or the existence of different isoforms of the receptor remains to be elucidated.

1.99.10   Thromboxane A2 (TXA2) Receptors

Thromboxane A2 (TXA2), a labile metabolite of arachidonic acid, is a powerful inducer of platelet activation and a potent smooth muscle constrictor (Samuelsson et al., 1978). TXA2 interacts with a GCR, which is linked to phospholipase C (Saussy, 1985; Ushikubi, 1989; Arita et al., 1989). The activation of the receptor renders the production of inositol trisphosphate and diacylglycerol, leading to the elevation of intracellular calcium and activity of protein kinase C (Arita et al., 1989; Kawahara et al., 1983). In addition, TXA2 has also been reported to decrease cAMP accumulation although such effect may be mediated indirectly via other factors such as intracellular calcium (Gorman et al., 1978; Gresele et al., 1991; Sage et al., 1992). The exact type of G-protein coupled to the receptor is unclear, but is found to be resistant to pertussis- and cholera toxins (Brass et al., 1987; Houslay et al., 1986), and possibly belongs to Gq family (Shenker, 1991). Pharmacologically, a vast number of compounds have been identified as potent antagonists for TXA2 receptors, and some of them possess a dual function as TXA2 blocker as well as thromboxane synthase inhibitors (Hall, 1991; Gresele et al., 1991). The responses to TXA2 are regulated by multiple factors. At the receptor level, agonist-induced homologous desensitization has been demonstrated (Murray and FitzGerald, 1989). The loss of activity was attributed, at least in part, to receptor-G protein uncoupling and agonist-occupied receptor internalization (Murray and FitzGerald, 1989; FitzGerald, 1991; Dorn, 1991).

1.99.10.1   Human TXA2 Receptors

Biochemically, the human platelet TXA2 receptor has been purified to apparent homogeneity, and has a molecular mass of 57 kDa (Ushikubi 1989). Based on the partial amino acid sequence of the purified receptor, an oligonucleotide probe was synthesized, which led to the isolation of a cDNA clone from human placenta (Hirata et al., 1991). The TXA2 receptor clone encodes a protein of 343 amino acids, displaying a seven transmembrane topology. When expressed in COS-7 cells, the cloned receptor bound to [³H]S-145, a selective TA2 receptor antagonist, with a Kd of 1.2 n*M*, and exhibited a similar pharmacological profile as those of the platelet receptor (Hirata et al., 1991). Northern blot analysis suggests that an identical form of TXA2 receptor is expressed in platelets and vascular tissues (Hirata et al., 1991) although the existence of TA2 receptor subtypes has been proposed (Arita et al., 1989; FitzGerald., 1991).

1.99.10.11   Mouse TXA2 Receptors

The mouse TXA2 receptor has been cloned recently from a mouse lung cDNA library (Namba et al., 1992). The clone encodes a protein of 341 amino acid residues, which displays 76% sequence homology to the human receptor. mRNA of the receptor was detected in a wide range of tissues, and was especially abundant in thymus, spleen, and lung.

1.99.11   Yeast Mating-Factor Receptors

Alpha- and a-mating factors are small peptide pheromones produced in the budding yeast *Saccharomyces cerevisiae,* and are essential for initiating the mating process through the interactions with their specific receptors on the surface of two types (alpha and a) of haploid cells (Cross et al., 1988; Blumer and Thorner, 1991; Sprague, 1991). The alpha pheromone has been shown to bind to its receptor with a Kd of 6 n*M* (Jenness et al., 1986), and such high affinity interaction requires the functional coupling with the G protein as a ninefold decrease in binding affinity was observed in the presence of GTPgammaS (Blumer and Thorner, 1990). The receptor also undergoes internalization by endocytosis upon pheromone binding (Jenness and Spatrick, 1986). Although the precise second messenger for mating factor-mediated effects has not been identified, both receptors appear to be coupled to the same heterotrimeric G protein, encoded by GPA1, STE4, and STE18 genes in the yeast (Nakafuku et al., 1987; Dietzel and Kurjan, 1987; Whiteway et al., 1989), and may induce a phosphorylation cascade in their signaling pathways (Sprague 1991). Various genetic and biochemical studies suggest that the alpha-factor receptor is encoded by STE2 gene, and a-factor receptor is encoded by STE3 gene. Mutations in STE2 and STE3 disrupted responses to alpha- and a-factors, respectively (Hartwell 1980; Hagen et al., 1986). The binding of alpha-factor to strains of temperature-sensitive mutations of STE2 displayed a corresponding thermosensitive profile (Jenness et al., 1983, 1986). In addition, alpha-factor has been shown to be specifically crosslinked to STE2 protein (Blumer et al., 1988).

The molecular cloning and sequencing of STE2 and STE3 in *Saccharomyces cerevisiae* has been reported (Nakayama et al., 1985; Burkholder and Hartwell, 1985). The STE2 and STE3 encode a 431- and a 470-amino acid protein, respectively. Both receptor proteins reveal a putative seven transmembrane topology with a long hydrophilic carboxy terminus, but share only limited sequence homology to each other. Microinjection of synthetic STE2 mRNA into *Xenopus* oocytes conferred specific surface binding of [³⁵S]alpha-factor with an apparent Kd of 7 n*M*, consistent with the binding data in yeast cells (Yu et al., 1989).

The gene for alpha-factor receptor in *Saccharomyces kluyveri*, a different species of budding yeast, was also isolated by using STE2 DNA as a probe (Marsh and Herskowitz, 1988). The *S. kluyveri* receptor gene encodes a protein of 426 amino acid residues and shares 50% sequence identity to STE2. When introduced into *Saccharomyces cerevisiae* cells, the *S. kluyveri* gene is able to confer the receptor selectivity for *S. kluyveri* alpha-factor (Marsh and Herskowitz, 1988).

Table 1. Miscellaneous G Protein-Coupled Receptors

Receptor	Second messenger	Species cloned	Chromosomal location	a.a. Sequence	Accession number	Primary reference
C5a Analphylatoxin	+PI	Human	19	350	P21730	Gerard et al., 1991
		Canine		352	—	Perret et al., 1992
Human cytomegolovirus						
UL33		Viral		390	P16849	Chee et al., 1990
UL27		Viral		362	P09703	Weston et al., 1986
UL28		Viral		323	P09704	Weston et al., 1986
edg-1		Human		381	P21453	Hla et al., 1990
F-Met-Leu-Phe (FMLP)	+PI	Human		350	P21462	Boulay et al., 1990
FMLP-1		Human	19	353	P25089	Lu et al. (unpublished)
Neuropeptide Y	−AC, +PI	Human1		384	P25929	Larhammar et al., 1992
		Human2		352	L01639	Jazin et al., 1992
		Bovine		353	P25930	Rimland et al., 1991
		Rat		382	P21555	Eva et al., 1990
		Drosophila		449	P25931	Li et al., 1992
Neurotensin	+PI, −AC	Rat		424	P20789	Tanaka et al., 1990
Platelet Activating Factor (PAF)	+PI	Human		342	P25105	Ye et al., 1991
		Guinea pig		342	P25556	Honda et al., 1991
Slime mold						
cAR1	+AC, +PI +PI, −AC	Dicytostelium discoideum		392	p13773	Klein et al., 1988
Thrombin		Human		425	P25116	Vu et al., 1991
		Rat		432	P26824	Zhong et al., 1992
		Hamster		427	M80612	Rasmussen et al., 1991
Thromboxane A (TXA2)	+PI	Human		343	P21731	Hirata et al., 1991
		Mouse		341	D10849	Namba et al., 1992
Yeast Mating Factors						
kSTE2		Yeast		426	P12384	Marsh et al., 1988
STE2		Yeast		431	P06842	Nakayama et al., 1985
STE3		Yeast		470	P06783	Hagen et al., 1986
mam2		Yeast		348	X61672	Kitamura and Shimoda, 1991

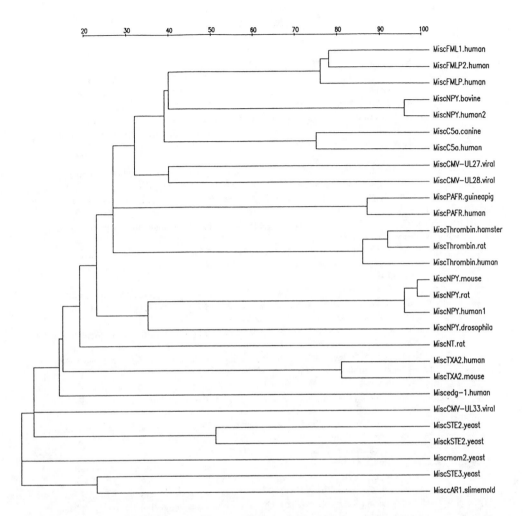

FIGURE 2. Phylogenetic tree of miscellaneous G protein-coupled receptors. The phylogenetic tree was constructed according to the method of Feng and Doolittle (1990). The length of each "branch" of the tree correlates with the evolutionary distance between receptor subpopulations.

Recently, another gene (mam2) from a genomic library of the fission yeast *Schizosaccharomyces pombe* was cloned (Kitamura and Shimoda, 1991). The mam2 contains an open reading frame composed of 348 amino acids and has about 26% sequence identity to STE2 product in *Saccharomyces cerevisiae*. The hydropathy plot analysis once again reveals seven potential hydrophobic membrane-spanning domains. Based on its sequence homology to alpha-mating factor receptor, and the observation that the mutant with disrupted mam2 gene failed to respond to P factor, a pheromone involved in mating process in *S. pombe*, mam2 has been proposed to encode the receptor for P factor (Kitamura and Shimoda, 1991).

REFERENCES

Arita, H., Hakano, T., and Hansaki, K. (1989) Thromboxane A2: its generation and role in platelet activation. *Prog. Lipid Res.* 28, 273–301.

Avdonin, P. V., Svitina-Ulitina, I. V., and Kulikov, V. I. (1985) Stimulation of high affinity hormone-sensitive GTPase of human platelets by 1-O-alkyl-2-acetyl-sn-glyceryl-3-phosphocholine (platelet activating factor). *Biochem. Biophys. Res. Commun.* 131, 307–313.

Bankier, A. T., Beck, S., Bohni, R., Brown, C. M., Cerny, R., Chee, M. S., Hutchison, C. A., III, Kouzarides, T., Martignetti, J. A., Preddie, E., Satchwell, S. C., Tomlinson, P., Weston, K. M., and Barrell, B. (1991) The DNA sequence of the human cytomegalovirus genome. *DNA Seq.* 2, 1–12.

Bao, L., Gerard, N. P., Eddy, R. L., Jr., Shows, T. B., and Gerard, C. (1992) Mapping of genesfro the human C5a receptor (C5AR), human FMLP receptor (FPR), and tow FMLP receptor homologue orphan receptors (FPRH1, FPRH2) to chromosome 19. *Genomics* 13, 437–440.

Blumer, K. J. and Thorner, J. (1991) Receptor-G protein signaling in yeast. *Ann. Rev. Physiol.* 53, 37–57.

Blumer, K. J., Reneke, J. E., and Thorner, J. (1988) The STE2 gene product is the ligandbinding component of the a-factore receptor of *Saccharomyces cerevisiae. J. Biol. Chem.* 263, 10836–10842.

Blumer, K. L. and Thorner, J. (1990) b and g subunits of a yeast guanine nucleotide-binding protein are not essential for membrane association of the a subunit but are required for reeptor coupling. *Proc. Natl. Acad. Sci. U.S.A.* 87, 4363–4367.

Bommakanti, R. K., Bokoch, G. M., Tolley, J. O., Schreiber, R. E., Siemsen, D. W., Klotz, K. N., and Jesaitis, A. J. (1992) Reconstitution of a physical complex between the N-formyl chemotactic peptide receptor and G protein. Inhibition by pertussis toxin-catalyzed ADP ribosylation. *J. Biol. Chem.* 267, 7576–7581.

Boulay, F., Mery, L., Tardif, M., Brouchon, L., and Vignais, P. (1991) Expression cloning of a receptor for C5a anaphylatoxin on differentiated HL-60 cells. *Biochemistry* 30, 2993–2999.

Boulay, F., Tardif, M., Brouchon, L., and Vignais, P. (1990a) The human N-formylpeptide receptor. Characterization of two cDNA isolates and evidence for a new subfamily of G-protein-coupled receptors. *Biochemistry* 29, 11123–11133.

Boulay, F., Tardif, M., Brouchon, L., and Vignais, P. (1990) Synthesis and use of a novel-N-formyl peptide derivative to isolate a human N-fromyl peptide receptor cDNA. *Biochem. Biophys. Res. Commun.* 168, 1103–1109.

Bradford, P. G. and Rubin, R. P. (1986) Quantitative changes in iniositol 1,4,5-triphosphate in chemoattractant-stimulated neutrophil. *J. Biol. Chem.* 261, 15644–15647.

Brass, L. F. (1992) Homologus desensitization of HEL cell thrombin receptors. Distinguishable roles for proteolysis and phosphorylation. *J. Biol. Chem.* 267, 6044–6050.

Brass, L. F. Shaller, C. C., and Belmonte, E. J. (1987) Inositol 1,4,5-triphosphate-induced granule secretion in platelets. Evidence that the activation of phospholipase C mediated by platelet thromboxane receptors involves a guanine nucleotide binding protein-dependent mechanism distinct from that of thrombin. *J. Clin. Invest.* 79, 1269–1275.

Brass, L. F., Manning, D. R., Williams, A. G., Woolkalis, M. J., and Poncz, M. (1991) Receptor and G protein-mediated responses to thrombin in HEL cells. *J. Biol. Chem.* 266, 958–965.

Burkholder, A. C. and Hartwell, L. H. (1985) The yeast a-factor recpetor: strucural properties deduced from the sequence of the STE2 gene. *Nucl. Acids Res.* 13, 8463–8475.

Chao, W., Liu, H., Hanahan, D. J., and Olson, M. S. (1989) Regulation of platelet-activating factor receptor in rat Kupffer cells. *J. Biol. Chem.* 264, 20488–20457.

Chee, M. S., Bankier, A. T., Beck, S., Bohni, R., Brown, C. M., Verny, R., Horsnell, T., Hutchison, C. A., III, Kouzarides, T., Martignetti, J. A., Preddie, E., Satchwell, S. C., Weston, P. K. M., and Barrell, B. G. (1990) Analysis of the protein-coding content of the sequence of human cytomegalovirus strain AD169. *Curr. Top. Microbiol. Immunol.* 154, 123–169.

Chee, M. S., Satchwell, S. C., Preddie, E., Weston, K. M., and Barrell, B. G. (1990) Human cytomegalovirus encodes three G protein-coupled receptor homologues. *Nature* 334, 774–777.

Chenoweth, D. E. and Hugli, T. E. (1978) Demonstration of a specific C5a receptor on intact polymorphonuclear leukocytes. *Proc. Natl. Acad. Sci. U.S.A.* 75, 3943–3947.

Colman, R. W. (1991) Receptors that activate platelets. *Proc. Soc. Exp. Biol. Med.* 197, 242–248.

Coughlin, S. R., Vu, T.-K. H., Huang, D. T., and Wheaton, V. I. (1992) Characterization of a functional trhombin receptor. *J. Clin. Invest.* 89, 351–355.

Cross, F., Hartwell, L. H., Jackson, C., and Konopka, J. B. (1988) Conjugation in *Saccharomyces cerevisiae. Annu. Rev. Cell Biol.* 4, 429–457.

Crowley, H. J., Yaremdo, B., Selig, W. M., Janero, D. R., Burghardt, C., Welton, A. F., and O'Donnell, M. (1991) Pharmacology of a potent platelet-activating factor antagonists: Ro 24–4736. *J. Pharmacol. Exp. Ther.* 259, 78–85.

Dana, C., Pelaprat, D., Vial, M., Brouard, A., Lhiaubet, A. M., and Rostene, W. (1991) Characterization of neurotensin binding sites on rat mesencephalic cells in primary culture. *Brain Res. Dev. Brain Res.* 61, 259–264.

De Wit, R. J. W. and Snaar-Jagalska, B. E. (1985) Folate and cAMP modulate GTP binding to isolated membranes of Dictyostelium discoideum. Functional coupling between cell surface receptor and G-proteins. *Biochem. Biophys. Res. Commun.* 129, 11–17.

Deveotes, P. N. and Sherring, J. A. (1985) Kinetics and concentration dependence of reversible cAMP-induced modification of the surface cAMP receptor in *Dictyostelium. J. Biol. Chem.* 260, 6378–6384.

Di Virgilio, F., Lew, D. P., and Pazzan, T. (1984) Protein kinase C activation of physiological processes in human neutrophils at vanishingly small cytosolic Ca++ levels. *Nature* 310, 691–693.

Didsbury, J. R., Uhing, R. J., Tomhave, E., Gerard, C., Gerard, N., and Snyderman, R. (1992) Functional high efficiency expression of cloned leukocyte chemoattractant receptor cDNAs. *FEBS Lett.* 297, 275–279.

Dietzel, C. and Kurjan, J. (1987) The yeast SCG1 gene: a Ga-like protien implicated in the a- and a-factor response pathway. *Cell* 50, 1001–1010.

Dolmatch, B. and Niedel, J. (1983) Formylpeptide chemotactic receptor: evidence for an active proteolytic fragment. *J. Biol. Chem.* 258, 7570–7577.

Dorn, G. W., II. (1991) Mechanism for homologous downregulation of thromboxane A2 receptors in cultured human chronic myelogenous leukemida (K562) cells. *J. Pharmacol. Exp. Ther.* 259, 228–234.

Dumont, Y., Martel, J.-C., Fourhier, A., St-Pierre, S., and Quirion, R. (1992) Neuropeptide Y and neuropeptide Y receptor subtypes in brain and peripheral tissues. *Prog. Neurobiol.* 38, 125–167.

Europe-Finner, G. N. and Newell, P. C. (1985) Inositol 1,4,5-triphosphate and calcium stimulate action polymerization in Dictyostelium discoideum. *Biochem. Biophys. Res. Commun.* 130, 1115–1122.

Eva, C., Keinanen, K., Monyer, H., Seeburg, P., and Spengel, R. (1990) Molecular cloning of a novel G prtoein-coupled receptor that may belong to the neuropeptide receptor family. *FEBS Lett.* 271, 81–84.

Feltner, D. E., Smith, R. H., and Marasco, W. A. (1986) Characterization of the plasma mebrane GTPase from rabbit neutrophils. I. Evidence for an N_i-like protein coupled to the formyl peptide, C5a, and leukotriene B4 leukotaxis receptors. *J. Immunol.* 137, 1961–1970.

Firtel, R. A., Van Haastert, P. J. M., Kimmel, A., and Devreotes, P. N. (1985) G protein linked signal transduction pathways in development: Dictyostelium as an experimental system. *Cell* 58, 235–239.

Firtel, R. A., Van Haastert, P. J. M., Kimmel, A. R., and Devreotes, P. N. (1989) G protein linked signal transduction pathways in development: Dictyostelium as an experimental system. *Cell* 58, 235–239.

FitzGerald, G. A. (1991) Mechanisms of platelet activation: thromboxane A2 as an amplifying siganl for other agonists. *Am. J. Cardiol.* 68, 11B–15B.

Fletcher, M. P., Seligmann, B. E., and Gallin, J. I. (1982) Correlation of human neutrophil secretion, chemoattractant receptor mobilization and enhanced functional capacity. *J. Immunol.* 128, 941–948.

Ganguly, P. (1974) Binding of thrombin to human platelets. *Nature* 247, 306–307.

Gerard, N. P. and Gerard, C. (1991) The chemotactic receptor for human C5a anaphylatoxin. *Nature* 349, 614–617.

Gerard, N. P., Hodges, M. K., Drazen, J. M., Weller, P. F., and Gerard, C. (1989) Characterization of a receptor for C5a anaphylatoxin on human eosinophils. *J. Biol. Chem.* 264, 1760–1766.

Gerisch, G. (1987) Cyclic AMP and other signals controlling cell development and differentiation Dictyostelium. *Annu. Rev. Biochem.* 56, 853–879.

Gierschik, P., Steisslinger, M., Sidiropoulos, D., Herrmann, E., and Jakobs, K. H. (1989) Dual Mg^{2+} control of formyl-peptide-recpetor-G-protein interaction in HL 60 cells. Evidence that the low-agonist-affinity receptor interacts with and activates the G-protein. *Eur. J. Biochem.* 183, 97–105.

Ginsburg, G. and Kimmel, A. R. (1989) Inositol trisphosphate and diacylglycerol can differentially modulate gene expression in Dictyostelium. *Proc. Natl. Acad. Sci. U.S.A.* 86, 9332–9336.

Goedert, M., Pinnock, R. D., Downes, C. P., Mantyh, P. W., and Emson, P. C. (1984) Neurotensin stimulates inositol phospholipid hydrolysis in rat brain slices. *Brain Res.* 323, 193–197.

Goldman, D. W., Enkel, H., Gifford, L. A., Chenoweth, D. E., and Rossenbaum, J. T. (1986) Lipopolysaccharide modulates receptors for leukotriene B4, C5a, and formyl-methionyl-leucyl-phenylalanine on rabbit polymorphonuclear leukocytes. *J. Immunol.* 137, 1971–1976.

Goldstein, I. M. (1988) Complement: biologically active products. In *Inflammation: Basic Principles and Clinical Correlates*, Gallin J. I., Goldstein I. M., and Snyderman R., Eds., Raven Press, New York, pp. 55–74.

Gorman, R. R., Fitzpatrick, F. A., and Miller, O. V. (1978) Reciprocal regualtion of human platelet cAMP levels by thromboxane A$_2$ and prostacyclin. *Adv. Cyclic Nucleotide Res.* 9, 597–609.

Gresele, P., Deckmyn, H., Nenci, G. G., and Vermylen, J. (1991) Tromboxane synthase inhibitors, thromboxane receptor antagonists and dual blockers in thrombotic disorders. *Trends Pharamcol. Sci.* 12, 158–163.

Gupta, S. K., Diez, E., Heasley, L. E., Osawa, S., and Johnson, G. L. (1990) A G protein mutant that inhibits thrombin and purinergic receptor activation of phospholipase A2. *Science* 249, 662–666.

Hagen, D. C., McCaffrey, G., and Sprague, G. F. (1986) Evidence that yeast STE3 gene encodes a receptor for the peptide pheromone a factor: gene sequence and implications for the strucutre of the presumed receptor. *Proc. Natl. Acad. Sci. U.S.A.* 83, 1418–1422.

Halldorsson, H., Magnusson, M. K., and Thorgeisson, G. (1991) Different methanisms of homologus and heterologous desensitization of thrombin-induced endothelial prostacyclin production. *Eur. J. Pharmacol.* 208, 193–198.

Handley, D. A. (1990) Preclinical and clinical pharmacology of platelet-activating facotr receptor antagonists. *Med. Res. Rev.* 10, 351–370.

Hartwell, L. H. (1980) Mutants of *Saccharomyces cerevisiae* unresponsive to cell division control by polypeptide mating hormone. *J. Cell. Biol.* 85, 811–822.

Herbert, J. M., Bernat, A., Valetter, G., Gigo, V., Lale, A., Laplace M. C., Lespy, L., Savi, P., Maffrand, J. P., and Le Fur, G. (1991) Biochemical and pharmacological activities of SR 27417, a highly potent, long-actine platelet-activating factor receptor antagonist. *J. Pharmacol. Exp. Ther.* 259, 44–51.

Herzog, H., Hort, Y. J., Ball, H. J., Hayes, G., Shine, J., and Selbie, L. A. (1992) Cloned human neuropeptide Y receptor couples to two different second messenger systems. *Proc. Natl. Acad. Sci. U.S.A.* 89, 5794–5798.

Hla, T. and Maciag, T. (1990) An abundant transcript induced in differentiating human endothelial cells encodes a polypeptide with structural similarities to G-protein-coupled receptors. *J. Biol. Chem.* 265, 9308–9313.

Honda, Z., Nakamura, M., Miki, I., Minami, M., Watanabe, T., Seyama, Y., Okado, H., Tho, H., It, K., Miyamoto, T., and Shimisu, T. (1991) Cloning functional expression of platelet activating factor recpetor from guinea-pig lung. *Nature* 349, 342–346.

Houslay, M. D., Bojanic, D., and Wilson, A. (1986) Platelet activating factor and U44069 stimulate a GTPase activity in human platelets which is distinct from the guanine nucleotide regulatory proteins, Ns and Ni. *Biochem. J.* 234, 737–740.

Huang, R. S., Sorisky, A., Church, W. R., Simons, E. R., and Rittenhouse, S. E. (1991) "Thrombin" receptor-directed ligand accounts for activation by thrombin of platelet phospholipace C and accumulation of 3-phosphorylated phosphoinositides. *J. Biol. Chem.* 266, 18435–18438.

Hui, K. Y., Jakubowski, J. A., Wyss, V. L., and Angleton, E. L. (1992) Minimal sequence requirement of thrombin recpetor agonist peptide. *Biochem. Biophys. Res. Commun.* 184, 790–796.

Hung, D. T., Vu, T.-K. H., Nelken, N. A., and Coughlin, S. R. (1992a) Thrombin-induced events in non-platelet cells are mediated by the unique proteolytic mechanisms established for the cloned platelet thrombin receptor. *J. Cell. Biol.* 116, 827–832.

Hung, D. T., Vu, T.-K. H., Wheaton, V. I., Ishii, K., and Coughlin, S. R. (1992b) Cloned platelet thrombin receptor is necessary for thrombin-induced platelet activation. *J. Clin. Invest.* 89, 1350–1353.

Hwang, S.-B., Lam, M. H., and Pong, S.-S. (1986) Ionic and GRP regulation of binding of platelet-activating factor to receptors and platelet-activating factor-induced activation of GRPase in rabbit platelet membranes. *J. Biol. Chem.* 261, 532–537.

Hwang, S. B. (1988) Identification of a second putative receptor of platelet-activating factor from human polymorphonuclear leukocytes. *J. Biol. Chem.* 263, 3225–3233.

Hwang, S. B. (1990) Specific receptors of platelet-activating factor, receptor heterogeneity, and signal transduction mechanisms. *J. Lipid Mediators* 2, 123–158.

Hwang, S. B. (1991) High affinity receptor binding of platelet-activating factor in rat peritoneal polymorphonuclear leukocytes. *Eur. J. Pharmacol.* 196, 169–175.

Jamieson, G. A. (1988) The activation of platelets by thrombin: A model for activation by high and moderate affinity receptor pathways. *Prog. Clin. Biol. Res.* 283, 137–158.

Janssens, P. M. W. and van Haastert P. J. M. (1987) Molecular mechanisms of transmembrane signal transduction in *Dictyostelium disoideum*. *Microbiol. Rev.* 51, 396–418.

Jenness, D. D., Burkholder, A. C., and Hartwell, L. H. (1983) Binding of alpha-factor pheromone to yeast a cells: chemical and genetic evidence for an alpha-factor receptor. *Cell* 353, 521–529.

Jenness, D. D. and Spatrick, P. (1986) Down regulation of the alpha-factor pherome recpetor in *Saccharomyces cerevisiae*. *Cell* 46, 345–353.

Jenness, D. D., Burkholder, A. C., and Hartwell, L. H. (1986) Binding of alpha-factor pheromone to *Saccharomyces cerevisiae* a cells: dissociation constant and number of binding sites. *Mol. Cell. Biol.* 6, 318–320.

Johnson, R. and Chenoweth, D. E. (1985) Labeling the granulocyte C5a receptor with a unique photoreactive probe. *J. Biol. Chem.* 260, 7161–7164.

Johnson, R. L., Van Haastert, P. J. M., Kimmel, A. R., Saxe, C. L., III, Jastorff, B., and Devreotes, P. N. (1992) The cyclic nucleotide specificity of three cAMP receptors in Dictyostelium. *J. Biol. Chem.* 267, 4600–4607.

Johnson, R. L., Vaughan, R. A., Caterina, M. J., Van Haastert, P. J. M., and Devreotes, P. N. (1991) Overexpression of the cAMP receptor 1 in growing Dictyostelium cells. *Biochemistry* 30, 6982–6986.

Kanaho, Y., Kanoh, H., Saitoh, K., and Nozawa, Y. (1991) Phospholipase D activation by platelet-activating factor, leukotriene B4, and FMLP in rabbit neutrophils: phospholipase D activation is involved in enzyme release. *J. Immunol.* 146, 3536–3541.

Kanba, K. S., Kanba, S., Okazaki, H., and Richelson, E. (1986). Binding of [^{3}H]neurotensin in human brain: properties and distribution. *J. Neurochem.* 46, 946–952.

Kasckow, J. and Nemeroff, C. B. (1991) The neurobiology of neurotensin: focus on neurotensin-dopamine interactions. *Regulatory Peptides* 36, 153–164.

Kawahara, Y., Yamanishi, J., Furuta, Y., Kaibuchi, K., Takai, Y., and Fukuzaki, H. (1983) Elevation of cytoplasmic free calcium concentration by stable throboxane A2 analogue in human platelets. *Biochem. Biophys. Res. Commun.* 117, 663–669.

Kesbeke, F. and Van Haastert, P. J. M. (1985) Selective down-regulation of cell surface cAMP-binding sites and cAMP-induced responses in Dictyostelium discoideum. *Biochim. Biophys. Acta* 847, 33–39.

Hirata, M., Hayashi, Y., Ushikubi, F., Yokota, Y., Kageyama, R., Nakaniishi, S., Narumiya, S. (1991) Cloning and expression of cDNA for a human thromboxane A2 receptor. *Nature* 349, 617–620.

Kitabgi, P. and Vincent, J. P. (1986) Effect of cations and nucleotides on neurotensin binding to rat synaptic membranes. In *Neural and Endocrine Peptides and Receptors*, Moody, T. W., Ed., Plenum Press, New York, pp. 313–319.

Kitabgi, P., Checler, F., Mazella, J., and Vincent, J.-P. (1985) Pharmacology and biochemistry of neurotensin receptors. *Rev. Clin. Basic Pharmacol.* 5, 397–486.

Kitabgi, P., Rostene, W., Dussaillant, M., Schotte A., Laduron, P. M., and Vincent. J.-P. (1987) Two populations of neurotensin binding sites in murine brain: discrimination by the antihistamine levocabastine reveals markedly different radioautographic distribution. *Eur. J. Pharmacol.* 140, 285–293.

Kitamura, K. and Shimoda, C. (1991) The *Schizosaccharomyces pombe* mam2 gene encodes a putative pheromone receptor which has a significant homology with the *Saccharomyces cerevisiae* Ste2 protein. *EMBO J.* 10, 3743–3751.

Klein, P., Vaughan, R., Borieis, J., and Devreotes, P. N. (1987) The surface cyclic AMP receptor in Dictyostelium. Levels of ligand-induced phosphorylatioin, solubilization, identification of primary transcript, and developmental regulation of expression. *J. Biol. Chem.* 262, 358–364.

Klein, P. S., Sun, T. J., Saxe, C. L., III, Kimmel, A. R., Johnson, R. L., and Devreotes, P. N. (1988) A chemoattractant receptor controls development in *Dictyostelium discoideum. Science* 241, 1467–1472.

Koo, C., Lefkowitz, R. J., and Snyderman, R. (1982) The oligopeptide chemotactic factor receptor on human polymorphonuclear leukocyte membranes exists in two affinity states. *Biochem. Biophys. Res. Commun.* 106, 442–449.

Korchak, J. M., Vienne, K., Rutherford, L. E., Wilkenfeld, C., Finkelstein, M. C., and Weissmann, G. (1984) Stimulus response coupling in the human neutrophil. II. Temporal analysis of changes in cytosolic calcium and calcium efflux. *J. Biol. Chem.* 259, 4076–4082.

Krause, J., Eva, C., Seeburg, P. H., and Sprengel, R. (1992) Neuropeptide Y1 subtype pharmacology of a recombinantly expressed neuropeptide receptor. *Mol. Pharmacol.* 41, 817–821.

Krych, M., Atkinson, J. P., and Holers, V. M. (1992) Complement receptors. *Curr. Opin. Immunol.* 4, 8–13.

Kumagai, A., Pupillo, M., Gundreson, R., Kiake-Lye, R., Devreotes, P. N., and Firtel, R. A. (1989) Regulation and function of Ga protein subunits in Dictyostelium. *Cell* 57, 265–275.

Kunz, D., Gerard, N. P., and Gerard, C. (1992) The human leukocyte platelet-activating factor recpetor. cDNA cloning, cell surface expression, and construction of a novel epitope-bearing analog. *J. Biol. Chem.* 267, 9101–9106.

Lapetina, E. G. (1990) The signal transduction induced by thrombin in human platelets. *FEBS Lett.* 268, 400–404.

Larhammar, D., Blomqvist, A. G., Yee, F., Jazin, E., Too, H., and Wahlestedt, C. (1992) Cloning and functional expression of a human neuropeptide Y/peptide YY receptor of the Y1 type. *J. Biol. Chem.* 267, 10935–10938.

Leeman, S. E. and Carraway, R. E. (1982) Neurotensin, discovery, isolation, characterization, synthesis and possible phsiological roles. *Ann. N.Y. Acad. Sci.* 400, 1–16.

Levin, E. G. and Santell, L. (1991) Thrombin- and histamine-induced signal transduction in human endothelial cells. Stimulation and agonist-dependent desensitization of protein phosphorylation. *J. Biol. Chem.* 266, 174–181.

Li, X.-J., Wu, Y.-N., North, A., and Forte, M. (1992) Cloning, functional expression, and developmental regulation of a neuropeptide Y receptor from *Drosophila melanogaster. J. Biol. Chem.* 267, 9–12.

Luderus, M. E. E., Van der Meer, R. F., and Van Driel, R. (1986) Modulation of the interaction between chemotactic cAMP-receptor and N-protein by cAMP-dependnet kinase in Dictyostelium discoideum membranes. *FEBS Lett.* 205, 189–193.

Malech, H. L., Gardner, J. P., Heiman, D. F., and Rosenzwig, S. A. (1985) Asparagine-linked oligosaccharides on formyl peptide chemotactic receptors of human phagocytic cells. *J. Biol. Chem.* 260, 2509–2514.

Marsh, L. and Herskowitz, I. (1988) STE2 protein of *Saccharomyces kluyveri* is a member of the rhodopsin/β-adrenergic receptor family and is responsible for recognition of the peptide ligand a factor. *Proc. Natl. Acad. Sci. U.S.A.* 85, 3855–3859.

Mazella, J., Amar, S., Bozou, J. C., Kitabgi, P., and Vincent, J. P. (1987) Functional properties and molecular structure of central and peripheral neurotensin receptors. *J. Receptor Res.* 7, 157–165.

Mazella, J., Chabry, J., Zsurger, N., and Vincent, J. P. (1989) Purification of the neurotensin receptor from mouse brain by affinity chromatography. *J. Biol. Chem.* 264, 5559–5563.

McLeish, K. R., Klein, J. B., Schepers, T., and Sonnenfeld, G. (1991) Modulation of transmembrane signalling in HL-60 granulocytes by tumour necrosis factor-alpha. *Biochem. J.* 279, 455–460.

McNicol, A., Gerrard, J. M., and MacIntyre, D. E. (1989) Evidence for two mechanisms of thrombin-induced platelet activation: one proteolytic, one receptor mediated. *Biochem. Cell Biol.* 67, 332–336.

Meylon, C. B., Nickashin, A., Little, P. J., Tkachuk, V. A., and Bobik, A. (1992) Trombin-induced Ca^{2+} mobilization in vascular smooth muscle utilizes a slowly ribosylating pertussis toxin-sensitive G protein. Evidence for the involvement of a G protein in inositol trisphosphate-dependent Ca^{2+} release. *J. Biol. Chem.* 267, 7295–7302.

Michel, M. C. (1991) Receptors for neuropeptide Y: multiple subtypes and multiple second messengers. *Trends Pharmacol. Sci.* 12, 389–394.

Mills, A., Demoliou-Mason, C. D., and Barnard, E. A. (1988) Purification of the neurotensin receptor from bovine brain. *J. Biol. Chem.* 263, 13–16.

Miyamoto-Lee, Y., Shiosaka, S., and Tohyama, M. (1991) Purification and characterization of neurotensin receptor from rat brain with special reference to comparison between newborn and adult age rats. *Peptides* 12, 1001–1006.

Murphy, P. M., Gallin, E. K., and Tiffany, H. L. (1990) Characterization of human phagocytic cell receptors for F4a and platelet activation factor expressed in *Xenopus* oocytes. *J. Immunol.* 145, 2227–2234.

Murphy, P. M., Ozcelik, T., Kenney, R. T., Tiffany, H. L., McDermott, D., and Francke, U. (1992) A structural homologue of the N-formyl peptide receptor. Characterization and chromosome mapping of a peptide chemoattractant receptor family. *J. Biol. Chem.* 267, 7637–7643.

Murray, R. and FitzGerald, G. A. (1989) Regulation of thromboxane receptor activation in human platelets. *Proc. Natl. Acad. Sci. U.S.A.* 88, 124–128.

Nakafuku, M., Itoh, H., Nakamura, S., and Kaziro, Y. (1987) Occurrence in *Saccharomyces cerevisiae* of a gene homologous to the cDNA coding for the a subunit of mammalian G proteins. *Proc. Natl. Acad. Sci. U.S.A.* 84, 2140–2144.

Nakamura, M., Honda, Z.-I., Izumi, T., Sakanaka, C., Kiroyuki, M., Minami, M., Bito, H., Seyama, Y., Matsumoto, T., Noma, M., and Shimizu, T. (1991) Molecular cloning and expression of platelet-activating factor receptor from human leukocytes. *J. Biol. Chem.* 266, 20400–20405.

Nakayama, N., Miyajima, A., and Arai, K. (1985) Nucleotide sequences of STE2 and STE3, cell type-specific sterile genes from *Saccharomyces cerevisiae*. *EMBO J.* 4, 2643–2648.

Namba, T., Sugimoto, Y., Hirata, M., Hayashi, Y., Honda, A., Watabe, A., Negishi, M., Ichikawa, A., and Narumiya, S. (1992) Mouse thromboxane A2 receptor: cDNA cloning, expression and northen blot analysis. *Biochem. Biophys. Res. Commun.* 184, 1197–1203.

Newell, P. C., Eruope-Finner, G. N., Small, N. V., and Liu, G. (1988) Inositol phosphates, G-proteins and ras genes involved in chemotactic signal transduction of Dictyostelium. *J. Cell Sci.* 89, 123–127.

Ng, D. S. and Wong, K. (1986) GTP regulation of platelet-activating factor binding to human neutrophil membranes. *Biochem. Biophys. Res. Commun.* 141, 353–359.

Nicholas, J., Cameraon, K. R., and Honess, R. W. (1992) *Herpesvirus saimiri* encodes homologues of G protein-coupled receptors and cyclins. *Nature* 355, 362–365.

Nielson, C. P., Stutchfield, J., and Cockcroft, S. (1991) Chemotactic peptide stimulation of arachidonic acid relese in HL60 cells, an interaction between G protein and phospholipase C mediated signal transduction. *Biochim. Biophys. Acta* 1095, 83–89.

Ong, R. C., Yoo, T. J., and Chiang, T. M. (1991) Activation mechanisms of platelet-activating factor in U937 cells: possible involvement of protein kinase C. *Cell. Immunol.* 137, 283–291.

Ostermann, G., Lorenz, A., Hofmann, B., and Kertscher, H. P. (1991) Species-dependent differences of high affinity [^{3}H]PAF-binding to platelets from human and pig and its inhibition by selective antagonists. *J. Lipid Mediators* 4, 289–298.

Pai, J. K., Siegel, M. I., Egan, R. W., and Billah, M. M. (1988) Phospholipase D catalyzes phospholipid metabolism in chemotactic peptide-stimulated HL-60 granulocytes. *J. Biol. Chem.* 263, 12472–12477.

Paris, S., Magnaldo, I., and Pouyssegur, J. (1988) Homologous desensitization of thrombin-induced phosphoinositide breakdown in hamster lung fibroblasts. *J. Biol. Chem.* 263, 11250–11256.

Paulson, S. K., Wolf, J. L., Novotney-Barry, A., Cox, C. P. (1990) Pharmacologic characterization of the rabbit neutrophil receptor for platelet-activating factor. *Proc. Soc. Exp. Biol. Med.* 195, 247–254.

Perret, J. J., Raspe, E., Vassart, G., and Parmentier, M. (1992) Cloning and functional expression of the canine anaphylatoxin c5A receptor: Evidence for high interspecies variability. *Biochem. J.* 288, 911–917.

Pike, M. C. and Snyderman, R. (1982) Transmethylation recactions regulate affinity and functional activity of chemotactic factor recpetors on macrophages. *Cell* 28, 107–114.

Pinckard, R. N., Ludwig, J. C., and McManus, L. M. (1988) Platelet-activating factors. In *Inflammation: Basic Principles and Clinical Correlates*, Gallin, J. I., Goldstein, I. M., and Snyderman, R., Eds., Raven Press, New York, pp. 139–167.

Prescott, S. M., Zimmerman, G. A., and McIntyre, T. M. (1990) Platelet-activating factor. *J. Biol. Chem.* 265, 17381–17384.

Pupillo, M., Kumagai, A., Pitt, G. S., Firtel, R. A., and Devreotes, P. N. (1989) Multiple alpha subunits of guanine nucleotide-binding proteins in Dictyostelium. *Proc. Natl. Acad. Sci. U.S.A.* 86, 4892–4896.

Rasmussen, U. B., Vouret-Craviari, V., Lallat, S., Schlesinger, Y., Pages, G., Pavirani, A., Lecocq, J.-P., Pouyssegur, J., and Van Obberghen-Schilling, E. (1991) cDNA cloning and expression of a hamster alpha-thrombin receptor coupled to Ca^{2+} mobilization. *FEBS Lett.* 288, 123–128.

Rimland, J., Xin, W., Sweetnam, P., Saijoh, K., Nestler, E. J., and Duman, R. S. (1991) Sequence and expression of a neuropeptide Y receptor cDNA. *Mol. Pharmacol.* 40, 869–875.

Rollins, T. E. and Springer, M. S. (1985) Identification of the polymorphonuclear leukocyte C5a receptor. *J. Biol. Chem.* 260, 7157–7160.

Sage, S. O. and Heemskerk, J. W. (1992) Thromboxane receptor stimualtion inhibits adenylate cyclase and reduces cyclic AMP-mediated inhibition of ADP-evoked responses in fura-2-loaded human platelets. *FEBS Lett.* 298, 199–202.

Samuelsson, B., Goldyne, M., Granstrom, E., Hamberg, M., Hammarstrom, S., and Malmsten, C. (1978) Prostaglandins and thromboxanes. *Annu. Rev. Biochem.* 47, 997–1029.

Saussy, D. L., Jr., Mais, D. E., Knapp, D. R., and Halushka, P. V. (1985) Thromboxane A2 and prostaglandin endoperoxide receptors in platelets and vascular smooth muscle. *Circulation* 72, 1202–1027.

Saxe, C. L., III, Johnson, R. L., Devreotes, P. N., and Kimmel, A. R. (1991a) Expression of a cAMP receptor gene of Dictyostelium and evidence for a multigene family. *Genes Dev.* 5, 1–8.

Saxe, C. L., III, Johnson, R. L., Devreotes, P. N., and Kimmel, A. R. (1991b) Multiple genes for cell surface cAMP receptors in *Dictyostelium discoideum. Dev. Genet.* 12, 6–13.

Scarborough, R. M., Naughton, M. A., Teng, W., Hung, D. T., Rose, J., Vu, T. K., Wheaton, V. I., Turck, C. W., and Coughlin, S. R. (1992) Tethered ligand agonist peptides. Structural requirements for thrombin receptor activation reveal mechanism of proteolytic unmasking of agonists function. *J. Biol. Chem.* 267, 13146–13149.

Schwertschlag, U. S. and Whorton, R. (1988) Platelet-activating factor-induced homologous and heterologous desensitization in cultured vascular smooth muscle cells. *J. Biol. Chem.* 263, 13791–13796.

Seiler, S. M., Michel, I. M., and Fenton, J. W., II. (1992) Involvement of the "tethered-ligand" receptor in thrombin inhibition of platelet adenylate cyclase. *Biochem. Biophys. Res. Commun.* 182, 1296–1302.

Sheikh, S. P. and Williams, J. A. (1990) Sturctural characterization of Y1 and Y2 receptors for neuropeptide Y and peptide YY by affinity cross-linking. *J. Biol. Chem.* 265, 8304–8310.

Sheikh, S. P. (1991) Neuropeptide Y and peptide YY: major modulators of gastrointestinal blood flow and function. *Am. J. Physiol.* 261, G701-G715.

Shenker, A., Goldsmith, P., Unson, C. G., and Spiegel, A. M. (1991) The G protein coupled to the thromboxane A2 receptor in human platelets is a member of the novel Gq family. *J. Biol. Chem.* 266, 9309–9313.

Shukla, S. D. (1991) Inositol phospholipid turnover in PAF transmembrane signalling. *Lipids* 26, 1028–1033.

Shukla, S. D. (1992) Platelet-activating factor receptor and signal transduction mechanisms. *FASEB J.* 6, 2296–2301.

Shuman, M. A. (1986) Trombin-cellular interactions. *Ann. N.Y. Acad. Sci.* 485, 228–239.

Siciliano, S. J., Rollins, T. E., and Springer, M. S. (1990) Interaction between the C5a receptor and Gi in both the membrane-bound and detergent-solubilized states. *J. Biol. Chem.* 265, 19568–19674.

Small, N. V., Europe-Finner, G. N., and Newell, P. C. (1986) Calcium induces cyclic GMP formation in Dictyostelium. *FEBS Lett.* 203, 11–14.

Smith, C. D., Uhing, R. J., and Snyderman, R. (1987) Nucleotide regulatory protein-mediated activation of phospholipase C in human polymorphonucleur leukocytes is disrupted by phorbol esters. *J. Biol. Chem.* 262, 6121–6127.

Snaar-Jagalska, B. E., Jakobsm, K. H., and Van Haastert, P. J. M. (1988a) Agonist-stimulated high affinity GTPase in Dictyostelium membranes. *FEBS Lett.* 236, 139–144.

Snaar-Jagalska, B. E., Kesbeke, F., and Van Haastert, P. J. M. (1988b) G-proteins in the signal-transduction pathways of Dictyostelium discoideum. *Dev. Genet.* 9, 215–226.

Snyderman, R. and Goetzl, E. J. (1981) Molecular and cellular mechanisms of leukocyte chemotaxis. *Science* 213, 830–837.

Snyderman, R. and Pike, M. C. (1984) Chemoattractant receptors on phagocytic cells. *Annu. Rev. Immunol.* 2, 257–281.

Snyderman, R. and Uhing, R. J. (1988) Phagocytic cells: stimulus-response coupling methanisms. In *Inflammation: Basic Principles and Clinical Correlates*, Gallin J. I., Goldstein I. M., and Snyderman R., Eds., Raven Press, New York, pp. 309–323.

Snyderman, R., Pike, M. C., Edges, S., and Lane, B. (1984) A chemoattractant receptor on macrophages exists in two affinity states regulated by guanine nucleotides. *J. Cell Biol.* 98, 444–448.

Sprague, G. F., Jr. (1991) Signal transduction in yeast mating: receptors, transcription factors, and the kinase connection. *Trends Genet.* 7, 393–398.

Stewart, A. G., Dusting, G. J. (1988) Characterization of receptors for platelet-activating factor on platelets, polymorphonuclear leukocytes and macrophages. *Br. J. Pharmacol.* 94, 1225–1233.

Sun, T. J. and Devreotes, P. N. (1991) Gene targeting of the aggregation stage cAMP receptor cAR1 in Dictyostelium. *Genes Dev.* 5, 572–582.

Sunkel, C. E., de Casa-Juana, M. F., Santos, L., Gomez, M. M., Villaroya, M., Gonzalez-Morales, M. A., Priego, J. G., and Ortega, M. P. (1990) 4-Alkyl-1,4-dihydropyridines derivatives as specific PAF-acether antagonists. *J. Med. Chem.* 33, 3205–3210.

Tanaka, K., Masu, M., and Nakanishi, S. (1990) Structure and functional expression of the cloned rat neurotensin receptor. *Neuron* 4, 847–854.

Theibert, A. and Devreotes, P. N. (1986) Surface receptor mediated activation of adenylate cyclase in Dictyostelium: regulation by guanine nucleotides in wild-type cells and aggregation deficient mutant. *J. Biol. Chem.* 261, 15121–15125.

Theibert, A., Klein, P., and Devreotes, P. N. (1984) Specific photoaffinity labeling of the cAMP surface receptor in *Dictyostelium discoideum*. *J. Biol. Chem.* 259, 12318–12321.

Tollefsen, D. M., Feagler, J. R., and Majerus, P. W. (1974) The binding of thrombin ot the surface of human paltelets. *J. Biol. Chem.* 249, 2646–2651.

Turner, J. T., James-Kracke, M. R., and Camden, J. M. (1990) Regulation of the neurotensin receptor and intracellular calcium mobilization in HT29 cells. *J. Pharmacol. Exp. Ther.* 253, 1049–1056.

Underwood, S. L., Lewis, S. A., and Raeburn, D. (1992) RP 59227, a novel PAF receptor antagonists; effects in guinea pig models of airway hyperreactivity. *Eur. J. Pharmacol.* 210, 97–102.

Ushikubi, F., Nakajima, M., Hirata, M., Okuma, M., Fujiwara, M., and Narumiya, S. (1989) Purification of the thromboxane A2/prostaglandin H2 receptor from human blood platelets. *J. Biol. Chem.* 264, 16496–16501.

Van Haastert, P. J. M. (1983) Binding of cAMP and adenosine derivatives to *Dictyostelium discoideum* cells. Relationships of binding, chemotactic, and antagonistic activities. *J. Biol. Chem.* 258, 9643–9648.

Van Haastert, P. J. M. (1984) Guanine nucleotides modulate cell surface cAMP-binding sites in membranes from *Dictyostelium discoidum*. *Biochem. Biophys. Res. Commun.* 124, 597–604.

Van Haastert, P. J. M. and De Wit, R. J. W (1984) Demonstration of receptor heterogeneity and negative cooperativity by nonequilibrium binding experiments. The cell surface cAMP receptor of *Dictyostelium discoideum*. *J. Biol. Chem.* 259, 13321–13328.

Van Haastert, P. J. M. (1985) cAMP activates adenylate and guanylate cyclase of *Dictyostelium discoideum* cells by binding to different classes of cell-surface receptors. A study with extracellular Ca^{2+}. *Biochim. Biophys. Acta* 846, 324–333.

Van Haastert, P. J. M., De Wit, R. J. W., Jassens, P. M. W., Kesbeke, F., and DeGoede, J. (1986) G-protein mediated interconversions of cell surface cAMP receptors and their involvement in excitation and desensitization of guanylate cyclase in *Dictyostelium discoideum*. *J. Biol. Chem.* 261, 6904–6911.

Van Haastert, P. J. M., De Wit, R. J. W., Van Lookeren Campagne, M. M. (1985) Ca^{2+}- or phorbol ester-dependent effect fo ATP on a subpopulation of cAMP cell-surface receptors in membranes from *D. discoideum*. A role for protein kinase C. *Biochem. Biophys. Res. Commun.* 128, 185–192.

Van Haastert, P. J. M., Snaar-Jagalska, B. E., and Janssens, P. M. W. (1987) The regulation of adenylate cyclase by guanine nucleotides in *Dictyostelium discoideum* membranes. *Eur. J. Biochem.* 162, 251–258.

Vassallo, R. R., Jr., Kieber-Emmons, T., Cichowski, K., Brass, L. F. (1992) Structure-function relationships in the activation of platelet thrombin receptors by receptor-derived peptides. *J. Biol. Chem.* 267, 6081–6085.

Vaughan, R. and Devreotes, P. N. (1988) Ligand-induced phosphorylation of the cAMP receptor from *Dictyostelium discoideum. J. Biol. Chem.* 263, 14538–14543.

Verghese, M. W., Uhing, R. J., and Snyderman, R. (1986) A pertussis/Cholear Toxin-sensitive Nprotein may mediate chemoattractant receptor signal transduction. *Biochem. Biophys. Res. Commun.* 138, 887–894.

Vitkauskas, G., Showell, H. J., and Becker, E. L. (1980) Specific binding of synthetic chemotactic peptide to rabbit peritoneal neutactic peptide to rabbit peritonial neutrophils: effects on dissociability of bound peptide, receptor activity and subsequencet biologic responsiveness (deactivation). *Mol. Immunol.* 17, 171–180.

Volpi, M., Yassin, R., Tao, W., Molski, T. F. P., Naccache, P. H., and Sha'afi, R. I. (1984) Leukotriene B4 mobilizes calcium without the breakdown of polyphosphoinositides and the production of phosphatidic acid in rabbit neutrophils. *Proc. Natl. Acad. Sci. U.S.A.* 81, 5966–5969.

Vu, T.-K. H., Hung, D. T., Wheaton, V. I., and Coughlin, S. R. (1991) Molecular cloning of a functional thrombin receptor reveals a novel proteolytic mechanism of receptor activation. *Cell* 64, 1057–1068.

Vu, T.-K. H., Wheaton, V. I., Hung, D. T., Charo, I., and Coughlin, S. R. (1991a) Domains specifying thrombin-recpetor interaction. *Nature* 353, 674–677.

Weisbart, R. H., Golde, D. W., and Gasson, J. C. (1986) Biosynthetic human GM-CSF modulates the numebr and affinity of neutrophil f-Met-Leu-Phe receptors. *J. Immunol.* 137, 3584–3587.

Welch, A. R., McGregor, L. M., and Gibson, W. (1991) Cytomegalovirus homologs of cellular G prtoein-coupled receptor genes are transcibed. *J. Virol.* 65, 3915–3918.

Whiteway, M., Hougan, L., Dignard, D., Thomas, D. Y., Bell, L., Saari, G. C., Grant, F. J., O'Hara, P., and MacKay, V. L. (1989) THe STE4 and STE18 genes of yeast encode potential beta and gamma subunits of the mating factor receptor-coupled G protein. *Cell* 56, 467–477.

Ye, R. D., Cavanagh, S. L., Quehenberger, O., Prossitz, E. R., and Cochrane, C. G. (1992) Isolation of a cDNA that encodes a novel granulocyte N-formyl peptide receptor. *Biochem. Biophys. Res. Commun.* 184, 582–589.

Ye, R. D., Prossnitz, E. R., Zou, A. H., Cochrane, C. G. (1991) Characterization of a human cDNA that encodes a functional receptor for platelet activating factor. *Biochem. Biophys. Res. Commun.* 180, 105–111.

Yu, L., Blumer, K. J., Davidson, N., Lester, H. A., and Thorner, J. (1989) Functional expression of the yeast a-factor receptor in *Xenopus* oocytes. *J. Biol. Chem.* 264, 20847–29850.

Zhong, C., Hayzer, D. J., Corson, M. A., and Runge, M. S. (1992) Molecular cloning of the rat vascular smooth muscle thrombin receptor. Evidence for in vitro regulation by basic fibroblast growth factor. *J. Biol. Chem.* 267, 16975–16979.

Index